CONTEMPORARY IDEAS ON SHIP STABILITY

Elsevier Science Internet Homepage
http://www.elsevier.nl (Europe)
http://www.elsevier.com (America)
http://www.elsevier.co.jp (Asia)
Consult the Elsevier homepage for full catalogue information on all books, journals and electronic products and services.

Elsevier Titles of Related Interest

MOAN & BERGE
13th Int Ship & Offshore Structures Congress (ISSC 1997)
ISBN: 008-042829-0

VUGTS
BOSS '97, Behaviour of Offshore Structures (3 Volume Set)
ISBN: 008-042834-7

WATSON
Practical Ship Design
ISBN: 008-042999-8

YOUNG
Wind Generated Ocean Waves
ISBN: 008-043317-0

SRIVASTAVA
Structural Engineering World Wide 1998 (CD-ROM Proceedings with Printed Abstracts Volume, 702 papers)
ISBN: 008-042845-2

OWENS
Steel in Construction (CD-ROM Proceedings with Printed Abstracts Volume, 268 papers)
ISBN: 008-042997-1

GODOY
Thin-Walled Structures with Structural Imperfections: Analysis and Behavior
ISBN: 008-042266-7

FUKUMOTO
Structural Stability Design
ISBN: 008-042263-2

USAMI & ITOH
Stability and Ductility of Steel Structures
ISBN: 008-043320-0

GUEDES-SOARES
Advances in Safety and Reliability (3 Volume Set)
ISBN: 008-042835-5

DUBINA
SDSS '99 – Stability and Ductility of Steel Structures
ISBN: 008-043016-3

BJORHOVDE, COLSON & ZANDONINI
Connections in Steel Structures III
ISBN: 008-042821-5

CHAN & TENG
ICASS '96, Advances in Steel Structures (2 Volume Set)
ISBN: 008-042830-4

CHAN & TENG
ICASS '99, Advances in Steel Structures (2 Volume Set)
ISBN: 008-043015-5

FRANGOPOL, COROTIS & RACKWITZ
Reliability and Optimization of Structural Systems
ISBN: 008-042826-6

OHTSUBO & SUMI
Proceedings of the 14th International Ship and Offshore Structures Congress
ISBN: 008-043602-1

Related Journals

Free specimen copy gladly sent on request: Elsevier Science Ltd, The Boulevard, Langford Lane, Kidlington, Oxford, OX5 1GB, UK

Applied Ocean Research
Advances in Engineering Software
CAD
Coastal Engineering
Composite Structures
Computers and Structures
Construction and Building Materials
Engineering Failure Analysis
Engineering Fracture Mechanics
Engineering Structures
Finite Elements in Analysis and Design
International Journal of Solids and Structures
Journal of Constructional Steel Research
Marine Structures
NDT & E International
Ocean Engineering
Structural Safety
Thin-Walled Structures

To Contact the Publisher

Elsevier Science welcomes enquiries concerning publishing proposals: books, journal special issues, conference proceedings, etc. All formats and media can be considered. Should you have a publishing proposal you wish to discuss, please contact, without obligation, the publisher responsible for Elsevier's civil and structural engineering publishing programme:

Ian Salusbury
Senior Publishing Editor
Elsevier Science Ltd
The Boulevard, Langford Lane
Kidlington, Oxford, OX5 1GB, UK

Phone: +44 1865 843425
Fax: +44 1865 843920
E-mail: i.salusbury@elsevier.co.uk

General enquiries, including placing orders, should be directed to Elsevier's Regional Sales Offices – please access the Elsevier homepage for full contact details (homepage details at top of this page).

CONTEMPORARY IDEAS ON SHIP STABILITY

Edited by

D. Vassalos
Department of Ship and Marine Technology, University of Strathclyde, Scotland

M. Hamamoto
University of Naval Architecture and Ocean Engineering, Osaka University, Japan

A. Papanikolaou
Department of Naval Architecture and Marine Engineering, National Technical University of Athens, Greece

D. Molyneux
Institute for Marine Dynamics, Newfoundland, Canada

Co-edited by

K. Spyrou
Department of Naval Architecture and Marine Engineering, National Technical University of Athens, Greece

N. Umeda
University of Naval Architecture and Ocean Engineering, Osaka University, Japan

J. Otto de Kat
MARIN, The Netherlands

2000

ELSEVIER

AMSTERDAM · LONDON · NEW YORK · OXFORD · PARIS · SHANNON · TOKYO

ELSEVIER SCIENCE Ltd
The Boulevard, Langford Lane
Kidlington, Oxford OX5 1GB, UK

First edition 2000

Library of Congress Cataloging in Publication Data

Contemporary ideas on ship stability / edited by D. Vassalos ... [et al.].-- 1st ed.
p. cm.
ISBN 0-08-043652-8 (hardcover)
1. Stability of ships. I. Vassalos, D. (Dracos)

VM159 .C65 2000
623.8'171--dc21

00-052126

British Library Cataloguing in Publication Data
A catalog record from the British Library has been applied for.

ISBN: 0-08-043652-8

∞ The paper used in this publication meets the requirements of ANSI/NISO Z39.48-1992 (Permanence of Paper).
Printed in The Netherlands.

PREFACE

Widely publicised disasters serve as a reminder to the maritime profession of the imminent need for enhancing safety cost-effectively and as strong indicator of the existing gaps in the stability safety of ships and ocean vehicles. The problem of ship stability is so complex that practically meaningful solutions are feasible only through close international collaboration and concerted efforts by the maritime community, deriving from sound scientific approaches. Responding to this and building on an established track record of co-operative research between UK and Japan, a Collaborative Research Project (CRP) was launched on 1 April 1995, jointly supported by the British Council, the Japanese Ministry of Transport and the Japanese Science and Technology Agency, aiming to foster international co-operation to investigate systematically the stability, survivability and operational safety of ships and to evolve design and operational guidelines for reducing the risk of ship losses. The project was co-ordinated by Professor Dracos Vassalos of Ship and Marine Technology at the University of Strathclyde in UK and Professor Masami Hamamoto of Naval Architecture and Ocean Engineering at Osaka University in Japan with participating institutions including The Centre for Non-Linear Dynamics at University College London in UK and Osaka Prefecture University, Ship Research Institute and National Research Institute of Fisheries Engineering in Japan.

As part of the CRP, Professor Vassalos organised a two day workshop at Ross Priory of the University of Strathclyde in July 1995 by inviting international experts on ship stability to address the timely and sensitive issue of ship capsize and to formalise ways for accelerating developments in the future. Twelve countries were represented, including all major shipping nations, with experts covering the whole spectrum of ship safety. This provided the foundation for the formation of the International Stability Workshops aiming to address contemporary ideas related to the stability and operational safety of ships in depth by promoting (Round-Table) discussion by internationally recognised experts on a restricted number of invited papers that address specific problem areas of on-going front-end research, development and application. Furthermore, to provide an enabling platform for promoting international collaboration on ship stability and for nurturing a continuous dialogue with the International Maritime Organisation (IMO) to facilitate an effective transfer of theoretical advances to practical rules and design procedures and guidelines. Following the 1st workshop, three others have been organised and plans are in place to continue in the foreseeable future with the workshops as well as with the publications of additional volumes by adopting the same format as presented here.

This volume includes selected material from the first four workshops: 2nd in Osaka Japan, Osaka University, November 1996 by Professor Masami Hamamoto; 3rd in Crete Greece, Ship Design Laboratory of the National Technical University of Athens (NTUA-SDL),

October 1997 by Professor Apostolos Papanikolaou; and 4th in Newfoundland Canada, Institute for Marine Dynamics, September 1998 by David Molyneux. It contains 46 papers that represent all currently available expertise on ship stability, spanning 17 countries from around the world. The framework adopted for grouping the papers aims to cover broad areas of ship stability in a way that it provides a template for future volumes, namely: (1) *Stability of the Intact Ship*; (2) *Damage Ship Stability*; (3) *Special Problems of Ship Stability*; and (4) *Impact of Stability on Design and Operation.*

We would like to express our gratitude and sincere thanks to all the speakers, especially to speakers from the industry; to all the participants of the workshops; to the members of the International Standing Committee and International Advisory Board for the Stability of Ships and Ocean Vehicles; and to Dr. Ismail Helvacioglou for his assistance in the final compilation of this volume.

A special thanks is reserved for all sponsoring organisations for their contribution in making these workshops a memorable experience.

Professor Dracos Vassalos
Chairman of the International Standing Committee for the Stability of Ships and Ocean Vehicles (on behalf of the editorial committee)

CONTENTS

Preface v

1. Stability of the Intact Ship

Experimental investigation of ship dynamics in extreme waves 3
S. Grochowalski

A mathematical model of ship motions leading to capsize in astern waves 15
M. Hamamoto and A. Munif

A note on the conceptual understanding of the stability theory of ships 27
A.Y. Odabasi

The role and the methods of simulation of ship behaviour at sea including ship capsizing 33
V. Armenio, G. Contento and A. Francescutto

Geometrical aspects of the broaching-to instability 47
K.J. Spyrou

Application of nonlinear dynamical system approach to ship capsize due to broaching in following and quartering seas 57
N. Umeda

Broaching and capsize model tests for validation of numerical ship motion predictions 69
J.O. de Kat and W.L. Thomas III

Sensitivity of capsize to a symmetry breaking bias 89
B. Cotton, S.R. Bishop and J.M.T. Thompson

Some recent advances in the analysis of ship roll motion 101
B. Cotton, J.M.T. Thompson and K.J. Spyrou

Ship capsize assessment and nonlinear dynamics 111
K.J. Spyrou

The mathematical modelling of large amplitude rolling in beam waves 125
A. Francescutto and G. Contento

Characteristics of roll motion for small fishing boats 137
K. Amagai, K. Ueno and N. Kimura

Piecewise linear approach to nonlinear ship dynamics 149
V.L. Belenky

2. Damaged Ship Stability

The water on deck problem of damaged RO–RO ferries 163
D. Vassalos

Water-on-deck accumulation studies by the SNAME ad hoc RO–RO safety panel 187
B.L. Hutchison

An experimental study on flooding into the car deck of a RO–RO ferry through damaged bow door 199
N. Shimizu, R. Kambisseri and Y. Ikeda

Damage stability tests with models of RO–RO ferries – a cost effective method for upgrading and designing RO–RO ferries 213
M. Schindler

About safety assessment of damaged ships 225
R. Kambisseri and Y. Ikeda

Survivability of damaged RO–RO passenger vessels 237
B.C. Chang and P. Blume

Dynamics of a ship with partially flooded compartment 249
J.O. de Kat

RO–RO passenger vessels survivability – a study of three different hull forms considering different RO–RO-deck subdivisions 265
A.E. Jost and P. Blume

Simulation of large amplitude ship motions and of capsizing in high seas 279
A.D. Papanikolaou, D.A. Spanos and G. Zaraphonitis

On the critical significant wave height for capsizing of a damaged RO–RO passenger ship 291
T. Haraguchi, S. Ishida and S. Murashige

Exploration of the applicability of the static equivalence method using experimental data 303
A. Kendrick, D. Molyneux, A. Taschereau and T. Peirce

Modelling the accumulation of water on the vehicle deck of a damaged RO–RO vessel and proposal of survival criteria 315
D. Vassalos, L. Letizia and O. Turan

3. Special Problems of Ship Stability

Damage stability with water on deck of a RO–RO passenger ship in waves 341
S. Ishida, S. Murashige, I. Watanabe, Y. Ogawa and T. Fujiwara

A study on capsizing phenomena of a ship in waves 353
S.Y. Hong, C.G. Kang and S.W. Hong

Physical and numerical simulation on capsizing of a fishing vessel in head sea condition 365
T. Hirayama and K. Nishimura

The influence of liquid cargo dynamics on ship stability 377
N.N. Rakhmanin and S.G. Zhivitsa

Exploring the possibility of stability assessment without reference to hydrostatic data 389
R. Birmingham

Stability of high speed craft 401
Y. Ikeda and T. Katayama

Nonlinear roll motion and bifurcation of a RO–RO ship with flooded water in regular beam waves 411
S. Murashige, M. Komuro, K. Aihara and T. Yamada

Effects of some seakeeping/manoeuvring aspects on broaching in quartering seas 423
N. Umeda

Ship manoeuvring performance in waves 435
K. Kijima and Y. Furukawa

Stability of a planing craft in turning motion 449
Y. Ikeda, H. Okumura and T. Katayama

An experimental study on the improvement of transverse stability at running for high-speed craft 461
Y. Washio, K. Kijima and T. Nagamatsu

Water discharge from an opening in ships 473
S.M. Calisal, M.J. Rudman, A. Akinturk, A. Wong and B. Tasevski

4. Impact of Stability on Design and Operation

Passenger survival-based criteria for RO–RO vessels 487
D. Vassalos, A. Jasionowski and K. Dodworth

Nonlinear dynamics of ship rolling in beam seas and ship design 499
K.J. Spyrou, B. Cotton and J.M.T. Thompson

Ship crankiness and stability regulation 511
N.N. Rakhmanin and G.V. Vilensky

The impact of recent stability regulations on existing and new ships. Impact on the design of RO–RO passenger ships 523
M. Kanerva

A realisable concept of a safe haven RO–RO design 537
D. Vassalos

Design aspects of survivability of surface naval and merchant ships 553
A. Papanikolaou and E. Boulougouris

A technique for assessing the dynamic stability and capsize resistance of ships 565
M. Renilson

Probability to encounter high run of waves in the dangerous zone shown on the operational guidance/IMO for following/quartering sea 575
Y. Takaishi, K. Watanabe and K. Masuda

Ongoing work examining capsize risk of intact frigates using time domain simulation 587
K. McTaggart

Author Index 597

1. Stability of the Intact Ship

Contemporary Ideas on Ship Stability
D. Vassalos, M. Hamamoto, A. Papanikolaou and D. Molyneux (Editors)

EXPERIMENTAL INVESTIGATION OF SHIP DYNAMICS IN EXTREME WAVES

S. Grochowalski

Institute for Marine Dynamics, National Research Council Canada,
St. John's, NF A1B 3T5, Canada

ABSTRACT

The problems related to prediction of ship behaviour in extreme wave conditions, including capsizing, are discussed. The differences between a conventional seakeeping analysis and an investigation into transient phenomena are highlighted. An experimental approach developed for study of ship capsizing and other phenomena in extreme waves is presented. Hydrodynamic effects generated on a submerged part of the deck illustrate the capability of the presented technique. The need for a mathematical model of extreme waves is emphasized.

KEYWORDS

Ship dynamics, extreme waves, capsizing, model testing, experimental technique, ship motion analysis, deck-in-water effects.

INTRODUCTION

Numerous capsize disasters which happened in the recent several years caused a significant increase of interest in the problems of safe operations in extreme weather conditions, in particular, in ship capsizing.

The dynamics of ships in extreme waves include strong nonlinearities, transient effects, and some additional physical phenomena, not defined theoretically yet, which may lead to a ship capsize. The methods used in seakeeping analyses and predictions are not applicable to these cases. There are no established methods or procedures for reliable prediction of transient phenomena or behaviour of a vessel in extreme waves. In many cases, there are even no definitions. The most acute problem still awaiting its solution is the prevention of ship capsize, both in intact and in damage conditions. This situation creates a need for development of reliable, standard methods for both, experimental testing and mathematical modelling.

It seems that at present, neither model testing nor pure theoretical modelling alone could provide a reliable prediction of strongly non-linear and transient phenomena involved in ship capsizing. Instead, a combination of theoretical modelling with model experiments proved to be a very powerful and effective approach in studies of this kind of problems. Theoretical modelling will always require experimental verification of the results or delivery of reliable data not achievable theoretically. Numerical simulations and model testing are complementary and feed each other. This new approach requires a development of some specific experimental techniques and put very demanding requirements with regards to test accuracy and precision. Recent advance in computers and sensors technology secure that these requirements can be satisfied.

For many years, the Institute for Marine Dynamics of the National Research Council, Canada was involved in an intensive research of ship capsizing. Comprehensive model tests of ship capsizing in extreme waves were carried out at the SSPA facilities in Gothenburg, Sweden. The innovative methodology applied in the studies was driven by the considerations presented above. It yielded a lot of unique information, and confirmed validity of the philosophy adopted in the research.

The following are some results and thoughts coming out of the experimental program which sheds some light on the specific nature of investigating ship dynamics in extreme wave conditions.

MODEL TESTING OF SHIP DYNAMICS IN EXTREME WAVES

Demands for predictions of ship or structure dynamics in extreme weather conditions set a real challenge for the research centres. In particular, analyses of transient phenomena such as broaching-to, capsizing, water shipping on deck, etc., require specific theoretical and experimental approaches. It became obvious that the traditional way of testing and analysis used in seakeeping, and based mainly on some assumptions of linearity, is not relevant for investigation of non-linear phenomena in extreme waves.

The dynamic identification of ship behaviour in extreme waves can be achieved, if for any set of environmental conditions, the composition of externally exerted forces and the corresponding ship response are fully identified. It means that all the elements presented in Figure1 have to be identified in order to find the qualitative and quantitative relation between the input (waves) and the response (ship behaviour or individual phenomenon).

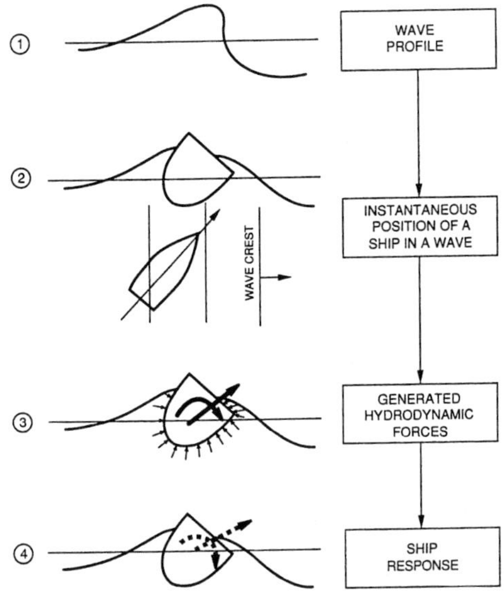

Figure 1: The elements of the "cause - result" chain in ship motions

In the traditional seakeeping testing only the wave parameters (input) and the final results (ship response) are measured. This is not sufficient for transient and non-linear phenomena. As the generated hydrodynamic forces depend on the instantaneous shape of the immersed body, the identification of the instantaneous position of the ship in the wave is essential.

For the validation purposes of numerical models used in time domain simulations, the measurement of the generated hydrodynamic forces is also needed, at least for some selected situations. Because of a strong influence of initial conditions on the result of numerical simulations, a good agreement between computed and experimental results may be achieved accidentally by changing initial conditions in the simulations, if the initial conditions in the selected fragment of the experiment record are not known. Obviously, this good agreement would be misleading. Thus, it is the validation against the hydrodynamic forces, and not only against the final ship response, which makes the mathematical/numerical model valid and the subsequent simulations reliable. The comparison of the computed hydrodynamic forces and free motions must be done for the same wave crest position with respect to the hull as it was in the experiment. Furthermore, the initial conditions in the simulations must be assumed the same as in the experiment.

The above mentioned requirements could be satisfied if the instantaneous position of the hull in the wave is continuously recorded in every experiment. In the IMD's capsizing model tests carried out at the SSPA, the instantaneous position of the model with respect to the acting wave was recorded continuously by video cameras. The time counter of the camera was synchronized with the time base of the main recording system. Through the analysis of frozen video pictures, frame after frame, it is possible to identify the time at which the wave crest

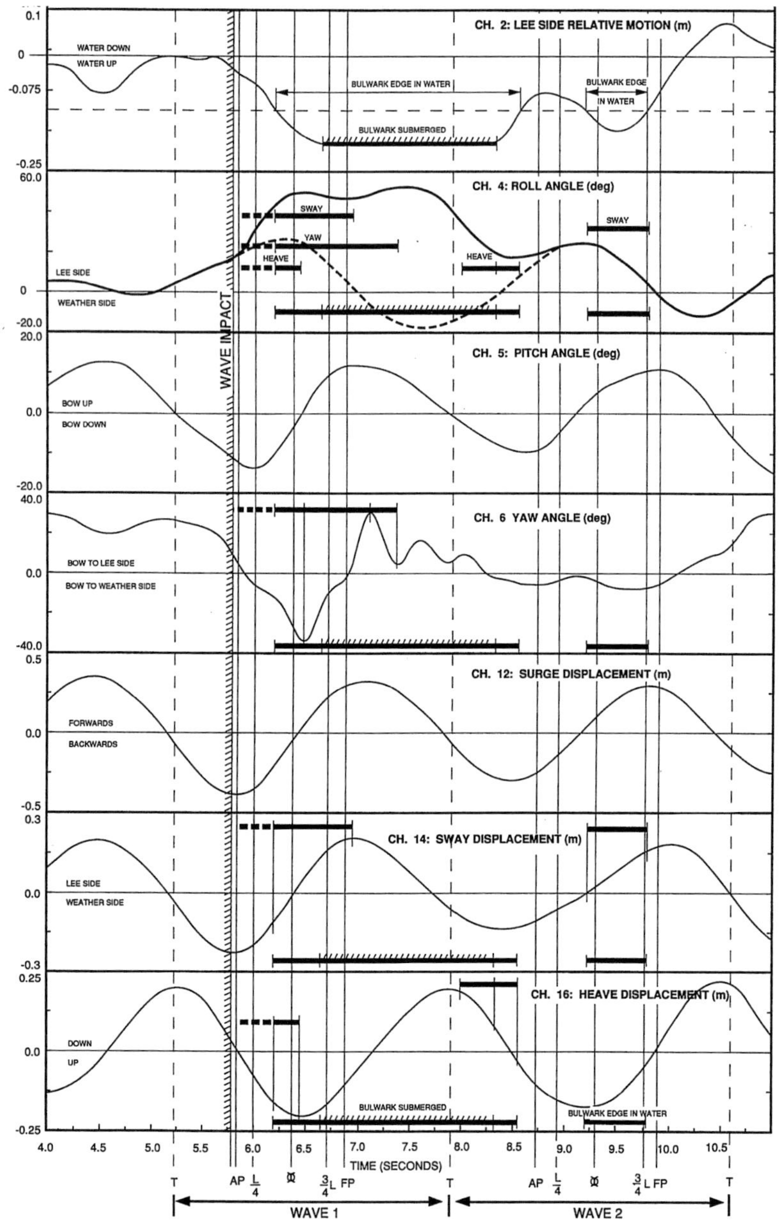

Figure 2: Analysis of a time history of model motions in extreme waves (Grochowalski, 1989)

reaches any considered point on the model. As the time base in the video-records is the same as in the recorded time histories, the identified time instances can be marked on the motion records.

Figure 2, taken from Grochowalski (1989), presents an example of such a detailed analysis. The time at which the model was in the wave trough (T), and when the wave crest reached the after perpendicular (AP), a quarter of the model length (L/4), etc., have been identified from the video-records and marked on the time histories in a form of vertical lines. This provide direct link between the instantaneous position of the hull in the wave (element 2 in Figure 1) and the resulting components of motions (element 4 in Figure 1).

The amount of information which can be obtained by use of this link is tremendous. For instance, all the components of ship motions, including velocities and accelerations, can be identified for any considered wave position (vertical lines in Figure 2). If this is done for various wave crest positions, a sequence of motion composition when the wave is passing along the hull can be identified. The Figure 3 presents an example of such a sequence for a case when the deck was not immersed in water. Bulwark submergence and water shipment on deck can also be identified by use of the same method.

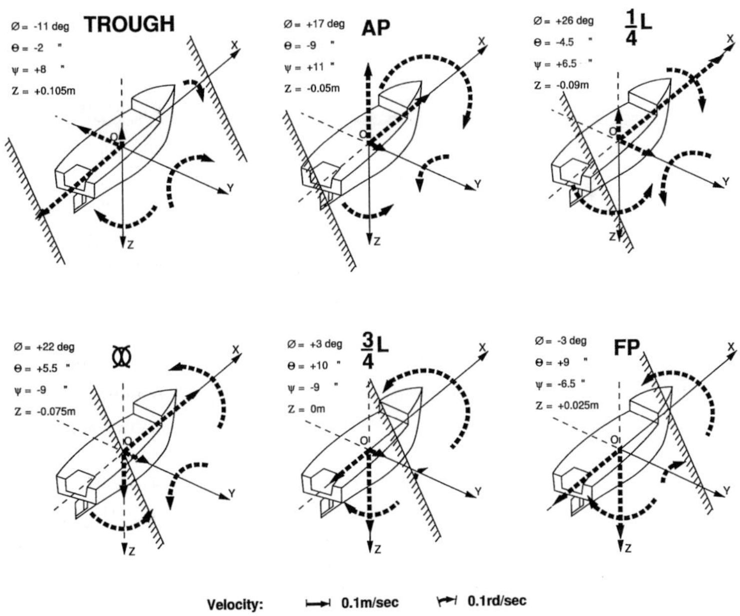

Figure 3: Motion components in quartering waves (Grochowalski, 1989)

If the same technique is applied to captive tests carried out in the same waves as the free-running tests, the hydrodynamic forces generated at various wave crest positions can be identified as well. This would provide the link, at least qualitatively, between the hydrodynamic forces (element 3 in Figure 1) and the resulting motions (element 4 in Figure

1), and thus all the elements of the chain in Figure 1 could be identified. An example of the composition of the hydrodynamic forces for various positions of the model in waves is presented in Figure 4. The solid vectors represent the measured forces, while the broken lines indicate the motions in the free modes.

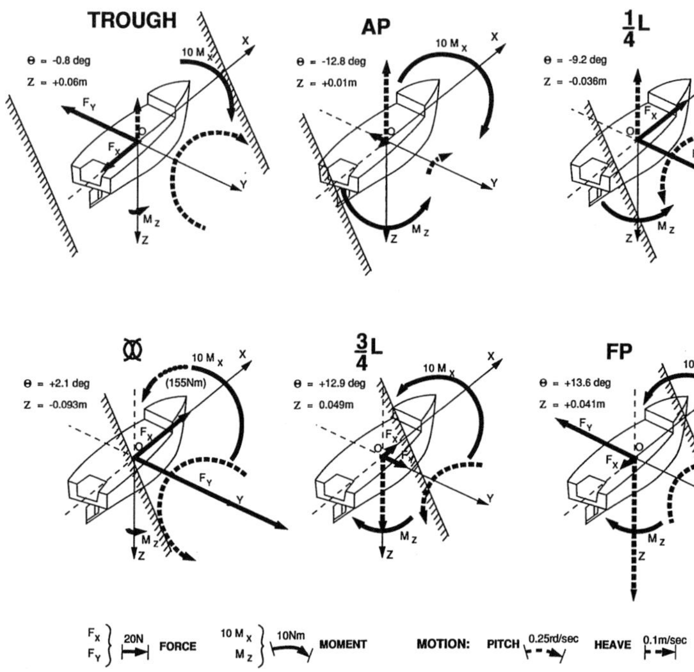

Figure 4: Forces and motions in a semi-captive model test (Grochowalski, 1989)

The results obtained through the identification of the position of the wave crest proves that this approach is a very powerful method in detailed analyses of complex behaviour of a ship in extreme waves. It provides a lot of insight into ship dynamics and the capsizing mechanism. A good example of the benefits brought by use of this method is the identification of additional hydrodynamic forces generated on the submerged part of a deck.

EFFECTS OF BULWARK AND DECK EDGE SUBMERGENCE

A ship advancing in quartering, extremely steep waves performs a very characteristic, complex composition of motions which is very unfavourable from the stability point of view. A typical one cycle of such motions is presented in Figure 3 (Grochowalski,1989).

The characteristic sequence of motions together with the corresponding position of the wave crest create possibility of submergence of the bulwark, in particular at the lee side. If this happens, and the submerged part of the deck is moving with a significant velocity relative to the surrounding water (Figure 5), a hydrodynamic reaction R is generated which constitutes water resistance to the movement of the submerged surface. This reaction introduces a restraint to ship motions and causes radical alterations in the roll. An additional heeling moment is generated which significantly reduces ship restoring capability, or causes a capsize (Grochowalski 1989, 1990, 1993).

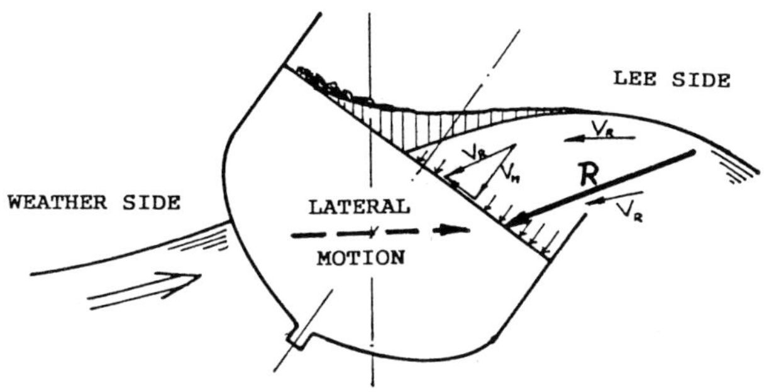

Figure 5: Hydrodynamic effects generated on the submerged deck in waves

The reaction R is a force additional to the static and Froude-Krylov forces calculated conventionally for the immersed hull surface. It has a dynamic nature and depends on a square of the relative velocity of water particles flowing to the deck. The experiments showed that a significant relative movement of the immersed part of the deck in waves take place if there is a large lateral motion caused by sway and yaw. Other components of ship motion may contribute, but their influence is insignificant if there is no lateral motion.

In order to explain the abnormal roll behaviour presented in Figure 2, the analysis of the instantaneous wave crest position was applied to the relative water surface motion, measured at midships at the model sides. It was found, that the deck at the lee side was deeply immersed from a certain time instance (see Figure 2 - Lee side relative motion). The time during which the bulwark at midships and part of the deck was in water is marked by the shadowed horizontal line. The analysis of the motion components revealed that during that time the direction of sway, yaw and heave was such that the submerged part of the deck was moving strongly relative to the adjoining water. The time when these motion components were conducive to generate the reaction R is marked by bold horizontal lines, and then all these indicators are put together on the time history of roll. It can be seen that the additional force on the deck was generated and it was responsible for the strange roll motion (solid line), radically different from the expected one (dotted line).

Thanks to the detection of the wave crest position with respect to the hull, the full identification of the immersed part of the deck is possible. The sequence of deck positions found from Figure 2 during action of "wave 1" and partially "wave 2" with the identified immersed deck surface (shadowed) is presented in Figure 6. The black spots indicate the area where the relative velocity of water was towards the deck surface, and thus was conducive to the generation of the additional force on the deck. The position of the wave crest is marked by the oblique line. As it can be seen, the deck at the stern was immersed when the wave crest was at AP (position "1" in Figures 6 and 7), well before the relative motion probe indicated the midships immersion (Figure 2).

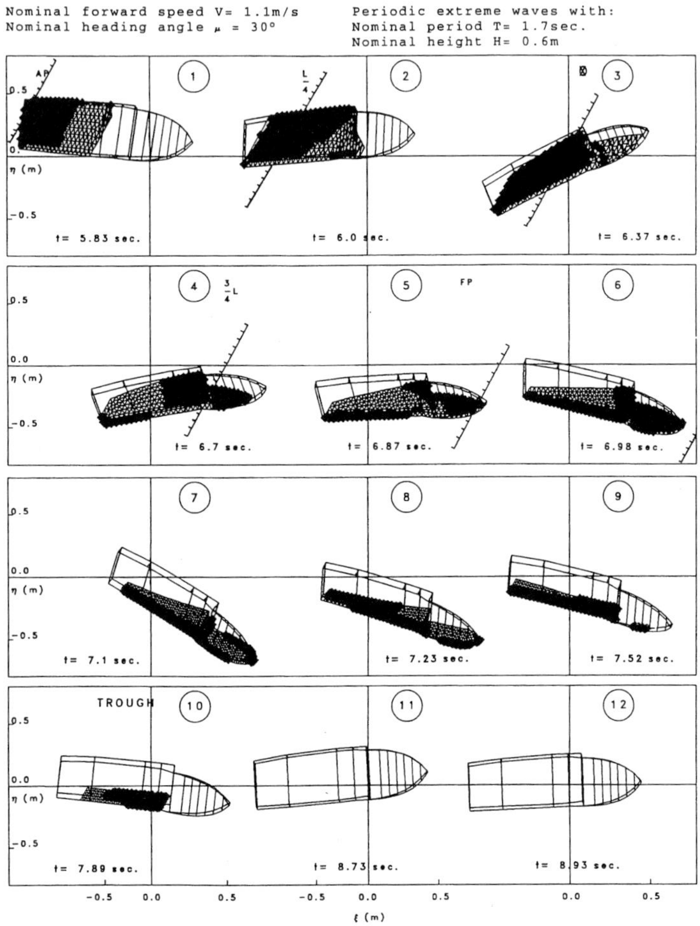

Figure 6: The history of deck immersion during the motions presented in Figure 2 (Grochowalski,1993)

Using the motion components from Figure 2, the force on deck and the corresponding additional moment was computed for the situations in Figure 6. The resulting additional heeling moment is shown in Figure 7/D (solid line). Also, the additional static load caused by mass of water shipped on deck is presented (broken line). The total additional heeling moment generated on the submerged part of the deck is presented in Figure 7/E (solid line) and collated with the conventional roll moment estimated from the partly captive tests (Grochowalski, 1993). The comparison of the additional heeling moment created by the deck immersion with the corresponding history of roll (Figure 7/C) clearly indicate that this moment was responsible for the abnormal rolling of the model.

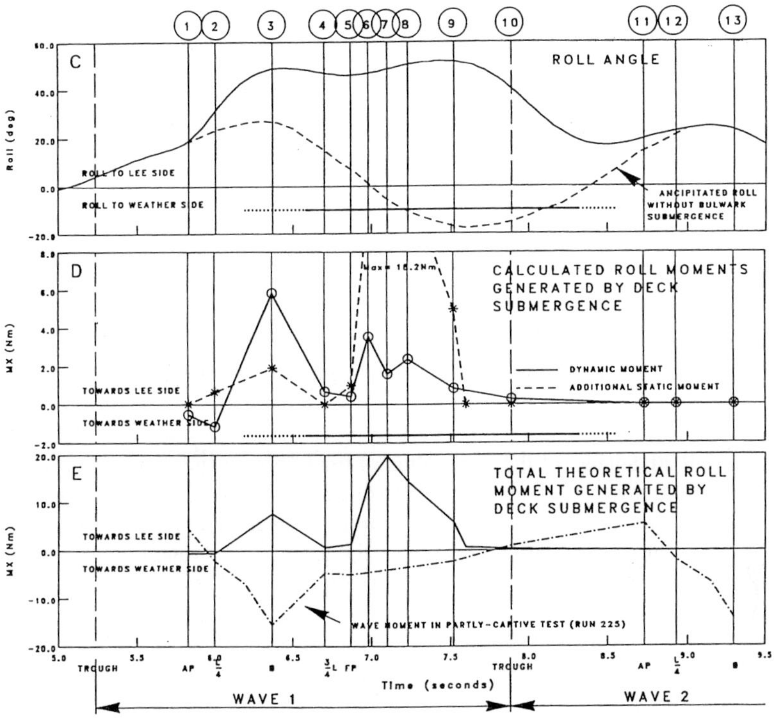

Figure 7: Comparison of the additional forces on deck with the corresponding roll motion (Grochowalski, 1993)

The analysis of the model tests proved that this phenomenon was responsible for capsizing of the model in the loading conditions in which, according to the existing criteria, the ship was considered as safe.

Detailed analysis of these hydrodynamic effects were presented by Grochowalski (1989, 1990, 1993). The phenomenon was never recognized before, and not considered in any studies of ship dynamics in extreme waves. Its dangerous effects have to be included in the stability safety analyses.

THE NEED FOR A MATHEMATICAL MODEL OF EXTREME WAVES

The methodology of prediction of seakeeping characteristics of ships in a seaway is well established and commonly accepted. It is based on the assumption that ship motions constitute a *steady state* process of a *linear* dynamic system, and the relationship between the excitement (waves) and the ship response (ship motions) is represented by the transfer function. Application of St.Denis-Pierson theory leads to spectral representation of irregular seas and the corresponding ship motions. This provides possibility of identification of the transfer functions and, as a result, prediction of all statistical characteristics of ship behaviour in irregular waves.

This approach, however, is not relevant to investigation of ship capsizing or other strongly non-linear and transient phenomena in extreme waves. Ship response to an action of an extreme wave, in particular transient behaviour, can not be obtained by superposition of linear steady responses to individual sinusoidal waves. Even though the profile of an individual large wave could be modelled by superposition of individual harmonics, the ship response to each wave component may not contain certain phenomena which appear during the action of an extreme wave. The hydrodynamic effects created on a submerged part of the deck could be the example of missing phenomena in the responses to individual components of the wave spectrum. Thus, the superposition of the responses to the individual components would not be the same as the real response to the extreme wave.

In order to take full advantage of the application of time-domain simulations to prediction of ship behaviour in extreme waves, an adequate mathematical model of such waves has to be developed. It can have a form of an individual extremely steep and periodic wave or a group of large and steep waves with defined profile of the group. The velocity potential of such waves must be defined.

Application of individual Stokes waves or W. Pierson's recent modelling of a group of steep waves by interaction of third order Stokes waves (Pierson, 1993) provide a good example of possible theoretical approaches.

Standard extreme (non-harmonic) wave representation is also needed for model testing. This is especially important if model testing become a mean for regulatory purposes, as it has been proposed recently (Ro-Ro safety, open top container ships, etc.) Model testing in large irregular waves generated to an assumed spectrum can be useful only if its purpose is just to examine ships behaviour in the particular conditions. There is no method yet to convert the result of recorded extreme or transient behaviour in one wave spectrum into another.

CONCLUSIONS

The examples presented in this paper lead to the following conclusions:

1. More attention should be paid to the research of ship dynamics in extreme waves, and to nonlinear and transient phenomena. Understanding of the physics involved is needed.

2. A combination of computer simulations with model experiments provides a very powerful and effective tool for investigating transient and strongly nonlinear phenomena of ship behaviour in extreme waves.

3. Measurement of the hydrodynamic forces acting on a ship in waves is required as the source for validation of numerical models.

4. Monitoring of the position of the model with respect to the acting wave is essential. Together with some selected measurements of the hydrodynamic forces, it will help to identify various transient effects, to get insight into the mechanism of ship capsizing, and will facilitate development of relevant mathematical models for numerical simulations.

5. There is an urgent need for development of an adequate mathematical model of extreme waves. Such a model should provide a relevant basis for numerical simulations and for standard model testing.

It is suggested that the International Towing Tank Conference (ITTC) committees take the lead and encourage the member organizations to work out the methodology and standard procedures for testing of ship dynamics in extreme waves. The unified way of testing would facilitate exchange of information, verification of computer simulations, and comparisons of results. Initiation of a review of currently used methodology and techniques in model testing of ship dynamics with large amplitudes and specific transient phenomena, would be a good beginning.

References

1. Grochowalski, S., Rask, I., and Söderberg, P., (1986). An Experimental Technique for Investigation into Physics of Ship Capsizing. *Proceedings, Third International Conference on Stability of Ships and Ocean Vehicles, STAB '86,* Gdansk, Poland.

2. Grochowalski, S., (1989). Investigation into the Physics of Ship Capsizing by Combined Captive and Free-Running Model Tests. *Transactions SNAME,* **vol.97,** 169-212, New York.

3. Grochowalski, S., (1990). Hydrodynamic Phenomenon Generated by Bulwark Submergence and its Influence on Ship Susceptibility to Capsizing. *Proceedings, Fourth International Conference on Stability of Ships and Ocean Vehicles, STAB '90,* Naples, Italy.

4. Grochowalski, S., (1993). Effect of Bulwark and Deck Submergence in Dynamics of Ship Capsizing. *Proceedings, US Coast Guard Vessel Stability Symposium '93,* New London, USA.

5. Pierson, W.J.Jr., (1993). Oscillatory Third-Order Perturbation Solutions for Sums of Interacting Long-Crested Stokes Waves on Deep Water. *Journal of Ship Research* **vol.37:4,** 354-383.

Contemporary Ideas on Ship Stability
D. Vassalos, M. Hamamoto, A. Papanikolaou and D. Molyneux (Editors)

A MATHEMATICAL MODEL OF SHIP MOTIONS LEADING TO CAPSIZE IN ASTERN WAVES

Masami HAMAMOTO [1)] and Abdul MUNIF [2)]

[1)] Fukui University of Technology
[2)] Graduate School of Engineering, Department of Naval Architecture and Ocean Engineering Osaka University

ABSTRACT

A reasonable mathematical model used in prediction of ship motions leading to capsize in astern waves was developed on the basis of a strip method. In this model, to realize ship capsizes, variations of metacentric heights in waves are taken into account. The variations are obtained from the balance condition of restoring moments up to an appropriate angle of inclination. Several simulations were conducted to predict the stability against capsizing of a container carrier 15000GT in severe waves due to parametric rolling for three metacentric heights, $\overline{GM}$ =0.3m, $\overline{GM}$ =0.6m and $\overline{GM}$ =0.9m. Finally stable and unstable areas leading to capsize of the ship running in severe astern seas, for the three metacentric heights, were presented

KEYWORDS

Ship Capsizes, Strip Method, Variation of Metacentric Height, Energy Balance, Parametric Rolling, Astern Waves

INTRODUCTION

As well known a linearized dynamic-hydrodynamic analysis of ship motion in waves has been successfully obtained in a strip method. The method now provides a workable design tool for predicting the average seakeeping performance of a ship in the early design process. Several cases, which have been successfully described by linear procedures, are ship motions, structural loads and even occurrence of seemingly nonlinear large amplitude phenomena such as the frequency of slamming and bow immersion. However, the method has been partially developed for predicting stability against capsizing due to the parametric resonance, pure loss of stability and broaching-to of a ship running through astern seas, because the strip method has been mainly concerned with small amplitude periodic motion in a higher frequency range. By taking into account the variation of metacentric height based on right arm curves of a ship in waves, it is useful to review some features of linear motion theory in hopes that its results may provide some guidance and insight into ship motion leading to capsizing. When a ship is running through waves with a constant forward speed U and encounter angle χ to waves, the linearized equations of motion with respect to heave displacement ζ_G, pitch angle θ, sway displacement η_G, yaw angle ψ and roll angle ϕ, are described by:

$$
\begin{aligned}
m\ddot{\zeta}_G &= Z(Rad) + Z(Dif) + Z(FK) + W \\
I_{yy}\ddot{\theta} &= M(Rad) + M(Dif) + M(FK) \\
m\ddot{\eta}_G &= Y(Rad) + Y(Dif) + Y(FK) \\
I_{zz}\ddot{\psi} &= N(Rad) + N(Dif) + N(FK) \\
I_{xx}\ddot{\phi} &= K(Rad) + K(Dif) + K(FK) - W\overline{GM}\phi
\end{aligned} \tag{1}
$$

Where m is the mass of ship, and, I_{yy}, I_{zz} and I_{xx} the mass moments of inertia of ship about y, z and x-axes as shown in Figure1. $Z(Rad)$ and $Y(Rad)$ radiation forces of heave and sway motions, $M(Rad), N(Rad)$ and $K(Rad)$ radiation moments of pitch, yaw and roll motions, $Z(Dif)$ and $Y(Dif)$ diffraction forces of incident waves, $M(Dif), N(Dif)$ and $K(Dif)$ the diffraction moments of incident waves, $Z(F.K)$ and $Y(F.K)$ the Froude-Krylov forces including the hydrostatic forces, $M(F.K), N(F.K)$ and $K(F.K)$ the Froude-Krylov moments including the hydrostatic moments, W the ship weight and $\overline{GM}$ the metacentric height.

In these linearized equations, the radiation, diffraction and Froude-Krylov forces acting on a section of the ship in the equilibrium position, can be computed by the ordinary strip method, but the linearized restoring moment computed for the equilibrium position can not be used for a ship with metacentric height varying with respect to the relative position of ship to waves and wave steepness. As pointed out by Kerwin (1955) and Paulling (1961) the variation of metacentric height is caused by the change of water plane area in the flare of fore and aft parts of ship hull plane with respect to the relative position of a ship to a wave. In order to take into account the effect of the variation, the linearized restoring moment of a ship in astern seas should be described by:

$$
W\overline{GM}\left[1 + \frac{\Delta\overline{GM}}{\overline{GM}}\cos(\omega_e t - k\xi_0)\right]\phi \tag{2}
$$

Instead of $W\overline{GM}\phi$ in the last equation of Equation (1). Where $\Delta\overline{GM}$ is the variation of metacentric height, ω_e the encounter frequency of ship to waves, k the wave number and ξ_0 the initial position of ship to waves.

The problem here is how to predict the variation of metacentric height depending on the wave height to length ratio, H/λ, wave length to ship length ratio λ/L, encounter angle of ship to waves χ and the geometry of ship hull. The purpose of this study is to investigate the insight of parametric resonance taking into account the variation of metacentric height of a container carrier running with constant forward speed *U*.

VARIATION OF METACENTRIC HEIGHT

In general the metacentric height $\overline{GM}$ can be obtained from the righting arm curve which is given by a nonlinear function of roll angle ϕ. When a ship is displaced in a regular wave with roll angle ϕ and encounter angle χ of ship to waves, the Froude-Krylov moment *K(F.K)* including the hydrostatic buoyancy with respect to the rolling about the center of gravity *G* is described as follows:

$$K(F.K) = -\int_L dx \iint \left[y\left(\frac{\partial p}{\partial z}\right) - (z - \overline{OG})\left(\frac{\partial p}{\partial y}\right) \right] dydz \tag{3}$$

Where:

$$\begin{aligned} \frac{\partial p}{\partial y} &= \rho g \sin\phi + \rho gake^{-k\sigma d} \sin\phi \cos k\Theta \\ &\quad - \rho gake^{-k\sigma d} \cos\phi \sin\chi \sin k\Theta \\ \frac{\partial p}{\partial z} &= \rho g \cos\phi + \rho gake^{-k\sigma d} \cos\phi \cos k\Theta \\ &\quad + \rho gake^{-k\sigma d} \sin\phi \sin\chi \sin k\Theta \\ \Theta &= \xi_G + x\cos\chi - (y\cos\phi - z\sin\phi)\sin\chi - ct \end{aligned} \tag{4}$$

ρ is water density, g the gravitational acceleration, a the amplitude of a regular wave, k wave number, ξ_G the position of ship to wave, c phase velocity of a wave, σ the sectional area ratio of x coordinate, t time, d draft in equilibrium, $\overline{OG}$ the position of center of gravity measured from the origin of body coordinate system $O-x,y,z$ in which the x is directed forward, the z axis directed downward and y axis directed to starboard as shown in Figure 1. In this computation, the integrals are taken over all volume up to the instantaneous submerged surface. The relative position of ship to wave is defined at $t=0$ by the ratio of ξ_G to the wave length λ.

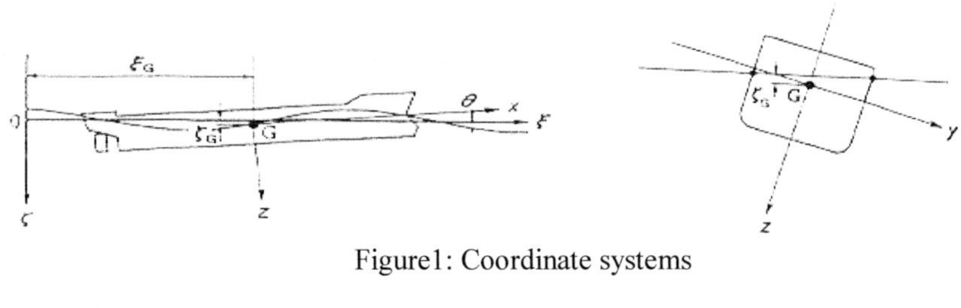

Figure1: Coordinate systems

And the displacement of submerged hull in waves must equal to the ship weight of the equilibrium condition in a still water and the pitch angle θ must be in the balance of pitch moment acting on the submerged hull. The Froude-Krylov moment $K(F.K)$ can be rewritten as:

$$\begin{aligned} K(F.K) = & -\rho g \int_L dx \iint (y\cos\phi - z\sin\phi) dydz \\ & - \rho g \overline{OG} \sin\phi \int_L dx \iint dydz \\ & - \rho gak \int_L e^{-k\sigma d} dx \iint (y\cos\phi - z\sin\phi)\cos k\Theta dydz \\ & - \rho gak \sin\chi \int_L e^{-k\sigma d} dx \iint (y\cos\phi + z\sin\phi)\sin k\Theta dydz \\ & - \rho gak \overline{OG} \sin\chi \int_L e^{-k\sigma d} dx \iint (\sin\phi\cos k\Theta - \sin\chi\cos\phi\sin k\Theta) dydz \end{aligned} \tag{5}$$

In this equation, the first and the second terms are righting moments due to the hydrostatic force acting on the submerged volume of ship hull in waves. According to the strip method, the righting arm $\overline{GZ}$ is defined by these two terms as:

$$W\overline{GZ} = \rho g \int_L dx \iint (y\cos\phi - z\sin\phi) dydz + \rho g \overline{OG} \sin\phi \int_L dx \iint dydz \tag{6}$$

In studying the large amplitude roll motion, the method of equivalent linearization has been utilized for describing a dynamic system in which large deviations from linear behavior are not anticipated. A reasonable approximation to the exact behavior of the real system, therefore, would be given by an equivalent linear system having linear coefficient appropriately selected. The $\overline{GZ}(wave)$ of container carrier as shown in Figure 2 increases at the wave trough amidship and decreases at the wave crest amidship in comparison with the righting arm $\overline{GZ}(still)$ in still water as shown in Figure 3.

When the ship is rolling in astern seas, the rolling angle can develop significantly. Therefore, the equivalent metacentric height should be determined on the basis of energy balance, that is the righting arm curve considered up to an appropriate angle of inclination as follows:

$$\begin{aligned} \int_0^{\phi_r} \overline{GM}(still)\phi\, d\phi &= \int_0^{\phi_r} \overline{GZ}(still)\, d\phi \\ \int_0^{\phi_r} \overline{GM}(trough)\phi\, d\phi &= \int_0^{\phi_r} \overline{GZ}(trough)\, d\phi \\ \int_0^{\phi_r} \overline{GM}(crest)\phi\, d\phi &= \int_0^{\phi_r} \overline{GZ}(crest)\, d\phi \end{aligned} \tag{7}$$

where ϕ_r is the vanishing angle, $\overline{GM}(still)$, $\overline{GM}(trough)$ and $\overline{GM}(crest)$ the equivalent linearized metacentric heights in still water, wave trough and wave crest respectively.

Items		Ship	Model
Length	L(m)	150	2.5
Breadth	B(m)	27.2	0.453
Depth	D(m)	13.5	0.225
Draught	d_f(m) d_a(m)	8.5 8.5	0.142 0.142
Block Coefficient	Cb	0.667	0.667
Metacentric height	GM(m)	0.3 0.6 0.9	0.051 0.077 0.112
Natural roll Period	T_ϕ(sec)	37.95 27.01 21.85	4.9 3.5 2.7
Model scale	-----		1/60

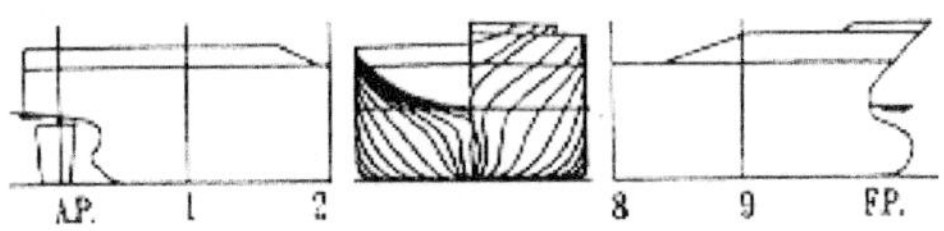

Figure 2: Principal particulars of container carrier

A further consideration is required to specify a reasonable expression of $\overline{GM}(wave)$ leading to a really equivalent solution. For this problem, an assumption is made here that the variation of metacentric height $\overline{GM}(wave)$ is sinusoidal and finally is given by the following form:

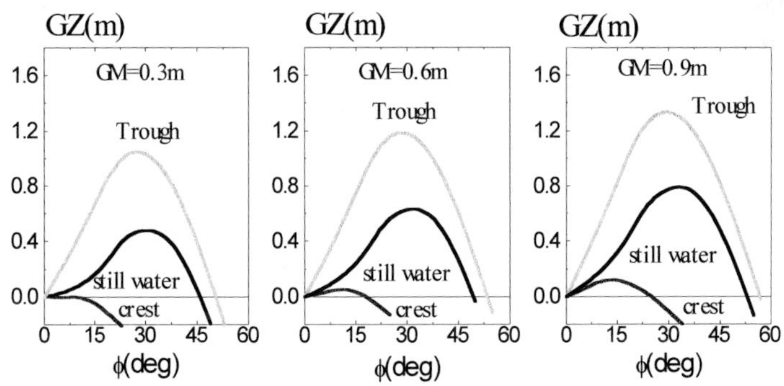

Figure 3: The righting arm curves of the ship at λ / L=1, H / λ=1/20 and χ=0

$$\overline{GM}(wave) = \overline{GM}(still)\left[1 + \frac{\Delta\overline{GM}}{\overline{GM}(still)}\cos(\omega_e t - k\xi_0)\right] \tag{8}$$

where

$$\frac{\Delta\overline{GM}}{\overline{GM}(still)} = \frac{\overline{GM}(trough) - \overline{GM}(crest)}{2\overline{GM}(still)} \tag{9}$$

The values of $\overline{GM}(still)$, $\overline{GM}(trough)$ and $\overline{GM}(crest)$ can be obtained by using the energy balance concept, their values depend on the wave steepness H/λ, the wave to ship length ratio λ/L, the encounter angle of ship to waves χ and the metacentric height $\overline{GM}$ in still water. For λ/L=1, H/λ=1/20, $\overline{GM}$=0.6m and χ=0 their values are shown in Figure 4, and the values of $\Delta\overline{GM}$ for several wave steepness and encounter angle χ are given in Figure 5.

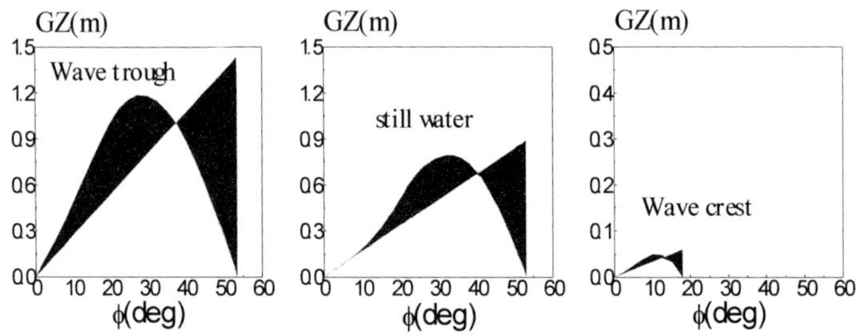

Figure 4: Equivalent linearized metacentric height for
λ/L=1, H/λ=1/20, $\overline{GM}$=0.6m and χ=0

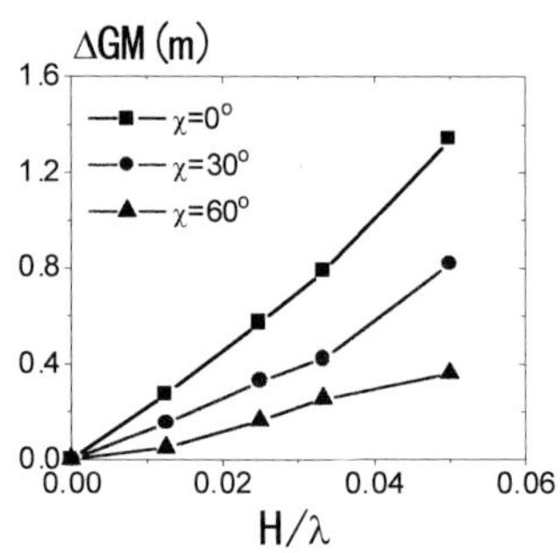

Figure 5: The variation of equivalent linearized metacentric height

MATHEMATICAL MODEL AND EXAMPLES OF NUMERICAL SIMULATIONS

According to the method mentioned in above section, equivalent linearized equations can be described in the following form:

Combined motions of heave and pitch

$$(m+m_z)\ddot{\zeta}_G + Z_{\dot{\zeta}_G}\dot{\zeta}_G + Z_{\zeta_G}\zeta_G + Z_{\ddot{\theta}}\ddot{\theta} + Z_{\dot{\theta}}\dot{\theta} + Z_{\theta}\theta = Z_C \cos\omega_e t + Z_S \sin\omega_e t$$

$$(I_{yy}+J_{yy})\ddot{\theta} + M_{\dot{\theta}}\dot{\theta} + M_{\theta}\theta + M_{\ddot{\zeta}_G}\ddot{\zeta}_G + M_{\dot{\zeta}_G}\dot{\zeta}_G + M_{\zeta_G}\zeta_G = M_C \cos\omega_e t + M_S \sin\omega_e t \qquad (10)$$

Combined motions of sway, yaw and roll

$$(m+m_y)\ddot{\eta}_G + Y_{\dot{\eta}_G}\dot{\eta}_G + Y_{\ddot{\phi}}\ddot{\phi} + Y_{\dot{\phi}}\dot{\phi} + Y_{\ddot{\psi}}\ddot{\psi} + Y_{\dot{\psi}}\dot{\psi} + Y_{\psi}\psi = Y_C \cos\omega_e t + Y_S \sin\omega_e t$$

$$(I_{zz}+J_{zz})\ddot{\psi} + N_{\dot{\psi}}\dot{\psi} + N_{\psi}\psi + N_{\ddot{\eta}_G}\ddot{\eta}_G + N_{\dot{\eta}_G}\dot{\eta}_G + N_{\ddot{\phi}}\ddot{\phi} + N_{\dot{\phi}}\dot{\phi} = N_C \cos\omega_e t + N_S \sin\omega_e t \qquad (11)$$

$$(I_{xx}+J_{xx})\ddot{\phi} + K_{\dot{\phi}}\dot{\phi} + W\overline{GM}\left[1 + \frac{\Delta\overline{GM}}{\overline{GM}}\cos(\omega_e t - k\xi_0)\right]\phi + K_{\ddot{\eta}_G}\ddot{\eta}_G + K_{\dot{\eta}_G}\dot{\eta}_G + K_{\ddot{\psi}}\ddot{\psi} + K_{\dot{\psi}}\dot{\psi} + K_{\psi}\psi = K_C \cos\omega_e t + K_S \sin\omega_e t$$

where the hydrodynamic and hydrostatic coefficients are obtained from the ordinary strip method and the metacentric height taking into account the variation of righting moment in waves is given by the equivalent linearization mention in section 2. It should be noted that the last equation in Eqn. (11) is a linear differential equation with respect to the roll angle ϕ although the unique feature of the equation is the presence of time dependent coefficient of the roll angle ϕ. Furthermore, this kind of equation has a property of considerable importance in ship rolling problem, for certain values of the encounter frequency ω_e, the solution is unstable. Physically, this implies that if the roll motion described by Eqn. (11) takes place in unstable region, the amplitude of rolling grows up. The unstable encounter frequency may be found from unstable solution of Mathieu's equation, in which unstable roll occurs when encounter frequency ω_e is equal to twice of the natural frequency ω_ϕ of roll. For this unstable condition ω_e=2 ω_ϕ, the encounter frequency ω_e is given by:

$$\omega_e = \sqrt{\frac{g}{L}}\left|\sqrt{\frac{2\pi L}{\lambda}} - Fn\left(\frac{2\pi L}{\lambda}\right)\cos\chi\right| \qquad (12)$$

and the natural frequency ω_ϕ is obtained from the natural roll period T_ϕ estimated by IMO resolution A 562 (1985) as follows:

$$\omega_\phi = \frac{2\pi}{T_\phi}$$
$$T_\phi = \frac{2B}{\sqrt{\overline{GM}}}[0.373 + 0.023(B/d) - 0.043(L/100)] \qquad (13)$$

where L is the ship length, B the breadth, d the draft, Fn the Froude number and λ the wave length.

By using these relations, it will be possible to specify the encounter frequency for the ship running with Fn and χ when the parametric resonance occurs.

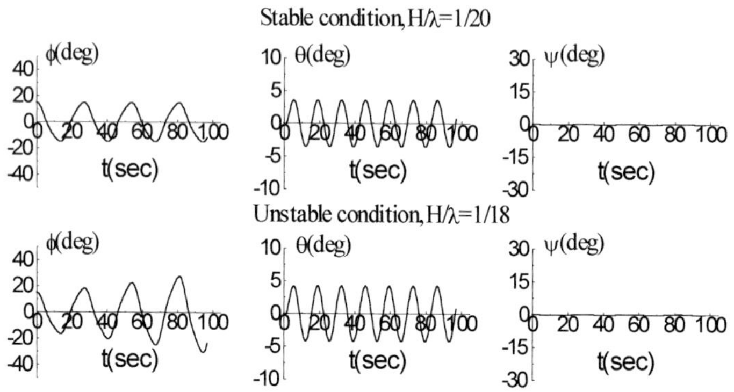

Figure6: Time history of roll, pitch and yaw in stable and unstable motions $\overline{GM}$ =0.6m, Fn =0.10935 χ =0

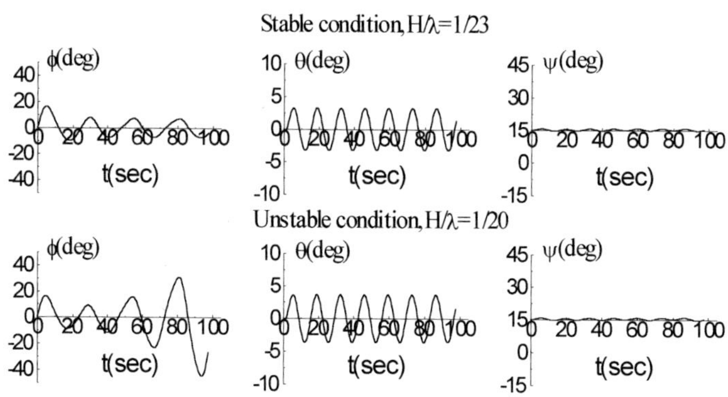

Figure 7: Time history of roll, pitch and yaw in stable and unstable motions $\overline{GM}$ =0.6m, Fn =0.11, χ =15

In general, the parametric resonance keeps a critical rolling of the constant amplitude when the energy due to the roll damping is balanced with the energy due to the variation of metacentric height. The roll angle grows up when the damping energy is smaller than energy due to the variation of metacentric height and it damps out when the damping energy is larger than the energy due to the variation of metacentric height. From the above physical point of view, several numerical simulations were carried out for the container carrier, which is running with constant speed U and encounter angle χ in astern seas. Three kinds of metacentric heights, $\overline{GM}$=0.3m, $\overline{GM}$=0.6m and $\overline{GM}$=0.9m of the container are selected to investigate the ship motion leading to capsize, the encounter angle is fixed at $\chi = 0^0, 15^0, 30^0, 45^0$ and 60^0. Figures 6, 7, 8, 9 and 10 are the time history of roll, pitch and yaw motions of the ship with metacentric height $\overline{GM}$=0.6m in stable and unstable conditions.

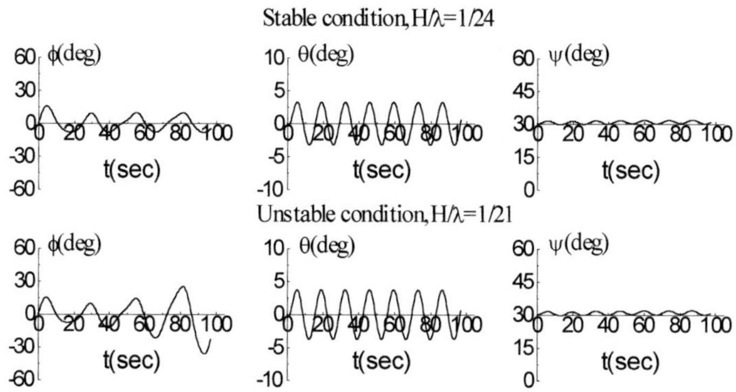

Figure 8: Time history of roll, pitch and yaw in stable and unstable motions $\overline{GM}$=0.6m, Fn=0.1263, χ=30

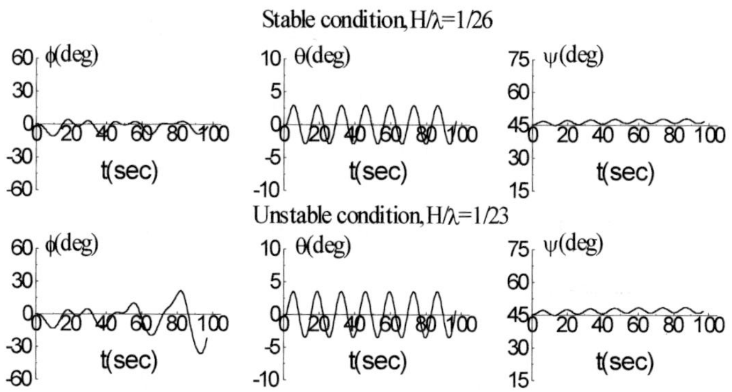

Figure 9: Time history of roll, pitch and yaw in stable and unstable motions $\overline{GM}$=0.6m, Fn=0.15 χ=45

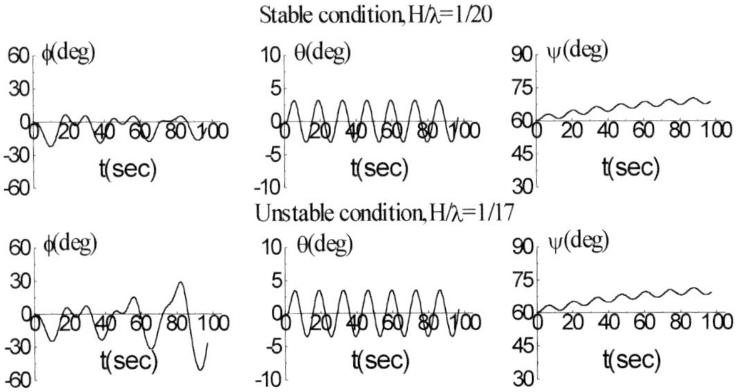

Figure10: Time history of roll, pitch and yaw in stable and unstable motions $\overline{GM}$=0.6m, Fn=0.22 χ=60

Finally, from the numerical simulations, it is possible to find out the stable and unstable area of roll motions in parametric resonance. Figure11 shows the waves steepness H/λ for encounter angle χ of the stable and unstable roll motions.

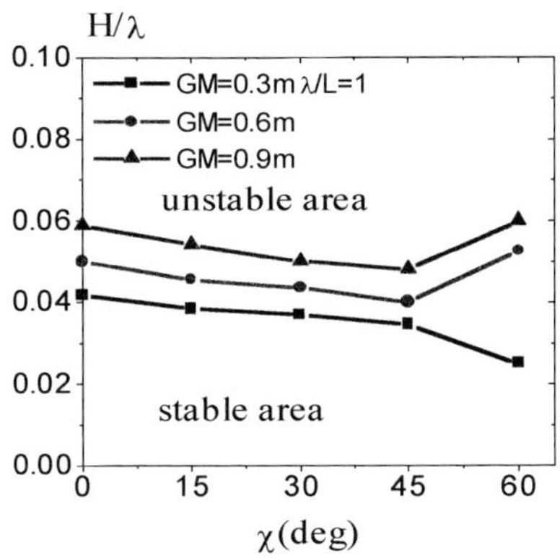

Figure 11: Stable and unstable areas of roll motions, H/λ Vs χ

CONCLUDING REMARKS

An analytical study of ship capsize phenomenon due to parametric resonance was conducted to investigate the stable and unstable areas of ship motions leading to capsize by making use of the ordinary strip method taking into account the variation of metacentric height with respect to relative position of ship to waves. The main conclusions are as follows:

1. The ordinary strip method taking into account the variation of metacentric height in waves is usable for predicting the occurrence of parametric resonance.
2. When the ship is running with large encounter angle such as $\chi = 45^0$, the roll angle deforms due to the wave excitation as shown in Figure 9 although the ship usually rolls with the natural roll period T_ϕ at the parametric resonance. This roll motion comes from the combination of the wave induced stability varying with the natural roll period and wave excitation varying with encounter period T_e.
3. For the design $\overline{GM}$=0.9m and operational $\overline{GM}$=0.6m, the most dangerous condition leading to capsize is at encounter angle approximately $45^{0.}$ However, for metacentric height, $\overline{GM}$=0.3m, the most dangerous condition is at encounter angle approximately to 60^0, because the Froude number becomes quite large to satisfy the condition of parametric resonance, then the ship is running with the velocity nearly equal to wave celerity.

ACKNOWLEDGEMENTS

This study was carried out under the scientific grant (No. 08305038) and RR 71 research panel of Shipbuilding Research Association of Japan. The authors would like to express their gratitude to the members of the RR 71, chaired by Prof. Fujino. The authors also wish to thank Prof. Saito at Hiroshima University for calculating the coefficients of radiation and diffraction.

References

Hamamoto,M., Enomoto, T., Sera,W., Panjaitan,J.P, Ito, H., Takaishi,Y., Kan, M., Haraguchi, T., Fujiwara, T. (1996). Model Experiment of Ship Capsize in Astern Seas (Second report). *Journal of the Society of Naval Architects of Japan* **179,** 77-87

Hamamoto,M., Panjaitan,J.P. (1996). Analysis on Parametric Resonance of Ships in Astern Seas. *Proceeding of Second Workshop on Stability and Operational Safety of Ships, Osaka*

Hamamoto,M., Sera,W., Panjaitan,J.P. (1995). Analysis on Low Cycle Resonance of Ship in Irregular Astern Seas. *Journal of the Society of Naval Architects of Japan* **178**, 137-145

IMO. (1985). *The intact Stability Criteria, Resolution A, 562*

Kerwin, J.E. (1955). Notes on Rolling in Longitudinal Waves. *International Shipbuilding Progress* **2:16**, 597-614

Lloyd A.R.J.M. (1989). *Seakeeping: Ship Behaviour in Rough Weather*, Ellis Horwood Ltd.

Paulling, J.R. (1961). The transverse Stability of a Ship in a Longitudinal Seaway. *Journal of Ship Research, SNAME* **4:4,** 37-49

Paulling,J.R., Oakley, O.H., Wood, P.D. (1975). Ship Capsizing in Heavy Seas: The Correlation of Theory and Experiments. *Proceeding of the International Conference on Stability of Ships and Ocean Vehicles, Glasgow*

Umeda N.,Hamamoto,M., Takaishi,Y., Chiba, Y., Matsuda A., Sera,W., Suzuki, S., Spyrou, K., Watanabe, K. (1995). Model Experiment of Ship Capsize in Astern Seas. *Journal of the Society of Naval Architects of Japan* **177,** 207-217

D. Vassalos, M. Hamamoto, A. Papanikolaou and D. Molyneux (Editors)

A NOTE ON THE CONCEPTUAL UNDERSTANDING OF THE STABILITY THEORY OF SHIPS

A. Yücel Odabasi

Faculty of Naval Architecture and Ocean Eng., Istanbul Technical University,
Maslak 80626, Istanbul, TÜRKIYE

ABSTRACT

Since initial efforts of Grim (1952) introducing the concepts of mathematical stability into the understanding of ship stability a considerable amount of research have been conducted and expressions like parametric resonance, bifurcation, domain of attraction, Lyapunov function, boundedness and simulation are being commonly used. However to many practicing naval architects both the terminology as well as their physical interpretation remains as illusive as ever. Therefore this paper is aimed at clarifying some of these concepts and suggesting a methodology for the relative uses of mathematical stability theory, numerical simulation and model experiments in the derivation of stability criteria.

KEYWORDS

Mathematical stability theory, Lyapunov function, stability criteria.

ASSESSMENT OF STABILITY AND STABILITY CRITERIA

From the practical point of view if a ship remains in an upright position it is stable. However since the environmental forces acting on the ship as well as ship's disposition with respect to sea will change in time, determination of a safe minimum amount of stability (i.e. stability criteria) becomes necessary.

In principle, for the stability assessment of a dynamic system, one needs to define two quantities; "norm" and "measure". Norm is the quantity to indicate the state of the system. For example, in the present IMO recommendations the initial metacentric height (=GM), the maximum righting arm (=GZ_{max}), and the area under the righting arm curve of a ship are examples of norms. Measure, on the other hand, defines the acceptable values of the norm elements. Again for the present IMO recommendations the set values, for example GM > 0.15 meters, GZ_{max} > 0.20 meters and the area under the righting arm curve up to 30° > 0.055 meter-radians are examples of measures.

If one follows the intuitive concept of stability it is natural to think that if the amplitudes of ship motions in every perceived combination of ship-environment conditions remain smaller than a predetermined safe value the ship should be considered stable. The mathematical counterparts of this definition are eventual stability and boundedness. However, determination of amplitudes of ship motions in varying ship and environmental conditions is not an easy task. Here three distinct approaches may be employed; scaled model tests, numerical simulation and direct methods.

Scaled model tests is based on geometric similarity and looks straightforward. However, they suffer from the viscous scale effect, which in turn significantly influence the ship response in linear and parametric resonance frequencies. As such, model testing is not a perfect substitute for the real ship behaviour. Comparison of model tests for F.P.V. 'Sulisker' (see Spouge & Collins (1986)) have proven this fact conclusively. Furthermore model tests alone cannot provide sufficient information on the selection of the elements of the norm, i.e. the critical form and nor can they provide the elements of measure, since one can only test a limited set of conditions which may or may not contain the critical ship condition-environment combinations.

Both the direct stability assessment methods and the numerical simulation require presence of an equation of motion. Although there are readily available equations of motion from the linear seakeeping theory, they are not particularly suitable for representing the large amplitude motions of ships. Therefore derivation of a representative equation of motion becomes the first and probably the most important step in the derivation of a stability assessment method if direct methods or numerical simulation are to be employed. Presence of an equation of motion is also a requirement for the analysis of model tests, especially for scale effects consideration.

While the theory of direct methods is quite general, explicit results are only available for relatively simple forms of equations. For more complex equations numerical techniques need to be employed and sometimes resulting criteria may be quite conservative. Numerical simulations provide time domain realisations of ship responses and may provide both qualitative information in different phenomena leading to capsize as well as quantitative information on response statistics. Their use as a method for stability assessment can only be justified if a design condition approach (similar to those used for off-shore platforms) is employed. However, such a choice will, no doubt, bring the associated risks and the consequential criticism.

A METHODOLOGY FOR CRITERIA DEVELOPMENT

Discussions presented so far may create an impression that the development of a rational stability criteria may not be possible. This however is not the case as one adopts a more pragmatic approach instead of a search for an all embracing criteria. Recommended steps for such a development plan is presented below.

Modelling

As noted before at present there is no comprehensive and justifiable model representing non-linear ship-wave-wind interactions and consequential ship responses. Therefore one needs identify different dangerous conditions which may lead to capsize. In such an identification not just the mode of capsize but also the dominant effects must be determined. Within this context scaled model tests and full-scale observation data are invaluable. When such a classification is made then a model equation of motion is easier to construct. For example in Haddara et al. (1971) an attempt was made to classify different modes of capsize into four categories. Grochowalski's (1989, 1993) experiments attempted to identify the significance of different factors, inclusive of deck and bulwark submergence as well as drift.

In such an approach one will end up with different sets of equations of motion addressing to different potential capsize mechanism. In this phase systematic model experiments are required to assist in the verification of equations of motion via simulation. Once the form of the equation of motion is known direct stability assessment methods should be utilised to determine the critical terms and functionals which appear in the resulting criteria. Mathematical modelling should be capable of producing reliable estimates of these terms and functionals. With these substantive measures it is possible to obtain representative set of equations addressing different eventualities.

Criteria Development

The basic premise of the methodology to be proposed is similar to the approach adopted in Odabasi (1977). Here, for each potential mode of capsize separate of criteria (or criterion) needs to be developed. During the development elements of the norm vector should be preferably be derived from the mathematical stability theory. Corresponding elements of the measure vector should be obtained from simulations or from other practical considerations. For example, Özkan (1982) considering mainly the initial portion of the GZ curve (up to the location of GZ_{max}) derived a stability criteria as

$$GM > \frac{C_1}{\Delta}\sqrt[3]{[E+WM]^2 e_1} \tag{1}$$

$$e_1 = \frac{\Delta}{\phi_m^3}\left(GM\phi_m - GZ_{max}\right) \tag{2}$$

where Δ is the displacement weight of the ship, E is the upper bound of the time varying forcing, WM is the wind heeling moment, GM is the initial metacentric height, GZ_{max} is the maximum righting lever, ϕ_m is the angle where GZ_{max} occurs, and C_1 is a constant (1.89 according to, Özkan (1982)). Leaving the obvious meaning of E and WM, e_1 is an interesting functional expression with the following interpretation:

- Larger the ϕ_m, lower the minimum GM requirement,
- Larger the deviation of GZ_{max} from the linear trend, larger the GM requirement.

Leaving the arguments about the actual values of the constant C_1, this expression (especially e_1 expression) is certainly worth investigating further since it is rationally derived, simple and

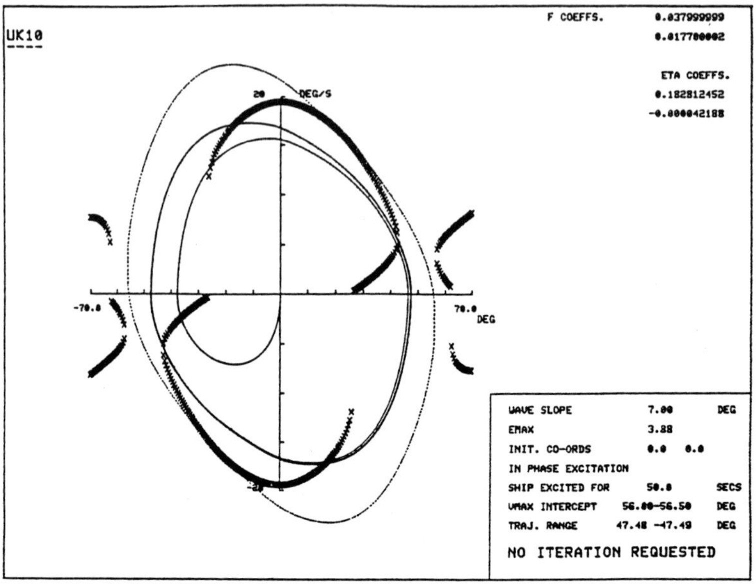

Figure 1 : Inner test simulation vs. stability bounds

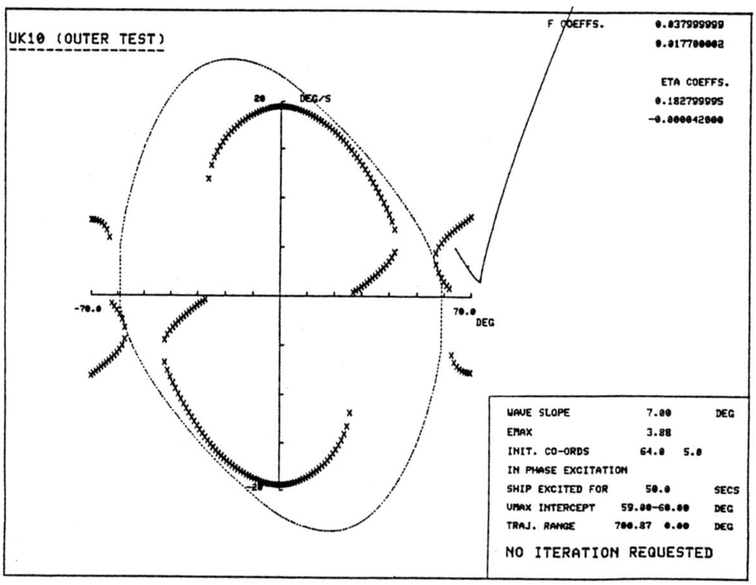

Figure 2 : Outer test simulation vs. stability bounds

practically meaningful. It is worth investigating further by simulations with a potential of becoming one of the elements of the criteria. Employing a different interpretation of the restoring curve (i.e. vanishing angle θ_v instead of ϕ_m and GZ_{max}) Thompson et al. (1990) indicated A_k/θ_v (A_k being the sustainable wave slope) and the roll damping as the determining parameters of ship stability. Here θ_v, being an angular scale, may easily be replaced with ϕ_m or any other representative angular scale and it is worth noting that A_k of Thompson and E of Özkan are closely related.

Effects of initial bias are brought home at Jiang (1996) as well as many others and almost all of the fractal analysis results (using similar state equations) indicated the significance of roll damping and nonlinearity coefficient (or its parametric representation) clearly into focus. In a much earlier paper Odabasi (1977) claimed that in nonlinear rolling resonance through phase capture may exist only in a certain range of roll damping coefficient.

Generalizing the results of Odabasi (1977) and Özkan (1982), Calderia-Saraiva (1986) proved a boundedness theorem using a Lyapunov function

$$V = \frac{1}{2}\left[\dot{\theta} + F(\theta) - h(\theta)\right]^2 + G(\theta) \tag{3}$$

for roll equation

$$\ddot{\theta} + f(\theta)\dot{\theta} + g(\theta) = e(t) \tag{4}$$

where

$$F(\theta) = \int_0^{\theta} f(s)\,ds, \quad G(\theta) = \int_0^{\theta} g(s)\,ds$$

and defined a method for the construction of $h(\theta)$ taking full consideration of a relatively general forcing term. The bounds determined by this method were tested by numerical simulation and were found to be not too conservative. Figure 1 and 2 illustrate the result of a ship named UK10 for inner and outer tests. The inner tests check numerically that the motion starting inside the Lyapunov envelope will remain inside of the same envelope. 'x' marked lines represent the boundedness auxiliary function whereas dotted lines show the Lyapunov stability limit.

In summary, it can be stated that there are a large body of research results which may provide some of necessary elements of the stability norm vector and others may need to be derived through similar methods. With these in hand we may move to safer and more defendable grounds in the development of criteria through the use of simulation, testing and historical data analysis.

CONCLUDING REMARKS

Development of stability criteria as a part of safety measures in ships has historically followed major accidents and were usually performed in a manner to address the particular problems identified as contributory to those specific incidents without addressing full problem. Current studies on safety against flooding seem to follow the same trend. A very limited set of scaled model tests are being used to derive additional stability criteria (50 cm water on deck) instead of providing well planned verification data for mathematical modelling and simulation.

Within this context it may be worth repeating Norbert Wiener's quotation (see Wiener (1920)): *"...things do not, in general, run around with their measures stamped on them like the capacity of a freight car. It requires a certain amount of investigation to discover what their measures are...What most experimenters take for granted before they begin their experiments is infinitely more interesting than any results to which their experiments lead."*
It is with this thought in mind a methodology has been proposed where various contributors may play their correct roles.

References

Calderia-Saravia, F. (1986). The Boundedness of Solution of a Linéard Equation Arising in the Theory of Ship Rolling, *IMA J. of Appl. Maths.*, **36.**

Grim, O. (1952). Rollschwingungen, Stabilitat und Sicherheit, *Schifftechnik*, **1.1**.

Grochowalski, S. (1989). Investigation into the Physics of Ship Capsizing by Combined Captive and Free-Running Model Tests, *Trans. SNAME*, **97.**

Grochowalski, S. (1993). Effect of Bulwark and Deck Submergence in Dynamics of Ship Capsizing, *U.S. Coast Guard Vessel Stability Symposium'93, New London.*

Haddara, M. R. et al. (1971). Capsizing Experiments with a Model of Fast Cargo Liner in San Fransisco Bay, *U.S. Coast Guard Project No. 723411*

Jiang, C., Troesch, A.W. and Shaw, S.W. (1996). Highly Nonlinear Rolling Motion of Biased Ships in Random Beam Seas, *J. Ship Research*, **40,** 2, 125.

Odabasi, A.Y. (1977). Ultimate Stability of Ships, *Trans. RINA*, **119**.

Özkan, I.R. (1982). Lyapunov Stability of Dynamical Systems as Applied to Ship Rolling Motion, *Int. Shipbuilding Prog.*, **29**, 329.

Spouge, J.R. and Collins, J.P. (1986). Seakeeping Trials on the Fisheries Protection Vessel Sulisker, *Int.Conf. on the Safeship Project, RINA, London*

Thompson, J.M.T., Rainey, R.C. and Soliman, M.S. (1990). Ship Stability Criteria Based on Chaotic Transients from Incursive Fractals, *Phil. Trans. R. Soc. London A332*.

Wiener, N. (1920). A New Theory of Measurement, A Study in the Logic of Mathematics, *Proc. London Math. Soc., Series 2*, **19**.

THE ROLE AND THE METHODS OF SIMULATION OF SHIP BEHAVIOUR AT SEA INCLUDING SHIP CAPSIZING

V. Armenio, G. Contento and A. Francescutto

Department of Naval Architecture, Ocean and Environmental Engineering,
University of Trieste, Via A. Valerio 10, 34127 Trieste, Italy

ABSTRACT

In this paper the research lines currently under development at the University of Trieste in the field of ship stability and safety from capsizing are shortly described. The main results regarding the development of tools for large amplitude ship motion simulation either based on concentrated parameters and on fully hydrodynamic approaches are outlined together with some suggestions for future work.

KEYWORDS

Nonlinear Dynamics, Roll Motion, Simulation, Ship Stability, Sloshing, CFD.

INTRODUCTION

Safety of life at sea and the protection of marine environment is receiving a growing interest by the different relevant parties. This is due to the too high rate of tragedy-level casualties at sea occurred during the last decade. The actions taken in this sense follow the intense phase devoted to the development of Reliability based methods of assessment of the resistance of structures and on the parallel development of techniques for Quality Management and Assurance. These last are based on the application of ISO9000 Standard Series of rules for quality control and are rapidly spreading in the world of ship construction. In line with this, IMO in his resolution A741 (18) of 4 November 1993 adopted the International Safety Management Code which constitutes a new approach to safety at sea *(Chauvel, 1994)*. Discovery, Learning, Understanding and Becoming are all practically new keywords of this new approach. Sentences as "Concern for safety is no longer focused on a product-centred response requiring technical improvements, but on human contribution and participation, in order to create a safety-conscious environment" are worth noting. In the following we shall use the word "safe" in the double sense of human life and environmental protection. Since 1980, the University of Trieste Ship Stability Research Group (SSRG), the authors belong to, has been involved in the investigation of large amplitude rolling motion. In the present paper recent results of the group in the framework of non linear formulation of the problem, including the application of CFD techniques are discussed. Finally the way the authors intend to pursue in the next future is outlined.

PRESENT WORK

The importance of Non-linear Dynamics

Rolling motion has been considered since long time the motion relevant to the hydrodynamic part of the ship safety *(Francescutto, 1993a)*. Great interest has been and is presently being paid to situations leading to large amplitude rolling, namely the rolling motion directly generated by the action of wind and waves as in the beam sea condition and the rolling motion indirectly generated by the action of waves in the following sea conditions.

Rolling motion is also crucial in the context of seakeeping and the related features of seaworthiness and seakindliness. In this perspective, large amplitude is not the only undesirable feature of roll motion where large accelerations, often obtained as a side effect of a roll stabilising system, are considered equally dangerous *(Sellars et al., 1992)*.

This twofold interest in ship rolling originates from the particular characteristics that this motion features for conventional ship forms. Roll motion is indeed the motion to which the ship opposes the minimum restoring moment and contemporaneously the minimum damping ability. As a consequence, large amplitudes can be experienced also in moderate sea states, provided that the sea spectrum is sufficiently narrow and the centerpeak frequency sufficiently close to the natural frequency of the ship *(Francescutto, 1991; Francescutto, 1993b)*. Unfortunately, these events are not so rare since the natural rolling frequency of the most part of existing ships falls within the range of frequencies of the most energetic part of the sea spectrum.

Finally, it is known that the rolling motion is a strongly non-linear phenomenon. Its intrinsic nonlinearities appear both in a qualitative and quantitative sense. Let's think for example to resonant frequency shifts and ultimately jump phenomena *(Francescutto et al., 1994a)* as a result of a non-linear restoring and/or to saturation of the amplitude at peak as a result of a strongly non-linear damping *(Contento et al., 1996)*. Biased oscillations may also occur in extreme situations and coupling effects may play dramatic roles.

Talking about all these features of the roll motion, typically experienced in Towing Tank tests and at sea, we implicitly refer to a simplified decoding of the physical problem: the traditionally adopted mathematical model (ODEs) allows us to approach different aspects of the fluid-body interaction problem. However the draft of a realistic equation of motion still stands as an open problem. Approximate roll motion equations are often used in the practice. These equations use mixed hydrodynamic/hydrostatic approaches and consider that linear or quasilinear hydrodynamic assumptions allow reliable descriptions well beyond their intrinsic validity limit.

Recently, the methods to study the complex dynamics of non-linear systems have received an extraordinary development. As a consequence, it is easier to study the possibilities of strange phenomena *(Falzarano et al., 1992)* (bifurcations, chaos, symmetry breaking, etc.) hidden in the rolling motion equation than to write down a correct non-linear equation of motion for rolling. This explains the huge amount of published papers on complex roll dynamics. The results obtained in this field are very interesting as they disclose a new world of possibilities, some of them being very dangerous. Despite that, the forecasting capability of these processes strongly depend on the reliability of the coefficients employed. In other words, even a mathematical model based on linear or quasilinear assumptions can often work well if used with 'ad hoc' parameter values, the possibility of the mentioned phenomena being often tied to very precise values for these coefficients. Moreover, bifurcations and chaos are usually studied in the deterministic case, i.e. in the presence of a regular excitation.

These coefficients may be obtained from experimental records of the motion post-processed with sophisticated parameter identification techniques. The fully theoretical calculation through analytical/numerical hydrodynamics is still to come *(Brook, 1990)* and the poor forecasting capability of the conventional seakeeping codes in the case of large amplitude motions witnesses this lack.

The growing capabilities of computers allow to face optimistically a fully hydrodynamic approach to the fluid-body interaction problem. Nevertheless simplified models based on non-linear ODEs are still needed in the perspective of a global/realistic shiphandling simulation system *(Francescutto, 1992; Francescutto, 1993a)*. The use of these 'simple' models, tuned on specific ship/sea conditions on the basis of more complicated 'pre-runs' of simulations/experiments, are undoubtedly on the side of human safety and environment protection.

With the aim to gain a deeper knowledge of the non-linear phenomena in ship rolling, according to the 20th ITTC Seakeeping Committee, campaigns of experiments on ship models in beam sea have been carried out and are presently in progress at the Hydrodynamic Laboratories of the University of Trieste *(Cardo et al., 1994)*. A numerical procedure similar to that proposed by Spouge *(1992)* for free decay tests has been developed at DINMA for the steady state oscillations in waves at constant incident wave slope and/or at constant incident wave frequency. Sophisticated models for the damping function, restoring and effective wave slope may be used, the results (coefficients) obviously being dependent on the limits of precision of the minimisation procedure and on the confidence range of the measurements *(Contento et al., 1996)*.

Several ship models of different typologies were subjected to a campaign of measurements of the steady state roll motion in beam sea *(Contento et al., 1996, 1999, Francescutto et al. 1998a,b Francescutto, 1999)*. The results are particularly interesting as regards the possibility of developing simulation tools for large amplitude rolling based on concentrated parameters mathematical models. For a description of main results, see (Francescutto & Contento, "The Mathematical Modeling of Large Amplitude Rolling in Beam Waves", in this book).

A fully hydrodynamic simulation of motions in waves - The development of a numerical Towing Tank

As mentioned before, the attractiveness of a simple mathematical model in ship motions is often frustrated by its own poor capability in describing properly the physical problem. The capability usually (but not necessarily) grows if detailed grids of experimental data, and consequently coefficients, are available. Full scale coefficients are in any case an 'a posteriori' option. As far as the theoretical predictions of loads/motions are concerned, they are fundamentally based on linearisation and on inviscid fluid assumptions apart from equivalent linearizations in the roll damping term. The non-linear effects from wave-floating body interaction are implicitly thought to be small if compared with the linear part and are therefore neglected. Kishev et al. *(1981)* have conducted a theoretical analysis up to the second order for the roll moment in forced oscillations in calm inviscid fluid. There they show that the traditional superposition of inertia, damping and restoring moment becomes inconsistent. As a consequence, in the presence of large amplitude incident waves and/or large amplitude motions the role of the excitation and of the response characteristics of the body is hardly distinguishable in the sense of the Non-linear Dynamics. Transient phenomena, slow drift motions, parametric oscillations, subharmonic responses, springing vibrations of the hull may therefore occur as a result of non-linear wave-body interaction. Recently, Contento et al. *(1996)* have conducted an extensive campaign of experimental measurements on the roll motion of a scale model of a RoRo vessel in regular beam sea and in free decay. On the other hand, conclusions similar to those of Kishev have been drawn in our analysis as shown in 2.1.

Recently, the so called 'body-exact approach' in numerical wave tanks has shown its attractiveness *(Faltinsen, 1977; Vinje et al., 1981, Isaacson, 1982; Dommermuth et al., 1987; Sen et al., 1989; Cointe et al., 1990; Sen, 1993; van Daalen, 1993; Zhao, 1993; Contento, 1996; Tanizawa, 1995; Contento and Casole, 1995)*. The main idea consists in simulating, without compromises on the amplitude of the incident wave and of the motion of the body, a 'physical' towing tank experiment in the time domain. A floating body in a closed domain is therefore subjected to an incoming wave train. Appropriate boundary conditions and/or a moving boundary allow to simulate an absorbing beach and the wavemaker respectively. The fully non-linear boundary conditions are applied both on the free surface and on the instantaneous wetted hull. Even if the inviscid fluid assumption is typically made, the method needs orders of magnitude of computational 'efforts' more than that required by linear frequency-domain solutions. In any case this matter doesn't justify the giving up of these methods when the need is stringent.

In the particular case of perfect fluid flow, after the appearance of the pioneering paper of Longuet-Higgins and Cokelet *(1976)* significant steps have been conducted in the direction of capturing second and higher order diffraction pressures on fixed structures *(Isaacson, 1982; Isaacson et al., 1991; Yeung et al., 1992; Kim et al. 1994)* or of calculating radiation/impact pressures on rigid bodies with prescribed motions in calm water *(Dommermuth et al., 1987; Zhao et al., 1993)*. Fully 3D computations seems to be a privilege of few *(Isaacson, 1982; Dommermuth et al., 1987; Tanizawa, 1995)* nevertheless several papers and applications have appeared in the 2D case *(Cointe et al., 1990; Contento, 2000; Faltinsen, 1977; Sen et al.,*

1989; Sen, 1993; Vinje, 1981; Zhao, 1993). An exhaustive review of the 'numerical wave tank approach' to the wave-body interaction problem was given by Kim *(1995)*.

Nowadays, non-linear wave loads predictions through the numerical wave tank approach are likely to become a standard procedure in ocean/coastal engineering. On the contrary, strong difficulties are encountered in ship motions computations. For example, the kinematic and dynamic characteristics of the incident waves generated by numerical wavemakers with a non-linear free surface, evidently affect the results, both as far as the pressures/loads are concerned and as regards the motion amplitudes in the floating body problem. Being obviously dependent on the amplitude and on the frequency of the motion of the wavemaker (boundary conditions), these characteristics have been shown to be dependent on computational features such as the size of the domain or the number of wavelengths in the tank *(Lee et al., 1987)*, the effectiveness of the non-reflective boundary *(Yeung, 1992; Jagannathan, 1988)* or damping sponge *(Israeli, 1981; Cointe, 1990)*, the regridding *(Dommermuth, 1987)* and the interpolation-extrapolation techniques *(Saubestre, 1991; Sen, 1993; Contento, 2000)*.

Operating at the moment in two dimensions, an accurate and robust algorithm has been developed and implemented at DINMA: the mathematical model for the wave generation and interaction with a fixed or free floating arbitrarily shaped body has been presented and deeply discussed *(Contento et al., 1996)*. Wave generation by a flap-type wavemaker and absorption by a non reflecting boundary condition are discussed in detail evidencing some physical and numerical aspects which respectively characterise and may affect the solution. In the free floating body problem, a stable and accurate procedure for the calculation of forces and moments is proposed and systematically used with good results *(Contento, 1995, 2000)*. Large amplitude motions can be simulated both prescribing the amplitude and frequency of the motion in calm water (radiation problem) and simulating free motions (decays in calm water and motions in waves). At present, a Sommerfeld radiation condition with 'numerical' celerity is enforced to allow a long time simulation *(Contento and Casole, 1995)*. Any other absorbing boundary condition can be easily implemented as well.

A flap-type wavemaker is chosen with axis of rotation at the bottom of the tank. Some significant quantities of the computed waves, such as the wave elevation, the potential and velocities, are monitored and plotted to detect the stability and accuracy of the scheme. Mass, inflow-outflow and energy-rate are systematically calculated during the simulation; moreover the Fourier analysis of the wave elevation is performed with reference to a fixed set of stations along the tank. Some results from the application of the numerical wave tank approach to the free floating body problem has been recently presented *(Contento, 1995)*. Prescribed motions in still water or in waves, free decays and motions in waves can be simulated. According to the 20th ITTC recommendations, an extensive campaign of numerical tests for internal consistency and validation against experimental data has been conducted *(Contento, 2000)*. The goodness of the comparison of numerical computation with experimental data from Vugts *(1967)* and linear theory *(Porter, 1960)* for the radiation and diffraction problem stands as a preliminary validation of the code. The intrinsic nature of the 'body exact approach' to ship motions avoids the use of traditional assumptions on the dynamics/hydrodynamics of the wave-body interaction problem (effect superposition, linear-quasi linear, ...). The results from the wave tank may however appear as a 'black box' so the application of a parameter identification technique to the records of the motion of the body from the simulations allows information about the nonlinear terms in the traditional ODEs. From an analysis like this, corroborated by Kishev et al. *(1981)*, it comes out that even a complicated non-linear ODE hardly simulates correctly free decays at moderate/large amplitudes. A computation carried out in the case of a scaled cross section after the bulb of a RoRo vessel with a pronounced flare, has shown that the immersion of the flare itself introduces sensible deviations from linearity in the decay record both in heave and in roll.

Finally, it has to be pointed out that the problem of accurate predictions of large amplitude motions for complicated geometries still remains open, mainly due to large impacts resulting in water jets and breaking waves at the liquid solid interface.

The hydrodynamic coupling between liquid sloshing and ship motions

One of the main challenges for safe ship design and operation is represented by the presence on board of liquids with free surface both desired as liquid cargo or consumable and undesired as water on deck or as a result of a flooding process. The motions connected with the presence on board of liquids are important in both aspects of ship safety: structural and hydrodynamic. The impulsive pressure peaks are important for structures, while dynamic pressures are relevant to transversal inclining moments and hence capsizing.
The introduction of double hull tankers as a response to the demand of pollution safe ships in case of grounding or collision, increases the importance of dynamic pressures since they are now relevant to structural safety too.

The importance attributed since long time to the possible danger represented by liquids with free surface on board is witnessed by the presence on intact ship stability rules of a specific regulation regarding first a correction to the initial metacentric height and successively the full curves of static and dynamic stability. In spite of the use of the term "dynamic", this is a static approach and is fully valid in this limit, i.e. in the case of infinitely slow or quasi-static inclinations of the ship.

Nowadays, it is completely understood that the hydrodynamic aspects of ship safety (and also part of the structural and operational ones) can only be treated as really dynamic phenomena, hence the consideration of sloshing motions and loads *(Francescutto, 1992; Francescutto, 1993a)*. This is particularly true when one realises that the loss of a ship at sea is generally a complex phenomenon involving the simultaneous action of different causes which superpose non linearly, to feedback phenomena and to the possible loss of structural integrity. Transient and large amplitude motions play the most relevant role. This is the reason why static, linear or quasi linear approaches cannot be used with sufficient reliability when capsizing is involved. They fail in the description of the phenomena, and therefore don't possess sufficient forecasting capability, in quantitative aspects, qualitative ones and often in both.

Previous statements are by no means questioned by some experimental observation *(Grochowalski, 1989)* reporting the correlation between actual behaviour at sea and complying to intact stability rules is quite good. In first place it is known that intact stability rules don't represent absolute safety. Furthermore, ships are an object with quite a high reliability, i.e. it is sufficient a small specimen of bad correlation between static stability and ship loss to explain the observed casualties at sea *(Francescutto, 1992)*.

From a mathematical point of view, the analysis of the roll motion of a ship with free surface liquids shipped on board constitutes a difficult task, due to the strong interaction between ship motions and liquid sloshing inside the tank. The problem as a whole can be split into two different subproblems. The first concerns the appropriate simulation of large amplitude motions of the ship, including the coupling between roll, sway and heave motion; the latter is related to the appropriate modelling of large amplitude liquid sloshing inside the partially filled tank.

In the past, several mathematical models have been proposed for the solution of such a problem. As a general rule, linear ship motion computer codes have been matched to algorithms which solve the partial differential equations modelling liquid sloshing. When the liquid depth inside the tank is small enough, the shallow water equations have been considered *(Pantazopoulos, 1990; Armenio, 1992)*, whereas in the other cases the sloshing problem has been solved by the use of BEM techniques in the hypothesis of inviscid liquids *(Francescutto et al., 1994b)*.

The shallow water equations, hyperbolic in nature, are usually solved by means of 'shock capturing' techniques developed in the framework of gasdynamic. The 'Random Choice' method, *(Chorin, 1976)*, has been widely used in the past. The method allows to treat sharp discontinuities effectively, nevertheless it is not conservative both in mass end energy. The main effects of such gaps consist in a wrong prediction of the wave speed inside the tank and in the numerical variation of water volume during computations. For the above reasons a new powerful technique (CE-SE) *(Chang et al., 1992)* recently developed has been successfully applied *(La Rocca, 1994)* for the solutions of the Shallow Water equations.

In the meantime, in order to simulate accurately large amplitude liquid sloshing in arbitrary shaped tanks (for instance equipped with internal baffle), an improved MAC method (SIMAC) *(Armenio, 1994)* has been developed. The algorithm is able to simulate large amplitude free surface flows, and, at the same time, to solve viscous stresses accurately. This last circumstance is due to its own ability in treating effectively very stretched grids. In order to validate the previous mathematical models experimental tests have been carried out considering a 0.50 meter breadth rectangular tank in roll motion *(Armenio and La Rocca, 1996)*.

Summarising, it has been proved that the use of the shallow water equations allows accurate evaluations of the wave pattern of the wave speed and of the pressure distribution over the rigid walls for filling ratios (λ/b) up to 0.10-0.12. This circumstance is basically due to the nature of the shallow water approximation and it is independent on the algorithm used for the solution of them. A detailed discussion on this topic is in *(Armenio and La Rocca, 1995a)*.

The RANSe provide accurate evaluation both of the pressure distribution and of the wave patterns in the whole range of filling ratios investigated. Nevertheless, break down of computations have to be expected when the phenomenon becomes rather violent including large splashes, breaking waves and air inclusion. Then, the coupled problem, concerning the interaction between the ship motion and liquid sloshing has been dealt with by matching the computer code solving the RANSe by means of SIMAC and the SWe by CE-SE with the uncoupled non-linear roll motion equation. In order to validate the above mathematical model experimental tests considering the scale model of a fishing vessel equipped with a rectangular tank, in a regular beam sea, have been carried out at the Hydrodynamic Laboratories of DINMA, University of Trieste.

This study has provided very interesting physical and numerical considerations *(Armenio and La Rocca, 1995b; Armenio et al., 1995)* briefly summarised in the following. The system is similar to a slightly damped two degrees of freedom oscillating system. The most significant effect due to the presence of free surface liquids on board results in the appearance of two different resonance frequencies at which ship rolling or liquid sloshing can experience large amplitude motion. The values of the new resonance frequencies are strictly related to the characteristics and the amount of the liquid shipped on board, the tank geometry and its own position inside the ship. The presence of small quantities of liquid (shallow water case) provides the reduction of the maximum roll angle as compared with the case without liquids on board. Moreover, in shallow water cases, in the zone of the anti-symmetric resonance the ship roll motion can experience the maximum roll amplitude in the frequency domain, whereas the opposite is true by increasing the liquid depth inside the tank. As regards the numerical model developed, a general good agreement with experimental data has been observed in the whole range of encounter frequencies of practical interest. When SWe are used, large disagreement between numerical and experimental results are evidenced in the first resonance zone for filling ratios greater than 0.10. As previously pointed out this circumstance is due to the shallow water approximation, that reduce the pressure field to an hydrostatic one. Despite the analysis described in 2.1, the use of constant hydrodynamic coefficients as derived by the roll decay experiments has constituted a fairly good approximation. This is mainly due to the fact that the sloshing induced moment is more than one order of magnitude larger than the roll damping moment in the whole range of wave frequencies investigated.

According to *(Rakhmanin, 1995),* in the next future a more complete mathematical model solving the coupled large amplitude roll heave and sway motion of the ship should be considered. This improvement would overcome the restrictive hypotheses that considers fixed the roll axis during the ship motion. This issue is partly controversial at least in the case of non water on board with centre of gravity not far from ship's centre of gravity recent experiments showed, on the other hand, that the influence of motions different from rolling and in the non-shallow water condition (Francescutto & Contento, 1999).

FUTURE WORK

Short-term research: fully hydrodynamic approach to ship motions with free surface liquids on board

In the previous paragraphs, an analysis has been conducted on the reasons leading to the choice of CFD approaches to ship motions forecasting and to on board free surface liquids motions/loads computations. In principle, these procedures can (and at present they do) work separately. The motion of the ship is indeed governed by Newton law where the force/moment vectors summarise the whole *external fluid actions* without any distinction between excitation and own response. Nothing forbids us to add forces and moments, if any, from the interaction of the body with other systems like moorings, wind or sloshing. From the other side, the simulation of the liquid motion in a tank needs the instantaneous knowledge of the displacements, velocities and accelerations of the wall of the container, i.e. of the ship (body forces in a fixed to the tank frame of reference). So at a specific time instant the motion of the body induced by the external hydrodynamics acts as an excitation for the liquid sloshing which in turn acts as a feedback on the motion of the ship. By this fully coupled ship/sloshing systems, the simulation of the ship in waves with free surface liquids is free from any assumption on the position of the roll axis *(Hutchinson, 1991),* i.e. important effects from sway and heave (plus surge, pitch and yaw in 3D) are no longer neglected. Moreover the inflow-outflow inside the flooded tank can be easily simulated, providing an accurate tool for the evaluation of the transient flooding after damage.

Long-term research - A "must" for future: domain decomposition

The scenario given in 2.1 to 2.3 about non linear dynamics and CFD tools for 'state of art' ship motions computations with the optional presence of free surface liquids on board, does not exhaust the problem at all. In fact, viscous effects often can affect wave loads on the hull and in a inviscid towing tank approach they can be accounted only in a 'concentrated' form (vortex shedding techniques) *(Standing et al., 1992)* and only in special cases (sharp sections where the influence of the Reynolds number on the separation point is scarce). In the field of ship resistance, the problem is rather similar in the sense that wave and viscous effects are strongly coupled. The effectiveness of an inviscid or viscous numerical approach in terms of resolution, computing time, ... has been widely discussed in the dedicated literature. Both methods are efficient in the range of validity of their formulation/application. Hence the domain decomposition has been first introduced as the 'magic potion' to the solution of the ship resistance problem: 'viscous' methods are applied on the part of the fluid domain where viscous effects are expected, namely near the body, whereas 'inviscid' methods are applied in the outer zone where the vorticity is expected to be vanished. At present this kind of simulations of unsteady oscillating flows around a surface piercing body are not available in literature. It is our intention to pursue this fascinating approach.

A BRIEF REMARK

Computing time - Spending less rarely means saving money

High reliability and precision of codes for numerical simulation is unfortunately paid in terms of computing time. Each numerical simulation requires a few hours in a medium speed workstation. While this result could be optimised, we don't consider this order of magnitude of computer time a too great limitation to the practical application of the developed methods. Apart the specific results, it allows to obtain in terms of flexibility, possibility of parameter variation, absence of scale effects, possibility of tank geometry variation, and so on. In addition we know that computations of such complexity and even more, are considered very attractive in the analysis of the circumstances of a casualty *after* its occurrence, i.e. when the problem to be solved is to find which of the parties involved in the processes of ship design, construction, maintenance, certification, operation *has to pay* for the ship loss (and its consequences as regards human life and environment). Our hope, corroborated by IMO actions as the ISM Code above mentioned, is that these computations and other relevant to safety, become more and more attractive *before* the occurrence of casualties, i.e. passing from an *understanding and learning* to a *prevention* scheme.

ACKNOWLEDGEMENTS

This research has been developed in the frame of the national project "Advanced technologies for ship safety and environmental protection" with the financial support of Ministero dell'Università e della Ricerca Scientifica e Tecnologica.

References

Armenio, V. (1992). Dynamic Behaviour of Free Surface Liquids on Ships, *Ph. D. Thesis, Dept. of Naval Architecture, Ocean and Environmental Engineering, University of Trieste.*

Armenio, V. (1994). A New Algorithm (SIMAC) for the Solution of Free Surface Unsteady High Reynolds Flows, *Proceedings, 9th International Workshop on Water Waves and Floating Bodies, Kuju, Oita, Japan*, 3-6.

Armenio, V. and La Rocca, M. (1995a). Numerical and Experimental Analysis of Liquid Sloshing in Rectangular Containers, *Proceedings, Moving Boundaries'95*, Bled.

Armenio, V. and La Rocca, M., (1995b). Numerical and Experimental Analysis of the Roll Motion of a Ship with Free Surface Liquids on Board, *Proceedings, the International Symposium on Ship Hydrodynamics*, St. Petersburg, 250-256.

Armenio, V. and La Rocca, M. (1996). On the Analysis of Sloshing of Water in Rectangular Containers: Numerical Study and Experimental Validation, *Ocean Engineering*, Vol. 23, pp. 705-739.

Armenio, V., Francescutto, A, and La Rocca, M. (1995). On the Hydrodynamic Coupling between the Ship Roll Motion and Liquid Sloshing in a Flooded Tank, *Proceedings, Int. Symposium Ship safety in a Seaway*, Kaliningrad, Vol. 2.

Brook, A. K. (1990). Evaluation of Theoretical Methods for Determining Roll Damping Coefficients, *Trans. RINA*, Vol. 132, 99-115.

Cardo, A., Coppola, C., Contento, G., Francescutto, A. and Penna, R. (1994). On the Nonlinear Ship Roll Damping Components, *Proceedings, Int. Symposium NAV'94,* Roma, Vol. 1.

Chang, R.K.C. and To, W.M. (1992). A Brief Description of a New Numerical Framework for Solving Conservation Laws: The Method of Space-Time Conservation Element and Solution Element, *Proceedings, 13th Int. Conference on Numerical Methods in Fluid Dynamics*, Roma, 396-400.

Chauvel, A-M., (1994). ISM Code: A New Approach to Safety at Sea, *Bull. Techn. du Bureau Veritas*, Vol. 76, 5-16.

Chorin, A., J. (1976). Random Choice Solution of Hyperbolic Systems, *Journal of Computational Physics*, Vol. 22, 517-533.

Cointe, R., Geyer, P., King, B., Molin, B., and Tramoni, M.P. (1990). Nonlinear and Linear Motions of a Rectangular Barge in a Perfect Fluid, *Proceedings, 18th ONR Symp. on Naval Hydrodynamics*, Ann Arbor, MI, National Academy Press, Washington D.C., 85-99.

Contento, G. (1995). A Numerical Wave Tank for the Free Floating Body Problem, *Proceedings, 10th Int. Workshop on Water Waves and Floating Bodies,* Oxford, 2-5 April.

Contento, G. (2000). Numerical Wave Tank Computations of Nonlinear Motions of 2D Arbitrarily Shaped Free Floating Bodies, *Ocean Engineering*, Vol. 27, pp. 531-556.

Contento, G. and Casole, S. (1995). On the Generation and Propagation of Waves in a Numerical Wave Tank, *Proceedings, 5th Int. Conf. on Offshore and Polar Engineering, ISOPE'95',* The Hague, 10-16 June, Vol. 3, 10-18.

Contento, G. and Francescutto, A. (1999). Bifurcations in Ship Rolling: Experimental Results and Parameter Identification Technique, *Ocean Engineering*, Vol. 26, pp. 1095-1123.

Contento, G., Francescutto, A. and Piciullo, M. (1996). On the effectiveness of constant coefficients roll motion equations, *Ocean Engineering*, Vol. 23, pp. 597-618

Dommermuth, D.G. and Yue, D.K.P. (1987). Numerical Simulation of Nonlinear Axisymmetric Flows with a Free Surface, *J. Fluid Mech.*, Vol. 178, 195-219.

Faltinsen, O.M., 1977, "Numerical Solutions of Transient Nonlinear Free Surface Motions Outside or Inside Moving Bodies", *Proceedings, 2nd Int. Conf. Num. Ship Hydro.*, Berkeley, 347-357.

Falzarano, J. M., Shaw, S. W., Troesch, A. W., 1992, "Application of Global Methods for Analyzing Dynamical Systems to Ship Rolling Motion and Capsizing", *Int. J. Bifurcations and Chaos*, Vol. 2, 101-115.

Francescutto, A., 1991, "On the Probability of Large Amplitude Rolling and Capsizing as a Consequence of Bifurcations", *Proceedings, 10th International Conference on Offshore Mechanics and Arctic Engineering 'OMAE'*, Stavanger, Vol. 2, 91-96.

Francescutto, A., 1992, "Towards a Reliability Based Approach to the Hydrodynamic Aspects of Seagoing Vessels Safety", *Proceedings, 11th International Conference on Offshore Mechanics and Arctic Engineering - OMAE'92*, Calgary, Vol. 2, 169-173.

Francescutto, A., 1993a, "Is it Really Impossible to Design Safe Ships?", *Transactions Royal Institution of Naval Architects*, Vol. 135, 1993, pp. 163-173.

Francescutto, A., 1993b, "Nonlinear Ship Rolling in the Presence of Narrow Band Excitation", in *'Nonlinear Dynamics of Marine Vehicles', ASME/DSC*, Vol. 51, 93-102.

Francescutto, A., Contento, G. and Penna, R. (1994a). Experimental Evidence of Strong Nonlinear Effects in the Rolling Motion of a Destroyer in Beam Sea", *Proceedings, 5th International Conference on Stability of Ship and Ocean Vehicles STAB'94*, Melbourne, Florida, USA, Vol. 1.

Francescutto A. and Contento G. (1994b). An Experimental Study of the Coupling between Roll Motion and Sloshing in a Compartment", *Proc. ISOPE'94, Osaka, Japan,* 3, 283-291.

Francescutto, A. and Contento, G. (1999). An Investigation on the Applicability of Simplified Mathematical Models to the Roll-Sloshing Problems, *Int. J. Offshore and Polar Engng*, Vol. 9, pp. 97-106.

Francescutto, A. (1999). An Experimental Investigation of a Dangerous Coupling Between Roll Motion and Vertical Motions in Head Sea, *Proceedings 13th International Conference on Hydrodynamics in Ship Design and joint 2nd International Symposium on Ship Manoeuvring* "Hydronav'99 – Manoeuvring'99", Gdansk, pp. 170-183.

Grochowalski S. (1989). Investigation into the Physics of Ship Capsizing by Combined Captive and Free-Running Model Tests, *Trans. SNAME*, Vol. 97, 169-212.

Hutchinson, B.L. (1991). The Transverse Plane Motions of Ships, *Marine Technology,* Vol. 28, N°2, 55-72.

Isaacson, M.St Q. (1982). Nonlinear-Wave Effects on Fixed and Floating Bodies, *J. Fluid Mech.*, Vol. 120, 267-281.

Isaacson, M.St Q., and Cheung, K.F. (1991). Second Order Wave Diffraction around Two-Dimensional Bodies by Time Domain Method, *Appl . Ocean Res.*, Vol. 13, No. 4, 175-186.

Israeli, M. and Orszag, S.A. (1981). Approximation of Radiation Boundary Conditions, *J. Comp. Phys.*, Vol. 41, 115-135.

Jagannathan, S. (1988). Non-linear Free Surface Flows and an Application of the Orlansky Boundary Condition, *Int. J. Num. Meth. in Fluids*, Vol. 8, 1051-1070.

Kim, C.H. (1995). Recent Progress in Numerical Wave Tank Research: A Review, *Proceedings, 5h Int. Conf. Offshore and Polar Engineering,* The Hague, The Netherlands, June 11-16, Vol. 3, 1-9.

Kishev, R. and Spasov, S. (1981). Second-Order Forced Roll Oscillations of Ship-Like Contour in Still Water, *Proceedings, Int. Symp. SMSSH*, Varna, Vol. 2, 28.1-28.4.

La Rocca, M. (1994). On the Roll Motion of Ships with Free Surface Liquids on Board: Solution of the Hydrodynamic Problem and Experimental Validation, *Ph.D. Thesis,* D.I.S.C., III University of Rome.

Lee, J.F. and Leonard, J.W. (1987). A Time Dependent Radiation Condition for Transient Wave-Structure Interaction, *Ocean Eng.*, Vol. 14, 469-488.

Longuet-Higgins, M.S. and Cokelet, E.D. (1976). The Deformation of Steep Surface Waves on Water. I. A Numerical Method of Computation, *Proceedings, Royal Society of London*, A, Vol. 350, 1-26.

Pantazopoulos, M. S. (1990). Sloshing of Water on Deck of Small Vessels, *Proceedings, 4th Int. Conf. on Stability of Ships and Ocean Vehicles , STAB '90*, Napoli, 58-65.

Porter, W.R. (1960). Pressure Distribution, Added Mass and Damping Coefficients fir Cylinders Oscillating in a Free Surface, Institute of Engineering Research, University of California, Berkeley, Series 82, Issue 16.

Rakhmanin, N., N. (1995). Written Contribution to : Armenio, V. et al. 1995.

Saubestre, V. (1991). Numerical Simulation of Transient Nonlinear Free-Surface Flows with Body Interaction, *Proceedings, 10th Int. Conf. Offshore Mechanics and Artic Engin., OMAE,* Vol. 1-A, Offshore Technology, ASME, 281-290.

Sellars, F. H. and Martin, J. P. (1992). Selection and Evaluation of Ship Roll Stabilization Systems, *Marine Technology*, Vol. 29, 84-101.

Sen, D., Pawlowsky, J.S., Lever, J. and Hinchey, M.J. (1989). Two-Dimensional Numerical Modelling of Large Motions of Floating Bodies in Waves, *Proceedings, 5th Int. Conf. Num. Ship Hydro.*, Hiroshima, 257-277.

Sen, D. (1993). Numerical Simulation of Motions of Two-Dimensional Floating Bodies", *J. Ship Res.*, Vol. 37, No. 4, 307-330.

Spouge, J. R. (1992). A Technique for Estimating the Accuracy of Experimental Roll Damping Measurements, *Int. Shipb. Progress*, Vol. 39, 247-265.

Standing, R.G., Jackson, G.E. and Brook, A.K. (1992). Experimental and Theoretical Investigation into the Roll Damping of a Systematic Series of Two-Dimensional Barge

Sections, *Proceedings, Int. Conf. on Behaviour of Offshore Structures, BOSS'92* , Vol. 2, 1097-1111.

Tanizawa, K. (1995). A Nonlinear Simulation Method of 3D Body Motions in Waves: Formulation with the Acceleration Potential, *Proceedings, 10th Int. Workshop on Water Waves and Floating Bodies,* Oxford, 2-5 April.

van Daalen, E.F.G. (1993). Numerical and Theoretical Studies of Water Waves and Floating Bodies, *Ph.D. Thesis,* University of Twente, the Netherlands.

Vinje, T. and Brevig, P. (1981). Nonlinear Ship Motions, *Proceedings, 3rd Int. Conf. Num. Ship Hydro.*, Paris, 257-268.

Yeung, R.W. and Vaidhyanathan, M. (1992). Non-Linear Interaction of Water Waves with Submerged Obstacles, *Int. J. Num. Meth. in Fluids*, Vol. 14, 1111-1130.

Vugts, Ir. J.H. (1968). The Hydrodynamic Coefficients for Swaying, Heaving and Rolling Cylinders in a Free Surface, *Tech. Rep. No. 194,* Ship Hydromechanics Laboratory, Delft University of Technology.

Zhao, R. and Faltinsen, O. (1993). Water Entry of Two-Dimensional Bodies, *J. Fluid Mech.*, Vol. 246, 593-612.

GEOMETRICAL ASPECTS OF THE BROACHING-TO INSTABILITY

K. J. Spyrou

Centre for Nonlinear Dynamics and its Applications
University College London, Gower Street, London WC1E 6BT, UK

ABSTRACT

Recent developments towards the clarification of the dynamics of the broaching-to mode of ship instability are reported. A multi-degree nonlinear mathematical model of an automatically steered ship operating in astern seas is taken as the basis of the investigation. A basic novelty of the approach lies in the fact that it unifies contemporary methodologies of ship controllability and transverse stability studies, within the framework of modern dynamical systems' theory. Specific nonlinear phenomena are identified as responsible for the onset of broaching behaviour. Steady-state and transient responses are investigated and it is shown how capsize is incurred during the forced turn of broaching. A classification of broaching mechanisms has been developed, concerning frequencies of encounter near zero, where surf-riding plays the dominant role, as well as frequencies of encounter away from zero, where, the instability is inherent of the overtaking-wave periodic mode.

KEYWORDS

Ship, broaching, surf-riding, nonlinear, dynamics, homoclinic.

INTRODUCTION

The broaching-to mode of dynamic instability is considered as one of the most enigmatic types of unstable behaviour. Although there seems to exist a consensus about what constitutes broaching (sudden loss of heading, felt as quick increase of heading deviation, sometimes ending with capsize, in spite of efforts to regain control), until recently we were lacking a satisfactory description of the fundamental dynamical phenomena that underpin the onset of such an undesirable behaviour. Nevertheless, earlier research has offered some valuable insights about the horizontal-plane stability of steered or unsteered ships in astern seas, particularly at zero frequency of encounter (Rydill 1959, Wahab & Swaan 1964, Eda 1972, Motora et al 1982, Renilson & Driscoll 1982). It is known for example that, yaw instability is more likely to arise when the ship centre rests at the down-slope of a long wave and, in particular, nearer to the trough where the wave yaw moment tends to turn the ship towards the beam sea condition. We also know that around the crest it is likely to encounter instability in surge. The common denominator for many earlier approaches is the concentration on local stability something that, in the spirit of 50's or 60's, was basically the result of necessity rather than choice. However, in recent years there has been increased interest for what was seen as a daunting task in the past, the study of large-amplitude motions where the system's response is shaped by nonlinear effects. Consideration of motion nonlinearity can open new horizons in at least two different ways: At first, it can often explain the instability of the linearised system as the fingerprint of the presence of some bifurcation phenomenon. Secondly, it can reveal the existence of new, qualitatively different types of responses that were not accounted for by the linearised model. Perhaps the best known (and verified) examples of nonlinear behaviour in ship dynamics are, the roll response curve near resonance and the spiral curve of a directionally unstable vessel, see Fig. 1. It is felt that a similar, simple geometrical representation of system dynamics is what is urgently needed for the broaching problem.

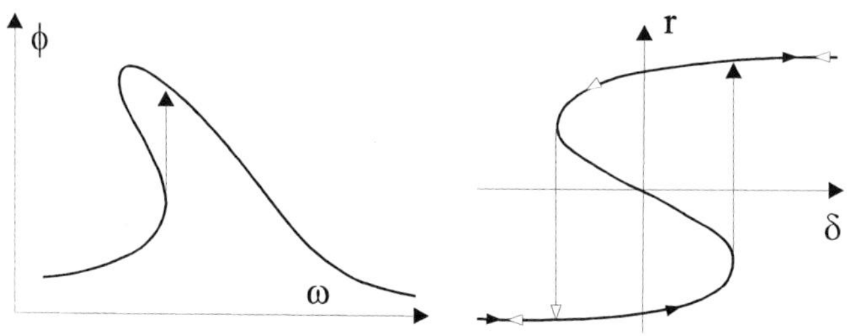

Fig. 1: The transition to resonant roll (left) and the phenomeno of directional instability of ships (right) are typica manifestations of nonlinear behaviour. For the latter linear analysis predicts instability when the rudder is se amidships. Nonlinear analysis shows in addtion th existence of a set of three states (two stable and on unstable in the middle) organized according to a classica cusp pattern (Spyrou 1990).

Geometrical considerations lie at the" heart" of a nonlinear dynamics approach and they can be also very helpful in the process of selecting a suitable mathematical model. If for example the main question asked is *what are the origins of broaching behaviour,* the answer should contain, in broad terms, a qualitative description of the underlying geometry that organises our system's state and control spaces. Such a description should be built primarily "around" specific bifurcation phenomena (notably, for typical configurations only a limited number of distinctive bifurcations can take place) and associated transient dynamic effects. On the other hand, the pursuit of quantitative accuracy will become meaningful only once a specific reference "landscape" has been established for our problem.So effects which do not alter significantly the character of system response may, at first instance, be neglected. This last observation is especially comforting for the development of a mathematical model given that still, little confidence exists about the accuracy of theoretical predictions of the hydrodynamic forces that act on moving ships in large waves.

In a series of recent papers, broaching was examined from such a nonlinear viewpoint with coosideration of multi-degree dynamics [Spyrou 1995a & 1995b, Spyrou & Umeda 1995, Spyrou 1996a, 1996b & 1996c]. Other contemporary approaches may be found for example in Rutgerson & Ottoson (1987), Umeda & Renilson (1992), Ananiev & Loseva (1994) and Umeda & Vassalos (1996). The emphasis on the multi-degree nature of the problem is very important; because from the outset it is known that broaching could involve (although not as a necessity) instabilities in at least three different directions: The first factor to deal with is the instability in surge that is connected with the surf-riding condition. Then we have the instability of yaw that triggers the uncontrolled turning motion; and finally, the instability in roll that leads to capsize. In studies of ship behaviour the systematic treatment of different instabilities within a single framework is rather unusual. However such an approach offers distinctive advantages, because it allows us to see how one instability is possible to be preparing the ground for the occurrence of the next one. By exploring the intimate connection between these dynamic effects, and also some others that are less obvious, we have developed a classification of broaching mechanisms organised in two groups (Spyrou 1996a): Broaching that involves surf-riding (basically at high Froude number); and broaching directly from the periodic motion. In the following we shall review some of the most important earlier results and furthermore, we shall present some more recent findings.

PHENOMENA OF INSTABILITY RELATED WITH SURF-RIDING IN QUARTERING SEAS

General features

It is rather well known that, a small ship sailing with high speed in an environment of large quartering waves with length at least equal to the ship length can be forced to advance with speed equal to the wave celerity , a mode of behaviour usually referred-to as, the surf-riding condition (Makov 1969; Kan 1990). The range of headings where surf-riding can arise covers a band around the purely following sea course and its onset is characterized by two speed thresholds: The first signals the existence of a pair of points of static equilibrium, that are "born " at the middle of the wave's down slope; with further increase of *Fn* they tend to move away from each other, towards the wave crest and trough. However the overtaking-wave oscillatory pattern remains in existence and it is a matter of the initial condition of the ship whether it will be engaged in the one or in the other type of behaviour. The second threshold flags the occurrence of a homoclinic connection, a bifurcation that is accountable for the abrupt disappearance of the oscillatory pattern, Fig. 2. Considering the arrangement in state-space, the limit-cycle correspoonding to the periodic motion tends to come nearer to the saddle of crest, driving the inset and outset of the saddle to be orientated almost parallel to the limit-cycle.

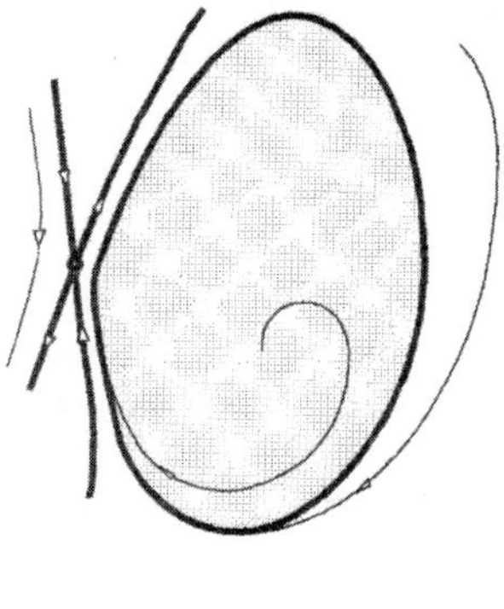

Fig. 2: Geometry of the homoclinic connection

As soon as the saddle "touches " the limit -cycle, the latter breaks and the oscillatory motion can no longer exist (from the previous description it becomes perhaps obvious why this bifurcation is also called "saddle-loop") . Thereafter a zone of headings emerges featuring stationary behaviour, due to the point of static equilibrium near trough that is stabilisable with proper selection of the proportional gain of the autopilot. The width of this zone increases at higher *Fn.* From certain initial conditions periodic motions may still be possible but the prescribed heading should lie at the outskirts of this zone, Fig. 3. At very high *Fn* the periodic pattern comes into existence again but in the present study we are not interested in that range of very high *Fn.*

In principle, the surf-riding states form a closed curve (Spyrou 1995a & 1995b). However diffraction effects have been shown to place heavy demand for larger rudder angles. Sometimes, the maximum rudder deflection is reached and, as a matter of fact, the curve cannot close (Spyrou & Umeda 1995, Spyrou 1996a). If the ship is steered with control law that icludes heading-proportional gain, the prescribed heading ψ_r is linked with the actual heading ψ with the relation $\psi_r - \psi = -\delta / a_\psi$ that involves only the rudder angle δ (in order to be found however, solution of the algebraic system of motion equations at steady state is required) and the gain a_ψ that is normally defined in advance. Obviously with a higher gain, ψ_r and ψ will tend to coincide although how exactly this happens depends on the specific form of motion equations that are "represented" through δ. The specific geometry of the $\psi_r - \psi$ relation with increasing a_ψ is shown in Fig. 4. However the relation between ψ_r and δ is independent of the gain value.

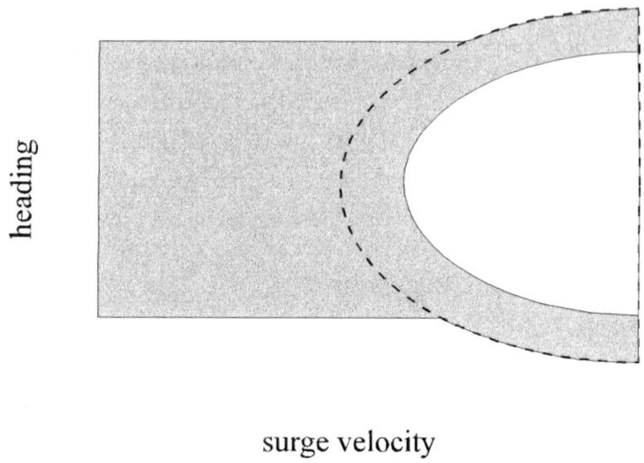

Fig. 3: Initial conditions' domain that is associated with surf-riding. The light grey area corresponds to a Froude number between the two surf-riding thresholds. The darker area shows a typical surf-riding domain where the higher surf-riding threshold is exceeded.

The condition of surf-riding should not be confused with the asymmetric oscillatory motion known as "large-amplitude surging" (Kan 1990) that precedes the first appearance of surf-riding. In terms of our system's state-space arrangement, as the limit-cycle that represents the oscillatory motion approaches the saddle of crest, the ship tends to create the impression that it remains for a while stationary at the crest of the wave (Spyrou 1996a). In a dynamical sense, this condition is significantly different from true surf-riding, where the ship remains stationary in the region of the wave trough.

Broaching at the encounter of surf-riding

Surf-riding can arise either with gradual increase of the nominal Froude number, or, as the result of a change concerning the wave parameters that represent the exogenous controls of our system. For "small" control parameter perturbations, a ship initially in overtaking-wave periodic motion will be eligible to "surf" only if it operates in the proximity of the higher surf-riding threshold; because only there the inset of the saddle of crest, that is the separatrix of stationary and periodic motion, is approached by the periodic orbit (Spyrou 1996a). Supposing that this threshold was, for some reason, exceeded, stable surf-riding will be realised only if the following two additional conditions are fulfilled: (a) at the prescribed heading the static equilibrium of the trough is stable, and, (b) the applied perturbation causes inward crossing of the boundary of the attracting domain of this stable equilibrium . When the condition (a) is not satisfied it is possible to experience surf-riding for limited time as a transient effect. As soon as the periodic motion disappears and given that stable surf-riding is not possible, the ship is left with no other option that turning towards the beam-sea, because no other stable steady-state (stationary or periodic) exists at the prescribed heading.

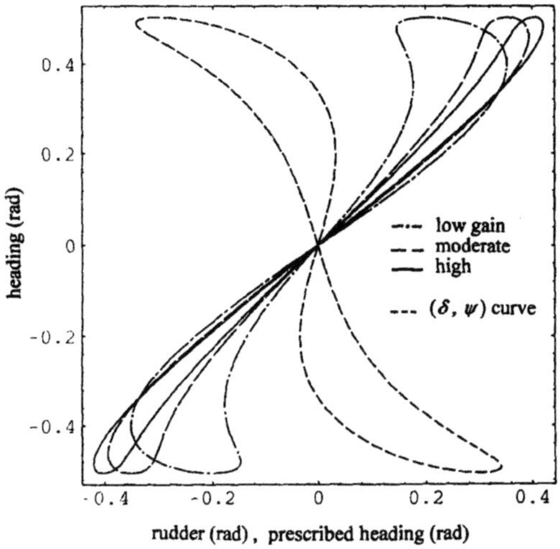

Fig. 4: Effect of proportional gain on surf-riding.

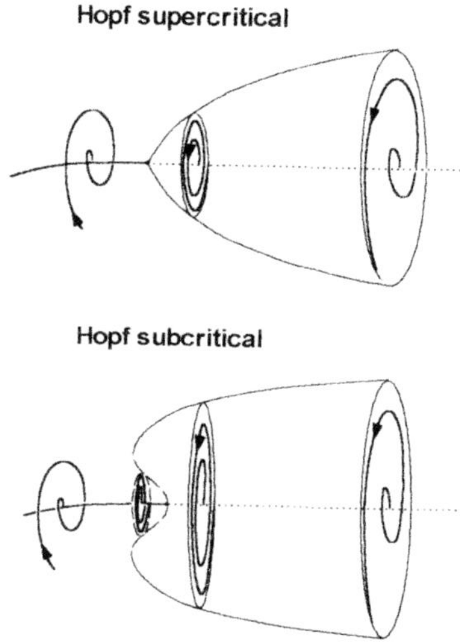

Fig. 5: The two types of Hopf bifurcation. The supercritical one was found in surf-riding

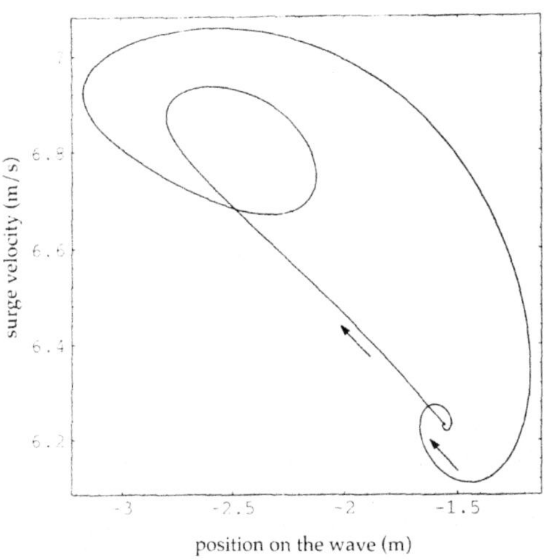

Fig. 6: The self-sustained oscillations can grow until they reach the saddle of trough where they disappear in a "blue-sky" situation.

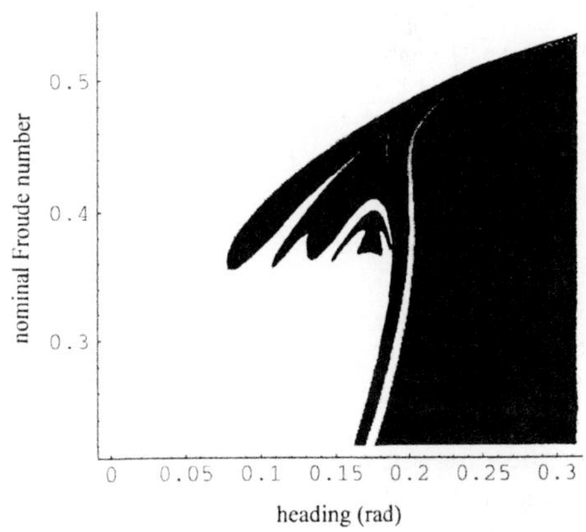

Fig. 7: Capsize (dark region) due to sudden reduction of propeller rate. The ship was initially at sterady surf-riding condition (initial $Fn = 0.56$, $\lambda/L = 2.0$, $H/\lambda = 1/20$, GM=*1.51* m.

Oscillatory surf-riding
Under certain circumstances, particularly when the heading is not very near to zero, it is possible to experience also an oscillatory-type surf-riding that is born due to a supercritical Hopf-bifurcation, Fig. 5. Moreover, the region of headings where oscillatory surf-riding is encountered can be host to period-doubling bifurcations, that lead sometimes to chaotic behaviour (Spyrou 1995b & 1996b). Generally, such behaviour was met only in very narrow ranges of control parameter values. Therein, the ship wanders at a slow rate on the down slope of a single wave, in an apparently erratic manner. Finally, there seem to exist different possible scenarios about the exit from oscillatory surf-riding. For example, the self-sustained oscillations can turn unstable at a fold, or, they can disappear in a blue-sky fashion due to a new homoclinic connection, if they collide with the saddles of trough, Fig. 6.

Voluntary escape
Another possibility is to attempt to escape voluntarily from surf-riding by setting a different propeller rate or heading, with desired destination the overtaking-wave periodic mode. For these manoeuvres, we derived the complete arrangement of domains of surf-riding, periodic motion, broaching and capsize, for the state/control parameters' plane (ψ, Fn). The characteristic layout of the capsize domain that corresponds to an abrupt reduction of propeller rate, is shown in Fig. 7. Here should be noted that, on the basis of the value ψ one could derive easily also the values of the other state-vector components.

BROACHING DIRECTLY FROM PERIODIC MOTIONS

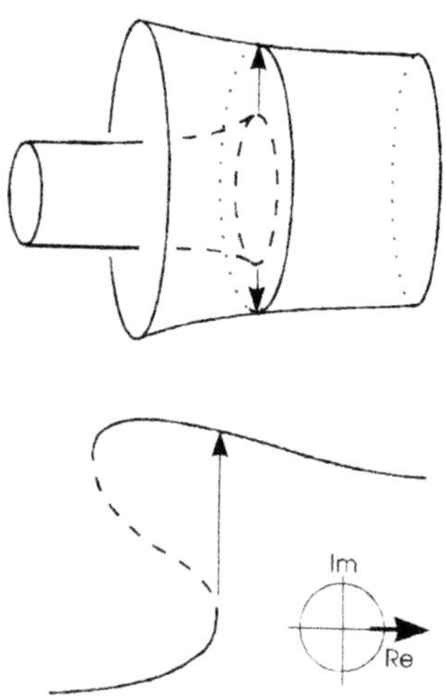

Fig. 8: Geometric representation of the jump associated with the cyclic fold.

Loss of stability of periodic motion
For larger vessels, surf-riding is rather unlikely to happen due to limitations of wave length and operational Froude number. However it is well known that broaching can occur also at frequencies of encounter that are not very near to zero (where the constraints set wave length and Froude number do not need hold). Often, this is the reflection of a change in the stability of the overtaking-wave periodic motion. It is quite common in dynamics, as periodic motions grow larger and nonlinear effects become more pronounced, to encounter phenomena such as, birth of new periodic states and exchanges of stability, coexistence of multiple periodic states, and, dynamic transitions from one periodic state to another. As discussed in detail in Spyrou (1997), in the presence of extreme excitation, periodic ship motions show a tendency to "fold" in the qualitative way shown in Fig. 8. This generates coexistence of two stable oscillatory-type motions; the one corresponds to what should be seen as the customary, relatively low amplitude, response, whereas the second corresponds to the condition of resonance,

Fig. 9. The autopilot gains, particularly the differential one, have a serious effect on the specific dynamics. It is interesting that, in many cases the resonant oscillation cannot exist in a practical sense because it extends beyond the usual rudder limits. If, with gradual increase of the prescribed heading, the state is reached where the ordinary oscillation loses stability, a transient arises that is felt as a progressive, oscillatory-type, build-up of yaw and rudder deviations. Broaching behaviour is, in this case, the manifestation of a classical transition to resonance.

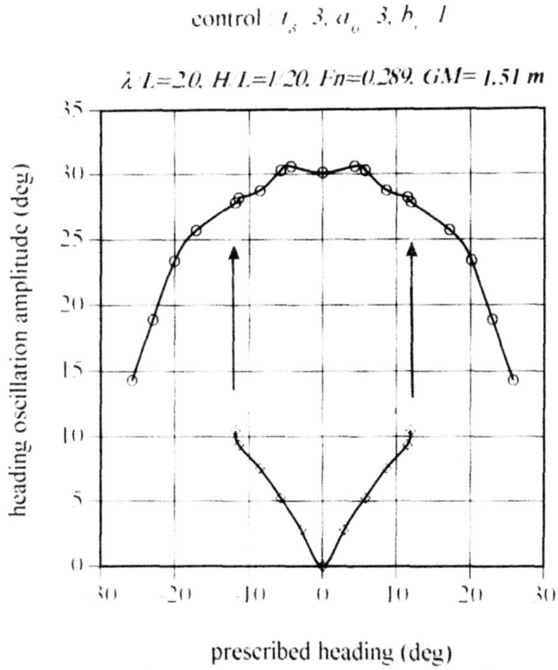

Fig. 9: Amplitude continuation for the two coexisting types of periodic motion showing how the jump from the ordinary towards the resonant oscillation comes about at a fold. With reduced *GM* the jump occurs at lower prescribed heading. The reference vessel is the same purse-seiner of our earlier studies.

Excessive rudder oscillation

This route is rather simple to perceive, and has been discussed to some length in Spyrou (1995b) and in Spyrou (1996a). At a certain combination of prescribed heading, wave length and wave height, the one end of the rudder's oscillatory motion reaches its physical limit. Thereafter, if the wave characteristics are fixed and higher heading, are prescribed (or. if the wave characteristics are altered in a sense that the yaw excitation on the ship is increased for the same prescribed heading), the preservation of the oscillatory motion requires rudder moments which however are not available due to system constraints. This scenario is meaningful when the rudder maintains considerable lift-producing capability up to large angles, something that often happens for fishing vessels .

CONCLUDING REMARKS

The association between well known nonlinear phenomena and broaching behaviour, together with the concurrent treatment of different, yet intrinsically connected instabilities, provide a new perspective for the study of a problem that has remained at the centre of research interest for nearly fifty years.

In earlier studies of broaching it was often tacitly assumed that, even in large waves, a small ship approaches the zero encounter-frequency condition in a smooth, basically "linear" manner. As we have seen however, this takes place in fact in a much more dynamic fashion where, at a certain stage, the ship is accelerated up to a speed as high as the wave celerity (a typical transition is from *Fn*=0. 35 or 0.40 to *Fn*=0.56 for wave length that is two times the ship length). Moreover, such transitions seem to concern only a limited range of headings "around" the following-sea course and there exists a sharp boundary that separates, in state-space, this from the domain of ordinary overtaking-wave response. The fact that this transition is abrupt in nature puts in question also the effectiveness of the customary examination of yaw-sway stability at zero frequency-of-encounter where, after all, properly selected autopilot gains are known to remove this type of instability . The problem however should be in reality considerably more complex since we ate dealing here with a change of state where the initial (periodic) and the final (stationary) state of the ship are dynamically unrelated. One question that immediately comes to mind is, whether the autopilot gains that seem adequate for maintaining stable periodic motion can be equally effective for the stationary mode (Spyrou 1997).

Of course, it cannot be forgotten that real ship operation will present several differences from the idealised system that was considered here. For example, long-maintained regular waves are rather rarely encountered in nature. Moreover, control systems and strategies may be considerably more complex from the simple model of control assumed here; and of course it is not unlikely that, for certain circumstances, some other important effects are not catered for by our mathematical model. Nevertheless, the understanding that a number of specific phenomena govern, to a large extent, the behaviour of the system in a physical sense, provides the solid basis on which future improvements can be specified and assessed.

It is believed finally that a similar know-how should be developed soon as regards experimental techniques since only through the physical verification of the predicted modes of behaviour we can expect that, one day, such knowledge will be integrated effectively in the customary design and operartional procedures of ships.

References

Ananiev, D.M and Loseva, L., (1994) Vessel's heeling and stability in the regime of manoeuvring and broaching in following seas. *Proceedings,* Fifth International Conference on Stability of Ships and Ocean Vehicles. Melbourne, Florida.

Eda, H. (1972) Directional stability and control of ships in waves, *Journal of Ship Research* **16/3**.

Kan, M. (1990) Surging of large amplitude and surf-riding of ships in following seas. *Selected Papers in Naval Architecture and Ocean Engineering* **28**, The Society of Naval Architects of Japan,.

Makov Y (1969) Some Results of Theoretical Analysis of Surf-Riding in Following Seas (in Russian). *Transaction of Krylov Society*, Vol. 126, pp.124-128.

Motora, S., Fujino, M. and Fuwa, T. (1982) On the mechanism of broaching-to phenomena. *Proceedings*, Second International Conference on Stability of Ships and Ocean Vehicles", Tokyo.

Renilson M.R. and Driscoll, A. (1982) Broaching - An investigatiom into the loss of directional control in severe following seas. *Trans.* RINA **124**.

Rutgerson, O. and Ottoson, P . (1987) Model tests and computer simulations - An effective combination for investigation of broaching phenomena". *Trans.* SNAME **95**.

Rydill, L.J. (1959) A linear theory for the steered motion of ships in waves", *Trans.* RINA **101**.

Spyrou, K.J. (1990) A new approach for assessing ship manoeuvrability based on dynamical systems' theory". *PhD Thesis*, Univ. of Strathclyde, Dept. of Ship and Marine Technology, Glasgow.

Spyrou, K. (1995a) Surf-riding, yaw instability and large heeling of ships in following/quartering waves. *Ship Technology Research/Schiffstechnik* **42/2**.

Spyrou, K.J., (1995b) Surf-riding and oscillations of a ship in quartering waves. *Journal of Marine Science and Technuiogy* **1/1**.

Spyrou, K.J. and Umeda, N. (1995) From surf-riding to loss of control and capsize: A model of dynamic behaviour of ships in following/quartering seas. *Proceedings*, Sixth lnternational Symposium on Practical Design of Ships and Mobile Units, PRADS '95, Seoul.

Spyrou, K.J . (1996a) Dynamic instability in quartering waves: The behaviour of a ship during broaching. *Journal of Ship Research* **40/1**.

Spyrou, K.J. (1996b) Homoclinic connections and period doublings of a ship advancing in quartering waves. *CHAOS* **6/2**.

Spyrou, K.J. (1996c) Dynamic instability in quartering waves: Part II - Analysis of ship roll and capsize for broaching. *Journal of Ship Research* **40/4**.

Spyrou, K.J. (1997) Dynamic instability in quartering seas - Part III: Nonlinear effects on periodic motions. *Journal of Ship Research* **41/3**.

Umeda N and Renilson, M R (1992): Broaching - A Dynamic Behaviour of a Vessel in Following Seas -, In: Wilson, P.A. (editor) Manoeuvring and Control of Marine Craft, Computational Mechanics Publications, Southampton, pp 533-543.

Umeda, N. and Vassalos, D. (1996) Nonlinear periodic motions of a ship running in following and quartering seas. *Journal of the Society of Naval Ardl itects of Japan*, 179.

Wahab, R. and Swann, W.A. (1964) Coursekeeping and broaching of ships in following seas. *Journal of Ship Research*, 7/4.

Contemporary Ideas on Ship Stability
D. Vassalos, M. Hamamoto, A. Papanikolaou and D. Molyneux (Editors)

APPLICATION OF NONLINEAR DYNAMICAL SYSTEM APPROACH TO SHIP CAPSIZE DUE TO BROACHING IN FOLLOWING AND QUARTERING SEAS

N. Umeda*

National Research Institute of Fisheries Engineering,
Ebidai, Hasaki, Ibaraki, 314-0421, Japan

ABSTRACT

This paper describes a method for applying a nonlinear dynamical system approach to capsize due to broaching of a ship running in regular following and quartering seas. First, a reasonable but simple mathematical model is presented for a coupled surge-sway-yaw-roll motion of an automatically steered ship with low encounter frequency. Second, methods for obtaining steady states, such as, equilibrium points and periodic orbits, of the nonlinear dynamical system described by the mathematical model are provided. Then numerical testing technique utilising a sudden change of control parameters is proposed to identify the final state where the ship is captured.

KEYWORDS

capsizing, broaching, dynamical system, sudden change concept, following seas, quartering seas, fixed point, periodic orbit

NOMENCLATURE

a_H	interaction factor between hull and rudder
A_R	rudder area
$B(x)$	sectional breadth

* Address correspondence to: N. Umeda, Department of Naval Architecture and Ocean Engineering, Osaka University, 2-1 Yamadaoka, Suita, Osaka, 565-0871, Japan

c	wave celerity
$d(x)$	sectional draught
D_p	propeller diameter
F_n	nominal Froude number
g	gravitational acceleration
GZ	righting arm
H	wave height
I_{xx}	moment of inertia in roll
I_{zz}	moment of inertia in yaw
J	advance coefficient of propeller
J_{xx}	added moment of inertia in roll
J_{zz}	added moment of inertia in yaw
k	wave number
K_p	derivative of roll moment with respect to roll rate
K_r	derivative of roll moment with respect to yaw rate
K_R	rudder gain
K_T	thrust coefficient of propeller
K_v	derivative of roll moment with respect to sway velocity
K_w	wave-induced roll moment
K_δ	derivative of roll moment with respect to rudder angle
K_ϕ	derivative of roll moment with respect to roll angle
L	ship length between perpendiculars
m	ship mass
m_x	added mass in surge
m_y	added mass in sway
n	propeller revolution number
N_r	derivative of yaw moment with respect to yaw rate
N_v	derivative of yaw moment with respect to sway velocity
N_w	wave-induced yaw moment
N_δ	derivative of yaw moment with respect to rudder angle
N_ϕ	derivative of yaw moment with respect to roll angle
OG	vertical distance between centre of gravity and waterline
p	roll rate
r	yaw rate
R	ship resistance
S(x)	sectional area
t	time
T	propeller thrust
T_D	time constant for differential control
T_E	time constant for steering gear
u	surge velocity
v	sway velocity
w_p	effective propeller wake fraction

x_H	longitudinal position of centre of interaction force between hull and rudder
x_R	longitudinal position of rudder
X_w	wave-induced surge force
Y_r	derivative of sway force with respect to yaw rate
Y_v	derivative of sway force with respect to sway velocity
Y_w	wave-induced sway force
Y_δ	derivative of sway force with respect to rudder angle
Y_ϕ	derivative of sway force with respect to roll angle
z_H	vertical position of centre of sway force due to lateral motions
z_{HR}	vertical position of centre of effective rudder force
z_R	vertical position of centre of rudder
δ	rudder angle
ε_R	wake ratio between propeller and hull
κ_p	interaction factor between propeller and rudder
λ	wave length
Λ	rudder aspect ratio
ξ_G	longitudinal position of centre of gravity
ρ	water density
$\rho S_y(x)$	sectional added mass in sway
$\rho S_y l_\eta(x)$	sectional added moment of inertia in roll
ϕ	roll angle
χ	heading angle from wave direction
χ_c	desired heading angle for auto pilot
ω	wave frequency
ω_e	averaged encounter frequency
ζ_w	wave amplitude

INTRODUCTION

A nonlinear system can possess several coexisting attractors. Thus, it is very difficult to clarify a global feature of the system by only repeating time domain simulations with a limited number of initial conditions. In particular, safety problems in engineering require us to exclude all potential danger in advance. For this purpose, the nonlinear dynamical system approach is the most suitable. It investigates recurrent behaviours of trajectories in a phase space. For a dissipative system, this approach has produced many and useful fruits because trajectories are attracted to subsets of the phase space. It is found that the forced Duffing equation, one of the simplest models that this approach was applied, has several steady states including even chaos attractors. Since the phase space of this system is three dimensional, several geometrical methods are very effective. (Thompson & Stewart, 1986)

In a seakeeping theory, a motion of a ship drifting in beam seas can be described with a coupled equation for sway and roll. If a wave length is much longer than the ship breadth, this equation can be simplified to a one degree of freedom equation in roll. Because, the roll radiation moment due to sway cancels out the roll diffraction moment. (Tasai, 1965) The nonlinear restoring moment can be represented with a third order polynomial or its equivalent. As a result, this roll equation is regarded as a kind of the forced Duffing equation. Thus, the nonlinear dynamical system approach has been directly applied to the roll motion and capsizing of a ship in beam seas. (e.g. Kuo & Odabasi, 1975; Thompson, 1990)

On the other hand, the IMO (International Maritime Organisation) stability criteria, especially weather criteria, almost succeed to prevent capsizing of a ship drifting in beam seas, except for capsizing due to a large scale breaking wave. It is also true, however, that the ship complying with the criteria may easily capsize when she runs in following and quartering seas. (e.g. Umeda et al., 1999) The Doppler effect can make the encounter period so long as the roll natural period or the time constant of manoeuvring motion. Thus, some instabilities may occur as coupled motions in surge, sway, roll and yaw. Capsizing due to broaching, a phenomenon that a ship cannot maintain her desired constant course despite the maximum steering effort and then suffer a violent yaw motion leading to capsizing, is one of the typical examples. For such phenomena, the above mentioned geometrical methods used for bean sea cases meet difficulties because increase of degrees of freedom prevents visualisation of the phase space. In the meanwhile, the ship master can avoid dangerous phenomena in following and quartering seas by changing ship speed or heading angle. Thus, it is important for upgrading the existing operational guidance at the IMO to comprehensively provide threshold for capsizing due to broaching by use of theoretical modelling.

Responding this practical requirement and expectation for the nonlinear dynamical system approach, the author proposes a method for applying the dynamical system approach to capsizing of a ship running in following and quartering seas. Here, as the first step, only ship motions in regular waves are discussed. Obviously effects of wave irregularity should be discussed in future.

MATHEMATICAL MODEL

Ship motions in waves are often described with a six degrees-of-freedom model as three-dimensional motions of a rigid body. On the other hand, it is desirable to use a simplest but still reasonable model for a nonlinear system because the increase in number of degrees-of-freedom makes nonlinear dynamical system analysis extremely difficult. When a ship runs in quartering seas with somewhat high speed, the ship has a certain possibility to suffer broaching. In this situation, the encounter frequency of the ship in waves becomes much smaller than the natural frequencies in heave and pitch. Surge, sway, roll and yaw motions, which have zero or very small restoring terms, significantly respond to such small encounter frequency. Therefore, heave and pitch motions can be reasonably approximated by simply tracing their stable equilibria. This approximation was well validated with a systematic comparison between captive model experiments and a strip theory in quartering waves with zero and very low encounter frequency. (Matsuda et al., 1997) Hence this investigation uses a 4 degrees-of-freedom model, surge-sway-yaw-roll model. Here it is noteworthy that all hydrodynamic terms should be obtained with the heave and pitch in a static equilibrium at zero encounter frequency.

Because of the low encounter frequency hydrodynamic forces acting on a ship, including wave-induced forces, mainly consist of hydrodynamic lift and buoyancy. Wave-making effects depending on the encounter frequency, which can be dealt with a strip theory, are negligibly small. On the other hand, vortices generated by hull sections immediately flow downstream with ship forward velocity and form trailing vortex sheets, which induce hydrodynamic lift forces. Therefore, a manoeuvring mathematical model focusing on hydrodynamic lift components (Hirano, 1980) can be recommended for broaching, while a seakeeping model focusing on wave-making components cannot be. Since wave steepness is much smaller than one, that is, at least up to 1:7, drift angle and non-dimensional yaw rate due to waves can be assumed to be as small as the wave steepness. Thus, wave effects on manoeuvring coefficients with respect to sway and yaw can be ignored as higher order terms. In addition, the longitudinal distance between the centre of the added mass and the centre of the mass is assumed to be negligibly small. Thus, a mathematical model used in this paper is based on a manoeuvring one with linear wave-induced forces but without nonlinear manoeuvring coefficients due to sway, yaw and waves.

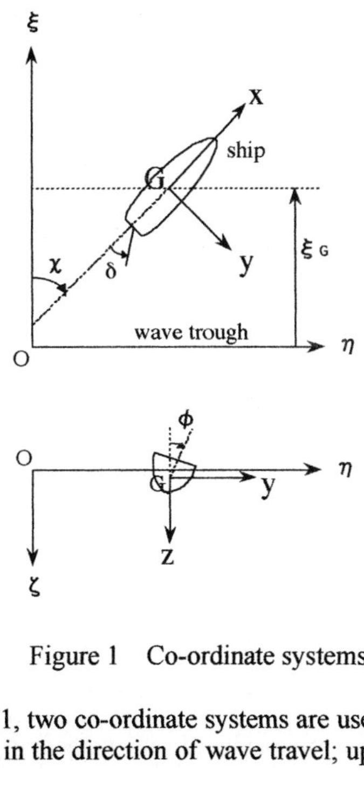

Figure 1 Co-ordinate systems

As can be seen in Figure 1, two co-ordinate systems are used: wave fixed with origin at a wave trough, the ξ axis in the direction of wave travel; upright body fixed with origin at the centre of ship gravity, the x axis pointing towards the bow, the y axis to starboard and the z axis downwards. The latter co-ordinate system is not allowed to turn about the x axis. (Hamamoto & Kim, 1993) The symbols and non-dimensionalisation are defined in the nomenclature.

The state vector, **x,** and control vector, **b,** of this system are defined as follows:

$$\mathbf{x} = \{\xi_G / \lambda,\ u,\ v,\ \chi,\ r,\ \phi,\ p,\ \delta\}^T \tag{1}$$

$$\mathbf{b} = \{n, \chi_c\}^T \tag{2}$$

The dynamical system can be represented by the following state equation:

$$\dot{\mathbf{x}} = \mathbf{F}(\mathbf{x};\mathbf{b}) = \{f_1(\mathbf{x};\mathbf{b}),\ f_2(\mathbf{x};\mathbf{b}),\ \cdots,\ f_8(\mathbf{x};\mathbf{b})\}^T \tag{3}$$

where

$$f_1(\mathbf{x};\mathbf{b}) = \{u\cos\chi - v\sin\chi - c\}\ /\lambda \tag{4}$$

$$f_2(\mathbf{x};\mathbf{b}) = \{T(u;\ n) - R(u) + X_w(\xi_G / \lambda,\ \chi)\}/(m + m_x) \tag{5}$$

$$\begin{aligned} f_3(\mathbf{x};\mathbf{b}) = \{&-(m + m_x)ur + Y_v(u;\ n)v + Y_r(u;\ n)r + Y_\phi(u)\phi \\ &+ Y_\delta(u;\ n)\delta + Y_w(\xi_G / \lambda,\ u,\ \chi;\ n)\}/(\mathrm{m} + \mathrm{m_y}) \end{aligned} \tag{6}$$

$$f_4(\mathbf{x};\mathbf{b}) = r \tag{7}$$

$$\begin{aligned} f_5(\mathbf{x};\mathbf{b}) = \{&N_v(u;\ n)v + N_r(u;\ n)r + N_\phi(u)\phi + N_\delta(u;\ n)\delta \\ &+ N_w(\xi_G / \lambda,\ u,\ \chi;\ n)\}/(\mathrm{I_{zz}} + \mathrm{J_{zz}}) \end{aligned} \tag{8}$$

$$f_6(\mathbf{x};\mathbf{b}) = p \tag{9}$$

$$\begin{aligned} f_7(\mathbf{x};\mathbf{b}) = \{&m_x z_H ur + K_v(u;\ n)v + K_r(u;\ n)r + K_p(u) \\ &+ K_\phi(u)\phi + K_\delta(u;\ n)\delta + K_w(\xi_G / \lambda,\ u,\ \chi;\ n) \\ &- mgGZ(\xi_G / \lambda,\ \chi,\ \phi)\}/(I_{xx} + J_{xx}) \end{aligned} \tag{10}$$

$$f_8(\mathbf{x};\mathbf{b}) = [-\delta - K_R(\chi - \chi_c) - K_R T_D r]/T_E \tag{11}$$

Since the external forces are functions of the surge displacement but not time, this equation is non-linear and autonomous.

The wave forces and moments can be predicted as the sum of the Froude–Krylov components and hydrodynamic lift due to wave particle velocity by a slender body theory, as shown in Appendix. Umeda et al. (1995) well validated this prediction method with a series of captive model experiments at a seakeeping and manoeuvring basin with an X-Y towing carriage. Even under the typical broaching conditions, namely, the runs with zero encounter frequency in extremely steep quartering waves, reasonably good prediction for wave-induced yaw moment in both amplitude and phase was reported. The resistance, propeller thrust and manoeuvring coefficients can be determined with circular motion tests. (e.g. Umeda & Vassalos, 1996) The added inertia terms, which cannot be obtained by circular motion tests, can be estimated theoretically in sway and yaw and empirically in surge.

An inertia co-ordinate system travelling with a mean ship velocity and mean ship course enables us to transform Eqn. (3) based on the wave fixed co-ordinate system to a nonlinear and nonautonomous model. In this model the ship motions are represented by periodic surge, $\tilde{X}_G$, sway, $\tilde{Y}_G$, roll, $\tilde{\phi}$, yaw, $\tilde{\chi}$, and rudder motion, $\tilde{\delta}$, around the inertia co-ordinate system travelling with a mean velocity and course. Here we do not assume that the surge and sway motions are small, because no restoring forces exist for these motions. When we consider the ship motions whose frequency is equal to the encounter frequency, the following van del Pol transformation is useful.

$$(u_1 \quad v_1)^T = P(\tilde{X}_G \quad \dot{\tilde{X}}_G)^T \tag{12}$$

$$(u_2 \quad v_2)^T = P(\tilde{Y}_G \quad \dot{\tilde{Y}}_G)^T \tag{13}$$

$$(u_4 \quad v_4)^T = P(\tilde{\phi} \quad \dot{\tilde{\phi}})^T \tag{14}$$

$$(u_6 \quad v_6)^T = P(\tilde{\chi} \quad \dot{\tilde{\chi}})^T \tag{15}$$

$$(u_7 \quad v_7)^T = P(\tilde{\delta} \quad \dot{\tilde{\delta}})^T \tag{16}$$

where

$$P = \begin{bmatrix} \cos\omega_e t & -1/\omega_e \sin\omega_e t \\ -\sin\omega_e t & -1/\omega_e \cos\omega_e t \end{bmatrix} \tag{17}.$$

Substituting Eqns.(12)-(16) to Eqn. (3) and averaging them over one period, that is,

$$0 < t < 2\pi/\omega_e \qquad (18),$$

the following averaged equation is obtained.

$$\dot{\mathbf{v}} = \mathbf{G}(\mathbf{v};\mathbf{b}) = \{g_1(\mathbf{v};\mathbf{b}),\ g_2(\mathbf{v};\mathbf{b}),\ \cdots,\ g_{10}(\mathbf{v};\mathbf{b})\}^T \tag{19}$$

where

$$\mathbf{v} = (u_1, v_1, u_2, v_2, u_4, v_4, u_6, v_6, u_7, v_7)^T \tag{20}$$

and the functions of $g_i(\mathbf{v};\mathbf{b})$ $(i = 1,\cdots,10)$ are shown in the separate publication (Umeda & Vassalos, 1996). Similarly we can find different averaged equations for several subharmonic motions.

STEADY STATE

In order to obtain steady states of the ship motions, the fixed points, $\mathbf{x}_0$ and $\mathbf{v}_0$, should be calculated by solving the following equations:

$$\mathbf{F}(\mathbf{x}_0;\mathbf{b}) = \mathbf{0} \tag{21}$$

$$\mathbf{G}(\mathbf{v}_0;\mathbf{b}) = \mathbf{0} \tag{22}$$

Then we examine its local stability by calculating eigenvalues of the Jacobian matrix for locally linearised equations at the fixed points.

Since $\mathbf{x}_0$ is a fixed point of the autonomous system on the wave fixed co-ordinate, it means that the ship runs with the wave celerity, a certain drift angle, a heading angle, a heel angle and a rudder angle. This is often called as surf-riding. If the auto pilot is appropriate, this fixed point can be stable at a wave down slope near a wave trough. If the auto pilot is not appropriate or heading angle is larger, this stable fixed angle easily becomes unstable. This unstable fixed point is usually saddle. (Umeda & Renilson, 1992; 1994)

$\mathbf{v}_0$ is a fixed point of the averaged equation. The averaging theorem (e.g. Guckenheimer & Holmes, 1983) indicates that, if an averaged equation has a hyperbolic fixed point, $\mathbf{v}_0$, the original equation possesses a unique hyperbolic periodic orbit of the same stability type as $\mathbf{v}_0$. Therefore, $\mathbf{v}_0$ means a periodic motion whose frequency is equal to the encounter frequency and its local stability can be examined with eigenvalues as the case of $\mathbf{x}_0$. Similarly we can investigate subharmonic motions.

TRANSIENT STATE

If the control parameter in the mathematical model changes slowly, the ship motion simply follows the above-obtained steady states. In some cases a fixed point may emerge or disappear. Its local stability may change from stable to unstable or from stable to unstable. Sometimes a periodic orbit may disappear and new fixed point may emerge. These are local or global bifurcations. As slowly-changing control parameters, the ship displacement, trim, mass distribution, wave height, wave length and so on can be pointed out. For these parameters, the analysis of steady states is enough for practical purposes.

On the other hand, the propeller revolution number, n, or the nominal Froude number, Fn, and the desired course for auto pilot, χ_c, cannot be assumed to be slowly changed. Here the nominal Froude number means the Froude number when the ship runs in otherwise calm water with that propeller revolution. These changes are rather sudden, as Figure 2, compared with the time constant of ship motion. Before this operational action for the propeller and rudder, the ship motion can be assume to exist in a stable steady state for the former control parameter set. When time tends to infinity after operational action, the ship must settle down to one of the stable steady states for the new control parameter set. However, it is not obvious which steady state is realised among several coexisting steady states. Therefore, we start to numerically integrate with time the state equation involving the new control parameters from the preceding steady state, as shown in Figure 3. If the preceding state is periodic, the phase lag between waves and the change of operational parameter is not unique although each phase lag of the ship motion to waves is unique. Thus, we have to repeat numerical integration with different phase lags between waves and the change of operational parameter. However, since the initial value set of the phase space is limited to be one dimensional, the procedure is still applicable for practical purpose. An example of this "sudden change" approach is shown in Figure 3. First the initial steady state was set to be a harmonic motion for the nominal Froude number of 0.1 and the auto pilot course of 0 degrees and then the control parameter was suddenly changed to the specified nominal Froude number and the auto pilot course of 10 degrees. Here the two steady states, a harmonic motion and capsizing due to broaching, theoretically coexist for the new control parameter set but this sudden change concept approach identified the boundary which state is practically realised between two.

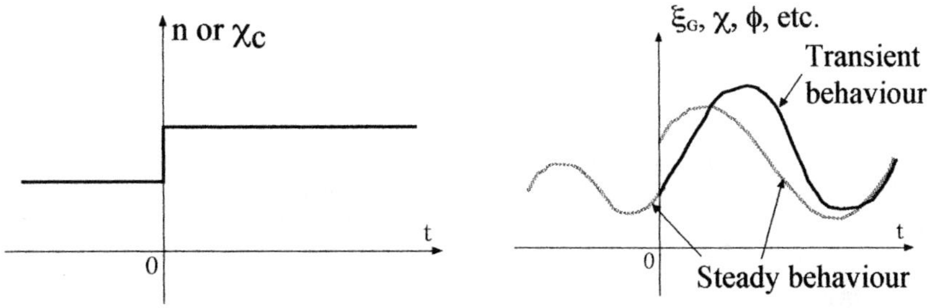

Figure 2 Sudden change concept

Although this procedure is quite straightforward, we can further reduce our effort for numerical simulation by a method in some cases. That is the analysis of invariant manifold. (e.g. Guckenheimer & Holmes, 1983) In case of uncoupled surge model in pure following seas, this analysis successfully identifies the initial value set, the domain of attraction, leading to a stable surf-riding. (e.g. Umeda, 1990) In case of coupled model in quartering seas, some attempt is recently reported for identifying the control parameter set leading to capsizing due to broaching. (Umeda, 1999) However, because of difficulty in visualisation of multi-dimensional phase space, this invariant manifold analysis is not always effective.

ACKNOWLEDGEMENTS

The work described in this paper was carried out at the University of Strathclyde during the author's stay there as a visiting research fellow from National Research Institute of Fisheries Engineering, Japan. This was supported by the Engineering and Physical Science Research Council in UK. The author expresses his sincere thanks to Professor

D. Vassalos from this university for his invitation and effective discussion.

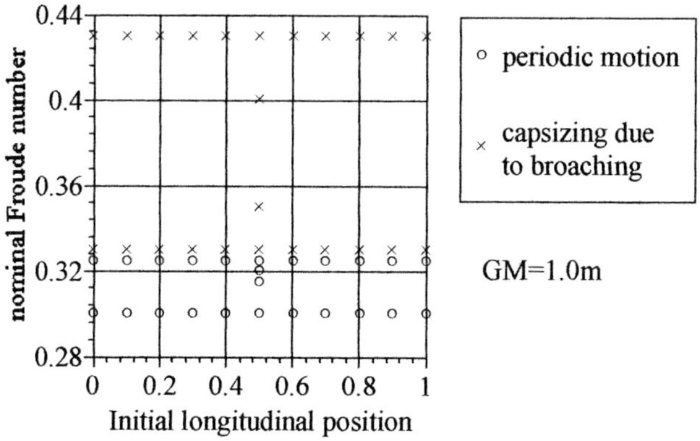

Figure 3 Results of numerical experiments based on sudden change concept for a fishing vessel. The system parameters for the experiment are $H/\lambda=1/9.2$, $\lambda/L=1.5$, $K_R=1$ and $T_D=0$. The operational parameters are changed from Fn=0.1 and $\chi_c=0$ degrees to the nominal Froude number specified at the ordinate and $\chi_c=10$ degrees. Here the initial longitudinal position indicates the phase lag between waves and the change of operational parameters divided by 2π.

REFERENCES

Guckenheimer J. and Holmes P. (1983). *Nonlinear Oscillations, Dynamical Systems, and Bifurcations of Vector Fields.* Springer-Verlag, New York, USA

Hamamoto M. and Kim Y.S. (1993). A New Coordinate System and the Equations Describing Manoeuvring Motion of a Ship in Waves (in Japanese). *Journal of the Society of Naval Architects of Japan* **173**, 209-220.

Hirano M. (1980). On Calculation Method of Ship Manoeuvring Motion at Initial Design Phase (in Japanese). *Journal of the Society of Naval Architects of Japan* **147**, 144-153.

Kuo C. and Odabasi A.Y. (1975). Application of Dynamical System Approach to Ship and Ocean Vehicle Stability. In: *Proceedings of the International Conference on Stability of Ships and Ocean Vehicles*, Uni Strathclyde, Glasgow, UK.

Matsuda A., Umeda N. and Suzuki S. (1997). Vertical Motions of a Ship Running in Following and Quartering Seas (in Japanese). *Journal of the Kansai Society of Naval Architects, Japan* **227**, 47-55.

Tasai F. (1965). On the Equation of Rolling of a Ship (in Japanese). *Bulletin of Research Institute for Applied Mechanics* **25**, 51-57.

Thompson J.M.T. (1990). Transient Basins: A New Tool for Designing Ships Against Capsize. In: Price, W.G. et al. (eds.) *Dynamics of Marine Vehicles and Structures in Waves*. Elsevier, Amsterdam, The Netherland, 325-331.

Thompson J.M.T. and Stewart H. B. (1986). *Nonlinear Dynamics and Chaos;*

Geometrical Methods for Engineers and Scientists. John Wiley & Sons Ltd., New York, USA.

Umeda N. (1990). Probabilistic Study on Surf-Riding of a Ship in Irregular Following Seas. In: *Proceedings of the 4th International Conference on Stability of Ships and Ocean Vehicles.* University Federico II of Naples, Naples, Italy, 336-343.

Umeda N. (1999). Nonlinear Dynamics of Ship Capsizing due to Broaching in Following and Quartering Seas. *Journal of Marine Science and Technology* **4:1**, (in press).

Umeda N, Matsuda A., Hamamoto M. and Suzuki S. (1999). Stability Assessment for Intact Ships in the Light of Model Experiments *Journal of Marine Science and Technology* **4:2**. (in press).

Umeda N. and Renilson M.R. (1992). Broaching - A Dynamic Behaviour of a Vessel in Following Seas -. In: Wilson P. A. (editor) *Manoeuvring and Control of Marine Craft.* Computational Mechanics Publications, Southampton, UK, 533-543.

Umeda N. and Renilson M.R. (1994). Broaching of a Fishing Vessel in Following and Quartering Seas. In: *Proceedings of 5th International Conference on Stability of Ships and Ocean Vehicles.* Florida Tech, Melbourne, USA, **3**, 115-132.

Umeda N. and Vassalos D. (1996). Non-Linear Periodic Motions of a Ship Running in Following and Quartering Seas. *Journal of the Society of Naval Architects of Japan* **179**, 89-101.

Umeda N., Yamakoshi Y. and Suzuki S. (1995). Experimental Study for Wave Forces on a Ship Running in Quartering Seas with Very Low Encounter Frequency. In: *Proceedings of International Symposium on Ship Safety in a Seaway.* Kaliningrad Tech Uni, Kaliningrad, Russia, 14:1-18.

APPENDIX

The wave forces and moments shown in Eqns.(5-6,8,10) are calculated as follows: (Umeda et al., 1995)

$$
\begin{aligned}
&X_w(\xi_G/\lambda,\chi) \\
&= -\rho g \zeta_w k \cos\chi \int_{AE}^{FE} C_1(x) S(x) e^{-kd(x)/2} \sin k(\xi_G + x\cos\chi) dx
\end{aligned}
\tag{A.1}
$$

$$
\begin{aligned}
Y_w(\xi_G/\lambda,u,\chi) &= \rho g \zeta_w k \sin\chi \int_{AE}^{FE} C_1(x) S(x) e^{-kd(x)/2} \sin k(\xi_G + x\cos\chi) dx \\
&+ \zeta_w \omega \omega_e \sin\chi \int_{AE}^{FE} \rho S_y(x) e^{-kd(x)/2} \sin k(\xi_G + x\cos\chi) dx \\
&- \zeta_w \omega u \sin\chi \left[\rho S_y(x) e^{-kd(x)/2} \cos k(\xi_G + x\cos\chi) \right]_{AE}^{FE} \\
&+ (1 + a_H) \frac{\rho}{2} A_R f_\alpha \varepsilon_R (1 - w_p) u \sqrt{1 + \kappa_P \frac{8K_T}{\pi J^2}} v_{WR}
\end{aligned}
\tag{A.2}
$$

$$
\begin{aligned}
N_w(\xi_G/\lambda, u, \chi) = {} & \rho g \zeta_w k \sin\chi \int_{AE}^{FE} C_1(x) S(x) e^{-kd(x)/2} x \sin k(\xi_G + x\cos\chi) dx \\
& + \zeta_w \omega\omega_e \sin\chi \int_{AE}^{FE} \rho S_y(x) e^{-kd(x)/2} x \sin k(\xi_G + x\cos\chi) dx \\
& + \zeta_w \omega u \sin\chi \int_{AE}^{FE} \rho S_y(x) e^{-kd(x)/2} \cos k(\xi_G + x\cos\chi) dx \\
& - \zeta_w \omega u \sin\chi \left[\rho S_y(x) e^{-kd(x)/2} x \cos k(\xi_G + x\cos\chi)\right]_{AE}^{FE} \\
& + (x_R + a_H x_H) \frac{\rho}{2} A_R f_\alpha \varepsilon_R (1 - w_p) u \sqrt{1 + \kappa_P \frac{8K_T}{\pi J^2}}\, v_{WR}
\end{aligned}
\tag{A.3}
$$

$$
\begin{aligned}
K_w(\xi_G/\lambda, u, \chi) = {} & \rho g \zeta_w k \sin\chi \int_{AE}^{FE} C_1(x) \,{}^{B(x)}\!/_2 \{d(x)\}^2 e^{-kd(x)/2} \sin k(\xi_G + x\cos\chi) dx \\
& - \rho g \zeta_w k^2 \sin\chi \int_{AE}^{FE} C_4(x) \{{}^{B(x)}\!/_2\}^3 e^{-kd(x)/2} \sin k(\xi_G + x\cos\chi) dx \\
& - \zeta_w \omega\omega_e \sin\chi \int_{AE}^{FE} \rho S_y l_\eta(x) e^{-kd(x)/2} \sin k(\xi_G + x\cos\chi) dx \\
& - \zeta_w \omega u \sin\chi \left[\rho S_y l_\eta(x) e^{-kd(x)/2} \cos k(\xi_G + x\cos\chi)\right]_{AE}^{FE} + Y_w(\xi_G/\lambda, u, \chi) \cdot \overline{OG} \\
& - (1 + a_H) z_{HR} \frac{\rho}{2} A_R f_\alpha \varepsilon_R (1 - w_p) u \sqrt{1 + \kappa_P \frac{8K_T}{\pi J^2}}\, v_{WR}
\end{aligned}
\tag{A.4}
$$

where the integrals are carried out from the aft end, AE, to the fore end, FE, of the ship, and

$$
C_1(x) = \frac{\sin(k\sin\chi \cdot B(x)/2)}{k\sin\chi \cdot B(x)/2}
$$

$$
C_4(x) = \{k\sin\chi \cdot B(x)/2\}^{-3}\left[2\sin\{k\sin\chi \cdot B(x)/2\} - k\sin\chi \cdot B(x)\cos\{k\sin\chi \cdot B(x)/2\}\right]
$$

$$
J = \frac{(1 - w_p)\, u}{nD_p} \qquad f_\alpha = \frac{6.13\Lambda}{2.25 + \Lambda}
$$

$$
v_{WR} = a\omega \sin\chi \exp(-kz_R) \cos(2\pi\xi_G/\lambda + kx_R\cos\chi).
$$

Here the sectional added mass and moment of inertia are calculated by solving the Laplace equation with the free surface condition for zero frequency as well as the hull surface condition.

Contemporary Ideas on Ship Stability
D. Vassalos, M. Hamamoto, A. Papanikolaou and D. Molyneux (Editors)

BROACHING AND CAPSIZE MODEL TESTS FOR VALIDATION OF NUMERICAL SHIP MOTION PREDICTIONS

J.O. de Kat[1] and W.L. Thomas III[2]

[1]Maritime Research Institute, Netherlands
[2]David Taylor Model Basin, NSWC, Carderock Division, USA

ABSTRACT

Model tests have been carried out with a frigate-type hull in waves leading to a variety of extreme motion events including capsizing. A specific requirement was the ability to perform tests in critical, stern quartering wave conditions at high speed and measure relevant parameters for validation of a large amplitude ship motion simulation program. The paper describes model testing techniques, test data and some comparisons with numerical simulations related to large amplitude rolling, surfriding, broaching and capsizing in following to beam waves. Tests comprised maneuvering (zigzag) tests, roll decay in calm water, and regular waves of moderate to extreme steepness for a range of GM values.

KEYWORDS

Capsize, broaching, surfriding, large amplitude rolling, model testing techniques, frigate-type hull, extreme motion.

INTRODUCTION

Since the cooperative research work presented by De Kat et al (1994), a second 4-year joint research effort on ship stability started in 1995 under sponsorship from the Cooperative

Research Navies group. This CRNAV group, which comprises five navies (from Australia, Canada, Netherlands, United Kingdom and United States), U.S. Coast Guard and MARIN, focuses its current activities on the dynamic stability assessment of intact and damaged ships using numerical simulations.

The applied numerical tools should be subjected to proper verification and validation. An extensive database is available with seakeeping and manoeuvering data for intact ships. Suitable validation data concerning large amplitude ship motions of intact ships (including broaching and capsizing) are available to a very limited extent. To make the validation database more complete in terms of extreme conditions, tests were carried out in 1997 with a free-running frigate model.

This paper describes these model tests in detail, with the objective to provide insights into the physics of extreme motion events observed in the tests and discuss validation issues from a model testing perspective. The tests cover the most critical wave directions concerning dynamic stability during ship operations: following, stern quartering and beam seas were tested at different ship speeds with Froude number ranging from Fn = 0.1 to 0.4. Observed events include surfriding, broaching associated with surfriding, extreme rolling, and capsizing in a variety of modes.

EXPERIMENTAL SETUP

Fulfilment of the extreme motion objectives required a large basin capable of generating moderate and steep waves under arbitrary heading angles. A large test basin was required to maximize run length. The capability to generate steep waves allowed a test matrix to be developed, which would ensure the occurrence of extreme events. In addition to the tests in waves, the requirements included zigzag tests and roll decay tests at different speeds in calm water.

The test matrix for runs in waves required:

1. High speed runs in beam seas through following seas of suitable run lengths to allow capsizing, broaching, and surfriding.
2. Large amplitude regular waves having typical steepnesses (H/λ of 1/20, 1/15, 1/10.
3. wave length to ship length ratios (λ /l) between .75 and 2.5

Basin

The capsize model tests and the calm water experiments were carried out in the Maneuvering and Seakeeping (MASK) Basin of the Carderock Division, Naval Surface Warfare Center. The MASK is an indoor basin having an overall length of 110 m, a width of 73 m and a depth of 6.1 m, except for a 10.7 m wide trench parallel to the long side of the basin. The Basin is spanned by a 115 m long bridge supported on a rail system that permits the bridge to transverse half the width of the basin and to rotate up to 45 degrees from the longitudinal centerline. Models can be tested at all headings relative to the waves, with the wave makers located along two adjacent sides of the basin. A towing carriage is hung from the bridge. The carriage has a maximum speed of 7.7 m/s.

Model Description

The model chosen for this study has a conventional frigate hull form; a 1/36th scale was chosen for the fiberglass model. See Figure 1. The 3 meter model was large enough for accurate seakeeping measurements, yet was small enough to take advantage of the steeper waves produced in the MASK. The model was assigned the number 9096 by the Carderock Division, Naval Surface Warfare Center (CDNSWC).

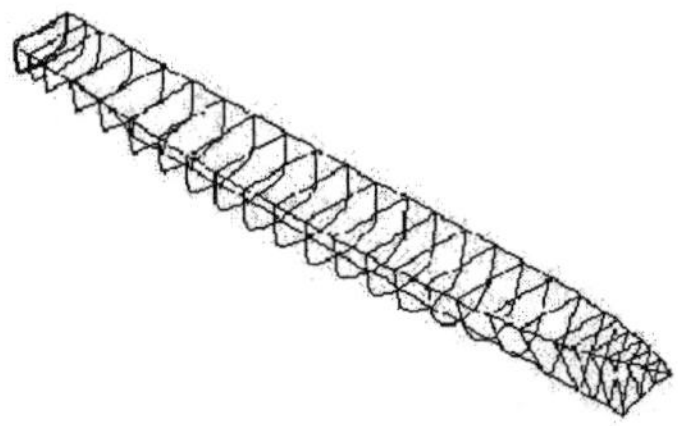

Figure 1: Isometric sketch of frigate (Model 9096)

The model was outfitted to be self-propelled using an autopilot for heading control. This autopilot uses a simple PID controller algorithm such that the desired rudder angle is:

$$\delta_{d\psi} = C_{1D}(\psi - \psi_d) + C_{2D}\dot{\psi} \tag{1}$$

where C_{1D} is the yaw gain and C_{2D} is the yaw rate gain.

The model was outfitted with bilge keels and twin rudders. Two four-bladed fixed pitch propellers (CDNSWC numbers 1991 and 1992) were installed on the model for inboard rotation. A Plexiglas deck was installed to protect the instrumentation and machinery systems inside the hull from water damage during capsize events. The model was painted to enhance the display of the model during video recordings.

The outfitted model was tethered to the carriage by a cable bundle that contained control, power, and data signals. The cable was looped to minimize the effect of the cable tension on the model. The cable was connected to an overhead boom via a pulley that could be reeled in and out by a boom operator who stood on a platform mounted on the carriage. The boom could be yawed as desired by the boom operator. During the calm water and regular wave experiments, the carriage followed the model and the boom operator ensured that cable tension did not affect the model. The boom operator did this by pointing the boom so that it stayed above the model while reeling in or out the overhead pulley so that the cable remained slack. See Figure 2.

Before the start of a test in waves, the wave maker was turned on and the model held in position until a sufficient number of waves had passed. Subsequently the model would start slowly under its own power (at low RPM) and once it was on course properly, the propeller RPM was set to the required level associated with a calibrated speed in calm water. The boom and trolley system followed the model during the test.

Figure 2 : Model tether arrangement

Instrumentation

The carriage was equipped with a Ship Motion Recording (SMR) system developed by the CDNSWC Seakeeping Department. This system was connected to the sensors used in the model test as listed in Table 1.

The primary sensor used in the model test was a KEARFOTT T16 *Miniature Integrated Land Navigation System* (MILNAV^{TM}). This inertial navigation system (INS) included a monolithic 3-axis Ring Laser Gyroscope and three, single axis linear accelerometers. The MILNAV had the capability to measure extreme pitch, roll, and yaw angles, rates and accelerations. It was also used to determine surge, sway, and heave displacements, velocities, and accelerations near the center of gravity of the model. These motions were translated to the ship's center of gravity reference for each load condition.

More details have been described by De Kat and Thomas (1998). Model speed was measured in three orthogonal directions using the MILNAV^{TM} inertial navigation system. Propeller speed was measured using a tachometer which was mounted to the propeller shafting. Rudder angle was measured using the rudder feedback unit which as mechanically linked to the rudder tiller arm. Longitudinal and transverse rudder forces were measured using block gauges mounted on a free-floating plate that was attached to the rudder posts.

TABLE 1
CAPSIZE MODEL TEST DATA CHANNELS

Measurement	Source
Basin Wave Height From Carriage (cm)	Sonic Transducer
Basin Wave Height on shore (cm)	Sonic Transducer
Carriage Speed (m/s)	Tachometer
Propeller RPM (rev/minute)	Tachometer
INS Hull Speed (m/s)	MILNAV™
INS Azimuth (°T)	MILNAV™
Pitch (deg)	MILNAV™
Pitch rate (deg/s)	MILNAV™
Roll (deg)	MILNAV™
Roll Rate (deg/s)	MILNAV™
Yaw Rate (deg/s)	MILNAV™
Rudder Angle (deg)	Autopilot
Rudder force, X direction (Newtons)	Block Gauge
Rudder Side force (Newton)	Block Gauge
Vertical Acceleration Forward (g)	Accelerometer
Longitudinal Acceleration Forward (g)	Accelerometer
Transverse Acceleration Forward (g)	Accelerometer
Vertical Acceleration Starboard (g)	Accelerometer
Longitudinal Acceleration Starboard (g)	Accelerometer
Transverse Acceleration Starboard (g)	Accelerometer
Relative Bow Motion, Port and Starboard (cm)	Capacitance Probe
Relative Stern Motion, Port and Starboard (cm)	Capacitance Probe

Model Hydrostatics

The model was tested in three load configurations. These configurations comprised:
1. Full Load condition
2. Marginal GM condition
3. Failed GM condition

The load conditions were chosen with first priority given to the validation of the time domain simulation program FREDYN developed by the CRNAV group. Since the capsizing characteristics of this frigate were previously unknown, it was deemed necessary to choose a load condition to ensure that capsizing would occur. As an additional consideration, it was desired that one load condition resemble a realistic operating condition of the ship. Thus, the first and most conservative condition tested closely resembled the typical frigate Full Load operating condition. The two remaining load conditions were achieved by the raising of weights above the deck of the model to reduce transverse GM. This procedure ensured that displacement and freeboard remained constant for the three load conditions. Thus, the second load condition called "Marginal GM" represented a decrease in GM such that the model is in marginal compliance with the U. S. Navy Criteria (NAVSEA, 1976), which include a weather criterion. The third load condition called "Failed GM" represented a load configuration that is in violation of the U. S. Navy Stability Criteria; as such this is not a realistic operating condition. A summary of the full scale load configurations tested is presented in Table 2. The GZ curve was measured for the full range from 0 to 180 degrees heel angle for the three loading conditions to provide comparison data as regards the computed hydrostatics.

TABLE 2

FULL SCALE LOAD CONDITIONS CORRESPONDING TO MODEL TEST

Parameter	Load condition		
	Full load	Marginal GM	Failed GM
L_{PP} (m)	106.68	106.68	106.68
B (m)	12.78	12.78	12.78
T (m)	4.73	4.73	4.73
GM_T(m)	0.78	0.68	0.43
T_ϕ(sec)	12.3	13.3	17.1

The transverse metacentric height (GM_T) was checked prior to each run series in waves by conducting an inclining experiment. The roll, pitch and yaw gyradii (k_{44}, k_{55} and k_{66}) were determined using the pendulum method by swinging the model in the air. For each load condition, the model was suspended by a pivot located above the COG of the model and allowed to oscillate in e.g. pitch or roll. The gyradii were calculated from the respective oscillation periods.

Test Conditions

The model test matrix was designed for runs in regular waves for the three load (GM) conditions for the purpose of identifying critical, extreme motion situations, including capsizing, broaching, harmonic resonance and surfriding. Regular waves were chosen instead of irregular waves to allow a better understanding of observed extreme events in terms of known wave length, steepness, and phase relation with respect to the model; an important factor was the ability to use the data for validating a time domain simulation model.

The critical wave length range involved waves with λ/L between 0.75 and 2.50 with associated wave steepness in the range of 1/20 and 1/10. Model speeds were chosen between Fn_0 = 0.1 and 0.35 in combination with the selected wave and heading conditions. The critical wave headings chosen included following seas and stern quartering seas predominantly. Some runs were performed in beam seas.

Runs were performed at selected critical headings and speeds, which were based on time domain simulations carried out *a priori*. In case of the "Failed GM" loading condition, the occurrence of a capsize event invoked procedures that isolated the capsize region at the particular wave length l/L. In general this meant that follow-up runs were made in higher and lower wave steepnesses and at lower (or higher) speeds until no capsizes were found. Adjacent headings were also tested at the capsize steepness to identify the effects of heading variations.

Table 3 provides an overview of the test runs for the ship with the Failed GM condition; χ represents the nominal heading angle. Test conditions for the other two loading conditions were in principle similar.

EXTREME MOTION EVENTS

The following provides a description of some extreme motion events, including:

- Surfriding (and periodic surging)
- Broaching
- Extreme rolling
- Capsizing

Surfriding

A ship in following seas can experience large speed fluctuations (at low encounter frequencies) about its mean forward speed. If the ship speed is sufficiently high, i.e. the speed that would be attained in calm water at a given propeller RPM and thrust, a wave may capture the ship and propel it at wave phase speed. The resulting speed can be significantly higher than the calm water ship speed.

Once wave capture takes place, instead of immediately attaining the wave phase speed, the ship can reach speeds well beyond the (steady) wave phase speed for an extended period before reaching a steady-state condition. To illustrate this, we consider the frigate running in following seas (zero degrees heading angle) of different periods and heights. The loading condition corresponds to the Full Load case discussed above. At a propeller RPM setting for calm water Froude number of $Fn\_0 = 0.3$, the ship experiences periodic oscillations in forward speed of significant amplitude, which increases with increasing wave amplitude (De Kat and Thomas, 1998).

The model tests show that for the speed range $Fn\_0 \leq 0.3$ wave capture (and hence surfriding)does not occur in the wave conditions tested. For $Fn\_0 = 0.35$ a drastic change in

TABLE 3

TEST MATRIX FOR MODEL 9096 IN FAILED GM CONDITION (0 DEGREES IS FOLLOWING SEAS)

λ/L	H/λ	Fn	χ (deg)						
			0	15	30	45	60	75	90
0.75	1/15	0.1							
		0.2			X				
		0.3							
		0.35		X	X				
	1/10	0.1		X					
		0.2			X				
		0.3		X					
		0.35		X	X	X			

λ/L	H/λ	Fn	χ (deg)						
			0	15	30	45	60	75	90
1.00	1/20	0.1							
		0.2			X				
		0.3			X				
		0.35		X	X				
	1/15	0.1			X				
		0.2				X			
		0.3		X	X	X			
		0.35	X	X	X	X			
	1/10	0.1							
		0.2							
		0.3		X	X				
		0.35		X	X	X			

λ/L	H/λ	Fn	χ (deg)						
			0	15	30	45	60	75	90
1.25	1/20	0.1							
		0.2			X				
		0.3			X				
		0.35	?	X	X				
	1/15	0.1				X			
		0.2				X	X		
		0.3			X	X			
		0.35	X	X	X	X	X		
	1/10	0.2							
		0.2		X	X				
		0.3		X	X				
		0.35			X				

λ/L	H/λ	Fn	χ (deg)						
			0	15	30	45	60	75	90
1.50	1/20	0.1							
		0.2							
		0.3							
		0.35			X				
	1/15	0.1				X			
		0.2				X			
		0.3		X	X		X		
		0.35		X	X	X	X		
	1/10	0.1							
		0.2							
		0.3		X	X				
		0.35		X	X	X	X		

λ/L	H/λ	Fn	χ (deg)						
			0	15	30	45	60	75	90
2.00	1/20	0.1							
		0.2							
		0.3		X					
		0.35		X					
	1/15	0.1							X
		0.2							
		0.3							
		0.35		X	X				

λ/L	H/λ	Fn	χ (deg)						
			0	15	30	45	60	75	90
2.50	1/20	0.1							
		0.2							
		0.3							
		0.35			X				
	1/15	0.1			X				X
		0.2							
		0.3			X				
		0.35		X	X				

surge character occurs, for at this speed setting the frigate model does experience wave capture and surfriding events. In a number of tests where wave capture takes place in following waves, the ship is accelerated to a speed that lies well beyond the phase speed of the wave. Figure 3 provides an overview of measured maximum speeds for Fn_0 = 0.35. The maximum ship speed increases with increasing wave steepness; for the conditions with $\lambda/L = 1$ the steepness tested is $H/\lambda = 0.077$ and 0.097, while for $\lambda/L = 1.25$ the steepness is $H/\lambda = 0.051$ and 0.092.

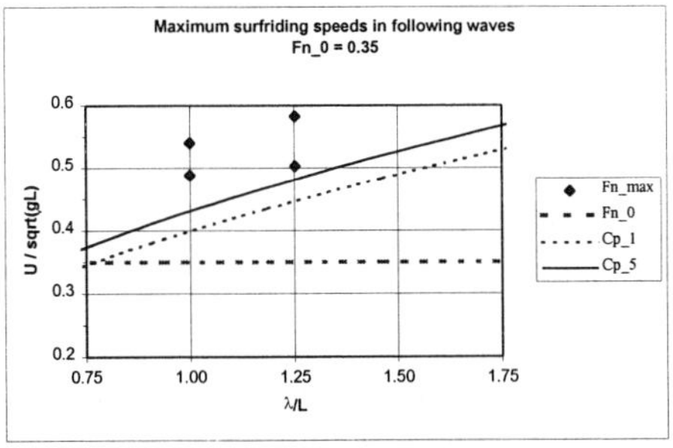

Figure 3 : Maximum attained ship speed (Fn_max) following wave capture in experiments. Cp_1 is the phase velocity based on linear wave theory; Cp_5 is the phase velocity for $H/\lambda=0.1$

For reference purposes in the figures presented in this paper, we use a definition for wavelength based on linear wave theory in deep water, i.e. $\lambda = gT^2/2\pi$, where T is the period. Cp_1 represents the phase speed according to linear wave theory, while Cp_5 is the phase speed determined according to Stokes 5th order theory (Fenton 1978); the phase speeds have been normalized by the square root of gL.

Figure 3 shows that especially for the steepest waves, the maximum ship speed reaches very high values. The time series describe the character of this behavior, as shown for run 231 in Figure 4, where $\lambda/L = 1.0$ and $H/\lambda = 0.1$.

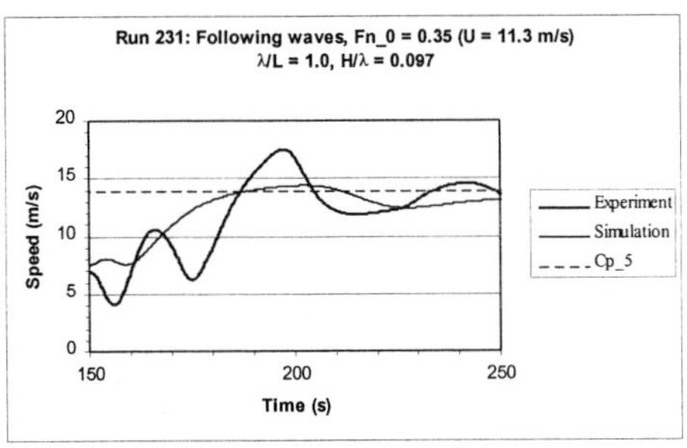

Figure 4 : Measured and predicted ship speed during surfriding events for run 231

This figure shows that as the ship reaches a critical speed level in the model test, wave capture occurs and the ship accelerates to a speed well beyond its calm water speed. The observed mechanism is as follows: at approximately t = 180 s a wave crest reaches the stern of the ship and causes the ship to surge forward. At around t = 185 s the crest has reached the aft quarter (i.e. it has overtaken the ship slowly until the ship speed equals the celerity), at which stage the ship is subjected to a large surge force. This causes the ship to accelerate and overtake the wave crest, which by t = 200 s is situated at the stern again; the maximum speed is 17.5 m/s. While the ship accelerates and overtakes the wave crest, it buries the bow in the back slope of the preceding wave.

The (complete) submergence of the bow increases the resistance and eventually causes the ship to decelerate; between t = 205 s and 235 s the speed decreases to below the wave celerity. As the crest overtakes the ship, the ship speed increases again to the wave celerity level. As a consequence, the speed drops to 11.9 m/s before accelerating again to wave celerity level.

Numerical simulations (De Kat and Thomas, 1998) predict similar trends in forward speed fluctuations, but the maximum simulated speed of 14.4 m/s lies well below the measured maximum speed of 17.5 m/s. As a consequence of the linear wave model in the simulation, the steady surfriding speed is equal to the linear wave celerity (Cp_1), which lies about 10% below the actual wave speed, Cp_5.

Prediction of transient forward speed above the wave phase speed is of relevance especially for irregular following seas, where (transient) steep waves will be able to push the ship speed to high values for some time and cause bow submergence, with broaching as a possible consequence.

Broaching

The model tests show that the frigate investigated can experience broaching under certain conditions at high Froude number only (Fn_0 = 0.35). In general the ship proved to have good course keeping qualities in following and stern quartering waves, i.e. broaching did not occur frequently in the conditions tested at high Froude number. Two modes of broaching were observed:

1. High speed broach preceded by surfriding, with rapidly and monotonically increasing heading deviation in following and stern quartering waves
2. Large amplitude, low-frequency yaw (heading) oscillations in stern quartering waves

Figures 5a and 5b depict the occurrence of the first broaching mode for the ship in the Marginal GM Condition. The nominal (desired) heading is 15 degrees. Figure 5a shows both the longitudinal and transverse ship speeds, where Us is defined in the horizontal plane along the x-axis of the yawed ship, and Vs is perpendicular to Us in the horizontal plane. Figure 8 shows the steadily increasing heading angle once the ship has been captured by the wave afterv t = 70 s; the ship experiences moderately large heeling angles during the event. As a consequence of the test set-up, the broach ended when a tight tether line limited the motions.

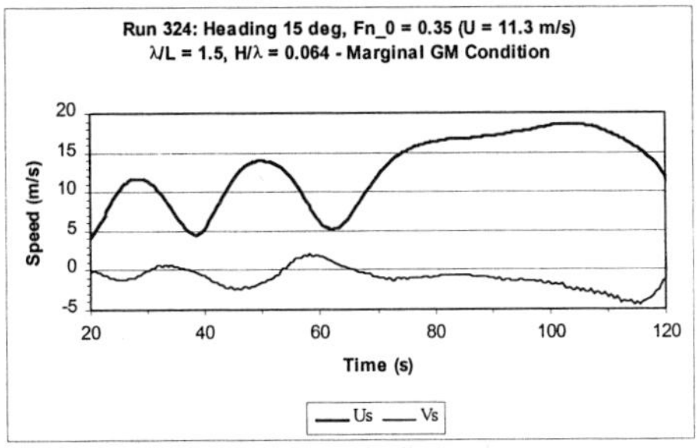

Figure 5a : Mode 1 broach: Measured longitudinal and transverse ship speeds for run 324

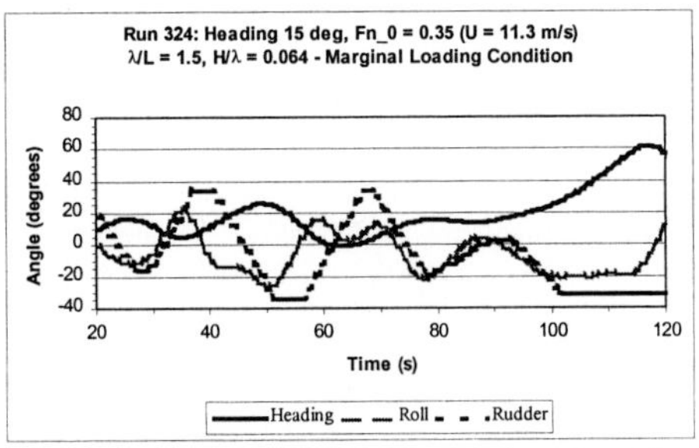

Figure 5b : Mode 1 broach: Measured heading, roll and rudder angles for run 324

Figures 6a and 6b depict the occurrence of the second broaching mode for the ship in the Full Load Condition, illustrating that the ship can experience large roll angles in this condition. Figure 8a shows that the ship has a significant mean negative drift velocity, i.e. it experiences a rather large drift speed to leeward while yawing. The highest transverse drift velocity occurs when the yaw angle (toward the wave) and forward speed increase while a wave crest is overtaking the ship (from aft to amidships). When the crest is in the midship area and the ship has reached its largest yaw deviation into the wave, the roll angle to leeward (negative sign) is largest; the reduction of the righting arm in the wave crest leads to asymmetric roll motions. In this case the ship experiences large roll angles, but it does not capsize.

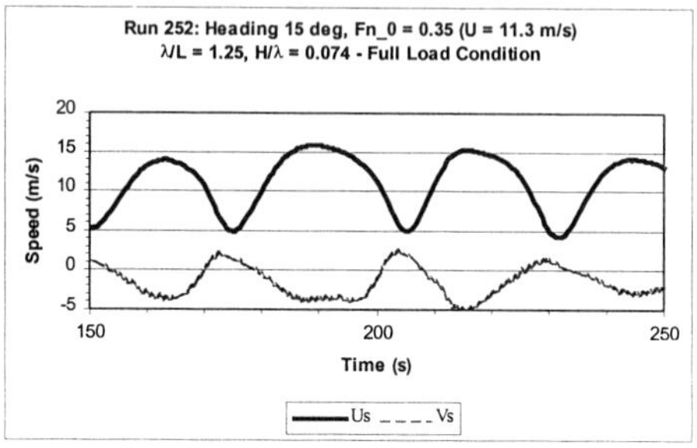

Figure 6a : Mode 2 broach: Measured longitudinal and transverse ship speeds for run 252

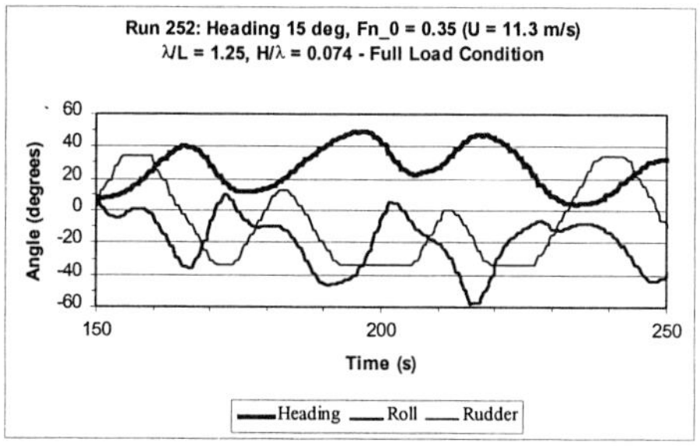

Figure 6b : Mode 2 broach: Measured heading, roll and rudder angles for run 252

The stable course keeping properties of the frigate in waves may be linked to its degree of directional stability in calm water. According to the simulations the ship is directionally very stable. The large skeg and twin rudders contribute to the frigate's directional stability.

Capsizing

The frigate did not capsize under any of the speed, heading and wave combinations tested in either the Full Load or Marginal Loading Condition. In the Failed Loading Condition, however, capsizes did occur frequently at the highest ship speeds ($Fn\_0 \geq 0.3$) in following and stern quartering waves; no capsizes occurred at $Fn\_0 < 0.3$, even though conditions included harmonic resonance in steep stern quartering and beam seas.

The following main modes of capsize could be distinguished from the experiments:

1. Loss of transverse stability in wave crest associated with surfriding or periodic surging
2. Dynamic loss of stability due to surge-sway-roll-yaw coupling
3. Broaching (mode 1 and 2)
4. Combinations of modes

The majority of observed capsizes were of mode 1 and 2. A mode 1 capsize is one where the ship capsizes while being overtaken slowly by a wave crest. For this capsize mode, typically the ship does not experience extreme roll motions before the final "half roll."

A mode 2 capsize involves significant rolling, and often the roll motion tends to build up in severity before capsizing occurs. We illustrate this mode for three experimental cases: run 414 with a short wavelength ratio (λ/L = 0.8), run 427 with λ /L = 1.25, and run 448 with a relatively long wave (λ /L = 2.5), as shown in Figures 7, 8 and 9, respectively.

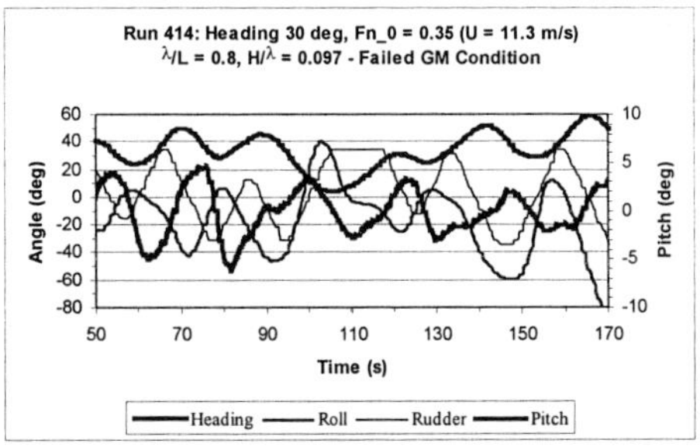

Figure 7a : Mode 2 capsize: dynamic loss of stability (run 414) with measured heading, roll, pitch and rudder angles

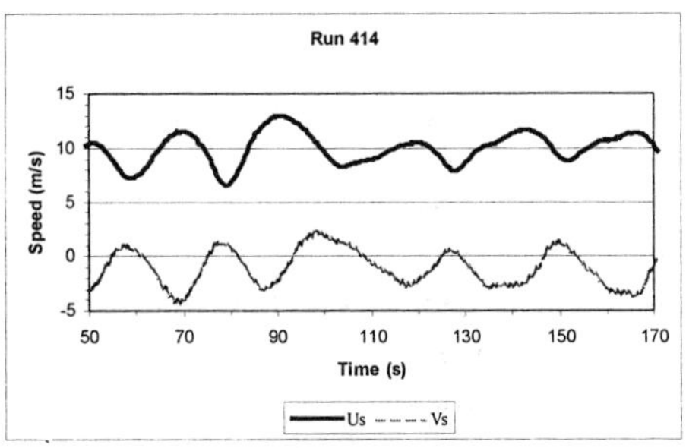

Figure 7b : Measured ship velocities associated with mode 2 capsize (run 414)

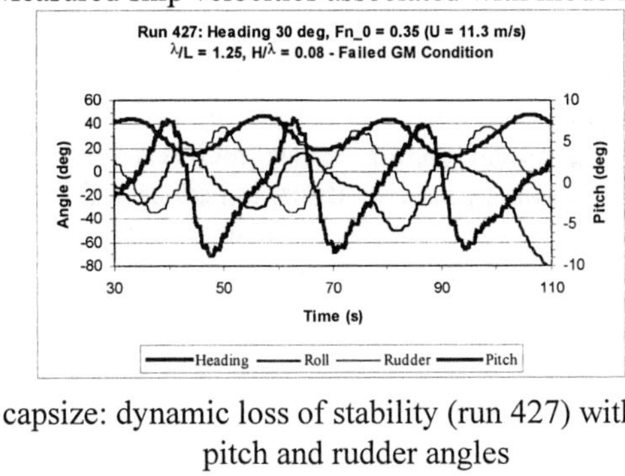

Figure 8. Mode 2 capsize: dynamic loss of stability (run 427) with measured heading, roll, pitch and rudder angles

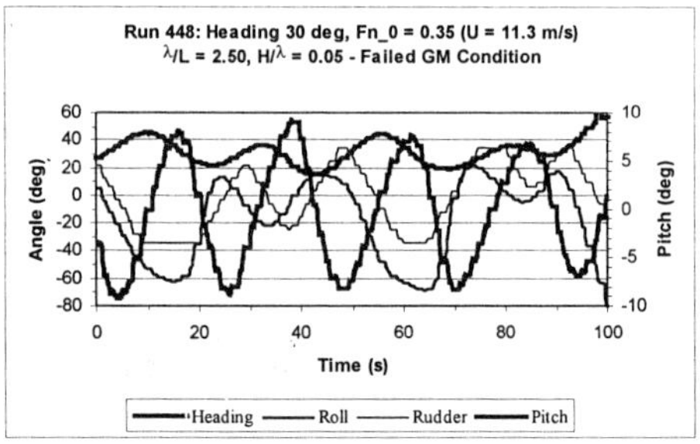

Figure 9a : Mode 2 capsize: dynamic loss of stability (run 448)

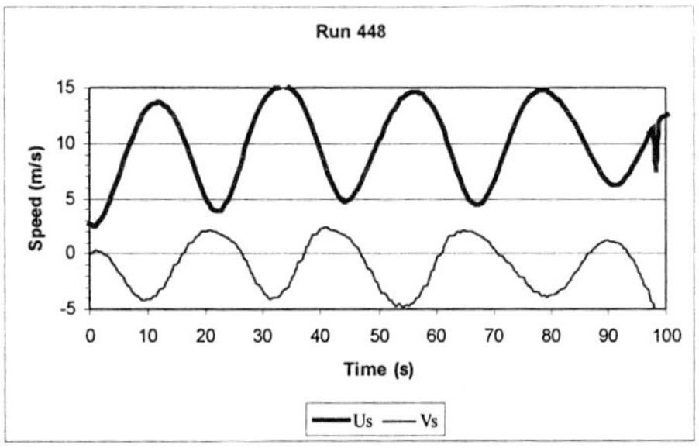

Figure 9b : Ship velocities associated with mode 2 capsize (run 448)

Although all three cases involve dynamically coupled rolling, the motion behavior differs significantly: for instance, the speed variations in run 414 are limited to relatively small fluctuations about the mean forward speed, the roll motion in run 427 shows increasing extreme roll angles to port (negative sign, to leeward), and the roll motion in run 448 has a double period. Concerning the latter, double period rolling ("period bifurcation") has been observed in some model tests with containerships in regular waves (Kan et al, 1994). In these dynamic (mode 2) capsize events the ship capsizes typically in the wave crest.

VALIDATION ASPECTS

This section discusses some model testing issues that bear relevance on validation of numerical simulations. In conjunction with the tests described above we will cover the following:

- Roll decay and influence of autopilot
- Measurement of ship velocities
- Measurement of ship position
- Wave height at ship-fixed reference point

Roll decay and inluence of autopilot

Roll decay tests in calm water provide information on roll period and damping as a funciton of ship speed. This type of testing is useful for validation against numerical predictions in the time domain.

Figure 10 provides an example of measured and predicted roll motion response in calm water at a Froude number Fn = 0.3 with an initial roll angle of 40 degrees. The two curves compare quite well for the frigate hull, but there is an offset in the model test data. The cause of the offset was found to be the autoilot: as the model heeled, a yaw moment was induced and the model started to veer off course. The ensuing rudder action demanded by the autopilot resulted in a (quasi-steady) heeling moment.

The recommendation for performing roll decay tests with large initial heel angles at forward speed is to switch off the autopilot.

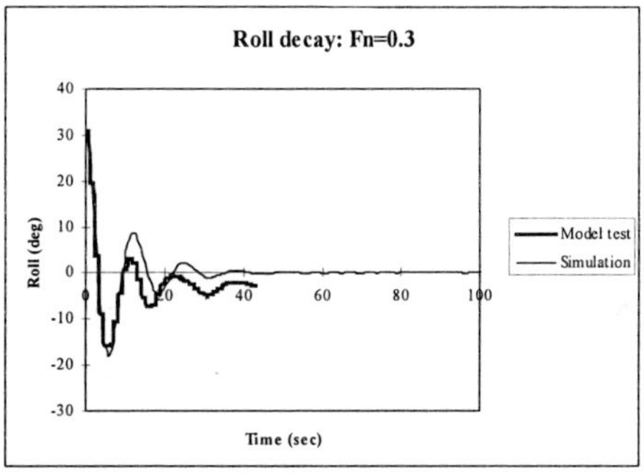

Figure 10 : Roll decay test at Fn = 0.3

Measurement of ship velocities

It is not standard practice in seakeeping or capsize model tests to measure ship speed in longitudinal and transverse directions; even the direct measurement of instantaneous forward speed is not common. Typically one knows the calm water speed associated with the propeller RPM, which is assumed to be the mean speed.

As some of the tests discussed above show, a ship can experience significant velocity fluctuations periodically in both forward and transverse directions. Quantitative knowledge of these speed variations will contribute to a better understanding of the physics in the validation process.

Let us consider a typical stern quartering condition case: Figure 11a represents motion measurements for run 337. Waves overtake the ship (with Marginal GM) slowly from the starboard quarter. Each passing wave captures the ship briefly while the crest passes the amidships area. The stability reduction experienced by the ship induces the ship to heel to port at a large angle of roll. As the ship is released by the wave, the ship rolls back to starboard and uprights itself in the wave trough, resulting in asymmetrical roll behavior. This cycle repeats with the next overtaking wave.

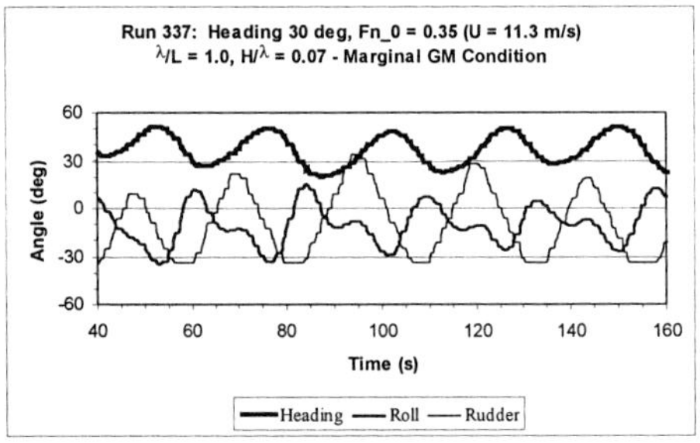

Figure 11a : Run 337: motions in stern quartering waves (Marginal GM condition, Fn_0 = 0.35)

As a consequence of the speed changes, the instantaneous drift angles can be substantial. To illustrate this aspect, we consider the velocity components Us and Vs associated with run 337 in Figure 13b. The instantaneous drift angle is given by

$$\beta(t) = \arctan\frac{Vs}{Us}$$

i.e., this is the drift angle with respect to the yawed x-axis of the ship in the horizontal plane.

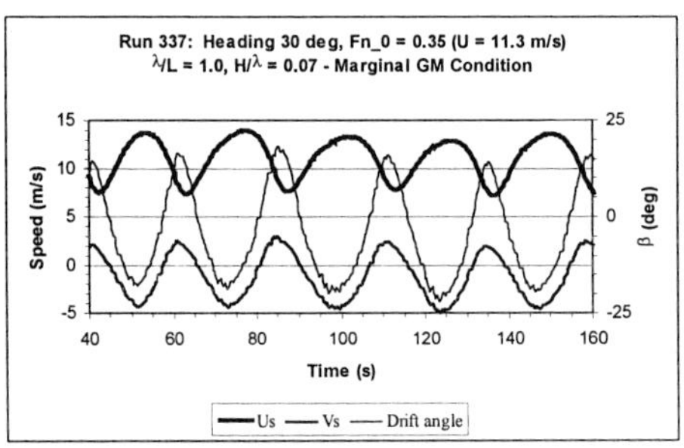

Figure 11b : Ship velocities and drift angle associated with run 337

Figure 11b shows the following:

- The transverse velocity can reach high values (up to -5 m/s to port) and it has a negative mean.
- The frigate undergoes large variations in instantaneous drift angle, with maxima of around 20 degrees to either side with respect to its longitudinal axis.
- The drift angle variations suggest a mean negative drift angle.

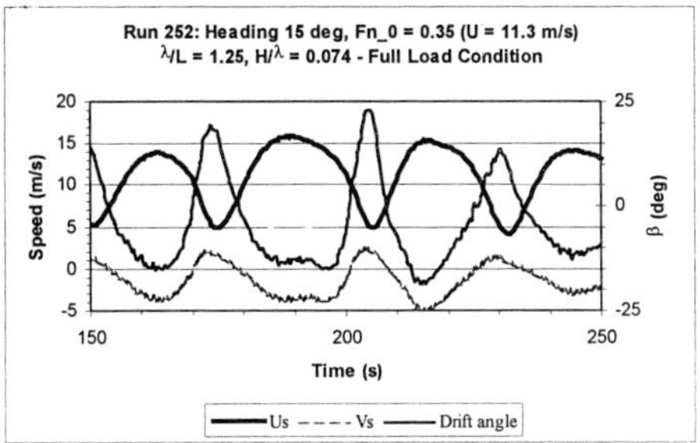

Figure 12 : Ship velocities and drift angle associated with run 252 (Broach shown in Fig. 6)

Figure 12 shows the drift angle for run 252 (Mode 2 broach, see Fig. 6). It is noted for run 252 and 337 that while the ship is undergoing these velocity and drift angle variations, it is at the same time undergoing large yaw (heading) changes. The large drift angles and drift velocities will influence the ship motions and course keeping behavior, noting that seakeeping and maneuvering are intricately intertwined and should not be modeled independently to predict motions in these conditions.

Measurement of ship position

As is the case for ship velocities, ship position is typically not measured in seakeeping and capsize tests. Yet knowing the earth-fixed position of the ship's CoG at each time instant, allows the comparison between measured and simulated ship track in space for validation purposes.

The ship track provides an indication of the amount of drift a ship may experience in stern quartering waves, for example. Figure 13 shows the measured track for run 252, referenced against the "no drift" track, which is the track the ship would have followed along its average heading (around 30 degrees) in the absence of mean drift effects as of the point in time at t = 150 s. Also shown is the actual heading angle, which corresponds to the one shown in Figure 8b for the time period between t = 150 and 250 s.

Figure 14 shows similar information for run 337 between t = 50 and 150 s. Here the "No drift" track is based on a mean heading of 35 degrees. Figures 15 and 16 illustrate the amount of drift a ship can experience in steep stern quartering waves, noting that the autopilot algorithm employed in the tests did not account for deviation from the desired path.

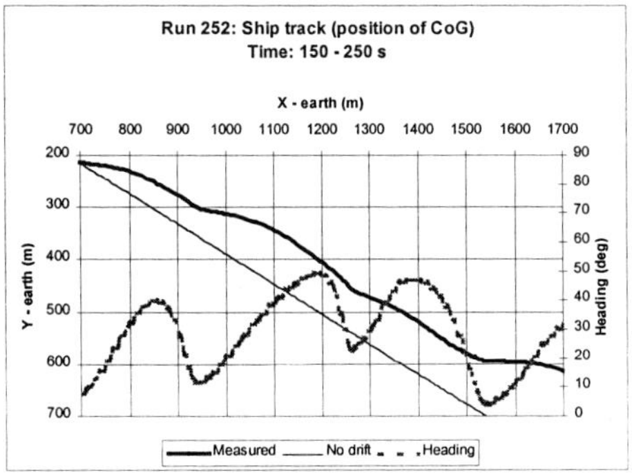

Figure 13 : Ship track and heading between associated with run 252 in stern quartering waves (waves travel along X-earth axis)

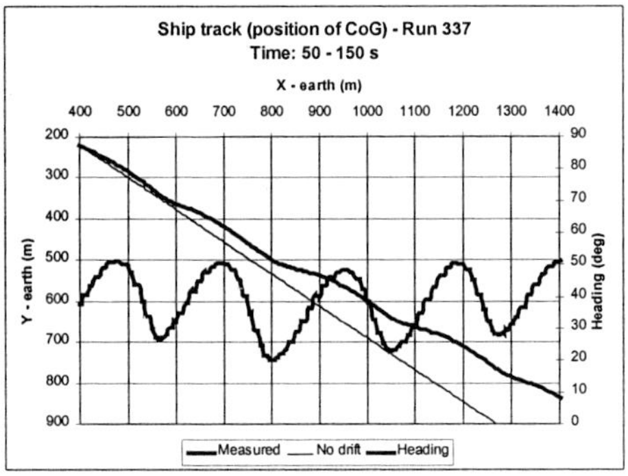

Figure 14 : Ship track and heading between associated with run 337 in stern quartering waves (waves travel along X-earth axis))

In addition to defining the ship track, earth-fixed measurements will allow the determination of the ship with respect to the (presumably known) wave system.

Wave height at ship-fixed reference point

To achieve simulation conditions that are similar to the model test, it is necessary to know the position of the ship in the wave at any time instant. This allows e.g. to start the simulation with the correct initial conditions and phasing with respect to the wave crest.

In the tests discussed here the wave elevation was measured at a basin-fixed location and at two locations on the towing carriage. Using the latter data and the measured ship motions (including positions), the time series of the wave height at the center of gravity could be generated. Figure 15 provides an illustration of the estimated wave elevation for run 252.

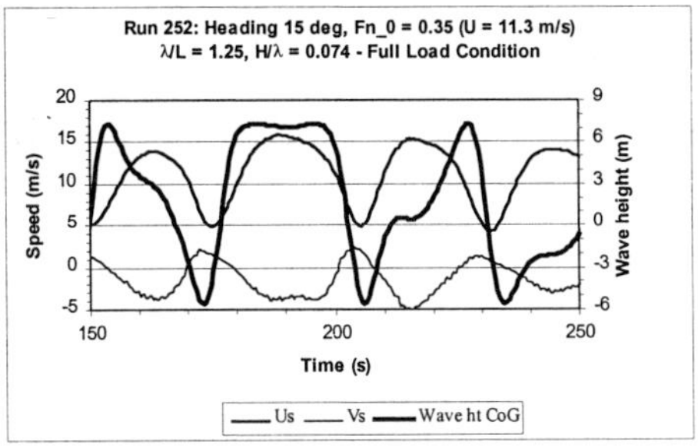

Figure 15 : Ship velocities and wave height at CoG for run 252 (Broach shown in Fig. 6)

Figure 15 shows that while the ship experiences a significant transverse velocity to port (negative sign), it has a high forward speed and runs with the wave crest at the CoG (i.e., amidships) for an extended period of time between t = 175 and 200 s. Comparison with Figure 8b shows that the roll motions to port are in phase with the wave crest coinciding approximately with the CoG position, leading to reduced stability and hence large roll angles to the leeside during the passage of the wave.

CONCLUSIONS

Model tests with an intact frigate-type hull form were carried out to obtain data on extreme motions and capsizing in critical wave conditions. To control the conditions leading to these events, all tests comprised regular waves of moderate to extreme steepness. A primary objective of these tests was to generate data that would be suitable for the vallidation of a time domain, large amplitude ship motion simulation program.
The paper discusses details of the experimental setup and test procedures, as well as physics associated with some observed extreme motion events in astern wave conditions:

- Periodic surging and surfriding
- Broaching
- Extreme rolling
- Capsizing

Test conditions leading to surfriding show that a ship could attain transient speeds that exceed not only its calm water speed, but also the wave phase speed; bow submergence causes a significant increase in resistance and subsequent deceleration. A numerical model, which makes use of linear wave theory, underpredicts wave celerity and hence surfriding speeds for steep wave conditions.

KG was varied to simulate three load conditions in the model tests: (1) typical Full Load condition, (2) Marginal GM condition that just meets the navy stability criteria, and (3) a Failed GM condition. No capsizes occurred for the first two load conditions. Tthe most extreme rolling and capsizing occurred at high speed (Fn $\geq$ 0.3) in stern quartering waves. An important observation is that a ship in steep stern quartering waves can experience large amplitude fluctuations in speed, both in longitudinal and transverse directions, and mean drift to leeward can be significant. Large variations in speed contribute to asymmetric rolling, as the ship spends more time (at higher speed and with reduced transverse stability) in a wave crest than in the wave trough.

From a validation perspective of a numerical simulation model, some of the conclusions for large amplitude motion tests in following and stern quartering waves are as follows.

- The autopilot should be switched off during roll decay tests at forward speed to avoid roll-yaw coupling.
- The measurement of ship velocity in longitudinal and transverse directions provides useful information on speed behavior and instantaneous drift angles.
- The measurement of earth-fixed ship position provides useful data on ship track and mean drift.
- Knowing the characteristics of the incoming wave system and earth-fixed ship position, enables one to determine the wave with respect to the ship model at any time instant.

Acknowledgements

The authors would like to express their gratitude to the Cooperative Research Navies Dynamic Stability group (navies from Australia, Canada, Netherlands, United Kingdom and United States, U.S. Coast Guard and MARIN) for their permission to publish this paper.

References

De Kat, J.O., Brouwer, R., McTaggart, K. and Thomas, W.L. (1994). Intact Ship Survivability in Extreme Waves: New Criteria from a research and Navy Perspective, *Proc. International Conference on Stability of Ships and Ocean vehicles STAB '94*, Melbourne, FL, Nov.

De Kat, J.O. and Thomas, W.L. (1998). Extreme Rolling, Broaching and Capsizing – Model Tests and Simulations of a Steered Ship in Waves, *Proc. Naval Hydrodynamics Symposium*, Washington, D.C, Aug.

Kan , M., Saruta, T. and Taguchi, H. (1994). Comparative Model Tests on Capsizing of Ships in Quartering Seas, *Proc. International Conference on Stability of Ships and Ocean vehicles STAB '94*, Melbourne, FL, Nov.

NAVSEA Design Data Sheet 079-1.(1976). Department of the Navy, Naval Ship Engineering Center, June.

Contemporary Ideas on Ship Stability
D. Vassalos, M. Hamamoto, A. Papanikolaou and D. Molyneux (Editors)

SENSITIVITY OF CAPSIZE TO A SYMMETRY BREAKING BIAS

B. Cotton, S.R. Bishop and J.M.T. Thompson

Centre for Nonlinear Dynamics and its Applications,
University College London,
Gower Street, London WC1E 6BT, UK

ABSTRACT

We discuss a mathematical model for ship roll motion which allows an investigation into the importance of symmetry in the system. A non-dimensional equation is derived and a family of potential wells discussed. A capsize diagram is plotted showing the steady state sensitivity to a symmetry breaking bias which considers a wide range of frequencies. The steady state dynamics of the system are discussed and bifurcation diagrams are numerically constructed. Finally, the importance of the flip bifurcation in the dynamics is assessed in relation to the alterations in the symmetry of the system.

KEYWORDS

Capsize, nonlinear, symmetry, bias, roll, steady-state, bifurcations

INTRODUCTION

A simple model of ship roll motion takes the form of a forced nonlinear oscillator (Wright & Marshfield 1980; Thompson 1997; Kan & Taguchi 1994) using the polynomial approximation of a nonlinear righting moment and an effective linear damping term. Thus the ship capsize model is simplified to that of escape from a potential well. This can then be used to explore the fundamental dynamics of the system.

The system studied here is that of a symmetric ship in beam seas, with the addition of a wind loading bias. Use can be made of an effective righting moment to incorporate wind loading as a fixed bias, generating the canonical escape equation (Thompson 1989; Thompson 1997), a simple, one sided potential well.

Extreme sensitivity to bias in the potential has been found for a system with righting moment parameterised to include varied bias (Macmaster & Thompson 1994). This takes the form of a sharp drop in the wave slope required to capsize the model ship for a small wind loading bias. Our aim here is to examine the dynamics underlying this sensitivity.

We follow previous work on the canonical escape equation (Falzarano 1994; Foale & Thompson 1991) and use numerical techniques to trace steady state solutions and the key bifurcations. By constructing such a bifurcation diagram for the varied bias system, the qualitative behaviour of the model ship as parameters are varied (e.g. wave forcing) can be predicted. The region of frequencies around linear resonance is studied, this being the expected worst case capsize zone.

THE PARAMETERISED ROLL MOTION EQUATION

A ship in deep water, on a long wavelength (compared to the ship width) beam wave can be modelled in its simplest form as a rotational oscillator, centered on the wave normal, (Thompson, Rainey & Soliman 1992). The roll motions are assumed to be uncoupled from other motions, with

$$I\theta'' + B(\theta') + C(\theta) = IAk\omega_f^2 \sin \omega_f \tau \tag{1}$$

where the prime denotes differentiation with respect to real (unscaled) time, τ, I is the rotational moment of inertia about the centre of gravity (incorporating any added hydrodynamic mass), θ is the roll angle relative to wave normal, $B(\theta')$ is the non-linear damping function, $C(\theta) = mgGZ(\theta)$ is the effective restoring moment (including wind loading etc), $GZ(\theta)$ is the GZ curve for the ship, Ak is the wave slope amplitude (A is the wave height and k the wave number) and ω_f is the wave frequency. We also write $\omega_n = \sqrt{mgGM/I}$ is the natural frequency of linearised ship motions where $GM = dGZ/d\theta(0)$.

The damping function $B(\theta')$ is dependent on a number of factors including the vessel's cross-section and its forward speed but will be assumed independent of roll frequency.

The restoring moment, $C(\theta)$, is highly nonlinear, dropping to zero at two angles of vanishing stability, θ_V and $-\theta_U$, where the potential energy $V(\theta)$ has a local maximum (for an unbiased ship $\theta_V = \theta_U = \theta_{V0}$). Here we are interested in the fundamental properties of the underlying dynamics and a suitable approach is to approximate $C(\theta)$ with a polynomial.

The restoring moment must also incorporate any added bias due to wind or cargo loading. For example we can model wind loading by subtracting $h\cos^2(\theta)$ from the unbiased GZ curve. This creates a heel angle θ_H and alters the effective angles of vanishing stability (originally $\pm\theta_{V0}$)

Now, taking $C(\theta) = mgGZ(\theta) = C_\theta(0)\theta(1-\theta/\theta_V)(1+\theta/\theta_U)$ we can characterise the shape of an effective GZ curve using the linear stiffness at equilibrium, $C_\theta(0)$, and the two angles of vanishing stability, θ_V and $-\theta_U$ (Macmaster & Thompson 1994). The key point here is that any cubic GZ curve with added cubic bias and two real angles of vanishing stability can be recast into the above form giving an effective restoring moment.

Equation (1) can be rescaled such that the effective GZ curve has unit slope at $\theta = 0$ and always passes through $x = 1$. This leads to a non-dimensionalised version of the equation,

$$\ddot{x} + b(\dot{x}) + c(x) = F\sin(\omega t) \tag{2}$$

where the dot now denotes differentiation with respect to scaled time and x is the scaled roll angle, θ/θ_V, t the scaled time $t = \omega_n \tau$, $b(\dot{x})$ the scaled nonlinear damping function, $c(x)$ the scaled nonlinear restoring function and $F = Ak\omega^2/\theta_V$ the scaled forcing amplitude

Now, defining $\alpha = \theta_V/\theta_U$ implies that $c(x) = x(1-x)(1+\alpha x)$ and by taking linear damping, $b(\dot{x}) = \beta\dot{x}$, one finally arrives at,

$$\ddot{x} + \beta\dot{x} + x(1-x)(1+\alpha x) = F\sin(\omega t) \tag{3}$$

which, following Thompson (1997), will henceforth be referred to as the HT_α equation.

The advantage of introducing this form of a general cubic potential is that the restoring curve shape can be varied without affecting other aspects of the system. By varying α, the shape of the curve is altered but the distance from $x = 0$ to the positive point of vanishing stability stays constant.

Note that the addition of bias to the symmetric restoring term has two effects on the effective restoring moment function, $C(\theta)$;

(a) it causes the original effective angle of vanishing stabilities to change by some heel angle, θ_H, with $\theta_V \approx \theta_{V0} - \theta_H$ and $\theta_U \approx \theta_{U0} + \theta_H$,

(b) the shape of the curve is altered as $\alpha = \theta_V/\theta_U$ is reduced from 1.

The parameterisation of the non-dimensional equation using α hides effect **(a)** by holding the point of vanishing stability at $x = 1$. This allows useful comparison between the dynamics seen for various α values, particularly the effect of bias as α is reduced from 1 .

It is useful to note that, using the approximations above, we can write

$$\alpha \approx \frac{1-H}{1+H} \tag{4}$$

where $H = \theta_H/\theta_{V0}$.

It is helpful to look at the set of non-dimensional potentials generated by the α parameterised restoring moment term, $c(x) = x(1-x)(1+\alpha x)$. Figure 1 shows the potential energy functions for $\alpha = 1$ (the symmetric well) $\alpha = 0.5$ and $\alpha = 0$ (the unsymmetric quadratic well). The two points of vanishing stability can be identified as the potential hilltops at $x = 1$ for all α and $x_\alpha = -1/\alpha$.

A bounded oscillation within the well corresponds to a ship rolling in beam seas. If $\alpha < 1$ then there is an additional wind loading on the model ship, although note that the effect of reduced θ_V is hidden, as previously discussed. Escape from the well corresponds to the capsize of the ship.

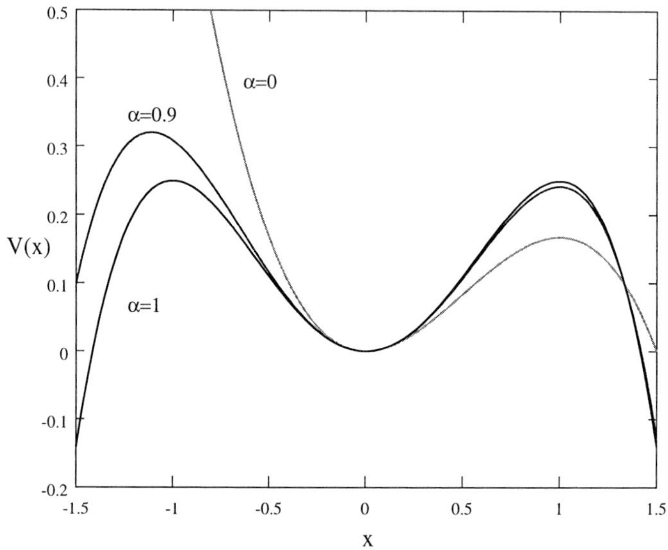

Figure 1: The potential wells for HT_α, with $\alpha = 0, 0.5, 1$

CAPSIZE SENSITIVITY TO A SYMMETRY BREAKING BIAS

Having recast the equations of motion into a suitable form, we are now interested in the conditions for escape as the shape of the potential well is altered. Of particular interest is the symmetry breaking that occurs in the system as α is reduced from unity.

Most previous studies have taken the forcing amplitude, F (at fixed ω), required for steady state escape to occur as a simple measure of the 'capsizability' of the ship. However this has no physical meaning so it is useful to introduce the concept of a *stability measure*. To derive such a measure for HT_α we can return to the original equations of motion and use the linearised equations to derive a simple design formula (Macmaster & Thompson 1994; Thompson 1997). Linearising equation (1) gives

$$I\theta'' + B_1\theta' + mgGM\theta = IAk\omega_f^2 \sin \omega_f \tau \qquad (5)$$

where $B_1 = dB/d\dot{\theta}$ at $\theta = 0$ is the linear damping coefficient and $GM = dGZ/d\theta$ at $\theta = 0$ is the metacentric height.

The steady state oscillation is given by the particular integral

$$\theta = \lambda \sin(\omega_f \tau) + \mu \cos(\omega_f \tau) \qquad (6)$$

It is possible to find θ_{max} by substituting this into equation (5), separating terms and solving to find λ and μ, arriving eventually at

$$\theta_{max} = \sqrt{\lambda^2 + \mu^2} = \frac{Ak\omega^2}{\sqrt{(1-\omega^2)^2 + (2\zeta)^2}} \qquad (7)$$

where $\omega_n = \sqrt{mgGM/I}$ is the natural frequency of the linearised system, $\omega = \omega_f/\omega_n$ and $\zeta = B_1\omega_n/(2mgGM) = \beta/2$ is the damping ratio, which is invariant under the scaling. Now, for tuned forcing $\omega = 1$ and by equating θ_{max} with θ_V, an approximate capsize criteria in terms of a critical wave slope, Ak, can be derived,

$$Ak = 2\zeta\theta_V \tag{8}$$

The concept of wave slope is a useful one since it is this quantity that is typically used to measure wave intensity (rather than amplitude, F) in an experimental set-up. In a numerical simulation, however, other factors such as θ_V and damping levels will be known so it is sensible to scale the wave slope to create a universal diagram. This criteria can be used to define a scaled, physical parameter, $J = Ak/(2\zeta\theta_V) = F/(\beta\omega^2)$ for a universal capsize diagram, (Thompson 1997). For tuned resonance, our simple design formula thus predicts that escape will occur at $J = 1$. A *stability measure* for a ship (with forcing at frequency ratio ω and linear equivalent damping factor β) can now be defined as,

$$J \text{ when capsize occurs} = J_{esc}(\omega, \alpha, \beta) = \frac{F_{esc}(\omega, \alpha, \beta)}{\beta\omega^2} \tag{9}$$

The word stability must be used carefully and not confused with the precise mathematical meaning. Here, it is used to descibe a ship's resistance to capsize, in terms of the wave intensity required (capsizability).

NUMERICAL DETERMINATION OF STEADY STATE CAPSIZE

To examine the stability of the HT_α oscillator as symmetry is broken by a bias, the response over a range of frequencies must be considered. To find $J_{esc}(\omega, \alpha)$ forcing amplitude is slowly increased from zero at constant ω until escape occurs. Further approximations are required to specify escape and steady state conditions. The system is then run for 16 forcing cycles before steady state is assumed to have been reached and F incremented by $\Delta F = 0.01$. Escape from the well is assumed to occur if the oscillations exceed $|x| = 1.2$. The forcing amplitude at which escape occured is recorded. A new frequency is then chosen and the process begun again. Figure 2 shows the J_{esc} lines for α varied in steps of 0.2 between 0 and 1. The value of the simple design formula $J = Ak/2\zeta\theta_V = 1$ as a worst case escape predictor is evident. The $J_{esc}(\omega, \alpha)$ lines are irregular throughout the frequency range, but trend behaviour is apparant. It is possible that some of the more minor irregularity may be eliminated by more stringent escape and steady state conditions. However, even then, rigorous criteria cannot be defined for the long transient motion associated with a chaotic orbit.

It is important to differentiate between irregularity and indeterminacy (Thompson & Soliman 1991). The symmetric restoring, $\alpha = 1$, escape boundary in particular, shows significant indeterminate regions between two trend escape lines for ω around worst case escape, one region of which is identified on the diagram. Indeterminacy is also seen at higher frequencies for $\alpha < 1$.

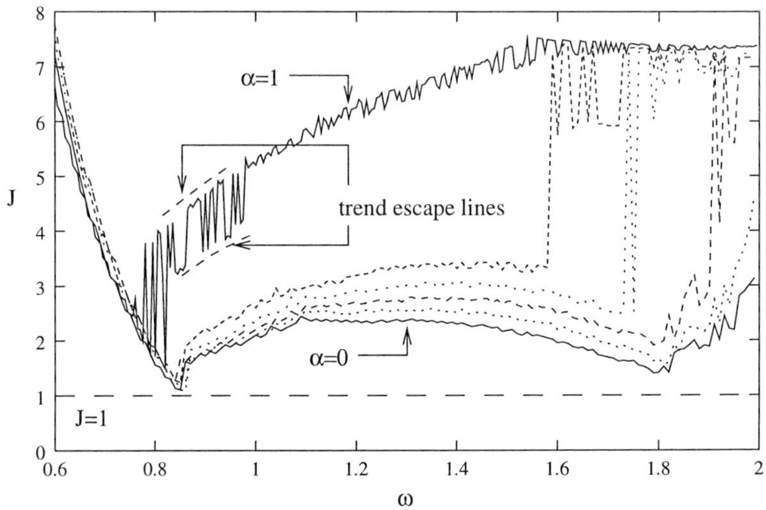

Figure 2: Here α is varied from 0 to 1 in steps of 0.2. The drop in J_{esc} due to a symmetry breaking bias occurs between frequencies $\omega_L(\alpha) \approx 0.84$ and $\omega_U(\alpha)$, such that $\omega_U(0.8) \approx 1.6$.

We now define two frequencies; $\omega_L(\alpha) \approx 0.84$, the worst case capsize point and $\omega_U(\alpha)$, the point at which $J_{esc}(\alpha)$ line moves back towards the $J_{esc}(1)$ line at higher frequencies (e.g. $\omega_U(0.8) \approx 1.6$). Escape from the well in the frequency region studied can now be roughly divided into 3 areas;

1 $\omega < \omega_L(\alpha)$, the low frequency regime where loss of symmetry has little effect on J_{esc},

2 $\omega_L(\alpha) < \omega < \omega_U(\alpha)$, mid-range frequencies where loss of symmetry leads to a significant reduction of J_{esc}. The J_{esc} lines are irregular in this region, but still show trend behaviour,

3 $\omega > \omega_U(\alpha)$, for higher freqencies J_{esc} is insensitive to loss of symmetry.

It is the behaviour in frequency region **2** that has serious implications for ship capsize. To illustrate the significance of these results we can use equation (4) to transform the information from the escape diagram into a physically meaningful form. In the sensitive frequency region J_{esc} is roughly halved when α is reduced from 1 to 0.9 (not plotted here). Now, $\alpha = 0.9 \Rightarrow \theta_H/\theta_{V0} \approx 0.05$. Hence we have a halving of the stability measure, J_{esc} when the heel angle is roughly 5% of the unbiased angle of vanishing stability, θ_{V0}. Note that for heel angles of this size, the effect of the stability drop due to reduced θ_V, is negligible when compared to the extreme symmetry breaking effect.

In reality there will always be some bias in a ship due to design or cargo imbalance. Thus, the dramatic sensitivity seen here will not occur should wind loading be increased. However the sensitivity will be important in an idealized mathematical model that might be used in the design process.

THE STEADY STATE DYNAMICS OF HT$_\alpha$

Solutions to the HT$_\alpha$ equation exist in a 3-dimensional phase space $(x, \dot{x}, t)$. However, the dynamics can be reduced to that of a 2-D map, $P(x, \dot{x})$ by taking a Poincaré section (Foale & Thompson 1991).

Since this is a periodically forced system a simple time T map can be used in which the phase projection $(x, \dot{x})$ is sampled every forcing period, from some t_0. Thus a steady state oscillation of period nT (n=1,2,3...) would be represented as a fixed point of P^n ($P^n(x, \dot{x}) = (x, \dot{x})$) of P. Here we are chiefly interested in the $n = 1$ fixed points and how parameter variation affects their stability and position in phase space.

To understand the drop in J_{esc} observed for certain frequency regions, the mechanisms of steady state escape need to be considered. We are interested firstly in the extreme cases $\alpha = 0$ or 1, the unsymmetric (canonical escape) and symmetric wells.

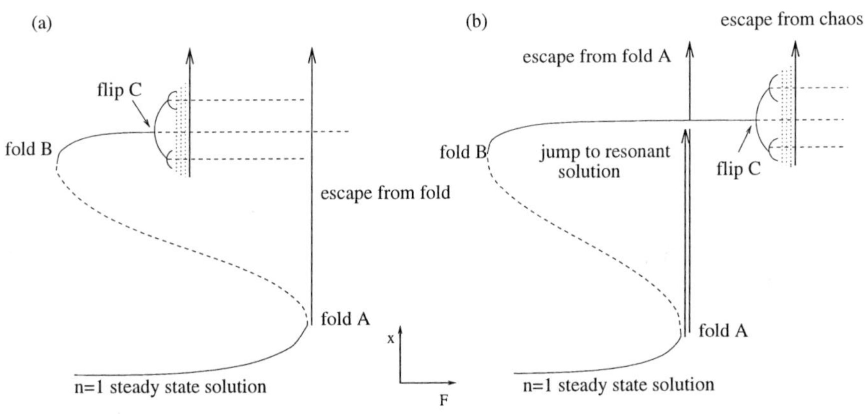

Figure 3: Schematic diagrams showing steady state paths (dashed = unstable) and escape routes. Each represents the x value in the Poincaré section plotted against forcing amplitude, F.

The unsymmetric well The unsymmetric case, HT$_0$ is an archetypal model for escape from a potential well (Thompson, Bishop & Leung 1987; Thompson 1989). Figure 3 shows schematic versions of the steady state solution paths as the parameter, F is varied.

For this system, as the forcing amplitude is increased, escape from the well can occur in two ways; **(a)**, the steady state being followed ceases to exist for higher F or **(b)** the steady state becomes unstable at higher F. Providing the current trajectory does not lie in the basin of attraction of another stable steady state, the oscillations will escape from the well if either of the above events takes place.

For the frequencies around linear resonance **(a)** occurs at the fold bifurcation, A, and **(b)** at the final crisis, E (after a period doubling cascade to chaos). The final crisis, E, is approximately located by the first period doubling bifurcation, C (or flip bifurcation in

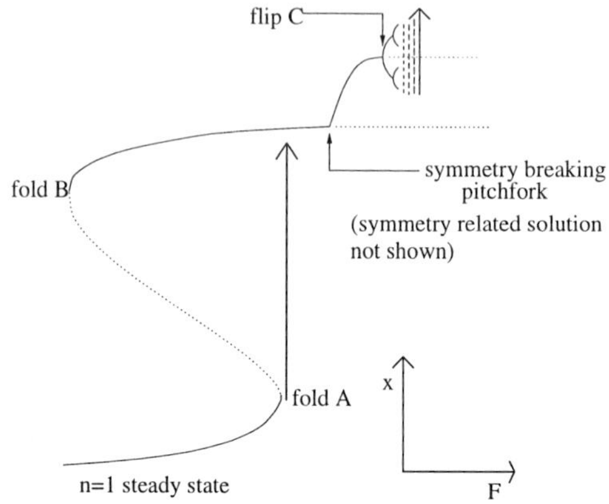

Figure 4: Schematic diagram showing solution paths for HT_1 in the frequency region where escape occurs from chaos

the Poincaré map). Note that solutions other than those shown here exist but can be safely ignored since they have very small basins of attraction (i.e. few initial conditions $(x_i, \dot{x}_i)$ will settle onto such steady states).

Winding up forcing amplitude from zero (attractor following) leads to escape by one of the above routes. For low forcing frequencies (figure 3(a)) the system will escape from fold A. As ω is increased fold A moves to lower F and the flip moves to higher F. At some ω the F values of the two bifurcations cross beyond which a stable $n = 1$ steady state exists when the system jumps from fold A (figure 3(b)). Whether escape from fold A occurs at these forcing frequencies depends on whether the steady state at this point exists in the basin of attraction of the resonant solution or not.

The symmetric well A similar picture was constructed for HT_1, figure 4, using a combination of attractor and path following techniques (Foale & Thompson 1991). The solution paths are similar to the unsymmetric case except that a symmetry breaking bifucation occurs just before the flip, as F is wound up. The symmetric $n = 1$ solution loses stability at the pitchfork bifurcation, where two unsymmetric, stable solutions branch off, losing stability after only a small further increase in F, at the flip bifurcation, C.

It is now useful to consider $\epsilon = 1 - \alpha$ as some small perturbation from symmetry. The paths for HT_α with $\alpha < 1$ can thus be seen as perturbations from that of HT_1. Crucially the pitchfork is replaced by a fold bifurcation for $\epsilon > 0$. The flip bifurcation that is taken as an approximation of the escape from chaos falls back to lower forcing amplitudes for a small increase in ϵ from zero.

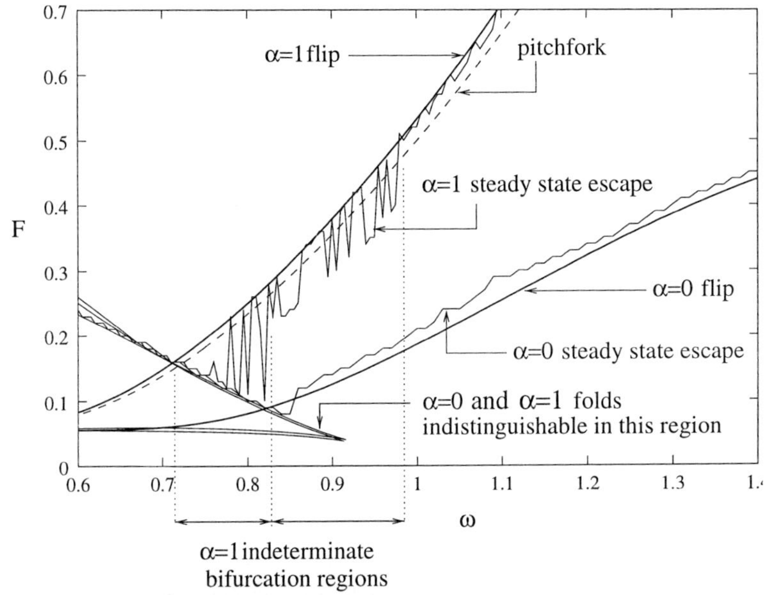

Figure 5: A bifurcation diagram for HT_1 and HT_0, showing the large difference in the flip lines and similar fold lines. The J_{esc} lines for each system are superimposed.

BIFURCATION DIAGRAMS

The positions of the fold and flip bifurcations have been shown to be vital in determining the parameter values at which escape occurs. A useful next step is to extend the path-following methods to generate these bifurcation lines in the 2-dimensional control space spanned by F and ω (a bifurcation diagram).

The bifurcation arcs are shown in figure 5 with the relevant J_{esc} lines superimposed. The significance of the fold and bifurcation lines in determining steady state escape is evident from figure 5. The bifurcation lines effectively mark escape for both systems over a range of frequencies. The main anomalies occur in the $\alpha = 1$ indeterminate regions (indicated on the diagram).

It is helpful to define $\omega_R(\alpha)$ as the frequency at which the fold and flip lines cross (e.g. $\omega_R(0) \approx 0.07$). For forcing frequencies lower than $\omega_R(\alpha)$, the steady state will escape from the fold A as F is increased. For higher ω the system can 'jump' to resonance and then escape from chaos (shortly after the flip bifurcation) or escape directly from the fold.

We can now identify the flip C, with the sensitivity to symmetry breaking bias in seen in figure 2. It is the change in the flip arc with α that causes the sensitivity to bias seen in J_{esc}.

Note also how the importance of indeterminacy in defining the worst case capsize point. The system continues to escape from the fold for frequencies higher than ω_R, in an indeter-

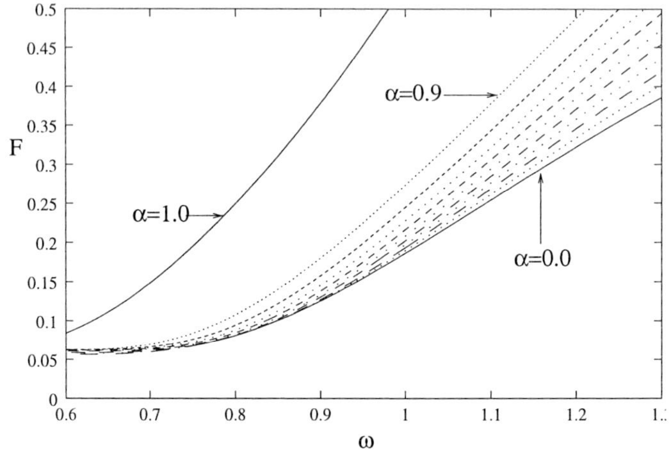

Figure 6: Flip arcs around worst case escape region with $\alpha = 0$ to 1 in steps of 0.1

minate manner, a phenomenon that is much more significant for the symmetric well.Further indeterminacy has been noted for $0.84 < \omega < 0.98$ for HT_1 and also at higher frequencies around $\omega_U(\alpha)$ for $\alpha < 1$. More details on the dynamics behind indeterminacy may be found in Thompson & Soliman (1991).

Having identified the main features of the stability diagram in terms of the bifurcation lines and indeterminacy, the sensitivity in the second frequency region is now considered in more detail. Focussing first on the worse case frequency region the high sensitivity of the flip line to symmetry breaking is evident in figure 6. The flip falls steeply back towards the HT_0 line as α is reduced from unity.

The flip sensitivity can be explained roughly in terms of the expected significance of the symmetry breaking for solutions of different oscillation amplitudes. From figure 1 we can see that the effect of symmetry breaking is greater away from the bottom of the well. Hence, it is sensible to expect oscillations of higher amplitudes (e.g. near to the flip) to be the most affected by variation of α.

We now use a relative sustainable wave slope to measure effects **(a)** and **(b)** together,

$$S = (Ak_{esc} \text{ with loading bias})/(Ak_{esc} \text{ without loading bias}) \tag{10}$$

which we can use with equation (4) to give S as a function of $H = \theta_H/\theta_{V0}$.

In figure 7 we plot S against H for the three main cases. The incorporation of the reduction in θ_V (effect **(a)**) can be seen in the linear relationship at higher H (remember $Ak \propto \theta_V$) where the effect **(b)** is less.

We can now compare this study with Thompson (1997) for which a diagram similar to that of figure 7 is plotted, for a damping level of $\beta = 0.2$. These results show similar sensitivity for the worse case frequency. However, the second frequency region shows a much greater sensitivity to bias. This is because the worse case capsize point is determined by the indeterminate bifurcation which occurs at lower amplitudes and is hence less sensitive to symmetry than the flip C.

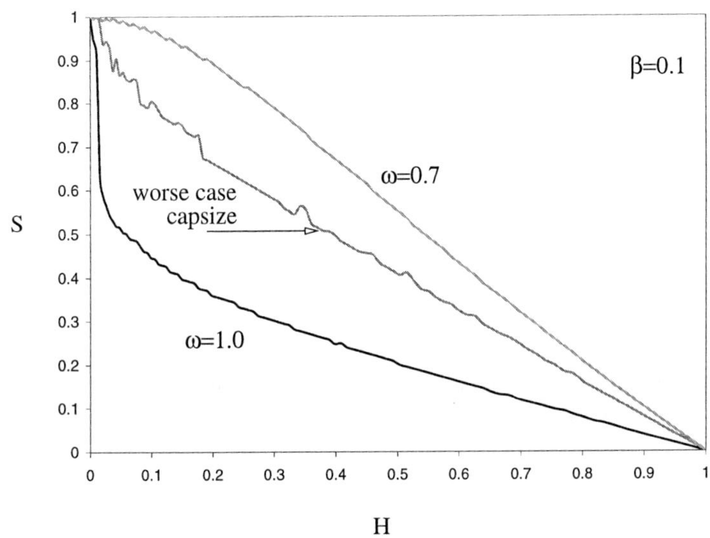

Figure 7: Relative sustainable wave slope, S, plotted against $H = \theta_H/\theta_V$. The $\omega = 1.0$ line shows much greater sensitivity than at worse case. At lower frequencies escape is insensitive to bias.

CONCLUSIONS

We have studied a model for varied bias on a symmetric ship in beam seas, considering in particular the effect of a symmetry breaking bias in the system. The relationship between potential well shape and escape for the HT_α system has been explored.

In order to study the escape with bias variation, a steady state escape diagram was plotted, using the stability measure, J_{esc}. From this three different escape regions were identified, the mid-frequency region of which showed extreme sensitivity to bias. This behaviour has great significance for ship safety criteria, implying that small biasing from symmetry can lead to a sudden reduction in the wave slope necessary to cause capsize. However the biasing required is extremely small and a real ship will always be biased beyond this point. We have therefore confirmed the canonical escape ($\alpha = 0$) to be a sensible model of ship roll motion.

Following previous work on the unsymmetric system, the bifurcation diagram for HT_α was constructed, giving a picture of the overall steady state dynamics. When compared to the stability diagram, the main areas of escape were identified with the possible routes displayed by the bifurcation diagram. The exisitence of an indeterminate bifurcation was used to explain the value of the worst case capsize frequency. The flip bifurcation line was shown to be highly senstitive to bias and this was identified as the mechanism behind the sensitivity of the escape conditions. An argument has been put forward to explain the relationship between the flip bifurcation and potential well shape in terms of the oscillation amplitudes in such regions of control space.

References

Falzarano, J.M. (1994). A combined approach to evaluate the effect of modelling approximations in predicting vessel capsizing. In J.M.T. Thompson and S.R. Bishop (Ed.), *Nonlinearity and Chaos in Engineering Dynamics*, pp. 408–410. Chichester: Wiley.

Foale, S. & Thompson, J.M.T. (1991). Geometrical concepts and computational techniques of nonlinear dynamics. *Computer Methods in Applied Mechanics and Engineering* **89**, 381–394.

Kan, M. & Taguchi, H. (1994). Ship capsizing and chaos. In J.M.T. Thompson and S.R. Bishop (Ed.), *Nonlinearity and Chaos in Engineering Dynamics*, pp. 418–420. Chichester: Wiley.

Macmaster, A.G. & Thompson, J.M.T. (1994). Wave tank testing and the capsizability of hulls. *Proceedings of the Royal Society London* **446**, 217–232.

Thompson, J.M.T. (1989). Chaotic phenomena triggering the escape from a potential well. *Proceedings of the Royal Society London* **A 421**, 195–225.

Thompson, J.M.T. (1997). Designing against capsize in beam seas: Recent advances and new insights. *Applied Mechanics Reviews* **50**, 307–325.

Thompson, J.M.T., Bishop, S.R. & Leung, L.M. (1987). Fractal basins and chaotic bifurcations prior to escape from a potential well. *Physics Letters* **A 121**, 116–120.

Thompson, J.M.T., Rainey, R.C.T. & Soliman, M.S. (1992). Mechanics of ship capsize under direct and parametric wave excitation. *Philosophical Transactions of the Royal Society London* **A 338**, 471–490.

Thompson, J.M.T. & Soliman, M.S. (1991). Indeterminate jumps to resonance from a tangled saddle-node bifurcation. *Proceedings of the Royal Society London* **A 432**, 101–111.

Wright, J.H.G. & Marshfield, W.B. (1980). Ship roll response and capsize behaviour in beam seas. *Transactions of the Royal Institute of Naval Architects* **122**, 129–148.

Contemporary Ideas on Ship Stability
D. Vassalos, M. Hamamoto, A. Papanikolaou and D. Molyneux (Editors)

SOME RECENT ADVANCES IN THE ANALYSIS OF SHIP ROLL MOTION

B. Cotton, J.M.T. Thompson & K.J. Spyrou

Centre for Nonlinear Dynamics and its Applications,
University College London,
Gower Street, London WC1E 6BT, UK

ABSTRACT

In an effort to place our previous investigations of ship roll dynamics within physically based limits, we extend a numerical steady state analysis to higher frequency forcing. Working with a simple nonlinear roll model, a number of different phenomena are discussed at above resonant frequencies, including sub-critical flip bifurcations and a *second resonance region.*

We then discuss a highly generalised approach to roll decay data analysis that does not require us to predefine damping or restoring functions. The problem is approached from a local fitting standpoint. As a result the method has potential for further extension to more complex models of damping as well as restoring force curves.

KEYWORDS

Nonlinear, roll, capsize, second resonance, parameter estimation, damping

INTRODUCTION

Previous studies of beam sea roll models (Thompson 1989; Thompson 1997; Virgin 1988) have focussed on the resonant region, where linear theory would predict capsize to be most likely. Here, we explore the steady state dynamics at higher frequencies of forcing and discuss some new features of the control space. In particular we discuss capsizing wave slopes at high forcing frequencies. Interestingly, the capsizing slopes are of similar magnitude to those at resonance.

The derivation of accurate representations of damping functions as part of a ship roll model is highly desirable in the study of roll dynamics. Roll damping functions, however, are extremely difficult to obtain by theory or experiment. The tendency has been to remain with simple linear or low order nonlinear velocity dependent models (Dalzell 1978; Haddara & Bennet 1989). To test the validity of such approaches we must be able to obtain damping functions from experimental data efficiently and accurately. However the difficulty in separating parameters in any such analysis has hindered improvement on existing ideas. Here we approach the problem from a local fitting standpoint using linear approximations to reconstruct a globally nonlinear curve.

Although the approach discussed is applied over all the data, separating angle and velocity dependent terms remains a serious problem. We conclude by briefly discussing some ideas for improving our ability to deal with these difficulties.

HIGH FREQUENCY FORCING

During the design of roll experiments it is necessary to ascertain the forcing parameter ranges over which our nonlinear oscillator model is valid. In particular we need to consider two limits; the maximum wave slope and frequency. The former is a consequence of the nature of waves and simple to evaluate. The latter is a more subtle problem related to the fact that the beam of a ship must be small compared to the wavelength for the model to be applicable.

Firstly we write our roll equation as,

$$I\theta'' + B(\theta') + mgGZ(\theta) = IAk\omega^2 \sin(\omega_f \tau) \tag{1}$$

where the prime denotes differentiation with respect to real (unscaled) time, τ, I is the rotational moment of inertia about the centre of gravity (incorporating any added hydrodynamic mass), θ is the roll angle relative to the wave normal, $B(\theta')$ is the non-linear damping function, $GZ(\theta)$ is the roll restoring force, Ak is the wave slope amplitude (A is the wave height and k the wave number) and ω_f is the wave frequency. We also write ω_n as the natural frequency of linearised ship motions.

We then utilise a simple non-dimensionalised model for roll motion, the Helmholtz-Thompson equation (Thompson 1997; del Rio, Rodriguez-Lozano & Velarde 1992)

$$\ddot{x} + \beta\dot{x} + x - x^2 = F \sin \omega t \tag{2}$$

where, in terms of (1), our two parameters are $F = Ak\omega^2/\theta_V$ and $\omega = \omega_f/\omega_n$ with $x = \theta/\theta_V$. We also introduce the parameter $J = Ak/(2\zeta\theta_V) = F/\omega^2$ which is a scaled measure of wave slope based on a linear capsize analysis, (Macmaster & Thompson 1994). Here, θ_V is the angle of vanishing stability and ζ the effective linear damping coefficient. We also set $\beta = 2\zeta = 0.1$.

The first limit is a consequence of the nature of water waves. For a steepness above $H/\lambda \approx 1/7$ the wave will break and the use of a simple sinusoidal forcing is no longer valid. Thus, with wave slope $Ak = \pi H/\lambda$, we can write,

$$J^{max} \approx \frac{\pi}{14\zeta\theta_V} \tag{3}$$

The model assumes that the ship tries to follow the motions of the water particles in the wave and does not interfere with the pressures in the wave. This is only valid when the beam of the ship is small compared to the wavelength. We can thus write a minimum wavelength, λ^{min}, permissable in terms of the beam, b

$$\lambda^{min} = \epsilon b \tag{4}$$

where we take, as a first estimate, $\epsilon \approx 6$. This in turn gives us a maximum forcing frequency

$$\omega_f^{max} = \sqrt{\frac{2\pi g}{\lambda^{min}}} = \sqrt{\frac{2\pi g}{\epsilon b}} \tag{5}$$

leading to

$$\omega^{max} = \frac{\omega_f^{max}}{\omega_n} = T_n\sqrt{\frac{g}{2\pi\epsilon b}} \tag{6}$$

where ω_n and T_n are the natural roll frequency and period of the ship. Note that this second limit is due to the approximations of our roll model whereas the first is a feature of wave behaviour.

Substituting in two real ship values (a purse seiner (Umeda, Hamamoto, Takaishi, Chiba, Matsuda, Sera, Susuki, Spyrou & Watanabe 1995) and a container (Takezawa, Hirayama & Acharrya 1990)) for beam dimension and natural frequency we can find example limits, Table 1.

TABLE 1

Ship	θ_V [degrees]	T_n [s]	b[m]	ω^{max}	J^{max}
Purse Seiner	40	7.47	7.6	1.4	3.2
Container	-	19.4	25.4	1.9	-

Therefore as a first step we extend previous steady state analyses to frequencies up to $\omega \approx 2$ with the additional limit $J^{max} \approx 3$. Using numerical techniques we are able to plot the development of steady state oscillations whilst varying wave amplitude (or slope). This process is repeated for a range of forcing frequencies. For below resonance frequencies it has been shown (Thompson 1996) that for (2), as F is increased, escape (corresponding to capsize) occurs with a jump from a fold bifurcation. Above resonance the system escapes from a chaotic orbit after a period doubling cascade. For the latter case the initial flip bifurcation is often taken to be a sufficiently accurate indicator of capsize in the control space.

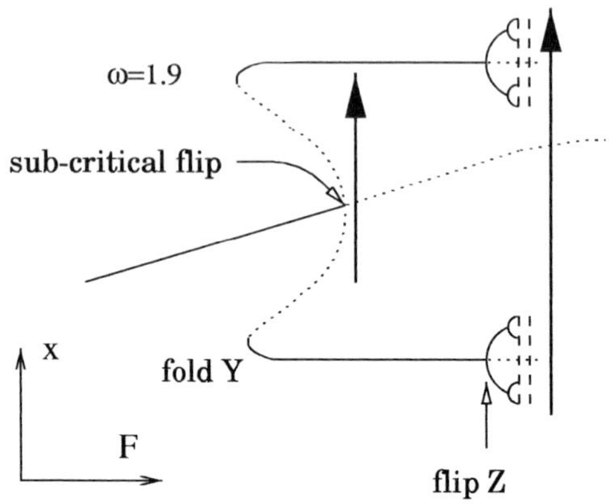

Figure 1: Schematic example of a high frequency capsize mechanism

Figure 1 shows a schematic example of a discontinuous jump found at higher frequency forcing. The solution path shows restabilisation after a sub-critical flip bifurcation onto a period 2 oscillation. Here we would see a sudden increase in roll amplitude. In this case the flip bifurcation is not a good estimate of capsize. With further increase in F, the system undergoes a period doubling cascade to chaos, before escaping. Note that the fold Y and the subsequent flip Z are bifurcations of the period 2 oscillation.

We illustrate the high frequency bifurcations in a control space diagram, figure 2. The steady state capsize line show the wave slope at which capsize occurs when J is increased in small steps from zero. The ragged nature of this line is primarily due to the computational approximations required in the numerical procedure.

Importantly we find that the flip C is a good estimate of capsize only below $\omega \approx 1.8$. However, the discontinuous jump (at the sub-critical flip) for $\omega > 1.8$ must be considered a highly dangerous phenomenon.

Of further interest is the existence of an effective *second resonance region* at $\omega \approx 1.8$ which shows qualitative similarity to the 'wedge' at resonant frequencies. At this second resonance capsizability of the model (as measured by J rather than F) is comparable to that at resonance. Note that the use of the scaled wave slope, J, rather than the amplitude, F, gives the correct emphasis to capsize in this higher frequency region. A simple design formula (based on a linear analysis), (Thompson 1997), predicts capsize at $J = 1$, which is a reasonable lower bound in the above case. For higher damping, this $J = 1$ formula is found to be more accurate.

ROLL TIME SERIES ANALYSIS

We have recently been considering whether we can extract the damping and restoring curves from simple roll decay data. In general, given a roll decay time series we can take two basic approaches to fitting our nonlinear model to the data; global or local. A global

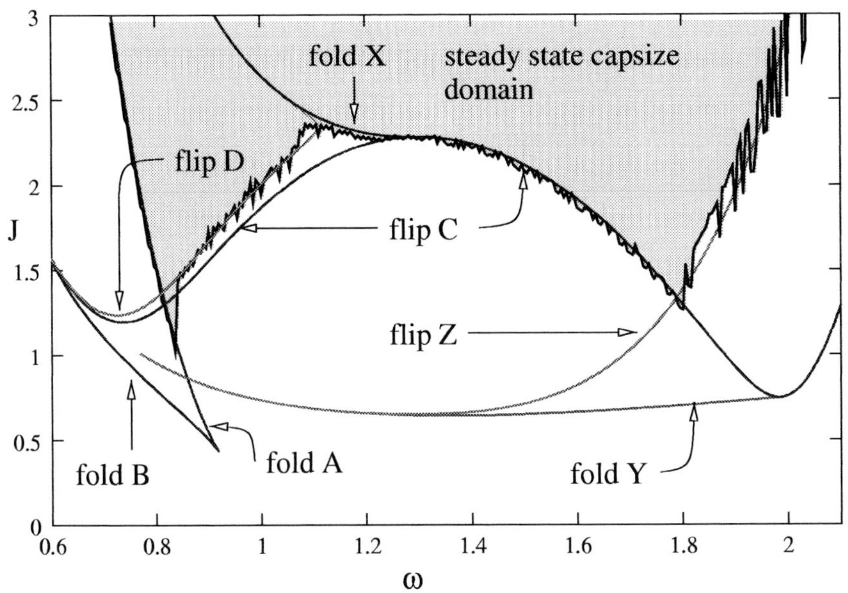

Figure 2: Bifurcation diagram for the capsize equation (2), extended to higher ω. The flip C is super-critical at large and small frequencies: it is sub-critical betweeen the codimension 2 events at which is meets fold X ($\omega \approx 1.3$) and fold Y ($\omega \approx 2$). Damping coefficient, $\beta = 0.1$.

approach predefines a polynomial to describe the damping (or restoring) functions. The predictions of such a model can thus be fitted to the data over some number of roll cycles. A local method does not require the pre-definition of these functions and instead fits local linear approximations over small sections of the data. These local approximations are then combined to reconstruct a global, nonlinear fit. Here we present the basic method and discuss its failings as well as their possible solutions.

The first step is to model the time series so that we can obtain estimates for its derivatives. At time τ_i the time series will have some value θ_i. Using the surrounding points we can also approximate $\dot{\theta}_i$ and $\ddot{\theta}_i$. We may employ a number of different methods to do this. Here we employ a Savitsky-Golay filter (Press, Teukolsky, Vettering & Flannery 1992) that we have succesfully used to obtain double derivatives from experimental roll decay data. We again use our roll motion model (1) and assume that we can write the two functions $B(\dot{\theta})$ and $GZ(\theta)$ as locally linear. We can now write our equation of motion *locally* as,

$$I\ddot{\theta} + B_0 + B_1\dot{\theta} + mg(\lambda + \mu\theta) = 0 \tag{7}$$

and

$$GZ(\theta)_{local} = \lambda + \mu\theta \tag{8}$$

$$B(\dot{\theta})_{local} = B_0 + B_1\dot{\theta} \tag{9}$$

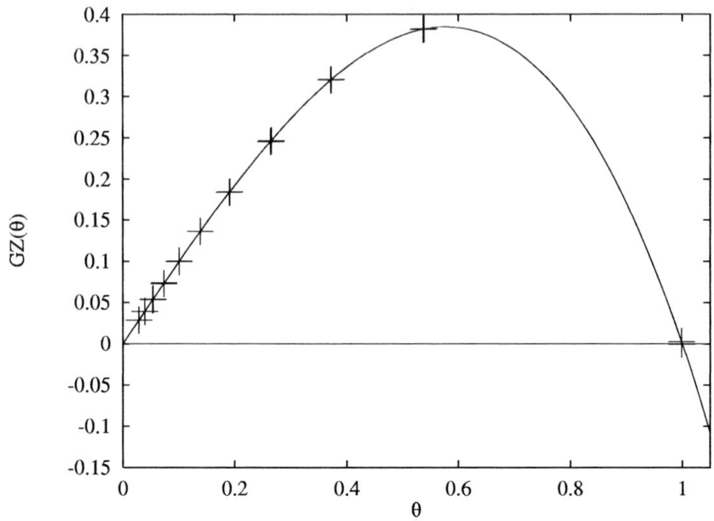

Figure 3: Reconstruction of restoring force curve for the symmetric escape equation, the reconstructed points are shown with the original curve

If we write $B_0 + mg\lambda = C$, we are left with three unknowns (B_1, μ, C) and thus require three equations to find these unknowns.

Therefore we simply need to sample the time series at three nearby points. Nearby here means that they must be close enough in phase space such that our local dynamical model is valid.

This gives the local slopes for $B(\dot{\theta})$ and $GZ(\theta)$ and the constant C. Since we cannot easily separate C we instead specify $GZ(0) = 0$ and $B(0) = 0$, and integrate over our local slopes to reconstruct the restoring and damping curves.

We then scan through our time series selecting three consecutive points every step and solving the equations to obtain locally fitted parameters over a wide range of phase space. We then reconstruct the curves by integrating over the local slopes.

EXAMPLES AND IMPROVEMENTS

As an example we have taken some numerically generated data from a model with known restoring and damping functions (the symmetric escape equation, (Thompson 1997), which is similar to (2) but with a restoring force of $x - x^3$). Here we have reconstructed damping and restoring simultaneously. Figure 3 show the reconstructed GZ curve.

Note that for this method velocity and angle dependent parameter separation remains a problem (the equations we are solving to find B_1, μ and C become ill-conditioned and much of the data series proves unusable for this method. Therefore we have applied the

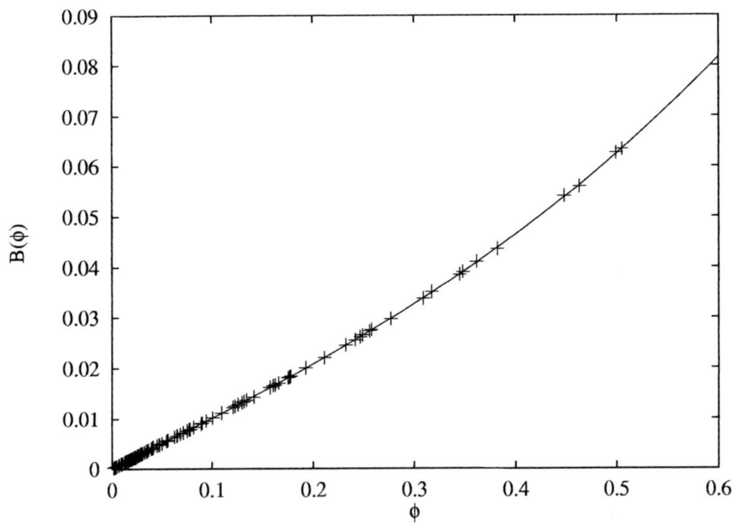

Figure 4: Reconstruction of a linear plus cubic damping curve with specified GZ

method carefully over parts of the data set for which it succeeds.

In figure 4 we plot a reconstructed nonlinear damping curve. Here the restoring function was pre-specified and the damping taken to be dependent only on velocity.

Therefore parameter separation was not a problem and all of the data was used. The routine has also been applied to some experimental roll decay data and was found to perform well in the presence of limited precision and noise. This experimental data was from a low angle decay test and so the restored functions were very close to linear. It was found that calculations of natural frequency using the reconstructed GZ gave results accurate to within 1% of the measured values.

We can improve our ability to deal with the parameter separation problem by employing singular systems analysis (Broomhead & King 1986), to provide us with more information on how and where the method fails. Treating the fitting as a matrix inversion problem we can rewrite our set of equations as,

$$\begin{pmatrix} \dot{\theta}_1 & \theta_1 & 1 \\ \dot{\theta}_2 & \theta_2 & 1 \\ \dot{\theta}_3 & \theta_3 & 1 \end{pmatrix} \begin{pmatrix} B_1 \\ mg\mu \\ C \end{pmatrix} = -I \begin{pmatrix} \ddot{\theta}_1 \\ \ddot{\theta}_2 \\ \ddot{\theta}_3 \end{pmatrix} \tag{10}$$

or

$$Ax = b \tag{11}$$

By expressing the problem is such a way, we are able to utilise singular value decomposition (SVD) which can be used to both solve for x and also provide information on separability of the parameters. When the data does not distinguish well between two or more parameters then A becomes ill-conditioned and this can be detected with SVD (Press, Teukolsky, Vettering & Flannery 1992).

The solution is obtained by decomposing A and then back-substituting given b (it is similar in application to solution by standard matrix decomposition methods). If A is ill-conditioned then SVD will provide the best approximation to a solution in the least squares sense. Thus we are able to go further than is possible with the simpler approach.

A further reason for employing SVD is that we can add additional rows to A and solve for x with a reduced likelihood of ill-conditioning. We can do this by simply selecting more nearby data points to provide local roll equations. A still more powerful addition is to include further rows representing energy balance equations for the sampled data points.

CONCLUSIONS

A steady state bifurcation analysis of a simple roll model has been extended to higher forcing frequencies. We have discussed a number of new phenomena, with particular reference to capsize mechanisms. The higher frequency region has been shown to bear qualitative similarities to that around resonance and we have identified a *second resonance region*. Capsizing wave slope at frequencies around $\omega \approx 1.8$ is found to be comparable to that at resonance, although the feasibilty of such conditions occuring must be considered. Furthermore we have shown that the usage of the flip bifurcation as a capsize estimate must be made carefully in this high frequency regime.

Secondly, we have applied a local fitting method to numerically generated roll decay data and succesfully recovered a nonlinear damping function. The method has been extended to the simultaneous reconstruction of restoring and damping curves, but in this case parameter separation problems remain. The basic difficulty is the separation of velocity and angle dependent terms over the whole data series. We have discussed the application of singular systems analysis to improve our ability to deal with this problem and sketched out how it may be applied.

References

Broomhead, D.S. & King, G.P. (1986). Extracting qualitative dynamics from experimental time data. *Physica D* **20**, 217–236.

Dalzell, J.F. (1978). A note on the form of ship roll damping. *Journal of Ship Research* **22**(3), 178–185.

del Rio, E., Rodriguez-Lozano, A. & Velarde, M.G. (1992). Prototype Helmholtz-Thompson nonlinear oscillator. *Review of Scientific Instruments* **63**, 4208–4212.

Haddara, M.R. & Bennet, P. (1989). A study of the angle dependence of roll damping moment. *Ocean Engineering* **16**, 411–427.

Macmaster, A.G. & Thompson, J.M.T. (1994). Wave tank testing and the capsizability of hulls. *Proceedings of the Royal Society London* **446**, 217–232.

Press, W.H., Teukolsky, S.A., Vettering, W.T. & Flannery, B.P. (1992). *Numerical Recipes in C, 2nd Edition.* Cambridge: Cambridge University Press.

Takezawa, S., Hirayama, T. & Acharrya, S. (1990, September). On large rolling in following directional spectrum waves. In *Fourth International Conference on Stability of Ships and Ocean Vehicles*, Volume 1, University of Naples, Italy, pp. 287–294.

Thompson, J.M.T. (1989). Chaotic phenomena triggering the escape from a potential well. *Proceedings of the Royal Society London* **42**, 195–225.

Thompson, J.M.T. (1996). *Nonlinear Mathematics and its Applications*, Chapter 1, pp. 1–47. Cambridge: Cambridge University Press. ed(Aston, P.J.).

Thompson, J.M.T. (1997). Designing against capsize in beam seas: Recent advances and new insights. *Applied Mechanics Reviews* **50**, 307–325.

Umeda, N., Hamamoto, M., Takaishi, Y., Chiba, Y., Matsuda, A., Sera, W., Susuki, S., Spyrou, K. & Watanabe, K. (1995). Model experiments of ship capsize in astern seas. *Journal of the Society of Naval Architects of Japan* **177**, 207–217.

Virgin, L.N. (1988). On the harmonic response of an oscillator with unsymmetric restoring force. *Journal of Sound and Vibration* **126**, 157–165.

Contemporary Ideas on Ship Stability
D. Vassalos, M. Hamamoto, A. Papanikolaou and D. Molyneux (Editors)

SHIP CAPSIZE ASSESSMENT AND NONLINEAR DYNAMICS

K. J. Spyrou

Centre for Nonlinear Dynamics and its Applications
University College London, Gower Street, London WC1E 6BT, UK

ABSTRACT

Certain aspects of ship stability assessment in beam and in following seas are discussed. It is argued that the use of detailed numerical codes of ship motions cannot solve alone the assessment problem. On the other hand, whilst simplified models can be very useful for acquiring a fundamental understanding of the dynamics of capsize, still a good number of theoretical obstacles need to be overcome. In respect to beam sea capsize, we begin by discussing the structure of the mathematical model and the types of excitation. Then we consider the mechanism of roll damping very near to capsize angles and we point out a very interesting connection that exists with the specification of predictors of capsize based on Melnikov's method. Finally we sketch out a constrained design optimization procedure which can be used in order to identify those ship parameters' values where resistance to capsize is maximized. In respect to the following sea, we show that if capsize is examined in a transient sense, it should be possible to have a unified treatment of pure-loss and parametric instability. We also present predictions of the qualitative effect on the stability transition curves coming from bi-chromatic waves.

KEYWORDS

Ship, capsize, nonlinear dynamics, Melnikov, parametric, design, model tests.

INTRODUCTION

Whilst one might think of many different methods for assessing the behaviour of a system, there is little doubt that the most reliable are those which are based on sufficient understanding of the system's key properties. For ship stability assessment however the application of this principle has been, so far at least, less than straightforward; because the behaviour of a ship in an extreme wave environment, where stability problems mostly arise, is often determined by very complex, hydrodynamic or ship dynamic, processes.

Ideally one would wish of course to have a full, meticulously developed and validated mathematical model of ship motions on which to carry out detailed analysis of dynamic behaviour and instability. Unfortunately this seems still to be well beyond our reach. But even if such a model were available, we would hardly know how to carry out in depth analysis for the nonlinear dynamical system in hand[1]. As a result, one can see two lines of research evolving, and it is essential that interaction between the two is encouraged: the first dealing with detailed mathematical modelling of the motion; whereas the second aiming to provide a better understanding of dynamic behaviour on the basis of simpler models that can capture however key features of the system's response. Areas of concern can be identified however in either of these directions: As the mathematical model of ship motion becomes larger, there is a cumulative effect from the uncertainties that often underlie the various assumptions and the unavoidable empiricism which lurks behind model development. On the other hand, when a simple model is used it is sometimes uncertain to what extent the observed behaviour corresponds to that of the real system.

The first direction represents essentially the extension of the traditional seakeeping approach from small towards larger amplitude motions. However the second is quite novel in naval architecture. Its importance is owed to the fact that nonlinearity can make large amplitude responses follow completely different patterns that their smaller-amplitude counterparts. As is nowadays increasingly realized, a ship, like many other dynamical systems, can exhibit a very rich envelope of large-amplitude behaviour which is sometimes very difficult to unravel. In order to understand the underlying principles of safety-critical behaviour one needs to have an effective methodology which will guide his search and here is where the techniques of nonlinear dynamics' can provide truly valuable inputs.

These techniques enable, at first instance, better focus during physical model testing. This is essential because in extreme seas comparisons between theory and experiment are non-trivial due to the fact that the number of unknowns involved is very large. But perhaps the more far-reaching implication is that they offer a potential for developing effective methods of stability assessment that can combine scientific rigour with practicality for better design and safer operation. This potential allows us to start thinking also about integrated stability assessment methods which will cover mechanisms of capsize associated with different environments and ship-wave encounters.

In the previous Workshop in Crete we have outlined some of our recent work along the above lines: We proposed a method of interfacing the findings of the nonlinear dynamics approach of ship capsize with design, in respect to the mechanism of capsize in resonant beam seas, Spyrou et al. (1997). Also we continued our investigations of the instabilities of the following/quartering sea, discussing the interesting parallel that exists between the yaw (related with broaching) and the roll dynamics (related with pure-loss and parametric instability), Spyrou (1997). The present paper consists of two parts: Firsly we discuss, very much in the spirit of this Workshop, some of the problems that exist in developing an effective stability assessment in beam-seas. Then we explain a practical assessment method for pure-loss and parametric instability in following seas.

[1] As is well known, the motion of a body on the surface of the sea entails partial differential equations (PDEs) for its description. As evidenced from approximations of PDEs from systems of ordinary differential equations (ODEs), an infinite number of ODEs is required for absolute equivalence. This corresponds to the well known fact that memory effects (or frequency dependence of hydrodynamic coefficients) render the system's state-space infinite dimensional.

BEAM-SEA CAPSIZE

A number of issues are currently under debate, such as the suitability of single roll or coupled models, the use of deterministic or stochastic-type of excitation; and the quantitative prediction of damping especially up to very large angles.

The suitability of the mathematical model

It is quite common, especially after Wright & Marshfield (1980) to model roll motion in regular beam waves by expressing the roll angle relatively to the local wave slope. A single-degree roll equation is then used to describe roll dynamics with nonlinearities in damping and in restoring. For ships with small beam compared to the wave length, it is often reasonable to assume that, in sinusoidal beam waves they experience a fluctuating "effective gravitational field" $\mathbf{g}_e$ where the centrifugal acceleration of the water particle is combined with the acceleration of gravity $\mathbf{g}$, Thompson et al. (1992). This says essentially that a small boat beam to long waves tends to follow the motion of the water particles and it allows direct use of the calm-sea restoring of the ship in the equation of relative roll. From an axes system tracking the motion of a water particle and having one axis always tangent to the wave surface the single roll model is then perfectly adequate. But if it is intended to carry out model experiments, the physical model should rather not be constrained rigidly in sway because then the model cannot follow the motion of the water particles and direct comparison between theory and experiment becomes difficult. On the other hand, if the model is not constrained at all, it is likely to yaw and to have also a mean drift which also hinders comparisons with theory.

There is of course the possibility also that the ship "cannot" follow the motion of the water particles. Then the coupled roll sway and heave need to be considered along with the type of wave excitation as the above single roll model has encountered its limits. This is even more evident if the effect of non-regular waves is under consideration. However, one must bear in mind here that, unlike some seakeeping studies where we examine performance degradation during a voyage, in intact-ship capsize we are only concerned about an almost momentary event which is usually the result of encountering a small number of steep, often quite similar, waves with which the ship cannot cope.

The nature of the excitation deserves however some further attention: In our capsize studies we are usually restricting our analysis, one might think unjustifiably, in excitations produced by steep but non-breaking waves. This is an idealization which can result in unsafe predictions; because in the extreme environments where we investigate capsize, wave breaking is quite common. The nature of such excitations, a combination of smooth and impacting, and their magnitude can be very conducive for capsize.

But even if we assume that the structure of the conventional mathematical model is satisfactory, at least two further tasks need to be tackled: (a) To derive roll damping coefficients that can be applicable for near-capsize-angle motions; and (b) to identify capsize thresholds in terms of combinations of wave amplitude and frequency. Interestingly, the two tasks are, as shown below, in fact intrinsically connected.

Derivation of damping coefficients

Currently it is quite common to derive the damping coefficients from free roll decrement data under the assumption that the undamped roll would be basically harmonic. However, near capsize the nonlinearity of restoring is very strong rendering the response of a rather different type. This means that energy dissipation near capsize angles is not taken into account accurately when the coefficients are derived, although the values of these coefficients are critical in the theoretical investigation of capsize.

To explain these, let us consider a scaled equation of free roll with a quite general, quintic-type restoring curve:

$$\ddot{x} + D(\dot{x}) + x + \delta x^3 - (\delta + 1)x^5 = 0 \tag{1}$$

where $x = \dfrac{\varphi}{\varphi_v}$ with φ the real roll angle and φ_v the vanishing angle. Differentiation is carried out in respect to scaled time $\tau = \omega_0 t$ where ω_0 is the 'undamped' natural frequency and t is the real time. $D(\dot{x})$ is the damping function that normally includes a linear plus an absolute quadratic or cubic component of roll velocity; and δ parametrizes the whole family of quintic restoring curves and therefore through δ we can establish a correspondence with the real (GZ) of our ship.

As has been shown by Spyrou & Thompson (2000), if damping is neglected we can obtain the following exact "Hamiltonian" solution for large amplitude relative free roll (assuming that the "ship" was released with zero initial velocity):

$$x = x_0 \frac{\mathrm{cn}(u,k)}{\sqrt{1 - \lambda^2 \,\mathrm{sn}^2(u,k)}} \tag{2}$$

where x_0 is the initial angle at $\tau = 0$; λ is a function of x_0 and δ; cn, sn are the so called Jacobian elliptic functions (respectively elliptic cosine and elliptic sine) with argument $u = wt$, and modulus k; w is also a function of x_0 and δ. We note that when $k \to 0$ we have the linear case and the solution (2) becomes harmonic; whereas for $k \to 1$ we obtain the hyperbolic solution that defines the boundary of the Hamiltonian safe basin.

In order to find damping coefficients appropriate for extreme roll angles we need to know how energy is dissipated at these angles which requires to know the trajectory in $(x, \dot{x})$ from one peak (that is, one crossing of the zero velocity line) to the next, see Fig. 1. For a linear roll equation such a solution is rather straightforward:

$$x = x_0 e^{-\zeta\tau} \sin\left(\sqrt{1 - \zeta^2}\,\tau - \theta\right) \tag{3}$$

Expressions for "mildly" nonlinear (GZ) can also be derived through a perturbation approach. But for the strongly nonlinear case, if damping is present, exact analytical solution cannot be obtained; and a perturbation-like approach (with damping's nonlinearity as small quantity) involving elliptic functions is extremely complex whilst the accuracy achieved may be doubtful.

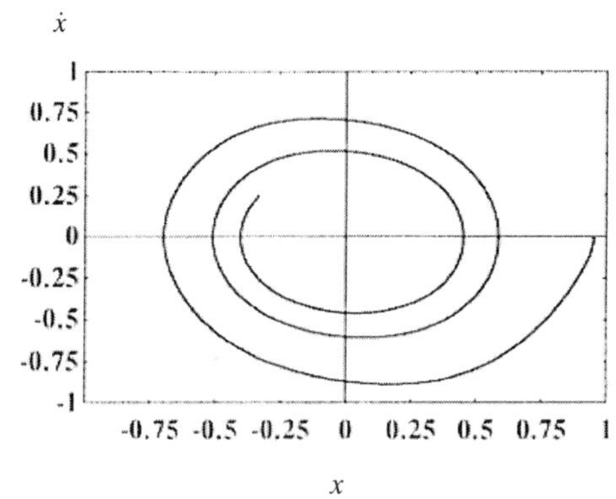

Fig. 1: Numerically derived roll decay for a quintic polynomial when $x_0 = 0.95$. The values of the nondimensionalised damping coefficients are $c_1 = 0.05$ (linear term) and $c_3 = 0.2$ (cubic).

Spyrou & Thompson (2000) have shown recently that it is possible to identify fully analytically the roll decrement per half-cycle for roll angles arbitrarily close to the vanishing angle by assuming that the roll trajectory constitutes a perturbation of the Hamiltonian solution. As will be shown next, this fits nicely with the Melnikov method of capsize assessment which is based on the same principle.

Predictors of capsize

Such predictors can be derived from an analysis either of steady-state or from transient roll responses, Spyrou et al. (1997). To resolve an issue which was raised in last year's Workshop, by "steady-state capsize" we mean the absence of stable steady-state solution in the vessel's response. If such a state does not exist at a certain level of wave forcing and damping, the ship simply cannot stay upright. On the other hand, by "transient capsize" we mean that although a stable state might exist, at the initial transient stage the response is such that capsize occurs. As is obvious, the threshold wave slope of transient capsize should be lower than that of steady-state capsize. For this reason it is more sensible to predict capsize on the basis of transients, Thompson (1996).

A good criterion of incipient transient capsize can be derived from the so-called Melnikov's method through which we can find an analytical approximation of the critical wave slope, given the frequency ratio, where manifold tangencies arise and the domain of bounded roll motion starts becoming fractal, triggering rapid loss of the "safe area" of state space. Melnikov's method has been applied both in a deterministic and in a stochastic context. It is very remarkable that the critical condition derived from the application of Melnikov's method can be interpreted also as an energy balance. Essentially, Melnikov's method "says" that in order to identify the critical wave slope, given the damping, the work done by the forcing should be balanced with the energy dissipated through damping *around the remotest orbit of bounded roll.* These special orbits are called in the literature heteroclinic or homoclinic, depending on whether a symmetric or a biased in roll ship is studied. What makes such an interpretation particularly interesting is that it provides a connection with the widely debated in the early eighties method of energy balance for capsize assessment. That method however relied on harmonic or nearly harmonic responses. Another observation on Melnikov is that it makes use of the perturbed Hamiltonian dynamics approach. This is the same fundamental assumption that has allowed, as discussed in the previous subsection, to find analytically the roll decrement during decay experiments for arbitrarily large initial roll.

From the above observations the conclusion may be drawn that the tasks of deriving damping coefficients and of predicting capsize are intrinsically connected and that, consistency in the followed approaches should be ensured.

Stability of symmetric and of biased ship

As has been pointed out by Thompson (1996), the presence of even a small amount of bias can reduce very considerably the critical wave slope where capsize occurs. It seems logical that a dynamic stability criterion should take into account this fact; but how much bias in needed in the assessment is very hard to define in a rational manner. As is well known, a ship can become biased as the result of wind loading or cargo imbalance; but what is further notable is that a ship shows a "preference" to capsize towards the wave; and that in large waves an initially symmetric ship may develop also some "dynamic" list towards the wave. This would possibly require consideration of sway and higher order wave effects to explain but whether these matters should be taken into account in a capsize assessment is a rather open question at this stage.

About the design problem

It is of course highly desirable the information produced from the analysis of dynamics to be linked with the design process. Unfortunately, until recently this problem had not even been addressed. Generally, there are two main problems that need to be solved: The first problem is how to maximize the critical wave slope where manifold tangencies arise, over a range of wave frequencies; and the second is how to generate practical hull shapes given some desirable form of the restoring curve, as identified from the first task. The latter is essentially the inverse of the conventional task of deriving the (GZ) curve given a hull. Here we shall discuss in further detail the first task.

Let's take the rather generic equation of roll with linearized damping and cubic-type (GZ) which has been thoroughly studied in the past:

$$\ddot{x} + 2\zeta\dot{x} + x - x^3 = F\sin(\Omega\tau) \tag{4}$$

$F = \dfrac{I}{I + \Delta I}\dfrac{Ak\Omega^2}{\varphi_v}$ and $\zeta = \dfrac{B}{2\sqrt{Mg(GM)(I + \Delta I)}}$ with B the dimensional equivalent damping, M the ship mass; $I, \Delta I$, respectively the roll moment of inertia and the added moment.

It can be noticed that in the expression of the equivalent damping ratio ζ, (GM) appears in the denominator which means that for our scaled equation increase of (GM) reduces ζ ! However, at the same time the forcing is reduced even more since $F \propto \Omega^2 = \dfrac{\omega^2}{\dfrac{Mg(GM)}{I+\Delta I}}$.

From Melnikov we find the critical forcing F_M to be:

$$F_M = \frac{4\zeta \sinh\left(\dfrac{\pi\Omega}{\sqrt{2}}\right)}{3\pi\Omega} \tag{5}$$

In order to understand the meaning of this we should go back to dimensional quantities in which case we can obtain the following expression of critical wave slope $(Ak)_M$:

$$(Ak)_M = \frac{2B}{3\pi} \frac{Mg}{I(I+\Delta I)} \varphi_v (GM) \frac{\sinh\left(\dfrac{\pi\sqrt{I+\Delta I}}{\sqrt{2Mg(GM)}}\omega\right)}{\omega^3} \tag{6}$$

where ω is the wave frequency.

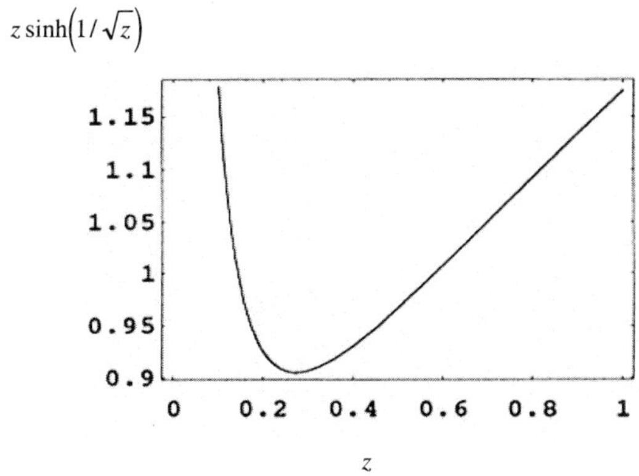

Fig.2: Basic trend of the dependence of (GM) on the critical wave slope

Increase of damping or of the vanishing angle are the typical ways to improve the resistance to capsize according to this mechanism, Thompson (1996). However some more intriguing observations are possible also on the basis of Fig. 2: Expression (6) allows for having a situation where very low (GM) can, under certain circumstances, be beneficial! It is essential therefore that the findings are not applied blindly but an understanding about the physical mechanisms involved is developed, and areas of practical validity are established.

As for any resonance mechanism, it can be dealt with by increasing damping and/or by detuning our system from the excitation. As the natural frequency of the ship depends on (GM), such detuning can be achieved not only by increasing but also by reducing (GM). In fact, it is possible that if the phase of rolling response is nearly opposite to the phase of the wave, then the "absolute" roll motion (the wave slope plus the relative to it roll angle) can be very little, giving the impression that ship is "insensitive" to the excitation. Of course, under no circumstances could be advised to set low (GM) for the ship because then capsize can easily happen from other reasons.

In a practical context it is sensible, rather than trying to establish the capsize limits of the ship, to set threshold absolute roll angles beyond which the ship is in grave danger of capsize due for example to cargo shift. In such a case however, we must be very careful in the interpretation of the output of a roll motion equation like (2). Because a small relative angle could mean a quite substantial one in absolute terms given the wave slope; and on the other hand, as hinted earlier, the phase between the roll response and the wave can make absolute rolling to be very large or very small.

As has been pointed out in Spyrou et al. (1997) it should be possible to combine an expression like (6) with an optimisation process, given certain ship parameter constraints obtained from existing stability standards. For example, for the considered simplest possible case of cubic restoring, the Naval Engineering Standard 109 would produce as far as (GM) and φ_v are concerned, the following constraints:

The area criteria for (GZ) up to 30deg, 40deg and between 30 and 40deg give:

$$\text{(a)} \quad (GM)\left(0.137 - \frac{0.019}{\varphi_v^{\;2}}\right) \geq 0.08 \tag{7}$$

$$\text{(b)} \quad (GM)\left(0.244 - \frac{0.059}{\varphi_v^{\;2}}\right) \geq 0.133 \tag{8}$$

$$\text{(c)} \quad (GM)\left(0.107 - \frac{0.04}{\varphi_v^{\;2}}\right) \geq 0.048 \tag{9}$$

Further constraints:

$$\max(GZ) \geq 0.3 \Rightarrow \quad \text{(d)} \quad 0.385(GM)\varphi_v \geq 0.3 \tag{10}$$

$$\text{(e)} \quad (GM) \geq 0.3 \tag{11}$$

$$\varphi_{(GZ)_{\max}} \geq 30\deg \Rightarrow \quad \text{(f)} \quad \varphi_v \geq 0.9064 \text{ rad} \tag{12}$$

In (f) the minimum φ_v is less than the recommended range of at least 70 deg .

The above lines are essentially sketching out an optimization process where Ak, expressed on the basis of (6), or preferably with a more detailed expression of the criterion taking better account of the hull, is sought to be maximized while making sure that realistic constraints like the above, are being satisfied.

It is very interesting that our concerns about the bias effects, expressed earlier, can be incorporated also into such a procedure. Let us consider the a- parametrized family of restoring curves with bias, where $a = 1$ means a symmetric system and $a = 0$ means a system allowing only one-sided escape, Thompson (1996):

$$\ddot{x} + 2\zeta\dot{x} + x(1-x)(1+ax) = f\sin(\Omega\tau) \tag{13}$$

In a recent MSc Thesis at UCL, Gurd (1997), it has been shown that it is possible to find analytically an expression for the critical (Ak) for small and for large bias, respectively as following[2]:

Perturbation of symmetric system (small bias):

$$F_M = \frac{\sqrt{2}\sinh\left(\frac{\Omega\pi}{\sqrt{2}}\right)\left(2\sqrt{2}\zeta - 1 + a\right)}{3\Omega\pi} \tag{14}$$

Strongly "one-sided" escape (large bias):

$$F_M = 2\zeta\sqrt{a}\,\frac{\left[3\alpha\beta^2\cosh^{-1}\left(\frac{\alpha}{\beta}\right) - \sqrt{\alpha^2-\beta^2}\left(\alpha^2+2\beta^2\right)\right]}{6\sqrt{2}\pi\mu\sqrt{\alpha^2-\beta^2}\sin\left[\mu\cosh^{-1}\frac{\alpha}{\beta}\right]} \tag{15}$$

Where: $$\alpha = \frac{2(2a+1)}{3a},\quad \beta = \sqrt{\frac{2(1-a)(2+a)}{3a}},\quad \mu = \frac{\Omega}{\sqrt{1+a}} \tag{16}$$

Again, the quantities will have to be expressed in dimensional form in order to be able to find the true critical relationship of ship parameters.

CAPSIZE IN A FOLLOWING SEA

As is well known, in a following sea a ship may capsize due to severe fluctuations of its righting arm. Capsize can occur either from a sudden divergent roll ("pure-loss") or from a more dynamic process ("parametric"), where roll is built-up in an oscillatory and gradual manner. Traditionally, the two mechanisms are considered independently. However, as they are both the result of time-dependence of the roll righting arm (in fact dependence on the

[2] These analytical results are of relevance to the papers of Jiang et al (1996) Kan (1992) where Melnikov's critical wave slope had been identified only numerically.

position of the ship on the wave), the propensity for capsize could be assessed more effectively if the two were treated in a unified manner.

Commonly, the parametric mechanism is examined on the basis of the principal and the fundamental resonance regions on the stability chart of a Mathieu-like equation. However, such a chart corresponds in fact to long term asymptotic behaviour which is rather unrealistic for a ship. This has created some controversy about the true relevance of the parametric scenario; because, although at realistic levels of ship roll damping the domain of the principal, and often of the fundamental resonance extend sometimes to feasible levels of restoring variation amplitude, this picture is correct if the considered number of wave cycles goes to infinity. Practically however, it is more important to know whether the instability becomes noticeable within a small number of wave cycles. But if the "allowed" number of wave cycles is small, the building-up of large roll requires very intensive variation of restoring which may, and one would indeed hope to, be unrealistic.

Another matter that needs to be taken also into account, more in respect to the pure-loss scenario, is the physical time required for capsize: At lower frequencies of encounter the ship may capsize more easily because it stays for longer time at unfavourable for stability regions of the wave. But because the ship is advancing very slowly relatively to the wave, the time for capsize can be excessively high. It is quite obvious in this case that for capsize assessment it becomes important where the ship was at $t=0$. One possible way to deal with this dependence on the initial phase is to assume that the ship, at $t=0$ is just entering the negative restoring region of the wave. For sinusoidal variation of (GM) this phase, say χ, is given from $\chi = -\arccos\left(\frac{1}{h}\right)$ where h is the amplitude of variation of (GM).

The major effect that the number of cycles has on the first resonances is shown clearly in Fig. 3 for a typical linear Mathieu-type roll equation which, on the basis of scaled quantities, takes the form:

$$\frac{d^2x}{d\tau^2} + \frac{2k\sqrt{a}}{\omega_0}\frac{dx}{d\tau} + a(1 - h\cos 2\tau)x = 0 \tag{17}$$

where $a = \frac{4\omega_0^2}{\omega_e^2}$, $x = \frac{\varphi}{\varphi_v}$ but this time $2\tau = \omega_e t$ (time nondimensionalized in respect to the encounter frequency ω_e). Also, ω_0 is the (dimensional) natural frequency and k is the equivalent damping factor ($2k = \frac{B}{I + \Delta I}$).

In Fig. 3 we examined whether the roll angle reaches the level of the vanishing angle within a prescribed number of wave cycles. It is noted that, when only four cycles are considered the h required is very high ($h = 2.1$, not shown in the graph). As the order of the resonance increases the required amplitude becomes less dependent on the number of wave cycles; but the practical relevance of these resonances for a ship is rather minimal. It is also noted that, the lower the number of cycles the more influential becomes the initial position of the ship on the wave.

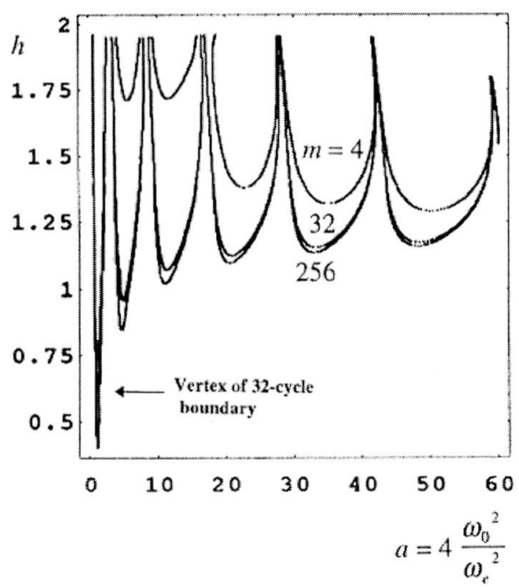

Fig. 3: Capsize regions in respect to the first six resonances, with parameter the considered number of encounter-wave cycles m. The initial heel was $x_0 = 0.01$ and as capsize was considered its 100-fold increase; $2k/\omega_0 = 0.025/0.144$.

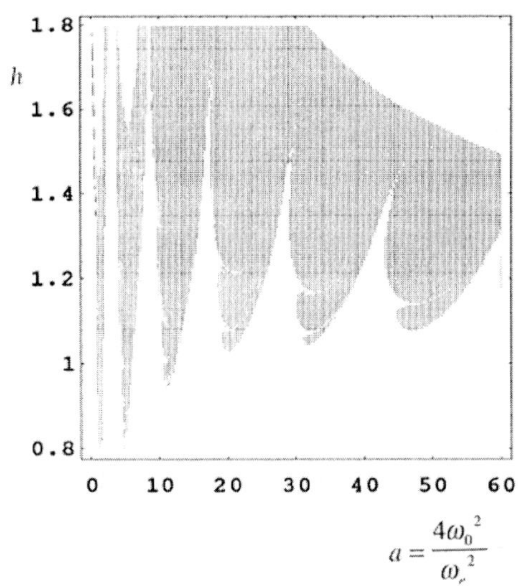

Fig. 4: Capsize regions for cubic-type restoring in less than 8 wave cycles and requiring less than 300 sec (natural frequency in calm sea $0.381\,\text{sec}^{-1}$). The dark regions correspond to capsize according to the parametric scenario. The white upper-right region is capsize in less than 50 sec and is according to the pure-loss mechanism. Quick capsizes ($t < 50$ sec) occur also in the first two resonances and it is notable that the required amplitude h is comparable with that of pure loss. The graph is drawn with $2k = 0.025$ and $x_0 = 0.1$.

Fig. 4 provides a combined view of the regions of pure-loss and parametric instability on the basis of cubic-type restoring (the nonlinear term was time independent). This unified assessment is allowed by the fact that behaviour is examined in a transient sense. Capsize occurrences are recorded if they happen in a small number of cycles and within a realistic limited time period. More details can be found in Spyrou (2000).

Of course, different hull forms will produce different laws of restoring variation. In turn, these will result in different arrangements of the capsize boundaries. At the moment, we are still lacking a systematic procedure for dealing with this fact. This is an area of research currently considered.

Behaviour in bi-chromatic seas

A possible extension of the traditional examination of parametric instability on the basis of sinusoidal variation of (GM), is to study the behaviour of a ship under the effect of a wave group containing at least two independent frequencies . We shall assume that, in a qualitative sense, this could bring about a quasiperiodically varying restoring which, for two frequencies present, results in the following roll equation:

$$\frac{d^2x}{d\tau^2}+\frac{2k\sqrt{a}}{\omega_0}\frac{dx}{d\tau}+a[1-rh\cos 2\tau-(1-r)h\cos 2\nu\tau]x=0 \tag{18}$$

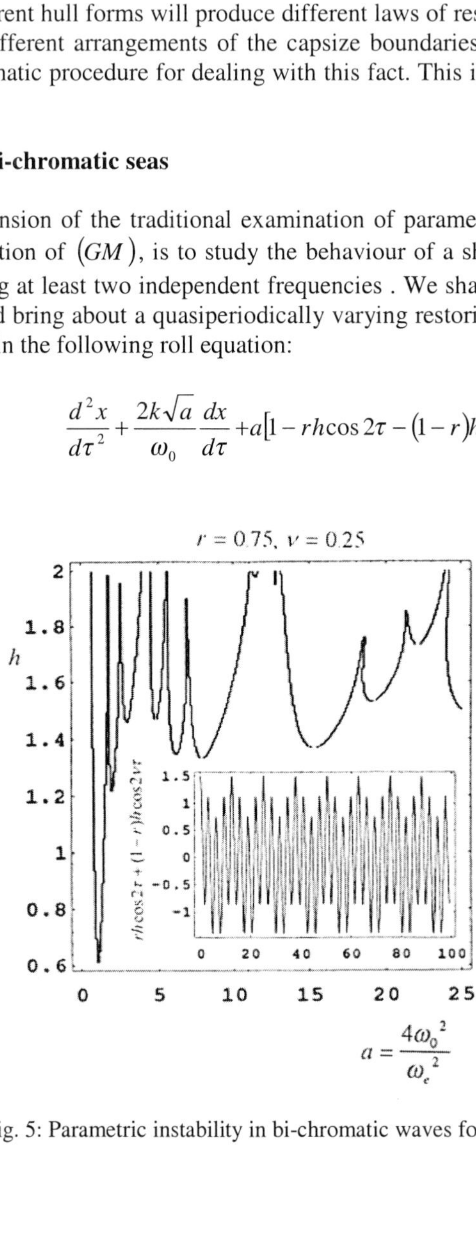

Fig. 5: Parametric instability in bi-chromatic waves for 32 wave cycles.

In (18) the parameters r and v represent respectively the relative strength of the basic frequency and the ratio of the second frequency to the basic. A general characteristic of the response is that a number of new "spikes" are growing on each primary resonant. However the effect of the extra frequency is not very influential on the principal resonance which extends at relatively low levels of required (GM) variation amplitude h.

References

Gurd, B.A. (1997): A Melnikov analysis of the Helmoltz-Thompson equation, *MSc Thesis*, Centre for Nonlinear Dynamics and its Applications, University College London, September.

Jiang, C., Troesch, A.W. & Shaw, S.W. (1996) Highly nonlinear rolling motion of biased ships in random beam seas, *Journal of Ship Research*, **40**, 2, 125-135.

Kan, M. (1992) Chaotic capsizing, *Proceedings*, The 20th ITTC Seakeeping Committee, Osaka, September 10-11.

Spyrou, K.J. (1997) The role of Mathieu's equation in the horizontal and transverse motions of ships in waves: Inspiring analogies and new perspectives, *Proceedings*, 3rd International Workshop on Theoretical Advances in Ship Stability and Practical Impact, Hersonissos, Crete, October, 14 pages.

Spyrou, K.J. (2000) Designing against parametric instability in following seas, *Ocean Engineering*, **27**, 625-653.

Spyrou, K.J. and Thompson, J.M.T.(2000) Damping coefficients for extreme rolling and capsize: an analytical approach. *Journal of Ship Research*, **44**, 1, 1-13.

Spyrou, K.J., Cotton, B. and Thompson, J.M.T. (1997) Developing an interface between the nonlinear dynamics of ship rolling in beam seas and ship design, *Proceedings*, 3rd International Workshop on Theoretical Advances in Ship Stability and Practical Impact, Hersonissos, Crete, October, 9 pages.

Thompson, J.M.T.(1996) Designing against capsize in beam seas: Recent advances and new insights, *Applied Mechanics Reviews*, **50**, 5, 307-325.

Thompson, J.M.T., Rainey, R.C.T. and Soliman, M.S. (1992) Mechanics of ship capsize under direct and parametric wave excitation, *Phil. Trans. R. Soc. Lond. A* **338**, 471-490.

Wright, J.H.G. and Marshfield,W.B.(1980) Ship roll response and capsize behaviour in beam seas, RINA *Trans.*, **122**, 129-148.

D. Vassalos, M. Hamamoto, A. Papanikolaou and D. Molyneux (Editors)

THE MATHEMATICAL MODELLING OF LARGE AMPLITUDE ROLLING IN BEAM WAVES

A. Francescutto and G. Contento

Department of Naval Architecture, Ocean and Environmental Engineering, University of Trieste, Via A. Valerio 10, 34127 Trieste, Italy

ABSTRACT

In this paper the effect of the excitation modelling on the fitting capability of the nonlinear roll motion equation to experimental data is studied. Several frequency dependent and constant effective wave slope coefficients are derived for five different scale models corresponding to different ship typologies by a Parameter Identification Technique. It appears that a mathematical modelling with constant damping parameters and frequency dependent excitation could give very good results. As regards the excitation parameters, a common trend for slender bodies is evidenced. Finally, the effects of the coupling roll/pitch in beam waves is analysed starting again from experiments in large amplitude waves. It appears again that a quadratic coupling term is needed in the pitch equation, but the introduction of a coupling term in the roll motion equation does not improve significantly its simulation capability.

KEY WORDS

Nonlinear Dynamics, Roll Motion, Simulation, Parameter Identification.

INTRODUCTION

The use of concentrated parameters mathematical models to simulate ship behaviour in waves, with particular regard to roll motion, is often at the centre of strong debates with some researchers claiming for the use of many degrees of freedom models instead of simulating the roll motion as an isolated motion. The aim of this paper consists in a contribution to these debates, showing that in beam waves the simulation of large amplitude roll motion is often possible with high efficiency by using a single degree of freedom model.

MATHEMATICAL MODELLING OF SHIP ROLLING AS AN ISOLATED MOTION

In this paper some results regarding a simplified mathematical modelling of ship rolling, possibly using as much as possible constant coefficients, are reported in synthesis. More details can be found in Francescutto, *et al.*(1998), Contento and Francescutto(1997, 1999), Penna, *et al.*(1997), and Francescutto and Contento(1998a, 1998b).
The incident wave is assumed long enough to be described by the local slope $\alpha(t)$.

Absolute angle description

$$(I_G + \delta I)\ddot{\varphi} + N\dot{\varphi} + D_1|\dot{\varphi}|\dot{\varphi} + D_2\dot{\varphi}^3 + ... + \Delta\overline{GZ}(\varphi) = E(t) \tag{1}$$

The excitation E(t) following Blagoveshchensky (1962) can be written:

$$E(t) = \chi_\varphi{}^{I}\Delta\overline{GM}\alpha + \chi_\varphi{}^{II} I_V\ddot{\alpha} + \chi_\varphi{}^{III} N\dot{\alpha} + \chi_\varphi{}^{IV}\delta I\ddot{\alpha} \tag{2}$$

where I_V is the inertia of the 'liquid hull' and the $\chi_\varphi{}^j$ $(0 \leq \chi_\varphi{}^j \leq 1)$ terms included to account for the hull shape/coefficients and for the ratio between wave parameters and hull breadth and draught.
In normalised form one has:

$$\ddot{\varphi} + 2\mu\dot{\varphi} + \delta_1|\dot{\varphi}|\dot{\varphi} + \delta_2\dot{\varphi}^3 + ... + \omega_0{}^2\varphi + a_3\varphi^3 + a_5\varphi^5 + ... = e(t) \tag{3}$$

with

$$e(t) = \pi s_w\left[\left(\alpha_1\omega_0{}^2 - \alpha_2\omega^2\right)\cos(\omega t) - 2\mu\alpha_3\omega\sin(\omega t)\right] \tag{4}$$

where α_2 accounts for both terms $\chi_\varphi{}^{II} I_V$ and $\chi_\varphi{}^{IV}\delta I$
The last contribution in e(t), proportional to the linear damping, is of order O(1/ω) with respect to the former and can be neglected, so that:

$$e(t) = \pi s_w\omega_0{}^2\left[\alpha_1 - \alpha_2\left(\frac{\omega}{\omega_0}\right)^2\right]\cos(\omega t) \tag{5}$$

Each term in e(t) has been derived under the Froude-Krilov hypotheses and long wave approximation. The presence of the hull on the incident flow is accounted for only in some so called 'effective wave coefficients' $\chi_\varphi{}^j$. In this way an important feature of wave load is lost when the wave length becomes comparable with the transversal dimension of the body, i.e. in the diffraction regime. A tentative formulation is here proposed as follows:

$$e(t) = \pi s_w\omega_0{}^2 e^{-\left[\omega/\alpha_1\right]^{\alpha_2{}^2}}\cos(\omega t) \tag{6}$$

Often, no explicit dependence of the amplitude of the excitation on the wave frequency appears. In this case e(t) reads as follows:

$$e(t) = \pi s_w \omega_0^{\;2} \alpha_1 \cos(\omega t) \tag{7}$$

Relative angle description

Assuming the same hydrodynamic model for the damping and the same notations as in the previous section, the equation of roll motion in the relative angle approach can be written following Wright-Marshfield (1980):

$$I_G \ddot{\varphi} = -\delta I(\ddot{\varphi} - \ddot{\alpha}) - N(\dot{\varphi} - \dot{\alpha}) - M_R(\varphi - \alpha) \tag{8}$$

Introducing the relative angle $\vartheta = \varphi - \alpha$ and dividing by $I_G + \delta I$, we obtain

$$\ddot{\vartheta} + 2\mu'\dot{\vartheta} + \delta'_1 \left|\dot{\vartheta}\right|\dot{\vartheta} + \delta'_2\, \dot{\vartheta}^3 + \ldots + \omega_0^{\;2}\vartheta + \sum_{i=1}^{M} a_{2i+1}\vartheta^{2i+1} = -\frac{I_G}{I_G + \delta I}\pi s_w \omega^2 \cos(\omega t) \tag{9}$$

The two approaches are equivalent since they correspond only to a different grouping of the terms. Some differences are due to the fact that:

- in the absolute angle approach only the first term in the development of $\overline{GZ}(\varphi - \alpha) - \overline{GZ}(\varphi)$ is retained corresponding to the fact that the nonlinear dependency on the wave slope is neglected;
- because of the relative angle approach, the inertia loads are accounted for only by the term δI

Moreover, in the assumed relative angle approach only one "unknown" coefficient

$$\alpha_0^{\;*} = \frac{I_G}{I_G + \delta I} \tag{10}$$

is left to account for the "reduction" of the wave slope effectiveness. In this sense the relative angle model is equivalent to the "constant" wave slope reduction.

Overall capability of the proposed excitation models

The application of an efficient Parameter Identification Technique to a large series of experiments conducted at the University of Trieste on several ship models in different loading conditions allowed to obtain the following results:

Absolute angle

As far as the form of the excitation is concerned, the results reported in Contento and Francescutto (1997), and Francescutto and Contento (1998a) indicate that frequency dependent effective wave slope coefficients work better than the constant. This becomes particularly evident outside the peak zone where the difference between estimated and measured values can exceed 100% even if at these frequencies the absolute roll amplitudes are quite small (few degrees). In Fig. 1 and Fig. 2 the results obtained from a highly non-linear restoring, leading to bifurcation, and from a mildly non-linear one respectively are reported.

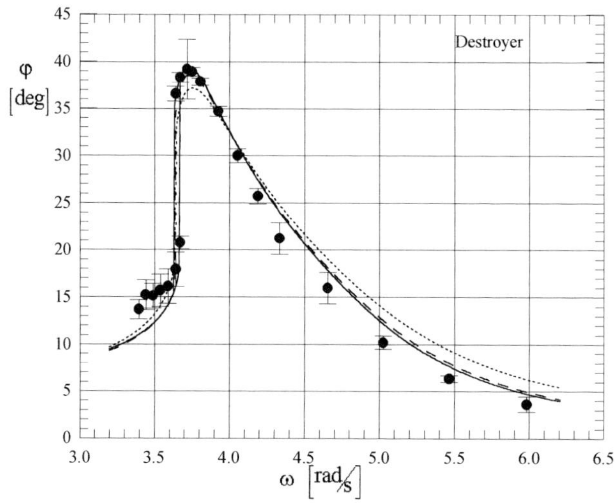

Figure 1: Steady roll amplitudes of a scale model of a destroyer in a regular beam sea. [●experimental data; ——— (eqns 3+5); .... (eqns 3+7); ----- (eqns 3+6)].

Relative angle

Results quite similar to the absolute angle description, but with some shortcoming in the mathematical modelling: α_0^* is indeed quite small, so that the introduction of a double factor could be more adapt.

Identification of a common trend

An evident frequency dependence of α_0^* is observed (Fig. 3). A common trend is evidenced that could be used as a default, at least for slender bodies.

Identified damping

The identified values of the damping coefficients, while different for the different ship typologies, show great stability with respect to the excitation modeling.
Several mathematical models were tried. Basically the linear-plus-quadratic and linear-plus-cubic showed almost the same fitting capability, so that using one or the other is a matter of preference.

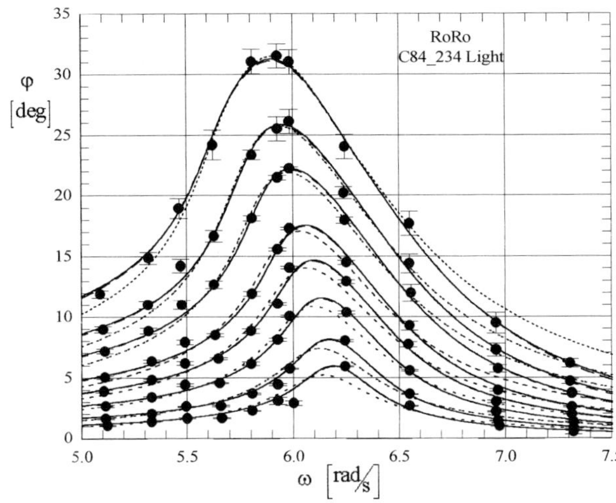

Figure 2: Steady roll amplitudes of a scale model of a RoRo (C84_234 Light) in regular beam sea. [● experimental data; ___ (eqns 3+5); .... (eqns 3+7); _ _ _ _ (eqns 3+6).

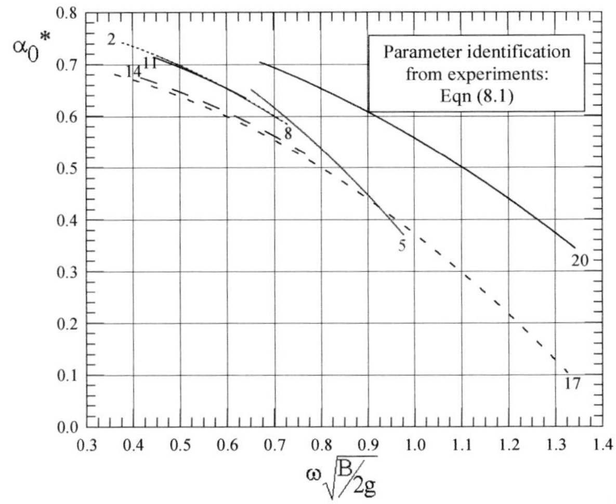

Figure 3: Effective wave slope coefficient α_0^* derived by PIT from eqn (3+5) versus the non-dimensional wave frequency for the different ship models (see Table 2 for curve labels).

MATHEMATICAL MODELING OF THE COUPLED ROLL AND PITCH MOTIONS IN BEAM WAVES

The scale model of a frigate was tested in beam waves with wave steepness $s_w = 2\zeta_a / \lambda = 1/20$. This severe sea condition led to very large roll amplitude at resonance and at a quite interesting pitch response with two peaks, one close to the roll peak frequency and the other close to the natural peak frequency. The results are reported in Fig. 4 with the roll amplitude normalised to the pitch peak for correlation analysis as will be explained later. The effective roll amplitude is reported in Fig. 5 together with the numerical simulation developed in the following. In Fig. 6 the experimental results regarding pitch motion are reported together with the position of the center of the oscillation. This indicates the presence of a negative bias (trim).

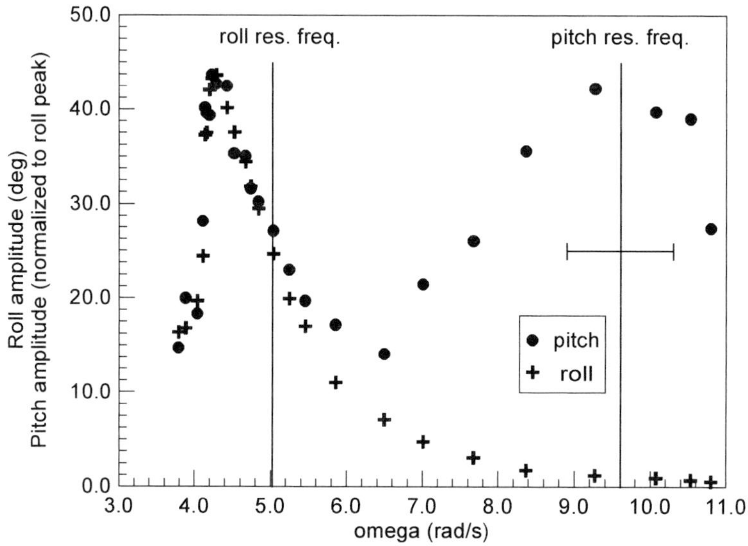

Figure 4: Roll and pitch amplitudes from beam sea experiments. The two oscillations have been scaled to the same peak value. The two vertical lines indicate the position and measurement uncertainty of roll and pitch natural frequencies.

The analysis of Fig. 4 and Fig. 5 indicates that roll motion exhibits typical non-linear features with considerable bending of the peak on the low frequency side in accordance with the strong less-than-linear behaviour of the righting arm curve. In particular, this leads to a pre-bifurcating roll response on the low frequency side of the resonance peak.

The analysis of Fig. 4 and Fig. 6 indicates that there is a strong correlation between pitch and roll at the low resonance pitch peak. Indeed:

- the two peaks follow closely each other, scale factor apart;
- the pitch bias, negligible out of this region, corresponds to a trim by-the-bow condition, typical of free trim transversal inclinations.

Mathematical model

The experimental results indicate a strong coupling between roll and pitch in beam sea also. As known from literature [Paulling and Rosenberg (1959), Nayfeh, *et al.*(1973) and Mook, *et al.*(1974)], a coupling between roll and pitch can be present in longitudinal (mainly following) sea. The two phenomena are quite different inasmuch as:

- the coupling in beam waves is predominantly of roll into pitch, so that there can be a significant pitch due to a large amplitude rolling motion, whereas roll motion exhibits negligible differences due to pitch:

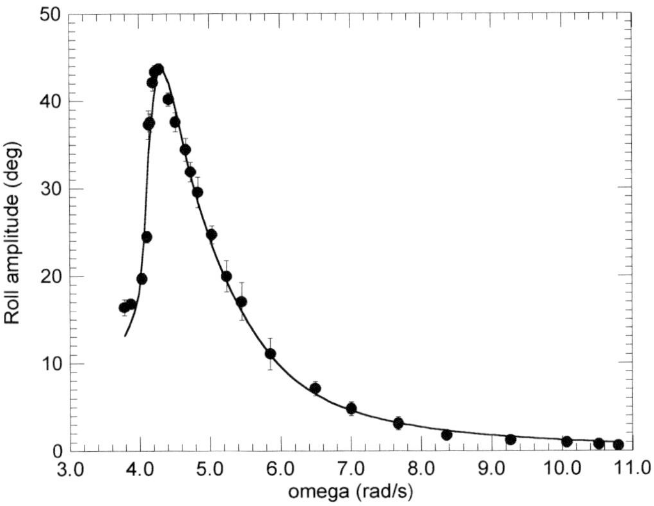

Figure 5: Measured roll amplitude as a function of excitation frequency. The solid line corresponds to the simulation obtained with mathematical model presented (eqn 3+5).

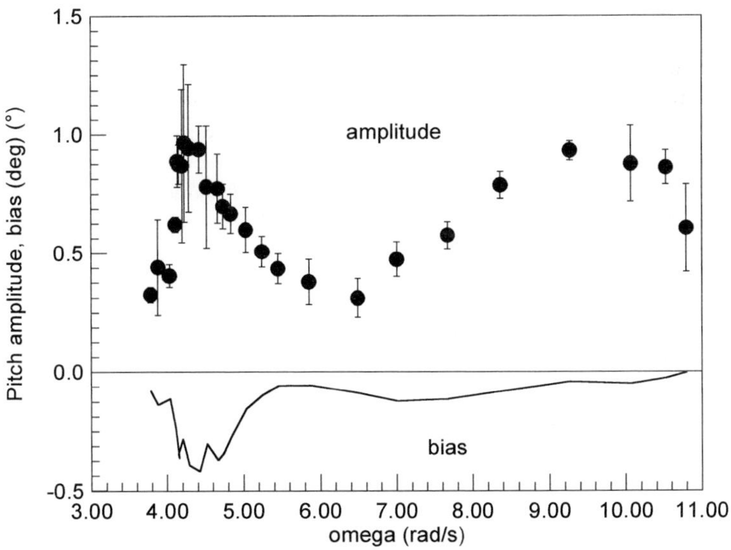

Figure 6: Measured pitch amplitude as a function of excitation frequency. In the lower part of the diagram the position of the center of the oscillation is reported.

- the coupling in longitudinal sea is predominantly of pitch (and heave) into roll, so that there can be a significant roll motion, whereas pitch is only indirectly affected by roll motion.
The difference in the effects is accompanied by some difference in the coupling mechanisms:
- in beam waves, the coupling acts via the difference fore-aft enhanced by the large amplitude transversal inclinations, so that it is effective *irrespectively* of the roll amplitude;
- in longitudinal waves, the coupling acts via variation of metacentric height due to pitch. Being a parametric excitation, it is effective *only above a threshold.*

On the other hand, the high resonance pitch peak is connected with the natural pitch excitation that is present even in transversal waves due to the fore-aft asymmetry.
Looking for a mathematical modelling of the coupled system of the two motions, it is thus natural to take into consideration two terms for the pitch equation:
- a natural forcing term connected with waves
- a term connected with rolling motion.
and
- a term for the roll equation.
In the commonly used sea-keeping approach, vertical motions and lateral motions are considered to be two uncoupled groups of coupled motions.
Here we propose a mathematical model based on the following system of differential equations:

$$\begin{cases} \ddot{\phi} + 2\mu_\phi\dot{\phi} + \delta\dot{\phi}|\dot{\phi}| + \omega_{0\phi}^2\phi + \alpha_3\phi^3 + \alpha_5\phi^5 + \alpha_7\phi^7 + \alpha_9\phi^9 = e_{w\phi}(\omega)\cos(\omega t) + \omega_{0\phi}^2 a_{\phi\theta}\phi\theta \\ \ddot{\theta} + 2\mu_\theta\dot{\theta} + \omega_{0\theta}^2\theta = e_{w\theta}\cos(\omega t) + a_{\phi^2}\phi^2 \end{cases} \quad (11)$$

where $e_{w\phi}(t)$ and $e_{w\theta}(t)$ are the roll motion forcing and the pitch motion forcing due to waves. While $e_{w\phi}(t)$ can be represented by eqns 5-7 above, little is known about $e_{w\theta}$, so that in the mathematical modelling this parameter was assumed to be a constant to be determined from experimental results.
A coupling term has been introduced in the RHS of the roll motion equation and one on the pitch equation. The starting point for the modelling of this last was an analysis of hydrostatic coupling introduced by isocarenic transversal inclinations with free trim. In Fig. 7 the trim angle required for hydrostatic equilibrium as a function of transversal inclination is reported. In agreement with Fig. 6, the trim is negative.
A look at the curve of Fig. 7 suggests that a quadratic dependence of pitch amplitude on roll amplitude could account for the hydrostatic part of the coupling. Hence, a term of the adopted type. The assumption of a term symmetric with respect to the instantaneous inclination is consistent with the physics of the phenomenon since the ship is symmetric port-starboard. On the other hand, a symmetric term constitutes an implicit pitch forcing with frequency double than that of the roll forcing (wave frequency). Assuming indeed, as it is reasonable to do as a first approximation, that the roll motion is harmonic with frequency ω: $\phi(t) \approx \phi_0\cos(\omega t + \psi)$, the proposed coupling term becomes:

$$a_{\phi^2}\,\phi^2 \approx a_{\phi^2}\frac{\phi_0^2}{2}\left[1+\cos(2\omega t+2\psi)\right] \quad (12).$$

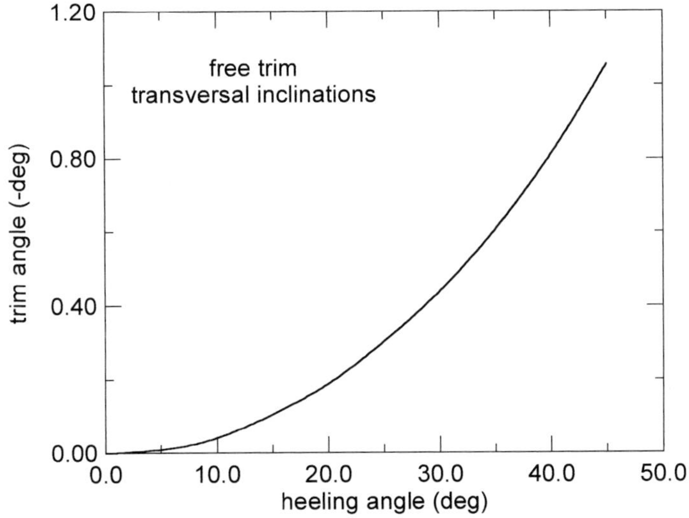

Figure 7: Trim angle as a function of transversal inclinations from free-trim hydrostatic calculations

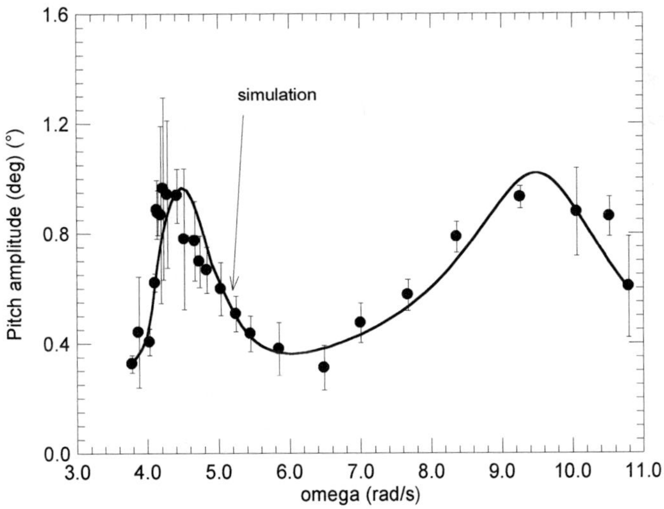

Figure 8: Comparison of predictions obtained by using different mathematical modelling of the term expressing the coupling of roll into pitch (eqn 11) with experimental results for pitch motion. Parameter Identification Technique was applied with respect to the coupling terms only.

The constant part of this function accounts for pitch bias (trim), while the second accounts for a forcing at double frequency which was effectively observed in the Fourier analysis of the experimental records.
The introduction of the coupling term in the roll motion equation only slightly improves the already good simulation obtained with roll motion alone.

Considerations on the coupling mechanisms

Pitch motion.

The differential equation of pitch motion, from eqn 11 is linear with respect to θ, so that we can apply the superposition principle and separate the effect of the two forcing contributions. Focusing on coupling with roll and assuming approximate solutions for roll motion, we obtain the roll-forced part of pitch oscillation as:

$$\theta \approx \theta_c + \theta_2 \cos(2\omega t + \varepsilon) \tag{13}$$

with the constant part given by:

$$\theta_c \approx \frac{a_{\phi^2}}{2\omega_{0\theta}^2}\phi_0^2 = b_{\theta_c\phi_0}\phi_0^2 \tag{14}$$

Eq. 14 expresses the constant part θ_c of the pitch oscillation, i.e. trim, as a quadratic function of roll amplitude ϕ_o. Assuming a quadratic form for the expression of trim as a function of transversal inclination from free trim hydrostatic computations (Fig. 7), one can easily see that the coefficients of the two expressions have values close each other, hence the principal part of the coupling of roll into pitch is hydrostatic.

Rolling motion.

The term $\omega_{0\phi}^2 a_{\phi\theta}\,\phi\theta$ expressing the influence of pitch into roll has the following behaviour in the two characteristic frequency regions:

- for $\omega \approx \omega_{\phi peak}$ one has $\phi\theta \approx \phi^3$

so that it is seen as a contribution to the cubic term $\alpha_3\phi^3$ of the polynomial representation of the restoring moment. The relevance of this part could probably be strongly varied or even disappear, by using a different approach to righting arm calculations (fixed trim/free trim);

- for $\omega \approx \omega_{0\theta}$ one has $\phi\theta \approx \phi^2$

in virtue of the linearity of pitch equation, the coupling term thus contributing to a small bias and a small modulation of the solution.

CONCLUSIONS

It appears that a single non-linear coupling coefficient is in most cases sufficient to have a satisfactory description large amplitude roll motion in beam waves. When the coupling with pitch is taken into account, it is seen that roll influences quite strongly pitch oscillation, while one coupling term helps to fit the roll motion at frequencies close to the roll peak, but it is not strictly necessary for a good simulation.

References

Blagoveshchensky, S.N. (1962). *Theory of Ship Motions*. Dover Publications, Inc., New York, Vol. 2.

Contento, G. and Francescutto, A. (1997). Intact Ship Stability in Beam Seas: Mathematical Modelling of Large Amplitude Motions, Proc. 3rd Int. Workshop on Theoretical Advances in Ship Stability and Practical Impact, Hersonissos, Crete.

Contento, G. and Francescutto, A. (1999). Bifurcations in Ship Rolling: Experimental Results and Parameter Identification Technique, Ocean Engineering, Vol. 26, 1999, pp. 1095-1123.

Francescutto, A. and Contento, G. (1998a). The Modelling of the Excitation of Large Amplitude Rolling in Beam Waves, Proc. 4th International Ship Stability Workshop, St.Johns, Newfoundland, September.

Francescutto, A. and Contento, G. (1998b). On the Coupling Between Roll and Pitch Motions in Beam Waves", Proc. 2nd International Conference on Marine Industry MARIND'98, P. A. Bogdanov Ed., Varna, Bulgaria, Vol. 2, pp. 105-113.

Francescutto, A., Contento, G., Biot, M., Schiffrer. L., and Caprino, G. (1998). The Effect of Excitation Modelling in the Parameter Estimation of Nonlinear Rolling, *Proc. 8th International Conference on Offshore and Polar Engineering - ISOPE'98*, Montreal, The Int. Society of Offshore and Polar Engineering, Vol. 3, pp. 490-498.

Mook, D. T., Marshall, L. R. and Nayfeh, A. H. (1974). Subharmonic and Superharmonic Resonances in the Pitch and Roll Modes of Ship Motions, J. Hydronautics, Vol. 8, pp. 32-40.

Nayfeh, A. H., Mook, D. T.. and Marshall, L. R. (1973). Nonlinear Coupling of Pitch and Roll Modes in Ship Motions", J. Hydronautics, Vol. 7, pp. 145-152.

Paulling, J. R. and Rosenberg, R. M. (1959). On Unstable Ship Motions Resulting from Nonlinear Coupling, J. Ship Research, Vol. 3, pp. 36-46.

Penna, R., Francescutto, A. and Contento, G. (1997). Uncertainty Analysis Applied to the Parameter Estimation in Nonlinear Rolling, Proc. 6th Int. Conf. on Stability of Ships and Ocean Structures - STAB'97, Varna, 1997, Vol. I, pp. 75-82.

Wright, J. H. G. and Marshfield, B. W. (1980). Ship Roll Response and Capsize Behavior in Beam Seas. *Trans. RINA*, Vol. 122, pp. 129-148.

AKNOWLEDGMENTS

This Research has been developed in the frame of EU Thematic Network "SAFER-EURORO".

D. Vassalos, M. Hamamoto, A. Papanikolaou and D. Molyneux (Editors)

CHARACTERISTICS OF ROLL MOTION FOR SMALL FISHING BOATS

K. AMAGAI[1], K. UENO[2] and N. KIMURA[1]

[1]Department of Marine Production System, Faculty of Fisheries, Hokkaido University,
3-1-1, Minato, Hakodate, 042-8611, Japan
[2]Department of Marine Production, Tokyo University of Fisheries,
4-5-7, Konan, Minato, 108-8477, Japan

ABSTRACT

In most of cases, it has been said that a small amount of water on deck acts as rather effective roll damper and roll of ship is less when a small amount of deck water exist than no deck water (Dillingham,1981). In our experiments, we reconfirmed the above effect of a small amount of water on deck. However, there were some cases that a small amount of water on deck did not act roll damper but increased the rolling motion. In addition, when damping effect by free water was enough, yaw was increased. This paper describes those cases from a view point of resonance.

KEYWORDS

Free water, sloshing, resonance, roll motion, small fishing boat, coupling

INTRODUCTION

The authors classified roughly into five types of characteristic behavior of shallow water on a ship's deck, based on tests using an oscillatory rectangular tank physically simulated with sinusoidal motion (Amagai et al., 1994; Ueno et al., 1994). Furthermore, the authors pointed out that the characteristics of roll damping of small fishing boats with projecting broadside and hard chine depend on rolling amplitude especially. Therefore, damping coefficient may depend upon rolling angle. As the improvement of damping term in the nonlinear equation of free rolling, the curve of damping was simulated by using the same parameters in the nonlinear equation of free rolling, the curve of damping was simulated by using the same coefficient for arbitrary rolling angle (Ueno et al., 1995; Ueno et al., 1997).

This paper describes comparison of rolling motions between the cases when small amount of deck water present and when no deck water. In many cases, it is said that a small amount of water on deck acts as rather effective roll damper and roll of boat is less when a small amount of deck water exist than no deck water (Dillingham,1981). In our experiments, we reconfirmed the above effect of a small amount of water on deck. However, there were some cases that a small amount of water on deck did not act as roll damper but increased the rolling motion. In addition, when damping effect by free water was enough, yaw was increased. This paper describes those cases from a view point of resonance.

EXPERIMENT

To clarify rolling motion due to the behavior of water on deck, tank tests were performed by using two ship models (see Table 1 and Figure 1) in beam sea, at near the resonant period. To evaluate the shipping water effect on ship's rolling motion, a tank was set on the upper deck of the ship model. The dimension of the tanks used in the experiment was shown in

TABLE 1
PRINCIPLE DIMENSIONS OF MODEL SHIP

	SHIP-A		SHIP-B	
	FULL SCALE	MODEL	FULL SCALE	MODEL
Lpp (m)	15.20	2.000	14.20	1.291
B (m)	3.80	0.500	3.63	0.330
D (m)	1.48	0.195	1.25	0.181
Disp. (ton)	60.14	0.137	24.69	0.019
GT (ton)	19.9		7.9	
GM (m)	0.51	0.067	0.33	0.030

SHIP-A: salmon drift net boats, SHIP-B: scallop-farming boats

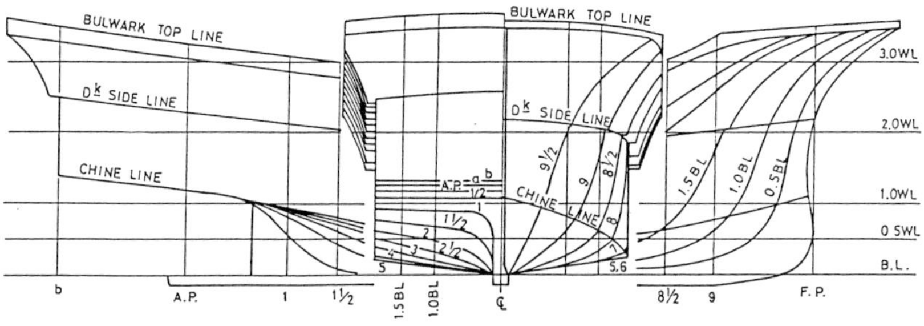

Ship-A

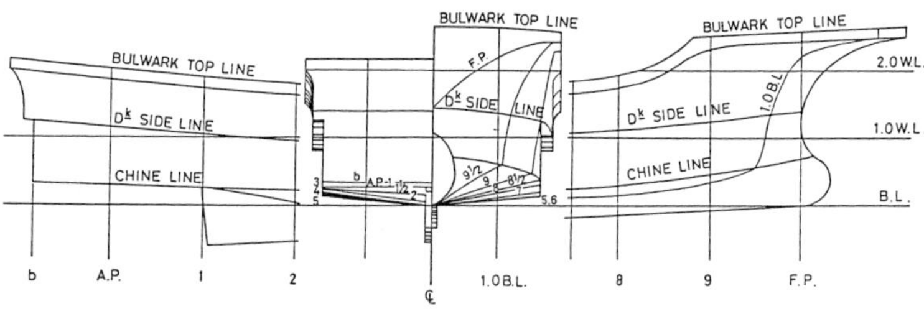

Ship-B

Figure 1: Lines of ships

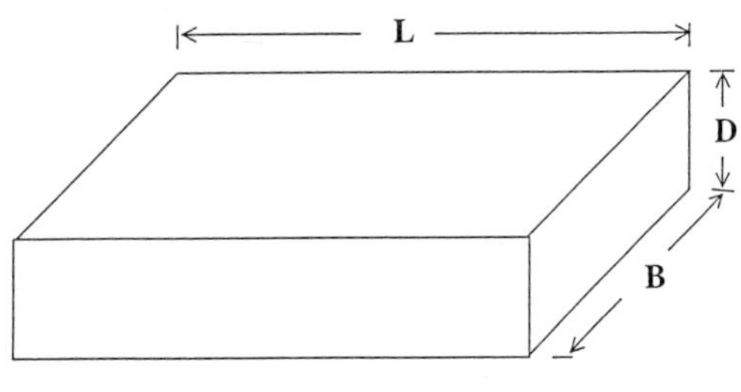

L=50cm, B=20cm, D=10cm (set on Ship-A)

L=31cm, B=22cm, D=11cm (set on Ship-B)

Figure 2: Dimension of tanks

Figure 2 and the locations of the tank on the ships were shown in Figure 3. The apparent rise of center of gravity due to the existence of water on deck is about 1.5 cm increase of Ship-A and 0.3 cm increase of Ship-B. The inclining angle of roll was measured by using Gyroscope system in free roll decay experiment. In experiment of forced rolling motion in regular beam waves, the angles of roll and yaw were measured without restriction.

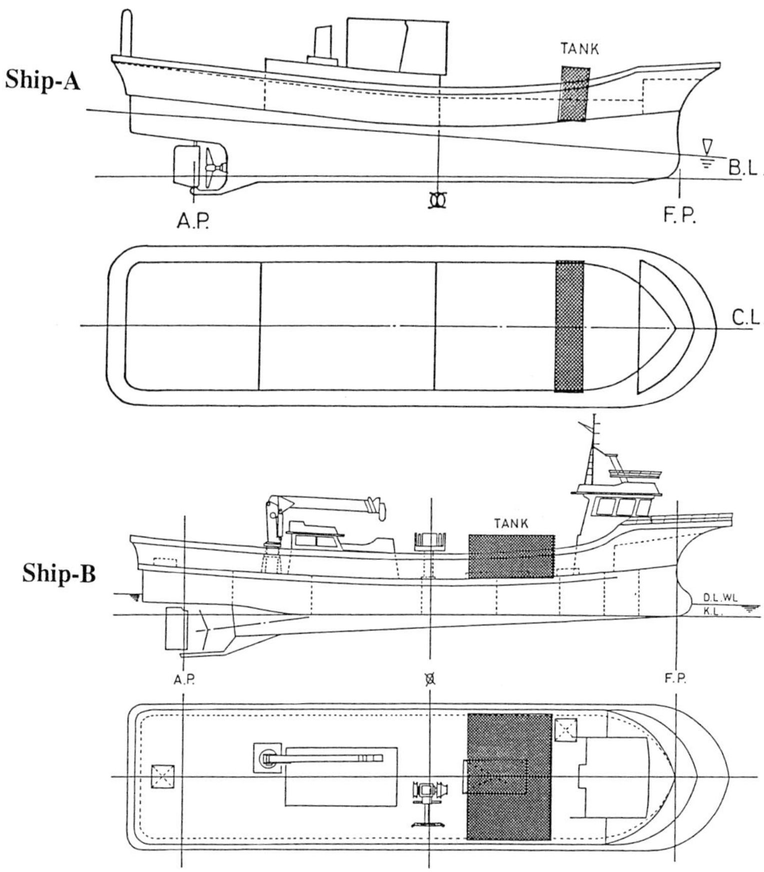

Figure 3: The location of the tank on the ship

CHARACTERISTICS OF ROLL DAMPING WITH WATER ON DECK

Considering the bulwark height, water depth of a tank on deck was set at in the range from 0.0 cm to 6.0 cm. The stability curves without free water are shown in Figure 4.

Figure 5 shows the results of free roll experiment when the damping effect on ship's rolling motion was the most notable in the case when the beginning rolling period was near natural period of water in a tank. It is evident as shown in Figure 5 that ship's rolling period tends to get shorter as roll angle decrease, in the case that natural period of water in a tank is longer than ship's rolling period (water depth h = 1cm, 2cm). On one hand side, when natural period is shorter than ship's rolling period (h = 4, 5, 6 cm), ship's rolling period tends to get longer. It is similar in the case that free water does not exist on deck (h = 0 cm).

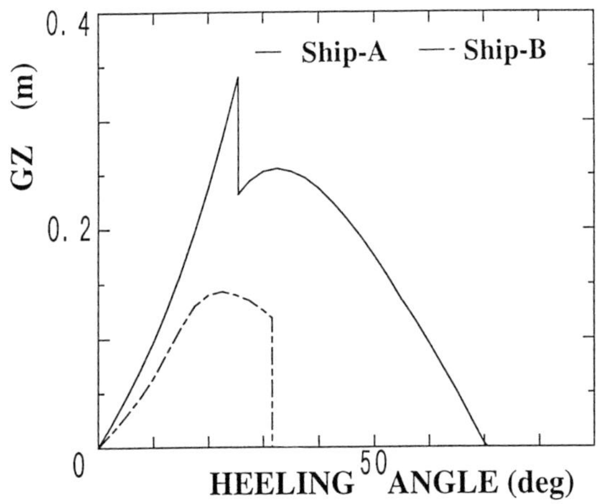

Figure 4: Stability curve

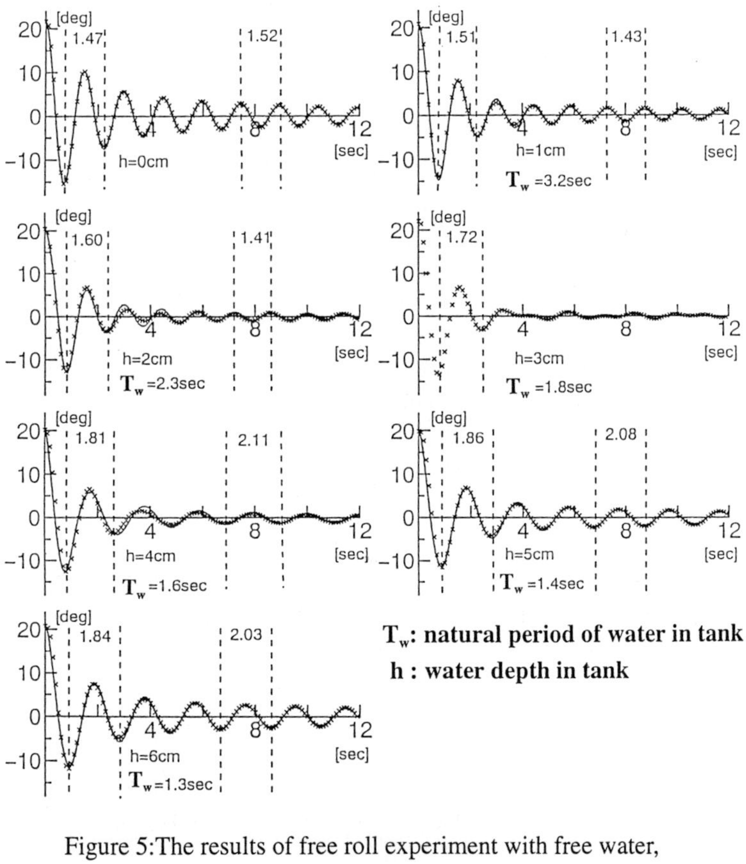

Figure 5:The results of free roll experiment with free water,
—; Estimate, ×; Experiment

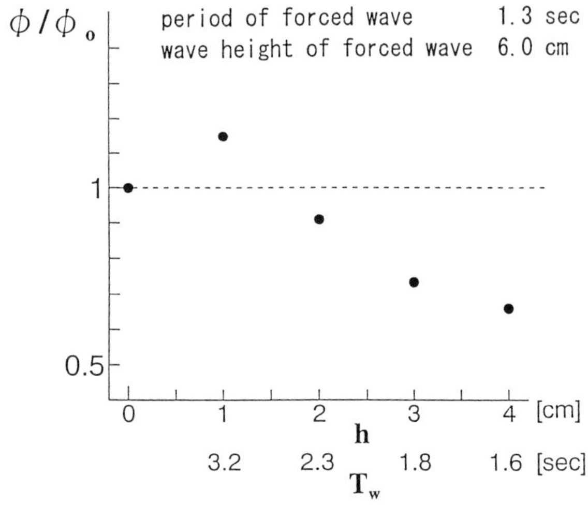

Figure 6: The comparison of roll amplitude between free water exist on deck and do not exist

THE EFFECT OF RESONANCE BETWEEN FREE WATER ON DECK AND FORCED WAVES

Figure 6 shows a comparison of roll amplitude between free water exist on deck and do not exist. Here, the horizontal axis means water depth h in tank and the vertical axis means the value which roll angle with free water ϕ divided by roll angle without free water ϕ_0. When ϕ / ϕ_0 is less than 1.0, the effect of damping due to free water can be seen and when ϕ / ϕ_0 is greater than 1.0, roll angle is increased. In many cases, the effect of damping due to free water can be seen as former report (Dillingham,1981) until now. However, it was confirmed that roll angle was promoted in the case that period of forced wave is a little shorter than the second resonant period (one half of natural period of water in a tank). The value of ϕ / ϕ_0 may be taken its maximum at the second resonant. An example in the case of water depth h = 1 cm is shown in Figure 7. The time series of roll angle corresponding to this case is shown irrespective of a phase in Figure 7. A typical behavior of water in a tank, which is occurred in condition with rolling angle ϕ = 5 deg, rolling frequency ω = 4.83 rad/sec, water depth h = 1cm and tank's length L = 50 cm, is shown in Figure 8. The continuous photographs in the left side are on a tank test and the figures on the other hand are the numerical simulation used by the Marker-and-Cell (**MAC**) method (Welch et al., 1968) and a Numerical Solution Algorithm for Transient Fluid Flow (**SOLA**) (Hirt et al., 1975). There were two transient waves which proceeded in opposite directions as in Figure 8. This behavior appears close to the second resonant period (Amagai et al., 1994; Ueno et al., 1997). Therefore, furious shock

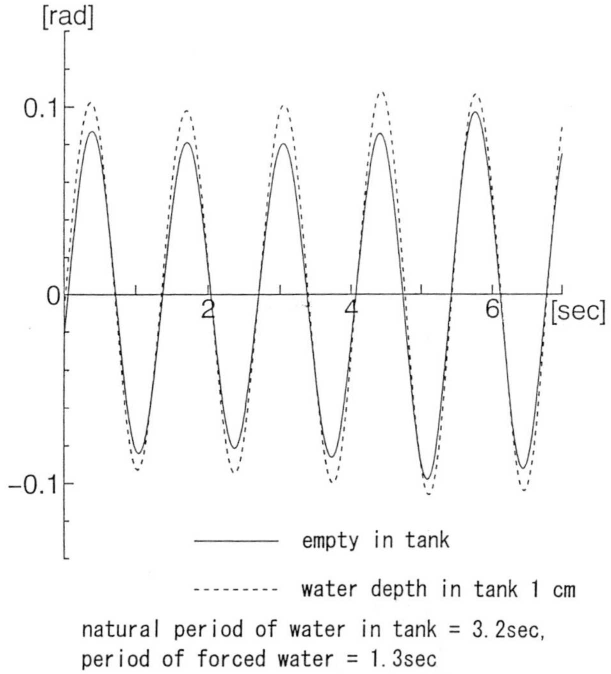

Figure 7: The time series of roll angle

wave was decreased and then the effect of damping due to free water was controlled.

Figure 9 shows the roll response $\phi / K\zeta$ by existence of free water on deck. Mark ● symbolizes the result in the case of empty tank and mark ■ symbolizes the result in the case of existence of free water in tank. K is wave number and ζ is wave amplitude. The roll response without free water is resonant at natural frequency ω_n = 4.13 rad/sec. The roll response with free water is greater than it without free water at two frequency (ω; 2.85 rad/sec, 5.86 rad/sec). When frequency of forced waves ω is 2.85 rad/sec, first resonance occurred because natural frequency of roll ω_n was changed into 2.85 rad/sec from 4.13 rad/sec by free water in tank. When ω is 5.86 rad/sec, this phenomenon is called one of subharmonic resonance and in this case it is second resonance (ω / ω_n = 2.05). Figure 10 shows this effect. It was confirmed that there were two components of natural period of roll and double the period of forced waves. As seen from these example, the existence of free water is sometimes dangerous for the stability of ship.

OCCURRENCE OF YAW DUE TO THE FREE WATER

When we consider coupling of roll and yaw, the effect of yaw is looked upon as slightness compared with the coupling of roll and sway. Therefore, yaw is sometimes neglected.

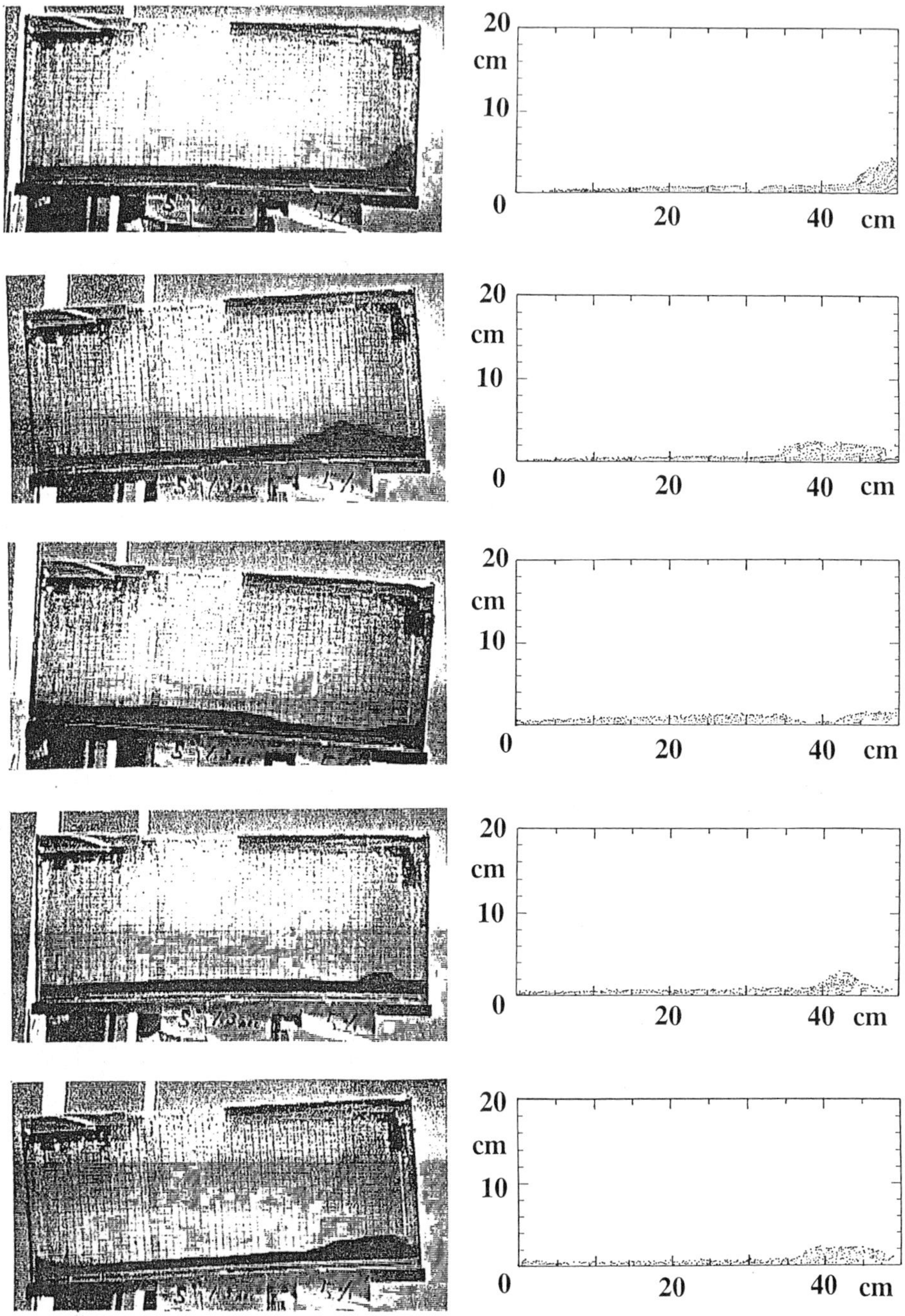

Figure 8: The behavior of water in tank

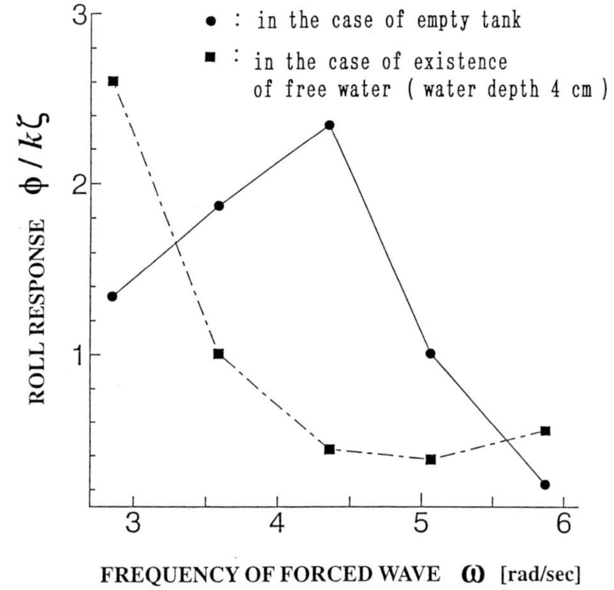

Figure 9: The roll response

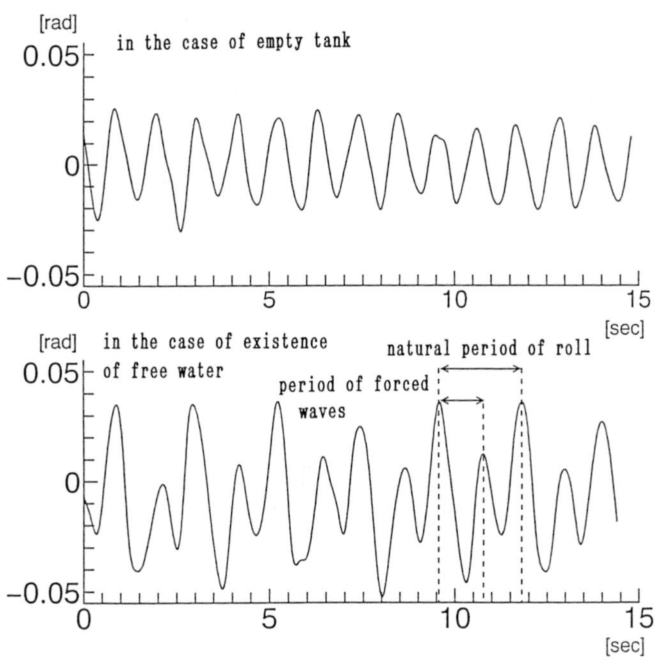

Figure 10: Time series of roll angle in the case of subharmonic resonance

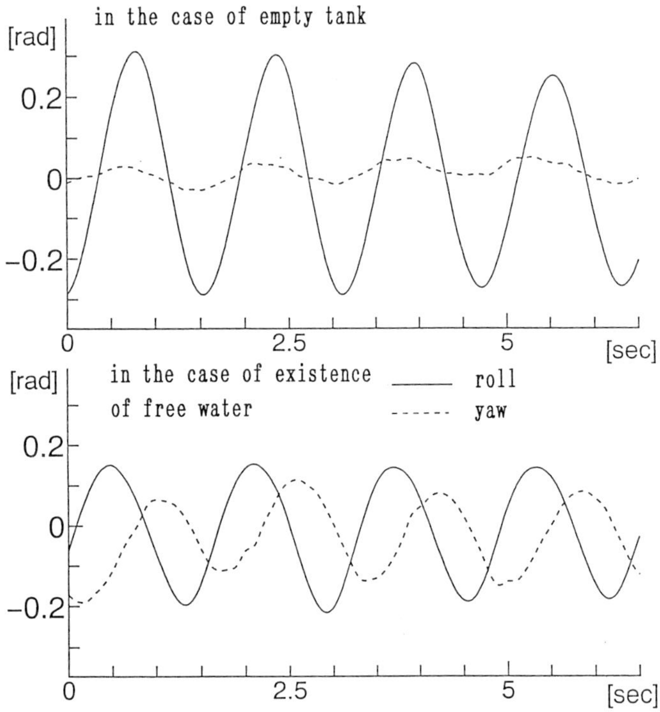

Figure 11: A typical example of occurrence of yaw

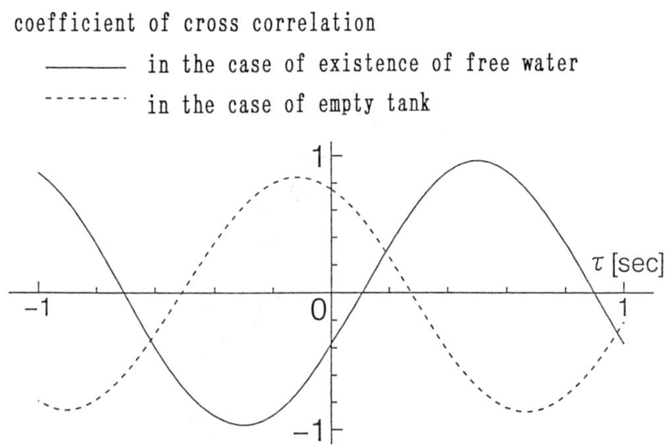

Figure 12: A coefficient of cross correlation between roll and yaw

However, it is recognized that the effect of yaw is not neglected when damping effect by free water is enough. A typical example is shown in Figure 11. The effect of yaw for roll is small and phase difference is short in the case of no free water. When free water exist in the tank even though roll is controlled yaw is increased and cannot be neglected as ship's behaviour. Still more the phase difference between roll and yaw become large. Figure 12 shows the cross correlation between roll and yaw. It became clear that roll goes ahead of yaw as time was plus. This increased yaw was caused by the existence of free water and its location and its behaviour.

CONCLUSIONS

(1) When natural period of free water is shorter than ship's rolling period, the free-decay data of rolling shows the characteristic like a soft spring. On the other hand, when natural period of water on deck is longer, the free-decay data of rolling shows the characteristic like a hard spring.

(2) It was confirmed that the roll response with free water is greater than it without free water at the first and the second resonant period of free water under following circumstances.

a) The second resonance between natural period of water on deck and the period of forced wave.
b) The second resonance between natural period of water on deck and the period of ship rolling motion.
c) The first resonance between natural period of forced wave and the period of ship rolling motion under the influence of water on deck.

(3) When free water exist in the tank even though roll is controlled, yaw is increased as compared with it without free water.

References

Dillingham, J. (1981). Motion studies of a vessel with water on deck, *Marine Technology*, **18:1,** 38-50.

Amagai, K., Kimura, N. and Ueno, K. (1994). On the practical evaluation of shallow water effect in large inclinations for small fishing boats, *Fifth International Conference on Stability of Ships and Ocean Vehicles*, USA, 3.

Ueno, K., Amagai, K., Kimura, N. and Hokimoto, T. (1994). Experimental and Numerical Studies on the Behaviour of Shallow Water in a Large Amplitude Oscillating Tank, *The Journal of Japan Institute of Navigation*, **90**, 201-213

Ueno, K., Amagai, K., Kimura, N. and Iwamori, T. (1995). On the Characteristics of Roll Damping and its Estimation for Small Fishing Vessels, *The Journal of Japan Institute of Navigation*, **93,** 149-161.

Ueno, K., Amagai, K., Kimura, N. and Iwamori, T. (1996). Characteristics of Roll Motion for

Small Fishing Vessels with Water on Deck, *The Journal of Japan Institute of Navigation*, **95**, 183-191.

Ueno, K., Amagai, K., Kimura, N. and Iwamori, T. (1997). On the Characteristics of Roll Motion for Small Fishing Vessels with Angle-Dependent Damping, *The Journal of Japan Institute of Navigation*, **97**, 121-129.

Welch, J.E., Harlow, F.H., Shannon, J.P. and Daly, B.J. (1968). The MAC Method – A Computing Technique for Solving Viscous, Incompressible, Transient Fluid–Flow Problems Involving Free Surfaces, Los Alamos Scientific Laboratory Report LA-3425.

Hirt, C.W., Nichols, B.D., and Romeo, N.C. (1975). SOLA – A Numerical Algorithm for Transient Fluid Flow, Los Alamos Scientific Laboratory Report LA-5852.

D. Vassalos, M. Hamamoto, A. Papanikolaou and D. Molyneux (Editors)

PIECEWISE LINEAR APPROACH TO NONLINEAR SHIP DYNAMICS

V.L. Belenky

National Research Institute of Fisheries Engineering, Ebidai, Hasaki-machi, Kashima-gun, Ibaraki, 314-0421, Japan

ABSTRACT

The paper considers piecewise linear dynamical system as a model of nonlinear rolling and capsizing of ship. Main advantage of the model is a possibility to describe capsizing directly: as a transition to oscillations near upside down stable equilibrium. Such a transition can be expressed in analytical functions that allow deriving symbolic solutions for both regular and irregular seas. Practical application of the model is estimation of ship capsizing probability per unit of time in beam seas.

The paper, however, does not deal with any stochastic matters. The following consequence was reproduced. Free undamped roll motions were studied; a dependence of free period vs. amplitude was derived. This figure was used as a backbone curve to obtain approximate solution for steady state forced roll motion by equivalent linearization. Then, using previous result as the first expansion an exact steady state solution was derived. The last figure allows analyzing motion stability; it was found that the system is capable for both fold and flip bifurcations. Deterministic chaos was observed as a result of period doubling sequence. Also it was found that safe basin of the piecewise linear system experiences erosion as conventional nonlinear system.

KEYWORDS

Capsizing, Nonlinear dynamics, Piecewise linear, Bifurcation, Chaos, Motion stability, Safe basin, Ship stability.

INTRODUCTION

There is a dual purpose for ship capsizing study: developing rational stability regulation and physical knowledge of phenomenon. The second one is necessary to be ready for new types of ship and ocean vehicles.

It seems that probabilistic approach might be the most important for future regulation development, having in mind stochastic character of wind / wave environment. Physical nature of capsizing as «a transition to motion near another stable (upside down) equilibrium» is a nonlinear phenomenon and can be studied by means of nonlinear dynamics. A capsizing model is an outcome of this study.

- Mathematical models of capsizing can be classified as follows:
- Energetic approach that is the background of weather criteria;
- Motion stability of steady state rolling, see Wellicom (1975), Ananiev (1981), Nayfeh and Khdeir, (1986), Virgin (1987).
- Classical ship stability definition or separatrix crossing model, see Sevastianov et al (1979), Umeda et al (1990).
- Safe basin or transient behavior approach, see Rainey at al (1990), Falzarano (1990).
- Piecewise linear approach, see Belenky (1989).

The recent development of the last one is a subject of our consideration.

BACKGROUND OF PIECEWISE LINEAR MODEL. GOALS OF THE STUDY

We consider the simplest model that contains capsizing as «...a transition to stable equilibrium, dangerous from practical point of view» see Sevastianov (1982, 1994), so we should have at least two stable equilibria: upright and upside down, see fig.1.

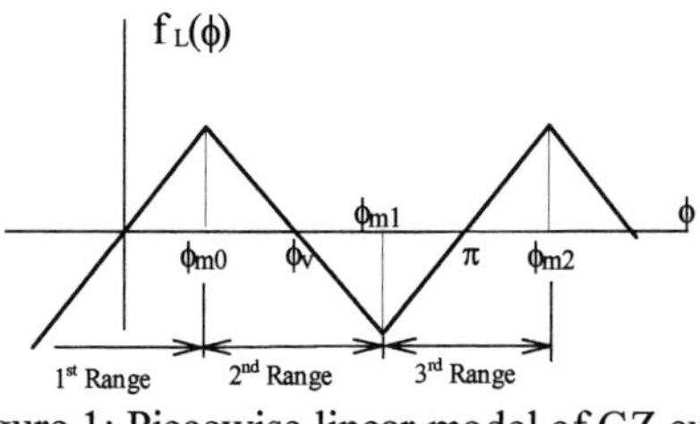

Figure 1: Piecewise linear model of GZ curve

A differential equation of ship rolling, corresponding to this model is expressed as follows.

$$\ddot{\phi} + 2\delta\dot{\phi} + \omega_\phi^2 f_L^*(\phi) = \alpha_E \omega^2 \sin(\omega t + \varphi_E) \tag{1}$$

Here: ϕ-roll angle, δ-damping coefficient, ω_ϕ-natural frequency, ω-excitation frequency, α_E-effective excitation amplitude, φ_E-initial phase angle. The motion is described by two linear solution linked by initial conditions at junction points:

$$\phi = \begin{cases} \phi_{0a} e^{-\delta t} \sin(\omega_0 t + \varepsilon) + q_a \sin(\omega t + \beta_q) \\ A e^{\lambda_1 t} + B e^{\lambda_2 t} + p_a \sin(\omega t + \beta_p) \end{cases} \tag{2}$$

Here A, B, $\phi_{0\alpha}$, ε - arbitrary constants depending on initial conditions at junction points, λ_1, λ_2 - eigenvalues, ω_0 - frequency of initial free damped roll motions. There is no way to simplify this model further, otherwise we shall get rid of capsizing as we defined it above. What is good about this model, besides its simplicity?

Early study of piecewise linear model of capsizing, see Belenky (1989, 1993), showed that capsizing will happen if the value of arbitrary constant A (if λ_1 is positive) becomes positive. The solution at the 2nd linear range becomes unbounded and reaches the 3rd linear range that describes oscillation near upside down position of equilibrium. If this value is negative, then the solution will be back to the 1st range, and capsizing will not happen at least at this semi-period of rolling. So the piecewise linear model offers clear criteria for immediate capsizing - sign of arbitrary constant A, which is determined by formula:

$$A = \frac{(\dot{\phi}_1 - \dot{p}_1) - \lambda_2(\phi_1 - p_1)}{\lambda_1 - \lambda_2}. \tag{3}$$

Here: $\phi_1, \dot{\phi}_1$ - are initial condition at junction point ϕ_{m0} and $p_1, \dot{p}_1$ are values of partial solution at the moment of crossing level ϕ_{m0}. Another important advantage of the model is a very easy way to apply probabilistic approach: probability of capsizing is just a probability of upcrossing (that is very well studied in theory of stochastic processes) with positive value of the arbitrary constant:

$$P_T(X) = P_T(\phi > \phi_{m0})P(A > 0). \tag{4}$$

The probability of upcrossing is connected with time (since upcrossings are Poisson flow), which makes entire probability of capsizing dependent on time of exposure. The last figure is essential for correct probabilistic approach to ship stability regulation see Sevastianov (1982, 1994).

Piecewise linear model can be used not for analytical study but as simulation tool as well. Algorithm contains the following steps:

1. Find the range where given point is;
2. Calculate next point with the given time step.
3. If the next point is within the same range, repeat step 2 until changing range or end of simulation.
4. If the next point in within the next range, find crossing time and crossing initial conditions. Then calculate the next point using the solution on the next range.
5. Repeat step 2 until changing range or end of simulation.

The algorithm has only one iteration procedure - crossing time search. It is numerical solution of nonlinear algebraic equation. Its accuracy can be checked easily - we just substitute crossing time into solution (2). All other steps involve calculation of elementary trigonomentric and exponential functions that can be done really accurate nowdays (at least error is known in advance). So, even working with high amplitudes, we are still able to control accuracy and error accumulation.

Practical using of the piecewise linear model for calculation of capsizing probability in beam seas and wind was found to be possible as well, see details in Belenky (1994, 1995). Practical applicability was reached by using of «combine» model of the GZ curve, see fig 2.

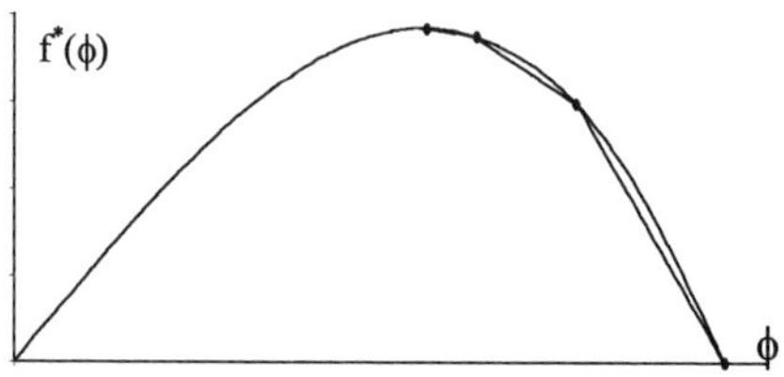

Figure 2: «Combined» model

Stability in beam seas is just a part of the problem In order to obtain a real practical solution it is necessary to take into account changing GZ curve in following and quartering seas. «Brute force» attempt is senseless: if linear coefficients of piecewise linear term are dependent on time harmonically (the simplest case of regular following seas), the expression (1) becomes Mathew equation range wise, that does not have general solution expressed in elementary functions. The problem can be solved using two-dimensional piecewise linear presentation, Belenky (1999).

Another problem is general adequacy. It is clear that behaviour of «combined system» (fig. 2) approximates real ship rolling. However, it is more complicated than just triangle. K. Spyrou proposed to use triangle presentation as piecewise linearization for real GZ curve, that might be fruitful especially for quartering seas to avoid additional complexity caused by large number of ranges. To do that we need a way to compare piecewise linear and conventional nonlinear dynamical system. Theoretically, piecewise linear system can be studied by nonlinear dynamic methods. M. Komuro (1988, 1992) carried out comprehensive math research on piecewise linear systems and found the way to derive bifurcation equations that allows analysing its nonlinear behaviour. However, there is no warranty that "triangle" behaves in the same way, conventional rolling equation does. To facilitate development of piecewise linearization we need methods to carry out conventional nonlinear study for piecewise linear rolling equation. This is the primary goal of the presented study.

We try to develop methods that are capable to reproduce the following results:

- Free rolling is not isohronic: period of free undamped roll motion depends on initial amplitude.
- Response curve of nonlinear forced rolling contains non-functionality: there is an area with several amplitudes corresponding to the same excitation frequency.
- Stability analysis of steady state nonlinear roll motion shows that some regimes are unstable see Wellicome (1975), Ananiev (1981).
- This instability leads either to fold (escape through positive real direction) or flip bifurcation (escape through negative real direction) and consecutive period doubling may lead to chaotic response see Nayfeh et al (1986), Virgin (1987).
- Dangerous combination of parameters of external excitation looks like erosion of safety basin area see Rainey et al (1990).

FREE MOTION

Let's consider first the simplest case of free motion: if there is no bias. If initial amplitude lies within the first range (see fig.1), we have pure linear oscillations: period does not depend on initial conditions - system is isohronic. If the initial amplitude is located within the second range, the period is described by formula from Belenky (1995-a):

$$T(\phi_a) = \frac{4}{\omega_\phi}\left\{\frac{1}{\sqrt{k_2}}\operatorname{arccosh}\frac{\phi_V - \phi_{m0}}{\phi_V - \phi_a} + \frac{1}{\sqrt{k_1}}\arctan\frac{\phi_{m0}\sqrt{k_1}}{\sqrt{k_2}\sqrt{(\phi_V - \phi_{m0})^2 - (\phi_V - \phi_a)^2}}\right\} \tag{5}$$

here k_1 and k_2 are angle coefficients of the first and the second range correspondingly, ϕ_V is angle of vanishing stability. As it could be clearly seen, from (5), the period depends on initial amplitude, so the system is not isohronic, if initial amplitude equals to angle of vanishing stability, the period becomes infinite.

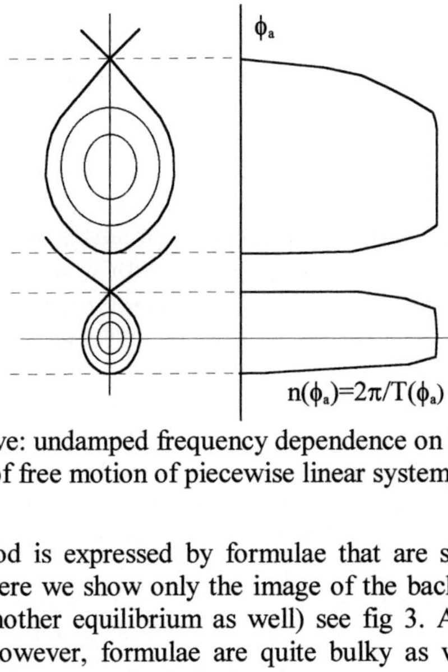

Figure 3: Backbone curve: undamped frequency dependence on initial amplitude and phase plane of free motion of piecewise linear system with bias

If there is bias, the period is expressed by formulae that are simple but rather bulky, see Belenky (1998, 1999). Here we show only the image of the backbone curve, along with the phase plane (build for another equilibrium as well) see fig 3. All phase trajectories can be expressed analytically, however, formulae are quite bulky as well and could be found in Belenky (1998, 1999-a). The system is not isohronic.

STEADY STATE FORCED MOTION

Since we have the backbone curve, the next step is evident, we can get approximate solution for steady state forced motion using equivalent linearization. It means that we substitute real system by linear one that has the same period of free oscillation. We can do the same procedure with piecewise linear system Belenky (1995-a)

$$\phi_a = \frac{\alpha_E}{\sqrt{\left[(\omega_\phi(\phi_a))^2 - \omega^2\right]^2 + 4\delta^2\omega^2}} \tag{6}$$

Appearance of the approximate response curve is show in fig 4. Phase curve can be calculated analogously.

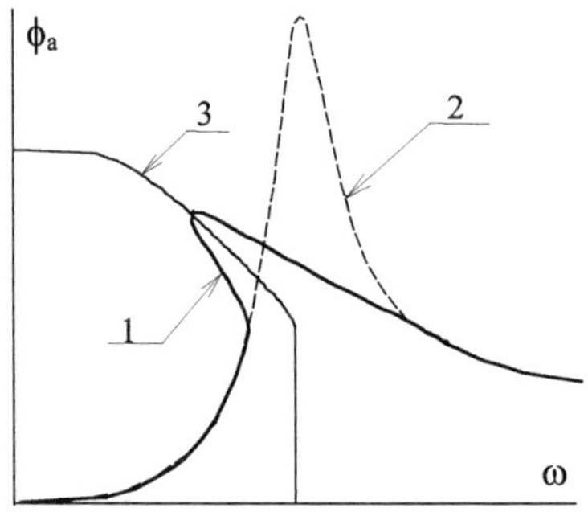

Figure 4: Response curve of piecewise linear system by equivalent linearization (1), response curve of linear system at the first range (2) and backbone curve (3) α_E=0.2, ϕ_{m0}=0.5, δ=0.1 s^{-1}, $k_1 = k_2$=1 s^{-2}, ϕ_v=1, bias 0.05

Steady state piecewise linear solution consists form fragments of linear solutions (2) as well as transition one. The only difference between transition and steady state solutions are crossing velocities and periods of time spent in different ranges of piecewise linear term. So if we find such figures that provide periodic solution with excitation frequency, steady state problem will be solved, see Belenky (1997). These conditions can be formalised as a system of simultaneous algebraic equations, if we look at unbiased case first, it is enough to consider just half of period:

$$\begin{cases} f_0(T_0, \dot{\phi}_0, \varphi_0) = \phi_{m0} \\ \dot{f}_0(T_0, \dot{\phi}_0, \varphi_0) = \dot{\phi}_1 \\ f_1(T_1, \dot{\phi}_1, \varphi_0 + \omega T_0) = \phi_{m0} \\ \dot{f}_1(T_1, \dot{\phi}_1, \varphi_0 + \omega T_0) = -\dot{\phi}_0 \\ T_0 + T_1 = \pi \cdot \omega^{-1} \end{cases} \tag{7}$$

Here functions f_0 and f_1 are solutions (2) at the first and second ranges of the piecewise linear term correspondingly, see also fig.5. The system (7) can be solved relative to unknown values T_0, T_1, $\dot{\phi}_0, \dot{\phi}_1$, and φ_0 using any appropriate numerical method. Results of equivalent linearization can be used for calculation of initial values of the unknown values that makes calculations more fast and simple. Biased case is more complicated. Main difficulty here is not only to consider entire period of motion, here we meet so-called «cross mode problem» Asymmetry caused by bias may affect on number of crossings per period: it is not necessary four as fig. 5 shows; period can contain two crossings as well. So considering biased steady state motions (that is especially important for further bifurcation analysis), we need to know in advance how much crossing will be hosted by one period. It can be done by searching frequency that provides «two crosses and one touch»; we use the system analogous to (7), that has excitation frequency as unknown value as well, see details in Belenky (1998,1999). Resulting response curve is shown in fig. 6.

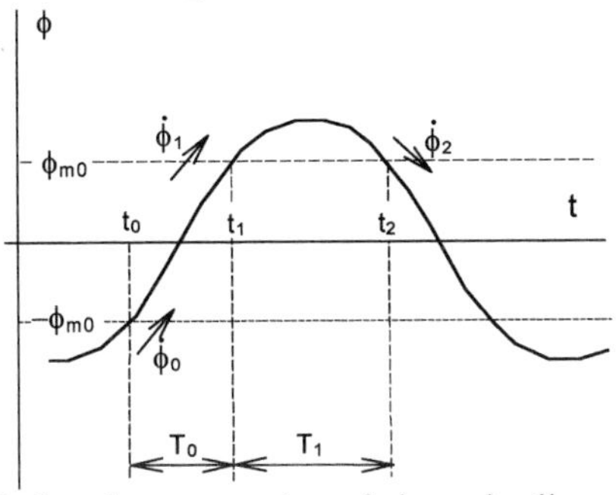

Figure 5: Steady state motion of piecewise linear system

As it could be seen from figure 6, response curve has quite conventional form, including hysteresis area, where three amplitudes corresponds to one excitation frequency. We call this steady state solution exact despite numerical method was used to calculate crossing characteristics; accuracy is still controllable: we always can substitute these figures into system (7) and check how solution turns equations into equalities. We can call this steady state solution analytical (or, at least semi-analytical), despite the numerical method was used; solution still is defined by formulae (2), so we still can manipulate it analytically.

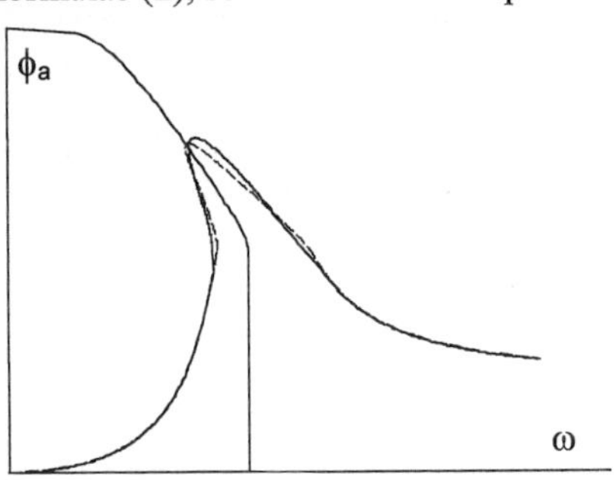

Figure 6: Exact response curve of piecewise linear system. (Dotted line shows response of linearized solution.) α_E=0.2, ϕ_{m0}=0.5, δ=0.1 s^{-1}, k_1= k_2=1 s^{-2}, ϕ_v=1, bias 0.05

MOTION STABILITY AND BIFURCATION ANALYSIS

The next conventional step is the motion stability determination. Analogous problem was considered in Murashige et al (1998) for piecewise nonlinear system. We also will search stability indicators as characteristics of Jacobian matrix (eigenvalues and trace-determinant) So we calculate Jacobian matrix for each range. The resulting Jacobian can be calculated as a product of the above, with regard on cross mode:

$$J_{4cross} = J_3 \cdot J_2 \cdot J_1 \cdot J_0 \qquad (8)$$

$$J_{2cross} = J_1 \cdot J_0 \tag{9}$$

Partial derivatives of Jacobian matrix should be calculated numerically. Analytical expressions for these figures are not available, because formulae (2) cannot be inverted in elementary functions. Results of motion stability calculation are shown in figure 7. It indicate presence of unstable steady state regimes, fold and flip bifurcations. Let's examine them more close. We get three responses in the hysteresis area, one of them is pure linear or trivial, so it is definitely stable. Two piecewise linear response were obtained from the same system of equation (like (7) depending on cross mode) using two different initial points. One of these initial points corresponds to high amplitude response of equivalently linearized solution; another one is from the middle branch. The middle solution is unstable, the high one - stable see fig 8.

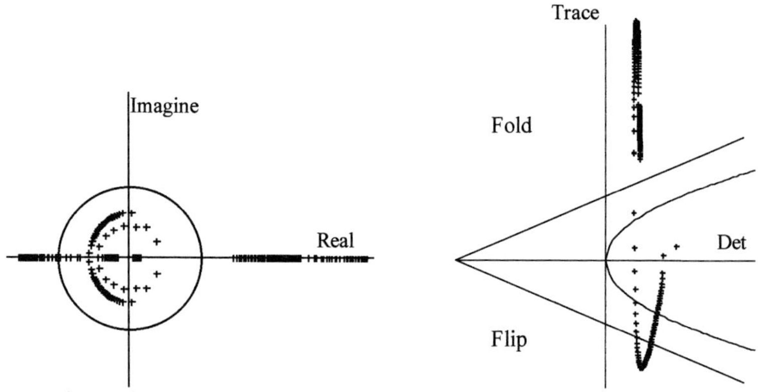

Figure 7: Eigenvalues of Jacobian matrix of biased piecewise linear system and trace - determinant plane of Jacobian matrix of biased piecewise linear system

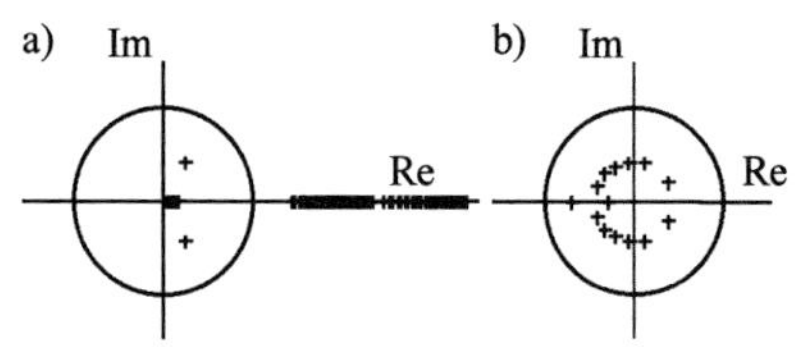

Figure 8: Middle (a) and high (b) amplitude response stability indexes

Eigen values escape unit circle through positive direction, what is an indication of fold bifurcation. To see the phase plane, we should reproduce unstable steady state regime and then disturb it in eigen vectors direction: the system will «jump» towards to stable mode, see figure 9. Another possible type of nonlinear behaviour is flip bifurcation: sequence of period doubling, see fig.10 leading to deterministic chaos, see fig 11.

Form of phase trajectories of piecewise linear system is very similar to conventional nonlinear ones: there is nothing that indicates piecewise linear origin of figures 9-11. Concluding motion stability and bifurcation analysis we can state that piecewise linear rolling equation qualitatively behaves in the same way as nonlinear one.

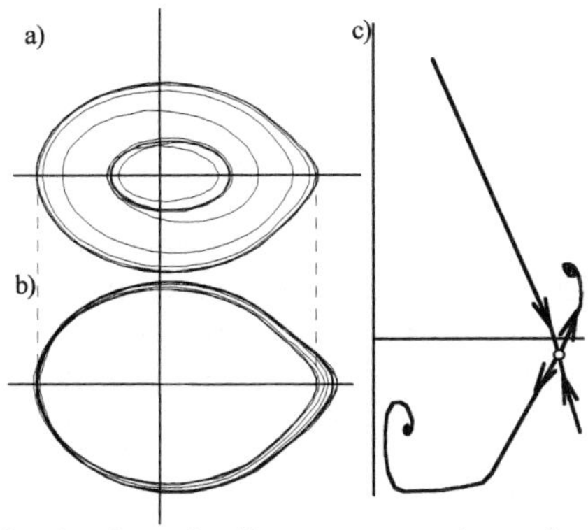

Figure 9: Fold bifurcation in piecewise linear system: «jump down» (a), «jump up» (b), invariant manifold (c), ω=0.77

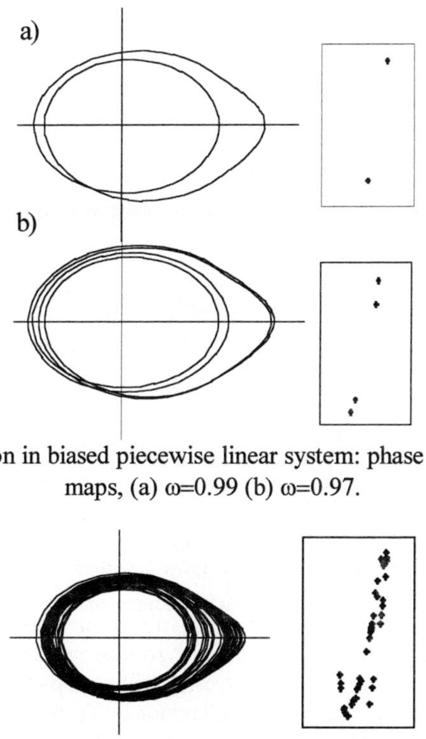

Figure 10: Flip bifurcation in biased piecewise linear system: phase trajectories and Poincare maps, (a) ω=0.99 (b) ω=0.97.

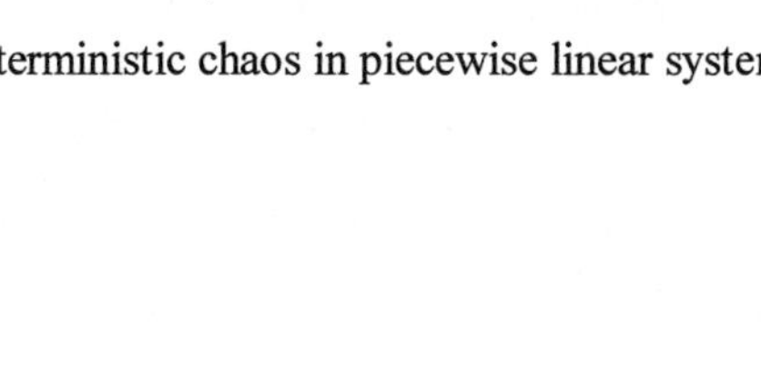

Figure 11: Deterministic chaos in piecewise linear system. ω=0.92439

SAFE BASIN ERROSION

Another important nonlinear quality of severe ship rolling is erosion of safe basin, see Rainey et al (1990), Falzarano (1990). Rainey et al (1990) considered relative area of the safe basin as stability criteria. We check if the piecewise linear system is capable for this type of behaviour. Calculations were carried out with the resolution 90x90 for a square: $\pm 1.5 \cdot \phi_V \times \pm 1.5 \cdot \phi_V \cdot \omega_\phi$. Excitation frequency was used as a control parameter. All other parameters were the same as for the above samples. Some of the results are shown on fig.12.

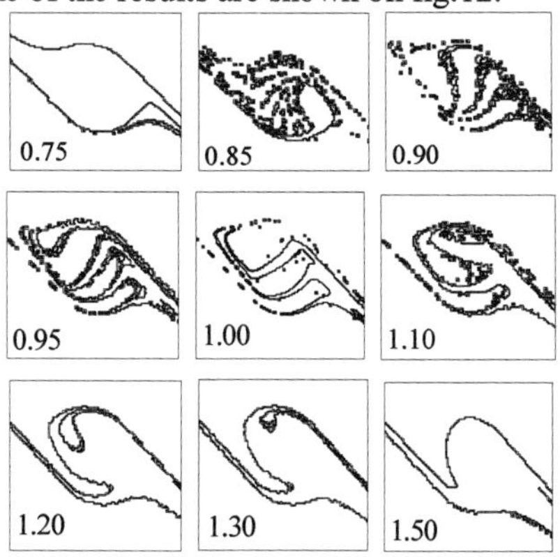

Figure 12: Erosion of safe basin of piecewise linear system

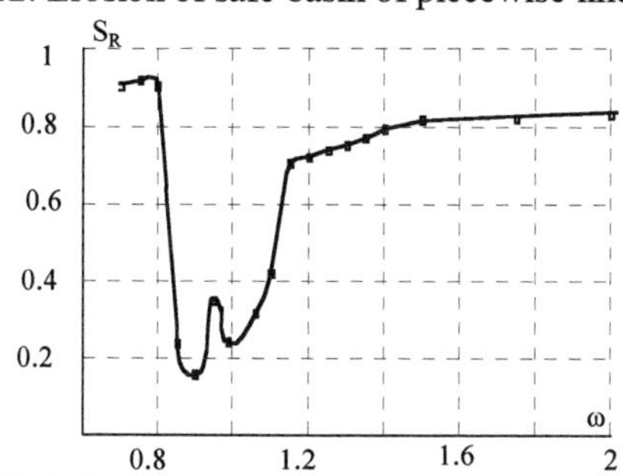

Figure 13: Relative safe basin area vs. excitation frequency

The results are summarised in a form of dependence of the safe basin relative area on excitation frequency:

$$S_R(\omega) = \frac{A_{SB}(\omega)}{A_{SB}\big|_{\alpha_E=0}} \tag{10}$$

Here: $A_{SB}(\omega)$ - area of the safe basin at given frequency, $A_{SB}\big|_{\alpha_E=0}$ - area of safe basin of free damped motion; see fig.13. As it could be clearly seen from figures 12 and 13, the safe basin of piecewise linear system experiences erosion that leads to decreasing of its area. The behaviour is similar to what is known on conventional nonlinear rolling equation.

CONCLUSIONS

The above study has shown that:

* Free motion of piecewise linear system is not isochronic; period of undamped free oscillations depends on initial amplitude (initial heel angle)
* Equivalent linearization can be applied to calculate approximated characteristics of steady state motion of the piecewise linear system.
* Exact characteristics of steady state motion can be calculated using the only assumption that this motion is periodic; exact steady state motion can be expressed by elementary functions, however, some of parameters are results of numerical solution of a system of simultaneous algebraic equations.
* Response curve of a piecewise linear system (both approximate and exact) contains non-functionality (or hysteresis): an area where there amplitudes correspond to one frequency;
* Conventional stability analysis can be applied to a piecewise linear system.
* Piecewise linear system is capable for fold and flip bifurcation; consecutive flip bifurcation leads to deterministic chaos in piecewise linear system.
* Safe basin of piecewise linear system experiences erosion, when waves become dangerous.

We have seen that qualitative behaviour of piecewise linear system with simplest "triangle" term coincides with conventional nonlinear dynamical system that represents rolling of ship in beam seas. So we can use "triangle" presentation of the GZ curve for further development of piecewise linearization.

ACKNOWLEDGEMENT

Contents of this work is partially based on research funded by Science and Technology Agency of Japan, STA fellowship Id No 269115 and carried out at National Research Institute of Fisheries Engineering (NRIFE). Help of Dr Naoya Umeda (NRIFE) was very fruitful and highly appreciated.

References

Ananiev D.M. (1981). On Stability of Forced Rolling Motion with Given Stability Diagram, *Trans of Kalningrad Technical Institute "Seakeeping of Ships"*, **93**, 17-25 (in Russian).

Belenky V.L. (1989). A New Method of Statistical Linearization in Severe Rolling and Capsizing Problem, *Proc. of 18th SMSSH*, Varna, Bulgaria.

Belenky V.L. (1993). A capsizing Probability Computation Method, *Journal of Ship Research*, **37:3**, 200-207.

Belenky V.L. (1994). Piece-Wise Linear Methods for the Probabilistic Stability Assessment for Ship in a Seaway. *Proc of 5th International Conference on Stability of Ships and Ocean Vehicles (STAB '94),* Melbourne, Florida, USA, **5**.

Belenky, V.L. (1995). Analysis of Probabilistic Balance of IMO Stability Regulation by Piece-wise Linear Method, *Marine Technology Trans. of Academy of Science*, Gdansk, Poland.

Belenky V.L. (1995-a). On the Dynamics of Piecewise Linear System, Proc. of Int. Symp. on Ship Safety in a Seaway: Stability, Manoeuvrability, Nonlinear Approach (SEVASTIANOV SYMPOSIUM), Kaliningrad, Russia, **1.**

Belenky V.L. (1997). Some problems of Stochastic Dynamics of Piecewise Linear and Nonlinear Systems. *Seminar at Naval Architecture Department*, University of Michigan, Ann Arbor, Michigan, USA, 13 March.

Belenky V.L (1998). Dynamics of Multiple Equilibria Piecewise Linear System. *Dynamical Brainics Seminar*, Aihara Laboratory, Department of Mathematical Engineering and Information Physics, University of Tokyo, Tokyo, Japan, 18 June.

Belenky V.L. (1999). Probabilistic Assessment of Ship Stability in Quartering Seas. *Proc. of 4th Japan-Korean Workshop on Ship Hydrodynamics (JAKOM'99)*, Fukuoka, Japan.

Belenky V.L. (1999-a). Piecewise Linear Approach to Nonlinear Dynamics of Ships, *Bulletin of National Research Institute of Fisheries Engineering*, Hasaki, Japan (to be printed).

Falzarano J.M. (1990). Predicting complicated dynamics leading to vessel capsizing. Ph.D. Dissertation, *Naval Architecture Department, University of Michigan*, Ann Arbor, Michigan, USA.

Komuro M. (1988). Normal Forms of Continuous Piecewise Linear Vector Fields and Chaotic Attractors. *Japan Journal of Applied Mathematics,* Part 1: **5:2**, 257-304, Part 2: **5:3**, 503-549

Komuro M. (1992). Bifurcation Equation of Continuous Piecewise Linear Vector Fields. *Japan Journal of Industrial and Applied Mathematics*, **9:2**, 269-312.

Murashige S. Komuro M. and Aihara K. (1998). Nonlinear Roll Motion and Bifurcation of a Ro-Ro Ship with Flooded Water in Regular Beam Seas. *Trans. R. Soc.* London, UK.

Nayfeh A.H. and Khdeir A.A. (1986). Nonlinear Rolling of Biased Ships in Regular Beam Waves. *International Shipbuilding Progress*, **33**.

Rainey R.C.T, Thompson J.M.T., Tam G.W. and Noble P.G. (1990) The Transient Capsize Diagram - a Route to Soundly-Based New Stability Regulations. *Proc. of 4th International Conference on Stability of Ships and Ocean Vehicles (STAB'90),* Naples, Italy.

Sevastianov N.B. and Fam Ngock Hoeh (1979) Boundary between the Domains of Stable and Unstable Free Ship Motion in Drift-Rolling Regime. *Trans. of Kaliningrad Technical Institute "Seakeeping of Fishing Vessels",* Kalininrgad, USSR, **81**, 17-25 (in Russian).

Sevastianov N.B. (1982). Probabilistic Stability Regulation as a Problem of Reliability Theory. *Trans. of Register of the USSR,* Leningrad, USSR, **12**, 94-100, (in Russian).

Sevastianov N.B. (1994). An Algorithm of Probabilistic Stability Assessment and Standards. *Proc. of 5th International Conference on Stability of Ships and Ocean Vehicles (STAB'94)* Melbourne, Florida, USA **5.**

Umeda N., Yamakoshi Y. and Tsichiya T. (1990) Probabilistic Study on Ship Capsizing due to Pure Loss of Stability in Irregular Quartering Seas. *Proc. of 4th International Conference on Stability of Ships and Ocean Vehicles (STAB'90),* Naples, Italy.

Virgin, L.N. (1987). The Nonlinear Rolling Response of a Vessel including Chaotic Motions Leading to capsize in regular seas, *Applied Ocean Research* **9:2**, 89-95.

Wellicome, J. (1975) An Analytical Study of the Mechanism of Capsizing, *Proc. of 1st International Conference on Stability of Ships and Ocean Vehicles (STAB'75),* Glasgow, Scotland, UK.

2. Damaged Ship Stability

D. Vassalos, M. Hamamoto, A. Papanikolaou and D. Molyneux (Editors)

THE WATER ON DECK PROBLEM OF DAMAGED RO-RO FERRIES

Dracos Vassalos

The Ship Stability Research Centre (SSRC), Department of Ship and Marine Technology
University of Strathclyde, Glasgow, UK

ABSTRACT

This paper addresses the research work undertaken by Strathclyde University Ship Stability Group in association with the ***Joint North West European Project***. The fundamental thinking behind the approaches adopted to tackling large scale flooding on the vehicle deck of Ro-Ro vessels is outlined and some of the promising early findings are highlighted, thus providing a suitable platform for discussions concerning the way forward to meaningfully safeguarding against this problem.

KEYWORDS

Early research on damaged Ro-Ro vessels, original numerical models, original concepts and applications

INTRODUCTION

The tragic accidents of the *Herald of Free Enterprise* and more recently of *Estonia* were the strongest indicators yet of the magnitude of the problem presented when water enters the deck of ships with large undivided spaces, such as Ro-Ro vessels. The ship loss could be catastrophic as a result of rapid capsize, rendering evacuation of passengers and crew impractical, with disastrous (unacceptable) consequences. These disasters have brought about a realisation that "***ship survival***" might have to be addressed separately from "***passenger survival***" in that the deterioration in the stability of such vessels, when damaged, could be "***catastrophic***" rather than one of graceful degradation. It would appear, therefore, that the approach to assessing realistically the damage survivability of passenger ships and indeed any ships, must derive from a logical framework and must, of necessity, offer the means of taking into consideration meaningfully both the operating environment and the hazards specific to the vessel in question. One of the tasks should be to quantify the probability of damage with water ingress in a given service area and, the second, to quantify the consequences of damage by identifying and analysing all the important factors using probabilistic methods. However, even though it is self-evident that reducing either the probability of damage or the consequences of damage or both can reduce the risks involved, there is a level beyond which consequences cannot be tolerated. In this case, reducing the probability of damage alone cannot reduce risk. The need arises, therefore, for a methodology whereby key questions are addressed and answers sought concerning definition of acceptable risks, definition and management of maximum tolerable consequences and procedures for dealing with residual risks. The Joint North West European Project has been set to address this need urgently, by bringing together all the available expertise in the relevant areas. The Strathclyde Group were responsible for Task 5 of this Joint R&D Project, pertaining to the development and validation of numerical tools for assessing the damage survivability of passenger/Ro-Ro vessels, leading to the development of survival criteria.

BACKGROUND

UK Ro-Ro Research Programme

In the wake of the *Herald of Free Enterprise* disaster, the need to evaluate the adequacy of the various standards in terms of providing sufficient residual stability to allow enough time for the orderly evacuation of passengers and crew in realistic sea states has prompted the UK Department of Transport to set up the Ro-Ro Research Programme comprising two phases. Phase I addressed the residual stability of existing vessels and the key reasons behind capsizes. To this end theoretical studies were undertaken into the practical benefits and penalties of introducing a number of devices, (BMT Report 1990), for improving the residual stability of existing Ro-Ro's. In addition, model experiments were carried out by the British Maritime Technology Ltd, (Dand 1990) and the Danish Maritime Institute, (DMI Report 1990) in order to gain an insight into the dynamic behaviour of a damaged vessel in realistic environmental conditions and of the progression of flood water through the ship. Phase II was set up with the following objectives in mind:

- To confirm the findings of Phase I in respect of the ability of a damaged vessel to resist capsize in a given sea state.
- To carry out damaged model tests, in which the enhancing devices assessed in Phase I would be modelled to determine the improvements in survivability achieved in realistic sea-going conditions.
- To confirm that damage in the region amidships is likely to lead to the most onerous situation in respect of the probability of capsize.
- To undertake theoretical studies into the nature of the capsize phenomenon, with a view to extrapolating the model test results to Ro-Ro passenger ships of different sizes and proportions.

The Department of Ship and Marine Technology at the University of Strathclyde was one of three organisations charged with the responsibility of developing and validating a theoretical capsize model which could predict the minimum stability needed by a damaged vessel to resist capsizing in a given sea state. This was subsequently to be used to establish limiting stability parameters that might form the basis for developing realistic survival criteria. Full details are given in Vassalos and Turan (1992).

Joint R&D Project

As the UK stood poised to share the findings from the Ro-Ro Research Programme with the rest of the world, the *Estonia* tragedy has once more shaken the foundations of shipping, forcing the profession to provide answers "immediately" and, in attempting to do so, to use the right expertise and experience to provide the right answers. The focus and effort in Ro-Ro and passenger ships capsize safety have suddenly reached the deserved and long overdue intensity. The Nordic countries reacted quickly in undertaking this responsibility leading to a wider-based project within a very short period. Taking onboard the fact that, in addressing the probability of a ship surviving a given damage, the problem of damage survivability does not end with quantifying the probability of damage and the consequences of damage. As indicated above, the *Estonia* disaster was the strongest indicator yet of the urgent need to define acceptable risks and maximum tolerable consequences as well as to identifying procedures for managing such consequences and dealing with the residual risks. To this end, the Joint R&D Project adopted the following framework:

A Framework for Rationalising the Probabilistic Approach

The risk of capsizing (or sinking) as a result of damage is given by

Risk = P(damage) x (1-A) ≤ AR, where

P(damage) = Probability of damage with water ingress (per year) in a given service area

A = Probability of surviving the said damage (Attained Subdivision Index)

AR = Acceptable risk

When AR is defined and P(damage) is known, then

$$A \geq 1 - AR / P(\text{damage}) \equiv R, \text{ where}$$

R = Required Subdivision Index

The Attained Subdivision Index is currently calculated as

A = $\sum$p.s, taken over all damage cases and combination of damage cases, where

p = Probability of damage calculated from damage statistics on damage location on the ship; length, height and penetration

s = Probability of surviving a given damage depending on vessel condition before damage, permeability of damaged compartments and vessel residual stability.

Currently the major defect in determining the factor "p" derives from the fact that the damage size is independent of the vessel's structural strength and the damage occurrence is also independent of the route, mode and area of service. To rationalise the probabilistic approach, these two deficiencies ought to be rectified. This forms Task 2.1 in the Joint R&D Project. Deriving from the success demonstrated by the Strathclyde Group during the UK Ro-Ro research and also during the two years following its completion, the Joint R&D Project decided to make full use of their mathematical model. The intention was, following a process of further development and validation, to apply it to different vessel types, forms, sizes and compartmentation and to representative damage scenarios and environments to verify its general applicability to assessing the capsize safety of a damaged ship in a given sea state leading to the development of generalised expressions for the factor "s" to be used in the determination of A. This will facilitate the way towards a Formal Safety Assessment methodology and help rationalise the probabilistic approach for assessing the damage survivability of ships. As mentioned above, this work forms Task 5 of the Project. The approach described in the foregoing will allow, in addition, the generation of knowledge for improving upon the design and operational practice of passenger ships.

Key aspects of this research are outlined in the following sections.

STRATHCLYDE APPROACH

General Remarks

Since the dynamic behaviour of the damaged vessel and the progression of the floodwater through the damaged ship in a random seaway are ever changing, rendering the dynamic system highly non-linear, the technique used, of necessity, is time simulation. The numerical experiment considered assumes a stationary ship beam on to the oncoming waves with progressive flooding taking place through the damage opening which could be of any shape, longitudinal and transverse extent and in any location throughout the vessel. As simulation begins with predefined initial conditions, the damaged ship starts moving under the action of random beam waves. Instantaneous water ingress is considered by taking into account the wave elevation and ship motions, which are also estimated at each time step. For each case under investigation, simulations are carried out for different loading conditions while the sea state used in the calculations is progressively increased to a limit where the ship capsizes systematically, thus allowing for a definition of survival boundaries. Having said this, the complexity of the problem at hand dictates that several simplifications are adopted in both the mathematical formulation of the damaged vessel motions and of the water ingress in order to derive engineering solutions. These are explained next before considering some key research findings.

Generalised Mathematical Models

As is commonly known, the static and dynamic stability of a ship depend on its heeling or rolling motion. The heel or roll angle is itself a criterion which is taken into account by intact and damage ship stability assessment procedures. However, in a real environment other motions could significantly affect the ship's stability and roll motion directly or indirectly. In studying extreme vessel behaviour one should clearly aim for a model that represents reality meaningfully. The strong hydrodynamic coupling of sway into roll and the non-linear hydrostatic coupling of heave into both significantly change the underwater volume of the ship in roll. Heave motion is also clearly important in affecting the rate of flooding through the ship and in influencing the roll motion itself. In addition, a vessel in beam seas will drift and this gives rise to additional forces acting on the ship and so would the sloshing motion of the floodwater. Therefore, the sway motion contribution to the ensuing vessel behaviour is expected to be significant. Furthermore, even in a beam seas situation, a vessel is generally expected to undergo a change in its heading relative to the waves, depending on the longitudinal distribution of the underwater volume and, as a result of this, pitch motion will be induced. To accommodate this situation and also the general situation of wave headings other than beam seas, a coupled six-degrees-of-freedom mathematical model of ship motions would be necessary.

In the majority of cases considered so far, however, it appears that a coupled sway-heave-roll model with instantaneous sinkage and trim will normally suffice. Considering the above, two generalised models are concurrently being pursued.

The UK Ro-Ro Research Model

This model was developed and used by the Strathclyde Group during the UK Ro-Ro Damage Stability Research Programme and formed the basis for the Joint R&D Project. It is a non-linear three-degrees-of-freedom model, coupled in sway-heave-roll motions and comprises the following:

$$\{[M(t)] + [A]\}\ \{\ddot{Q}\} + [B]\ \{\dot{Q}\} + [C]\ \{Q\} = \{F\}_{WIND} + \{F\}_{WAVE} + \{F\}_{WOD}$$

with, $[M(t)]$: Instantaneously varying mass and mass moment of inertia matrix.

$[A]$, $[B]$: Generalised added mass and damping matrices, calculated once at the beginning of the simulation at the frequency corresponding to the peak frequency of the wave spectrum chosen to represent the random sea state.

$[C]$: Instantaneous heave and roll restoring, taking into account ship motions, trim, sinkage and heel.

$\{F\}_{WIND}$: Regular or random wind excitation vector.

$\{F\}_{WAVE}$: Regular or random wave excitation vector, using 2D or 3D potential flow theory.

$\{F\}_{WOD}$: Instantaneous heave force and trim and roll moments due to flood water.

The latter is assumed to move in phase with the ship roll motion with an instantaneous free-surface parallel to the mean waterplane. This assumption is acceptable with large ferries since, owing to their low natural frequencies in roll, it is unlikely that floodwater will be excited in resonance and this is further spoiled as a result of progressive flooding. Indeed, when the water volume is sufficiently large to alter the vessel behaviour, small differences are expected between the floodwater and ship roll motions. During simulation, the centre of gravity of the ship is assumed to be fixed and all subdivisions watertight.

On-going Research Model

This model has been developed recently by the Stability Group and is currently undergoing validation. It allows for a vessel drifting with the centre of gravity updated instantaneously during progressive flooding. All the parameters in the model are updated instantaneously as a function of the vessel's mean attitude relative to the mean waterplane and her mean position relative to an earth-fixed system. It is a non-linear, coupled six-degrees-of-freedom model comprising the following:

$$\{[M]+[M_W(t)]+[A_\infty]\}\{\ddot{Q}\}+\{[\dot{M}_W(t)+[B]_{Viscous}\}\{\dot{Q}\}+\int_0^t[K(t-\tau)]\{\dot{Q}(\tau)\}d\tau$$
$$=\{F\}_{Wave}+\{F\}_{Drift}+\{F\}_{Wind}+\{F\}_{Current}+\{F\}_{Restoring}+\{F\}_{Gravitational}-\{F\}_{WOD}$$

with, $[M]$: Generalised mass matrix.

$[M_W(t)]$: Flood water moving independently of the vessel but with an instantaneous free surface parallel to the mean waterplane.

$[A_\infty]$: Generalised added mass matrix (asymptotic values).

$[\dot{M}_W(t)]$: Rate of flood water matrix (acting as damping).

$[B]_{Viscous}$: Non-linear damping matrix.

$\int_0^t[K(t-\tau)]\{\dot{Q}(\tau)\}d\tau$: Convolution integral representing radiation damping.

$\{F\}_i$: Various generalised force vectors comprising wave (1st and 2nd order), wind and current excitation as well as restoration and gravitational effects. All these are updated instantaneously as a function of the vessel attitude relative to the mean waterplane by using a database which spans the whole practical range of interest concerning heel, trim, sinkage, heading and frequency. The same applies to the hydrodynamic reaction forces. Excitation from shifting of cargo can also be considered.

$\{F\}_{WOD}$: This force vector is now comprised of dynamic effects of floodwater in contrast to its counterpart in the previous model which involves only gravitational effects.

The phase/amplitude difference between vessel roll and flood water motions could be determined, for example, by building a database through a systematic series of model experiments using a sway-heave-roll bench test apparatus. This undertaking is currently underway at Inha University in South Korea through a collaborative research arrangement supported by the British Council.

In case when the dynamic behaviour of the floodwater is considerable and could prove to be dominating or heavily influencing the vessel behaviour, the dynamic system of vessel-flood water must be treated as two separate worlds interacting, using CFD techniques to describe flood water sloshing. Considerable effort along these lines has already been expended at the University of Trieste in Italy with the Strathclyde Stability Group collaborating through yet another British Council supported research link.

Modelling the Water Ingress

This is indeed a very difficult phenomenon to model as it involves very complex hydrodynamic flows. Some degree of approximation is, therefore, expected in order to derive engineering solutions. In the approximate method adopted, water ingress is modelled as an intermittent probabilistic event based on the calculation of the relative position between wave elevation and damage location. The mode of flow is affected largely by the hydrostatic pressure head and the area of the damage hole but this is influenced by dynamic effects, edge effect, shape of opening, wave direction and profile, water elevation on either side of the opening and damage location. Considering, for example, damage below the bulkhead deck, a flooding scenario is depicted by the simplified picture shown in Figure 1 with the sea treated as a reservoir and the pressure distribution in the hold assumed hydrostatic. If Bernoulli's equation is applied at sections A and B, assuming that the total pressure head is maintained constant and the velocity is zero in the reservoir, the inflow velocity at point P can be calculated as follows:

$$h_{out} + \frac{P_{atm}}{\rho g} + 0 = h_{in} + \frac{P_{atm}}{\rho g} + \frac{v^2}{2g} \qquad \rightarrow \qquad v = \sqrt{2g(h_{out} - h_{in})}$$

The flow rate through the horizontal layer around P is then given by:

$$dQ = K\sqrt{2g(h_{out} - h_{in})}dA$$

The total flow rate can be found by integrating dQ over the damage opening height. This expression reduces to the general form of those used for free-discharging orifices and notches when either h_{out} or h_{in} is negative, if the following limits are set:

$$\begin{cases} h_{in} = 0 \text{ if } h_{in} \le 0 \\ h_{out} = 0 \text{ if } h_{out} \le 0 \end{cases}$$

This takes care of those situations in which water is present only on one side of the damage. Of course, when h_{out} is less than h_{in}, the flow becomes negative, and water is expected to flow out of the compartment and into the sea. To accommodate for this the *pressure head* equation is put into the form:

$$dQ = K.\text{sign}(h_{out} - h_{in}).\sqrt{2g|h_{out} - h_{in}|}.dA \text{, with the same limits as above.}$$

Considering that $|h_{out} - h_{in}|$ represents the instantaneous downflooding distance, which is relatively easy to compute, the whole problem of progressive flooding reduces to the evaluation of the coefficient K, and this is done experimentally.

Validation/Calibration of the Mathematical Model

In addition to the work undertaken during the UK Ro-Ro research, considerable effort is being expended in the Joint R&D Project to ensure the validity of the mathematical model on the whole range of possible applications, regarding vessel type and compartmentation (above and

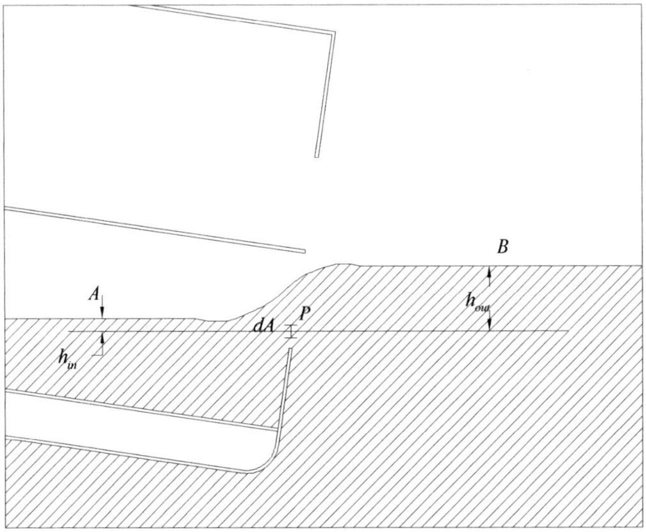

Figure 1: Water Ingress Main Parameters

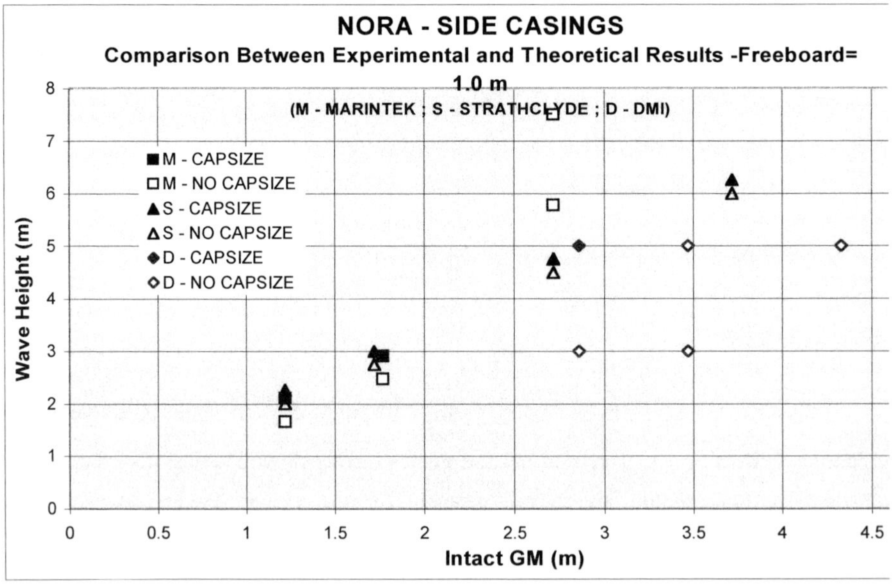

Figure 2: NORA Boundary Survivability Curve

below the bulkhead deck), loading condition and operating environment as well as location and characteristics of damage opening. The various aspects involved are being tackled on three fronts, involving two ship models tested in random wave conditions. Details of the vessels and test conditions are given in Tables 1 and 2 below.

TABLE 1
PRINCIPAL DESIGN PARTICULARS OF ST NICHOLAS (UK RO-RO RESEARCH VESSEL) & NORA (JOINT R&D PROJECT GENERIC VESSEL)

	St Nicholas	**NORA**
Model Scale	42.05	34.67
Length, L_{BP}	131.00m	130.00 m
Beam, **B**	26.00m	25.50m
Depth, **D**	7.80m	8.35m
Draught, **d**	6.12m	5.75m
Displacement, Δ	12,400.0 tonnes	12,000.0 tonnes
Block Coefficient, C_B	0.582	0.612
KM_{intact}	13.84m	14.26m
KM_{damage} (DL=13.5m; $F_{flooded}$ =1.02m)	13.88m	
KM_{damage} (DL=22.8m; $F_{flooded}$ =0.55m)	14.34m	
KM_{damage} (DL=28.7m; $F_{flooded}$ =0.21m)	14.74m	
KM_{damage} (DL=24.05m; $F_{flooded}$ =1.50m)		13.46m
KM_{damage} (DL=32.43m; $F_{flooded}$ =1.00m)		14.48m
KM_{damage} (DL=40.26m; $F_{flooded}$ =0.50m)		14.59m

TABLE 2
SEA STATES (JONSWAP SPECTRUM WITH γ=3.0)

Significant Wave Height, H_s (metres)	Peak Period, T_p (seconds)	Zero-Crossing Period, T_0 (seconds)
1.0	4.0	3.1
1.5	4.90	3.8
2.0	5.66	4.4
2.5	6.33	4.9
3.0	6.93	5.4
4.0	8.00	6.2
5.0	8.95	6.9

On the basis of the above, the following series of tests have been undertaken:

DMI Model Experiments (NORA model)

The DMI experiments were designed to investigate the water ingress phenomena, comprehensively. To this end, the water level inside the deck as well as the water elevation outside the damage opening are measured using an array of wave gauges, together with roll and pitch motions including static heeling and trim. The analysis of these results will also be used to calibrate the developed numerical water ingress model in a range of sea states, conditions and compartmentation as indicated below:

- Open Ro-Ro deck
- Centre Casing
- Side Casings
- Size of damage opening (25%, 100% and 200% SOLAS)
- Location of damage (amidships and forward)
- $Freeboard_{flooded}$ (0.5m, 1.0m and 1.5m)
- Loading conditions (KG ranging from 9.5m to 12.0m)
- Transverse Bulkheads (Partial and full height)
- Sea States (Hs=1.3m, 3.0m and 5.0m)
- Ro-Ro deck damage only

In addition to information pertaining to water ingress, valuable information will also be obtained concerning the survivability of NORA in these conditions. Video records of all the above were also obtained.

MARINTEK Model Experiments (St Nicholas and NORA models)

The MARINTEK tests were designed to test the capsizal resistance of both models in a range of loading conditions, sea states and compartmentation (Table 3), recording all relevant information pertaining to model motions and attitude, as well as wave characteristics, including again video recordings.

TABLE 3
CAPSIZE MODEL EXPERIMENTS AT MARINTEK

St Nicholas	NORA
Centre Casing	Side Casings
$Freeboard_{flooded}$ (0.21m, 0.55m and 1.02m)	Transverse Bulkheads (Partial and full height)
Loading conditions (10.0m ≤KG≥ 12.0m)	$Freeboard_{flooded}$ (0.5m, 1.0m and 1.5m)
Sea States (Hs=1.0m to 5.0m)	Loading conditions (11.5m ≤KG≥ 13.0m)
Forward Speed (5 and 10 knots in full scale)	Sea States (Hs=1.0m to 8.0m)
Wave Heading (30 and 60 degrees)	

An example comparison between experimental and theoretical boundary survivability curves is shown in Figures 2.

Strathclyde Marine Technology Centre Experiments (St Nicholas model)

Following suggestions by the management of the Joint Nordic Project, the St Nicholas model was brought to the University of Strathclyde for undertaking additional tests pertaining either to the recent recommendations by the IMO panel of experts or related to the project itself, particularly so tests relevant to the validation of the mathematical model. In relation to the above, the following modifications to the model were made:

- Decoupling of the car deck from the main hull and attaching on load-cells for continuous measurement of the water on deck. This is believed to be a more effective method for assessing inflow/outflow than the DMI method of using a number of capacitance probes inside the deck.
- Fitting arrangements allowing for the positioning of movable transverse (partial or full height) and longitudinal bulkheads/casings.

The suggested range of tests includes the following:
- Measurement of water accumulation in a range of sea states, loading conditions and freeboards
- Central Casing
- Partial height and full height bulkheads
- Varying number of transverse bulkheads
- Combinations of the above
- Testing damage survivability in a zero freeboard case (upright or inclined condition)
- Investigation of uncertainties that might arise through the correlation studies between experimental and numerical simulation results.

Representative results demonstrating the effectiveness of measuring water accumulation on the Ro-Ro deck as well as comparisons between theoretical and experimental results are shown in Figures 3 and 4.

APPLICATIONS OF THE MATHEMATICAL MODEL

Sensitivity study

In order to identify the most influential parameters for the stability and survivability of a damaged ship, a series of parametric studies have been carried out using the time simulation program. For this purpose a matrix that combines different damaged freeboards, vehicle deck subdivisions, loading conditions and sea states has been tested as shown in Table 4.

TABLE 4
SENSITIVITY STUDY TEST MATRIX FOR ST NICHOLAS

	Open deck			Central casing			Side casings			Open deck: Transverse BHDS (45m)		
KG (m)	**F_1**	**F_2**	**F_3**	**F_1**	**F_2**	**F_3**	**F_1**	**F_2**	**F_3**	**F_1**	**F_2**	**F_3**
9.0	X	X	X	X	X	X	X	X	X	X	X	X
10.0	X	X	X	X	X	X	X	X	X	X	X	X
11.0	X	X	X	X	X	X	X	X	X	X	X	X
12.0	X	X	X	X	X	X	X	X	X	X	X	X
13.0	X	X	X	X	X	X	X	X	X	X	X	X

As can be seen there are 60 conditions and for each condition a minimum of four different sea states has been considered. The sea states were tested to a resolution of 0.25m (i.e. sea states

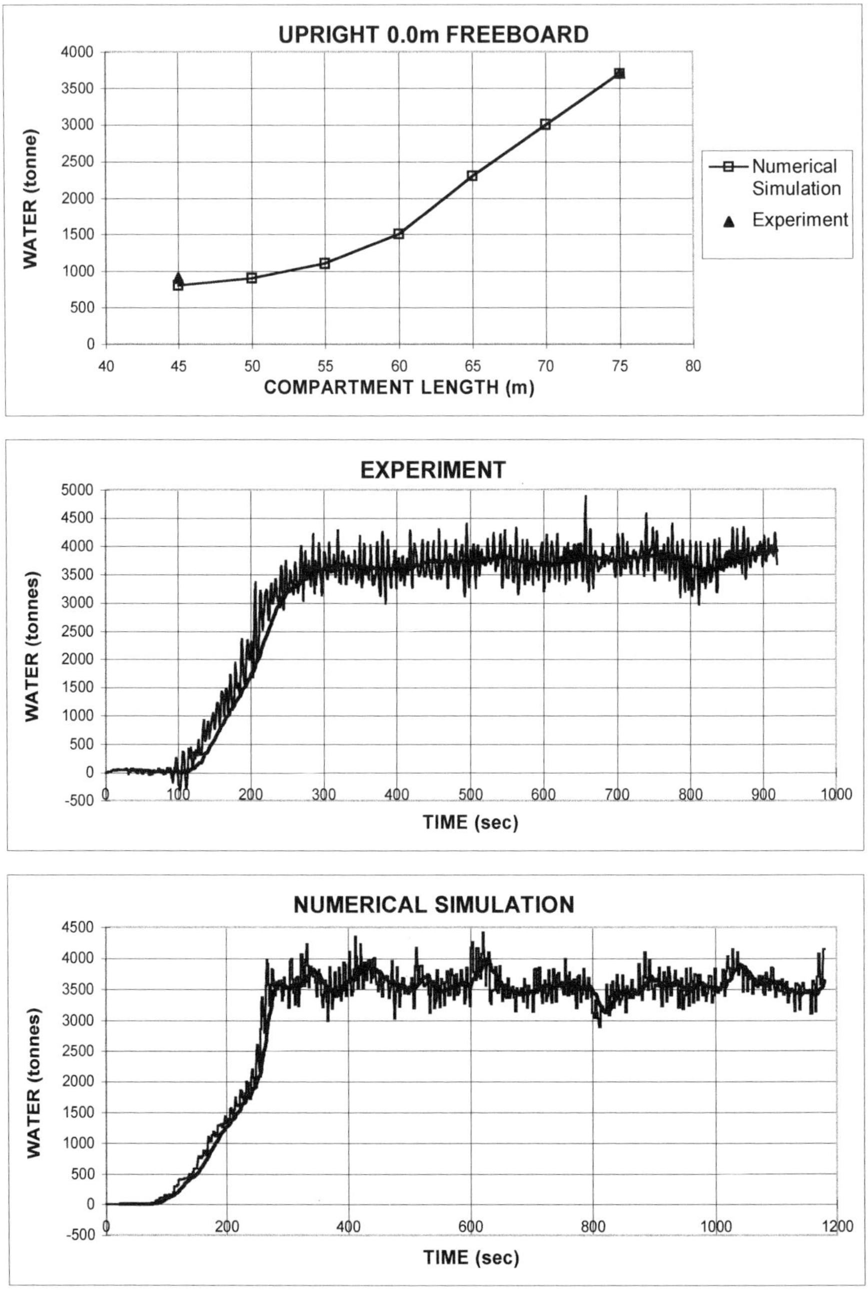

Figure 3: Water Ingress - Theory and Experiment
(St Nicholas, Damage Length = 75m)

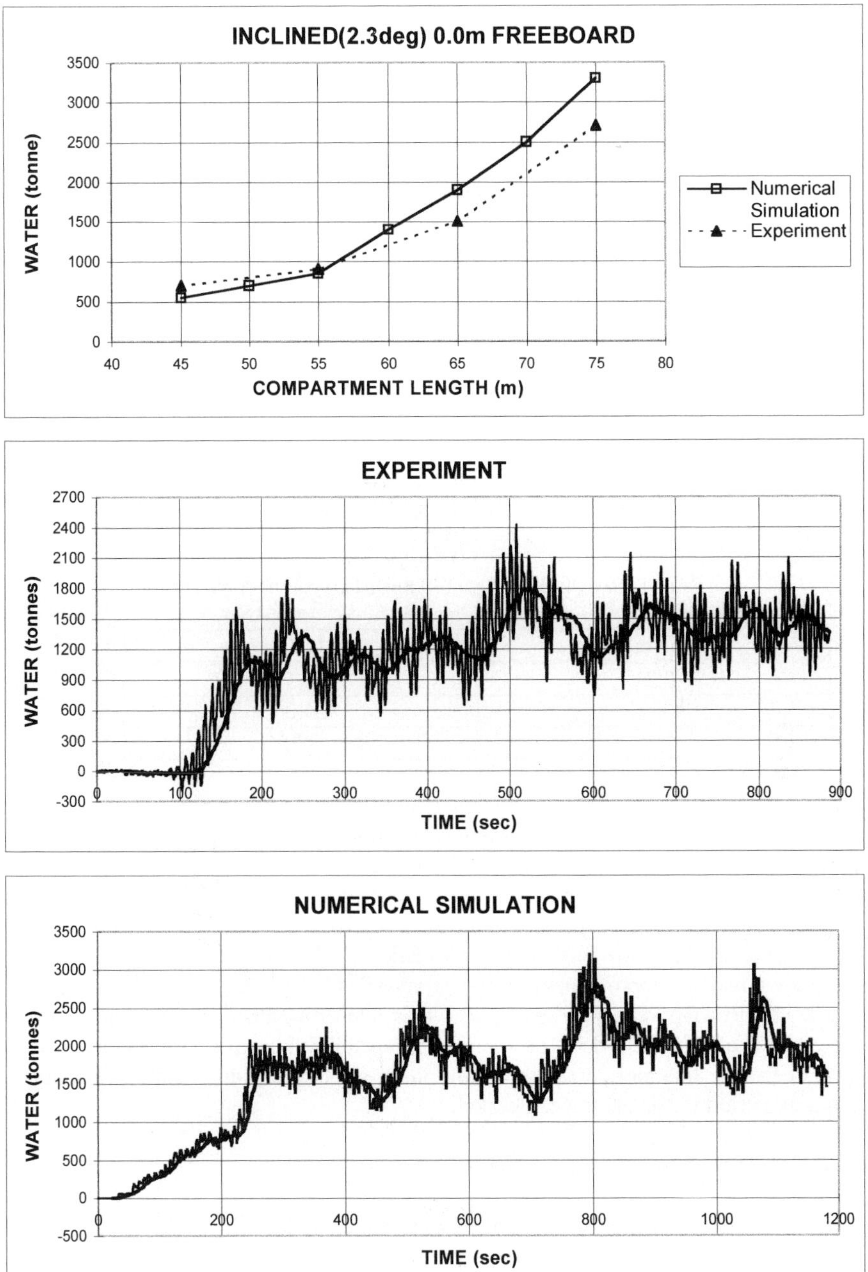

Figure 4: Water Ingress - Theory and Experiment
(St Nicholas, Damage Length = 65m)

were increased progressively by 0.25m intervals). Where necessary, several runs were carried out for the same conditions to ensure statistical consistency of the results. The damage conditions used and the corresponding details are as indicated in Table 1.

Results and Discussion

The results of the study are presented in the form of limiting boundary curves in the form of H_s v GM_f and are summarised in Figure 5. The damaged freeboards (F) and the corresponding metacentric heights (GM_f) refer to the final equilibrium following flooding of the compartment below the bulkhead deck.

Effect of Damaged Freeboard (F) on Survivability

The results clearly indicate that freeboard is one of the key parameters influencing stability and survivability of damaged ships. In this respect, it is interesting to note that the relationship between limiting sea states (Hs) and damaged freeboards (F) is not linear. Table 5, for example, shows the results corresponding to the open deck case and GM_f of 3.0m.

TABLE 5
RELATIONSHIP BETWEEN FREEBOARD AND SEA STATES

F (m)	Hs/F
0.21	5.5
0.55	4.0
1.02	3.25

From this it is clear that the use of Hs/F ratios in boundary survivability curves needs careful interpretation if one is not to be led to wrong conclusions. As shown in the table, a vessel with lower freeboard can survive at higher Hs/F but the actual sea state is in fact significantly smaller. It is also clear from Figure 5 that the open Ro-Ro deck and the central casing designs would need a damaged freeboard close to 1.5m to survive a sea state of 3-4 metres Hs, which is likely to be required by the forthcoming regulations. However, the results relating to side casings show a marked improvement on the survivability of the vessel, which appears now to be capable of surviving very high sea states. It is interesting also to note that, at very high GM_f, the effect of water on deck on damage survivability becomes less dominant as is the effect of freeboard, this particular ship rolling quite significantly due to the proximity of the spectral peak period to the natural roll period.

Effect of Vehicle Deck Subdivision on Survivability

The large open vehicle deck poses a great danger to the survivability of Ro-Ro type vessels if serious flooding of the vehicle deck takes place. Notwithstanding this, the majority of the existing designs have open deck or central casing as implementation of side tanks has been limited due to economical reasons. Thus, the clear and substantial benefit to be gained by a ship with side casings, as shown in Figure 6, has not been taken advantage of. In the example considered, the limiting boundary curves referring to the open deck and central casing are almost identical with the open deck showing a slight improvement which derives mainly from

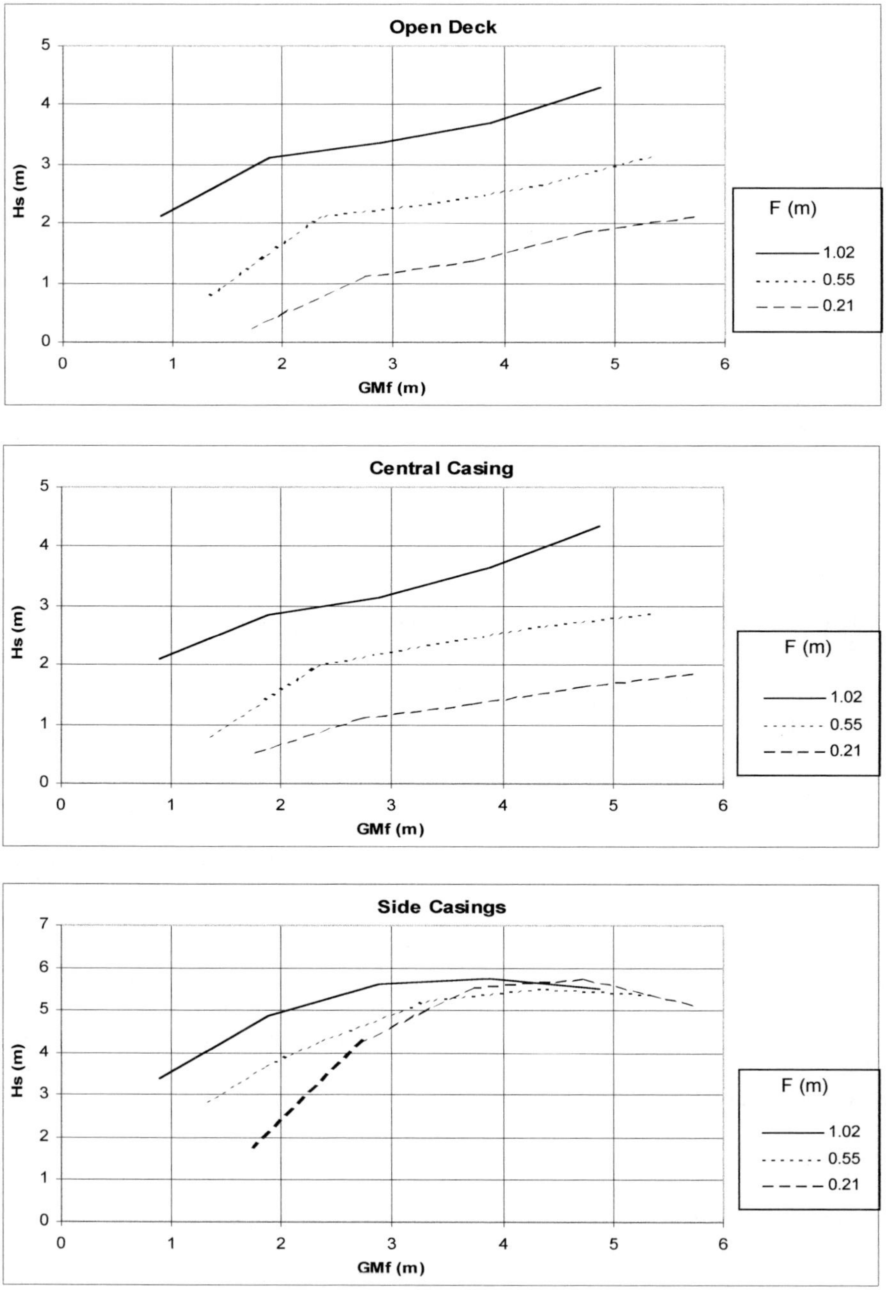

Figure 5: Effect of Damaged Freeboard on Survivability (St Nicholas)

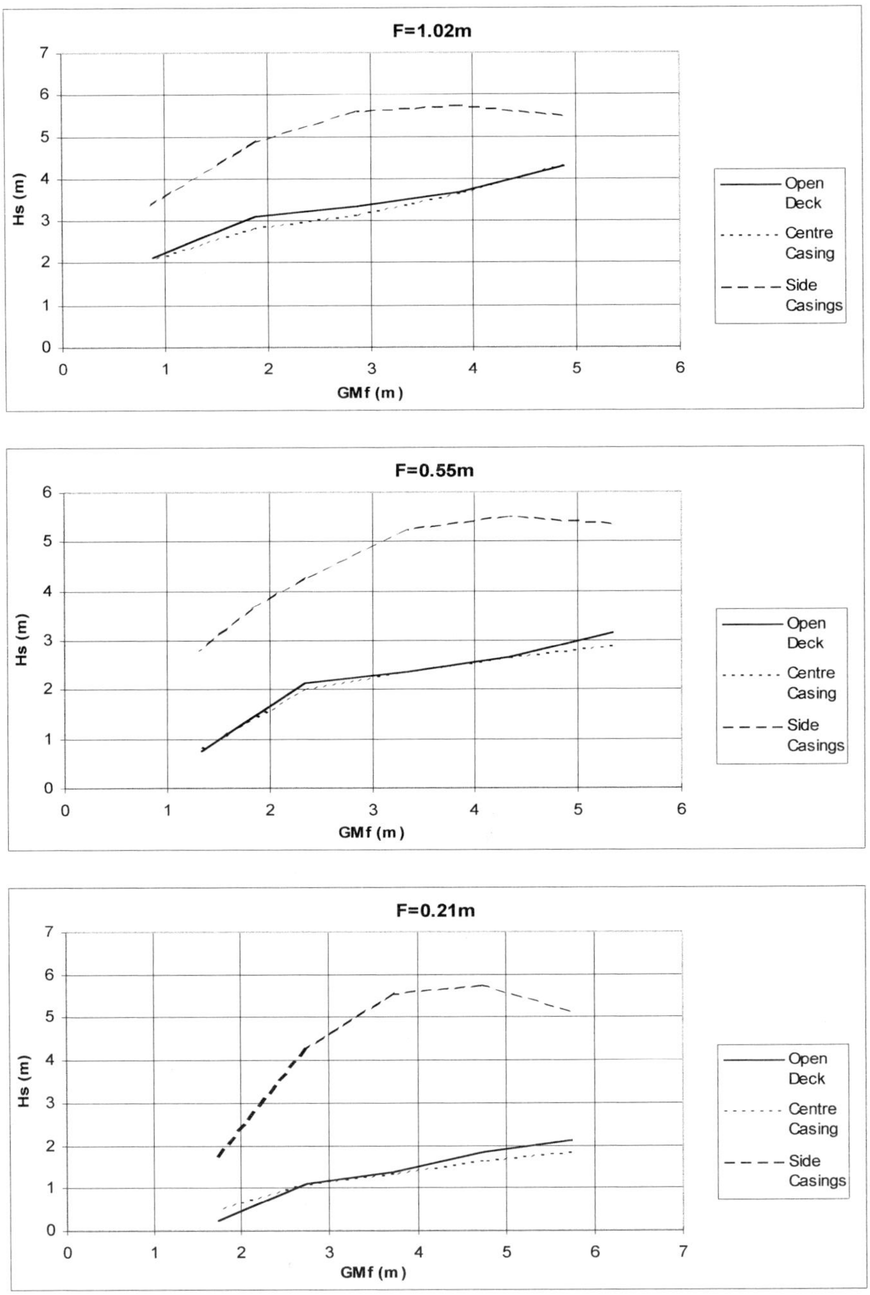

Figure 6: Effect of Vehicle Deck Subdivision on Survivability (St Nicholas)

the fact that, under certain conditions, the open deck Ro-Ro vessel may incline to the lee side, thus enhancing her chance of survival. The beneficial effect of side casings on ship survivability derives mainly from the following:

- Due to their location away from the centre of rotation, side casings increase substantially the roll restoring ability of the damaged vessel in addition to improving significantly the reserve buoyancy.
- For the same reason, side casings decrease the heeling moment resulting from flooding of the vehicle deck, as the body of floodwater moves closer to the centreline (roll centre). It is obvious that this beneficial effect increases as flooding progresses and the ship tends to return to the upright condition.

This effect, however can be outweighed by low damaged freeboards and small GM_f as shown in Figure 6.

Effect of Transient Flooding on Survivability

Depending on the damaged freeboard, a vessel with small GM_f may capsize due to transient heeling resulting from the flooding of the damaged compartment below the bulkhead deck. However, this depends critically on the direction of the initial heel. If this is to the lee side, asymmetric flooding of the compartment below the bulkhead deck will cause the vessel to incline to large angles, thus increasing her effective freeboard, water ingress on the vehicle deck is prevented and she survives. The ship may remain inclined or, due to the increasing amount of water in the compartment below the bulkhead deck she may return to the upright position. On the other hand, if the initial heel is to weather side, the asymmetric flooding of the compartment below the bulkhead deck will have the exact opposite effect on the survivability of the vessel. The above effects are demonstrated in Figures 7 and 8, which refer to the same vessel condition and sea state but to different wave realisations. The effect of transient flooding on survivability diminishes with increasing damaged freeboard or GM_f.

Sensitivity of Survivability on GM_f and other Residual Stability Parameters

If different subdivisions of the vehicle deck are contemplated, then clearly GM_f cannot be considered as a representative parameter to characterise the damage survivability of passenger/Ro-Ro vessels. This is demonstrated in Figure 6. This is not, however, the first time that GM has been dismissed as a parameter in assessing ship stability but GM in itself is the key opening the door to unarguably the most successful characteristic property to date of a vessel's ability to resist capsize in any condition and environment, namely, the restoring curve. Even if one does not support this view, any results that this route is likely to yield, offer two distinct advantages: *simplicity* and *applicability*. As explained earlier, the objective is to express the survival factor "s" as a function of residual stability characteristics, judiciously chosen (e.g. systematic parametric investigations, regression analyses, experiential judgement, etc.) to enable such a factor to be generalised for application to all vessel types and compartmentation. Parameters to be considered in such an investigation include:

- GZ_{max} at a certain angle
- Positive GZ range
- Area under the GZ curve
- Area under the GZ curve up to a certain angle

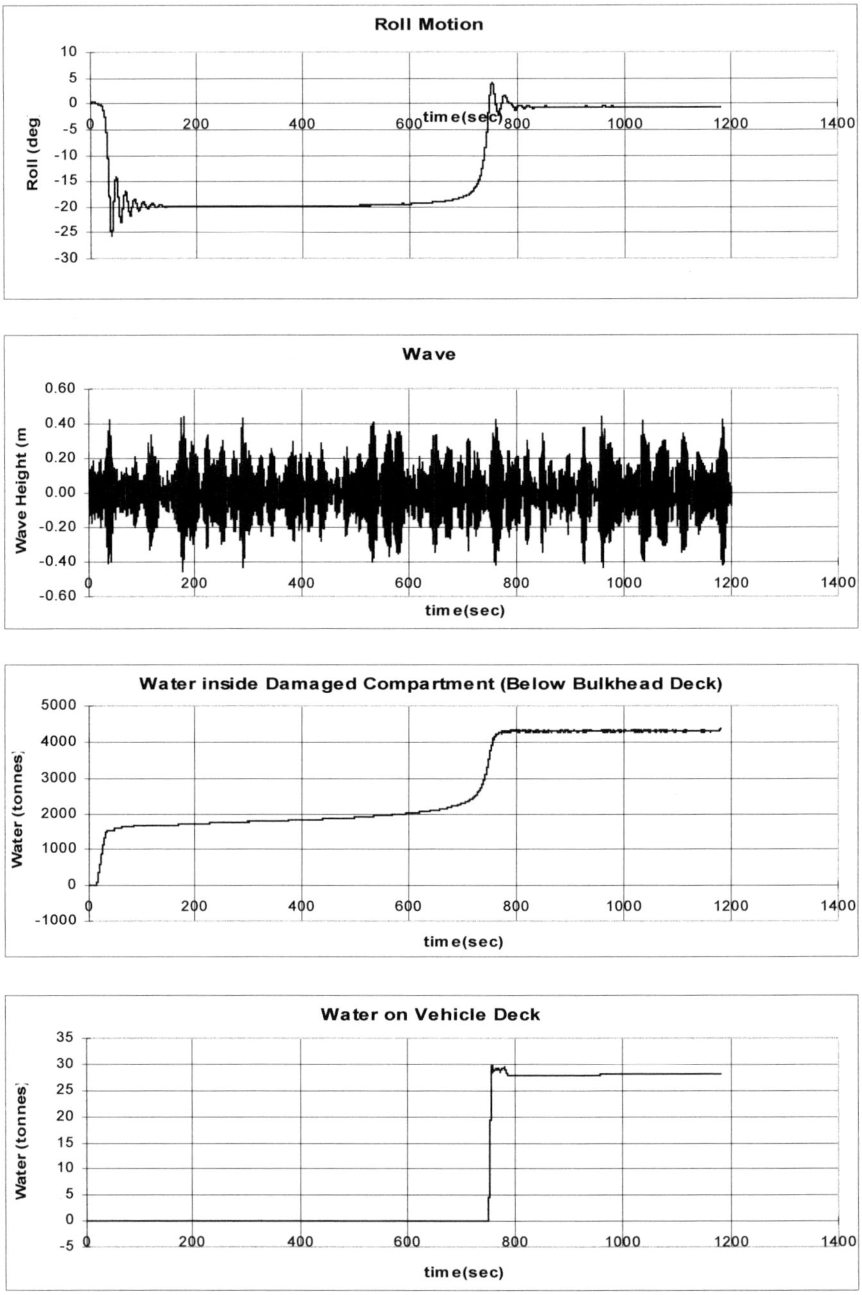

Figure 7: Beneficial Effect of Transient Flooding on Survivability (St. Nicholas)

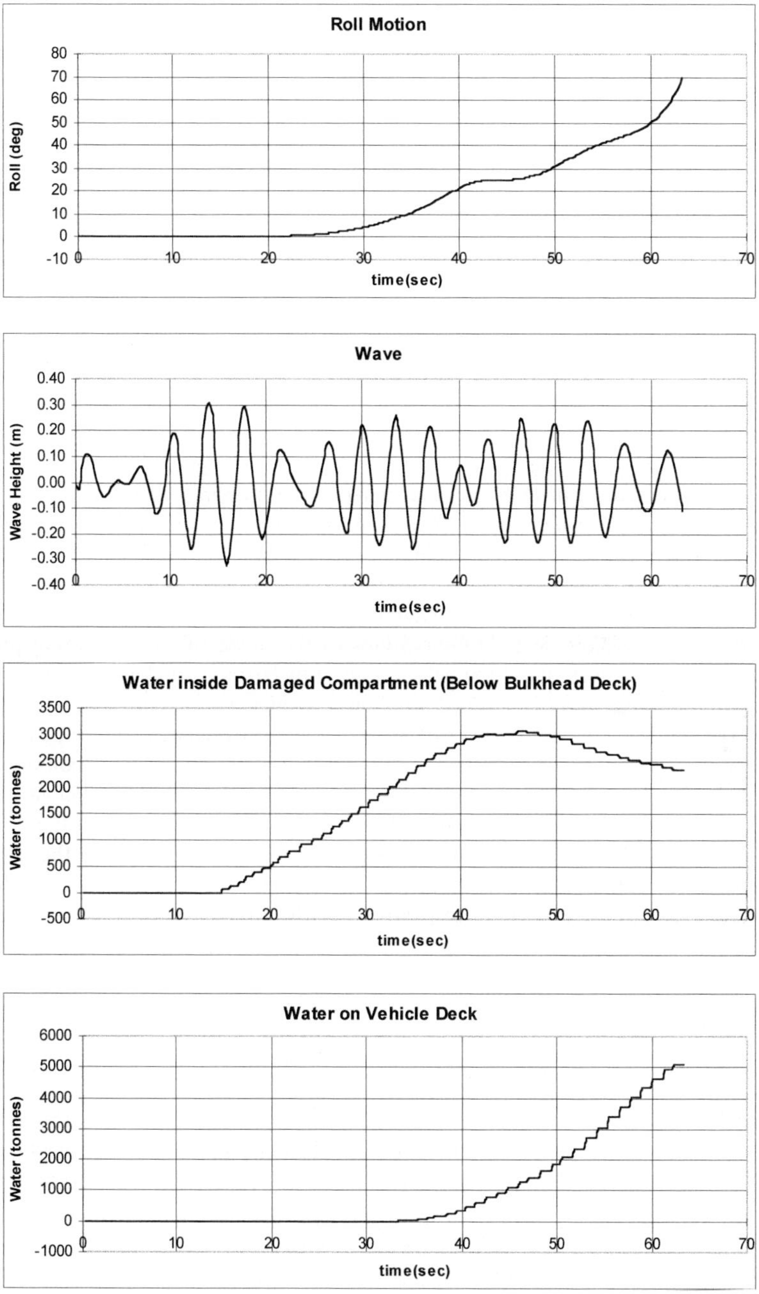

Figure 8: Adverse Effect of transient Flooding on Survivability (St Nicholas)

A first exploration in this direction met with a problem in need of careful thinking. Stability calculations for vessels damaged both above and below the bulkhead deck would require, according to IMO, that the water level in each damaged compartment open to the sea should be at the same level as the sea i.e., final equilibrium be reached. However, the GZ curves derived on the basis of this approach, simply fail to offer any useful information. The principal reason lies on the wrong assumption that water is free-flooding the deck in these calculations. Taking heed from this and from the fact that water on deck is a dominant parameter affecting damage survivability, as earlier experience amply demonstrated, it was decided to attempt to quantify the critical amount of water on deck as a matter of top priority. It would appear that this effort is likely to bear fruits and early results appear very promising. The first important step was to achieve a good understanding of what is meant by *critical amount of water on deck* and to develop a practical method of quantifying this.

Critical Amount of Water on Deck - "The Point of No-Return"

The effect of random waves on the rolling motion of the damaged ship appears to be rather small and for capsize to occur in a "pure" dynamic mode should be regarded as the exception rather than the rule. The main effect of the waves, therefore, is that they exacerbate flooding. In this respect, the effect of heave motion in reducing the damaged freeboard is as important as the roll motion. Model experiments and numerical simulations have clearly demonstrated that the dominant factor determining the behaviour of the vessel is the amount of floodwater accumulating on the vehicle deck. Observations of the mode of capsize during progressive flooding of the vehicle deck show the vessel motion to become subdued with the heel angle slowly increasing until a point is reached when heeling increases exponentially and the vessel capsizes very rapidly. This is *the point of no-return*. Put differently, the floodwater on the vehicle deck increases slowly, depending on the vessel and environmental conditions, until the amount accumulated reaches a level that cannot be supported by the vessel/environment and the vessel capsizes very rapidly as a result. The amount of floodwater when the point of no-return is reached is *the critical amount of water on deck*. In relation to this, two points deserve emphasis: This amount is substantially less than the amount of water just before the vessel actually capsizes but is considerably more than the amount required to statically capsize the ship. In this respect, the energy input on account of the waves help the vessel sustain a larger amount of water than what her static restoring characteristics appear to dictate. Because of the nature of the capsize mode when serious flooding of the vehicle deck takes place, it is not difficult to estimate the critical amount of water on deck at the point of no-return and this is demonstrated in Figures 9 and 10 using the generic vessel of the Joint R&D Project, NORA.

Careful examination of the numerical simulation results, such as those shown in Figure 10, has revealed that the point of no-return occurs at a heeling angle very close to ϕ_{max}, where a maximum of the GZ curve occurs. This observation led to the development of a Static Equivalent Method (SEM), which allows for the calculation of the critical amount of water on deck from static stability calculations. In this respect, a damage scenario is assumed in which the vessel is damaged only below the bulkhead deck with water added on the deck progressively until the ship assumes an angle of loll (angle of equilibrium) equalling ϕ_{max}. The amount of accumulated water on the deck when this angle is reached corresponds very closely to the critical amount referred to above. The simple way of calculating this most important

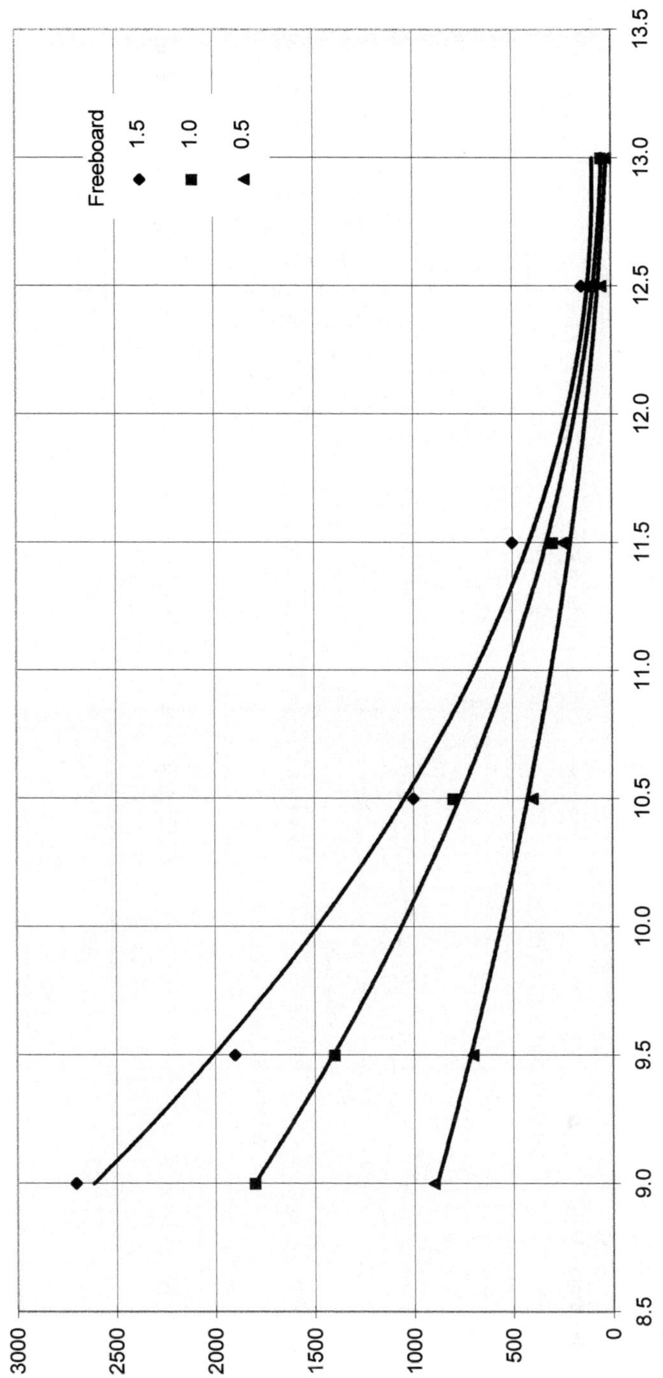

Figure 9: Critical Amount of Water on Deck at "The Point of No-Return" NORA, Open Deck, Deck Area = 3,000m^2)

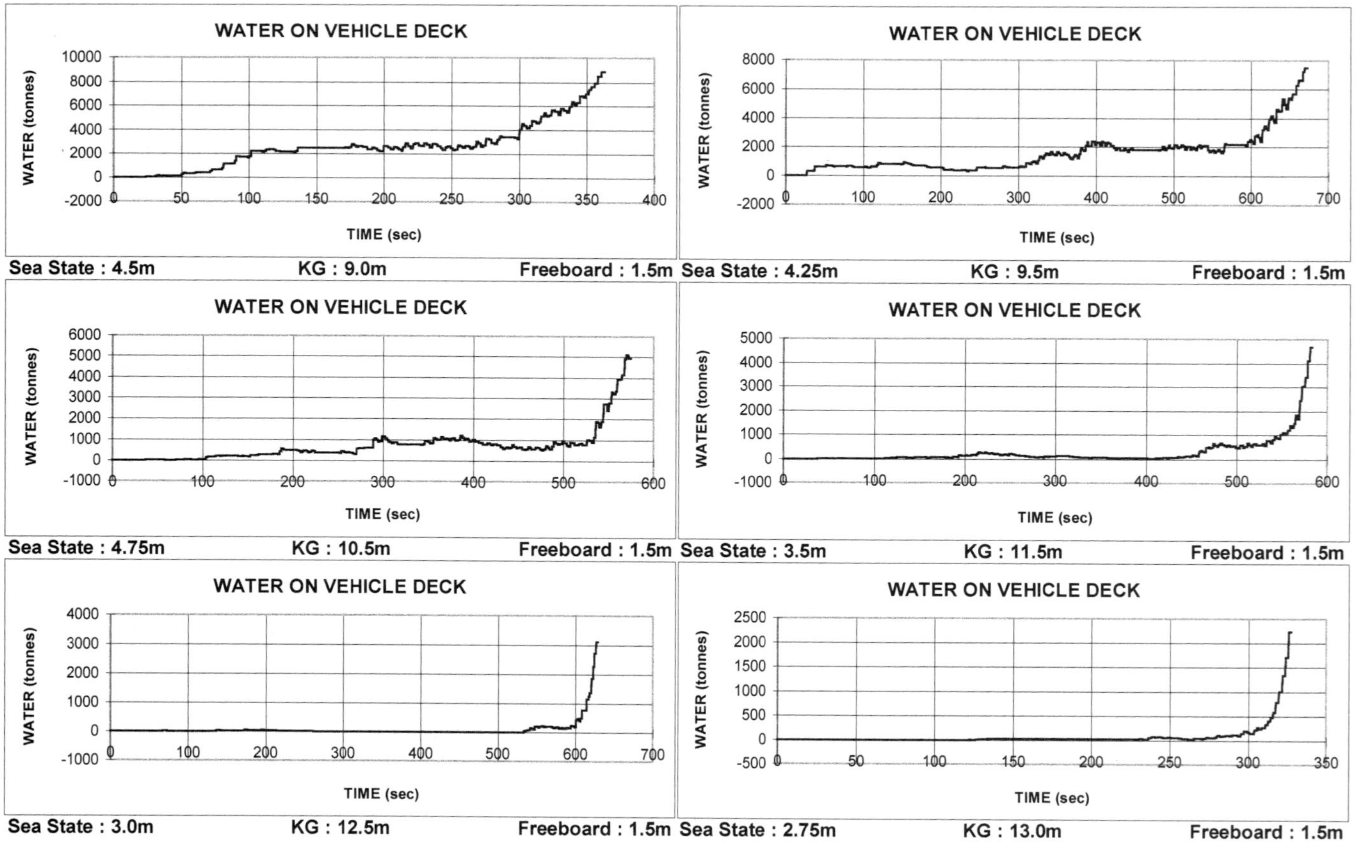

Figure 10: Evaluation of Critical Amount of Water on Deck
(NORA, Open Deck, Deck Area = 3,000m^2)

factor allows, in turn, for a plausible way of developing a rational method for assessing damage survivability. Efforts in this direction are currently under way.

CONCLUDING REMARKS

As the Joint R&D Project is drawing to its conclusion, discussions on the provision of meaningful criteria for ensuring the damage survivability of passenger/Ro-Ro vessels gather momentum, results from model experiments and numerical "tools" are being made available and confidence is slowly being built up that what was perceived to be an intractable problem, can in fact be tackled with sufficient engineering accuracy to yield solutions which by the nature of the problem are likely to have a profound effect on the way these vessels evolve. For this reason alone, the profession must take a step back and attempt to see the wider implications of the problem at hand. The commercial success of Ro-Ro's lies principally on the provision of large unrestricted enclosed spaces for the stowage of vehicles and cargo. When addressing the safety of Ro-Ro vessels, therefore, one has of necessity to focus on subdivision. In so doing, however, one is pointing a finger at the immediate problem rather than towards the required solution. ***One should not loose sight of this fact!***

ACKNOWLEDGEMENTS

The financial support of the UK Department of Transport and of the Joint R&D Project is gratefully acknowledged. I should also like to record my appreciation to all my colleagues in the Project and to the Stability Research Group members for their help, contribution and support in more ways than one.

References

British Maritime Technology Ltd. (1990). Research Into Enhancing the Stability and Survivability Standards of Ro-Ro Passenger Ferries: Overview Study, BMT Ltd., Report to the Department of Transport, March.

Dand I.W. (1990). Experiments with a Floodable Model of a Ro-Ro Passenger Ferry, BMT Project Report, for the Department of Transport, Marine Directorate, BMT Fluid Mechanics Ltd., February.

Danish Maritime Institute (1990). Ro-Ro Passenger Ferry Safety Studies, Model Tests for F10, Final Report of Phase I for the Department of Transport, DMI 88116, February.

Vassalos D. and Turan O. (1992). Development of Survival Criteria for Ro-Ro Passenger Ships - A Theoretical Approach, Final Report on the Ro-Ro Damage Stability Programme, Phase II, Marine Technology Centre, University of Strathclyde, December.

Contemporary Ideas on Ship Stability
D. Vassalos, M. Hamamoto, A. Papanikolaou and D. Molyneux (Editors)

WATER-ON-DECK ACCUMULATION STUDIES BY THE SNAME AD HOC RO-RO SAFETY PANEL

Bruce L. Hutchison

The Glosten Associates, Inc., Seattle, WA 98104-2224, USA

ABSTRACT

A mathematical theory is presented for the accumulation of water on the deck of a damaged RoRo passenger vessel. Excellent agreement is demonstrated between results obtained from extensive time domain simulations and corresponding results obtained from integrals in the probability domain. Comparison is also made to results obtained during free floating model tests in waves at National Research Council Canada, Institute for Marine Dynamics (IMD). The mathematical theory presented leads to a simple curvilinear relationship between the accumulated depth of water on deck, freeboard and significant wave height.

KEYWORDS

Water-on-deck, freeboard, significant wave height, *Estonia*.

THE SNAME AD HOC RO-RO SAFETY PANEL

The Society of Naval Architects and Marine Engineers (SNAME) Ad Hoc RO-RO Safety Panel was created in 1994 in response to the capsize and sinking of the *Estonia* and its final report was issued prior to dissolving in 1996. The Ad Hoc Panel was composed of owners and operators, regulators, designers, researchers and academics from both Canada and the United States. It provided input directly to the IMO Panel of Experts (also established following the *Estonia* tragedy) and provided advice to U.S. and Canadian delegations to IMO.

The SNAME Ad Hoc Panel agreed that it was advisable to develop requirements that address the hazard posed by water on the decks of vessels such as fully enclosed RoRo passenger ferries. The Panel believes that any proposal to address the water-on-deck hazard should be rationally based on:

- The operating environment
- The freeboard at the point of assumed damage
- The means to remove water from the vehicle deck

The SNAME Ad Hoc RO-RO Safety Panel addressed the problem of water accumulation on deck using time domain simulation and integral methods based on the Gaussian distribution of wave elevations. The research that resulted is the primary focus of this synoptic paper. In addition, but outside the scope of this paper, the Panel applied these same principles and methods to model the benefit of outflow through flow biased freeing ports. Readers interested in results of the freeing port investigation should consult SNAME Ad Hoc RO-RO Safety Panel Annex B (1995B) and Hutchison et al. (1995).

STATIONARY SHIP MODEL

The SNAME Ad Hoc Panel investigated a highly simplified model for the accumulation of water on the deck of a damaged RoRo vessel. A stationary ship was assumed, with a flat deck and side damage represented by a rectangular opening of unlimited vertical extent beginning at the deck. The assumed stationarity corresponds to no vessel motion in response to waves (no sinkage, trim or heel), resulting in a fixed elevation, f, of the deck at the point of assumed side damage. The treatment of essential fluid flow processes in the stationary ship model is two-dimensional.

A partial rationale for the stationary ship approach is that, once new rules are implemented, the burden of water on deck is supposed to be limited to a quantity that the vessel can survive without capsize. This argument helps to explain why it may be possible to ignore sinkage, trim and heel. It has not been established whether relative motion effects, hydrodynamic interaction between the hull and the waves, or internal dynamics of the accumulated water pool lead to excessive departures from the expectations based on the stationary ship concept. However, encouraging agreement has been found between predictions based on the stationary ship concept and model test data obtained using a free floating model at IMD.

The Two Phases of the SNAME Panel's Research

In order to appreciate the following presentation of results obtained by the SNAME Ad Hoc Panel, it is necessary to explain that its analytical work has proceeded through two phases. The first phase encompasses all work accomplished through 28 February 1995, culminating in the submission to the IMO Panel of Experts of SNAME Annexes A and B (1995A & 1995B). The second comprises that work accomplished since 28 February 1995.

The research of the first phase made use of a velocity superposition principle that permitted separation of inflow and outflow processes. This approximation resulted in significant simplifications of the analysis. The research of the second phase dispensed with these simplifications and determined an exact solution of the stated problem in terms of Gaussian integrals. Comparison of phase one and phase two solutions revealed that the approximation obtained from velocity superposition is quite good. Salient results from the first phase are presented here, albeit without mathematical development, which is covered in SNAME Annex A (1995A) and Hutchison (1995), Hutchison et al. (1995), and Hutchison et al. (1996). The exact solution obtained in phase two is presented in this paper.

Independent Parameters

Given the assumption of a stationary ship, the independent problem parameters are reduced to:

- A area of the deck subject to flooding
- f freeboard at the point of assumed damage
- W width of the damage opening measured normal to the direction of wave travel
- H_S significant wave height

Objectives

The SNAME Ad Hoc Panel's primary objective was to develop simple mathematical relationships for the following, as functions of the independent parameters:

- $\overline{D}$ asymptotic average water depth on deck
- $\overline{V}$ asymptotic average water volume on deck, (i.e., $\overline{V} = A\,\overline{D}$)

A secondary but important objective of the SNAME Panel's research was to determine out-flow credit functions for deck drains, freeing ports and active deck pumping systems. The development of these out-flow credits is too extensive for inclusion here. The interested reader could consult SNAME Annex B (1995B), Hutchison (1995), Hutchison et al. (1995) or Hutchison et al. (1996)

VELOCITY SUPERPOSITION RESULTS

Results were obtained both from time domain simulations and from probability domain integrals during the first phase of the SNAME analytical research. The probability domain integrals were based on the Gaussian distribution of wave elevation in an irregular sea. SNAME (1995B), Hutchison (1995), Hutchison et al. (1995) and Hutchison et al. (1996) provide greater detail regarding the simulation procedures and the development of the Gaussian integrals. A total of 252 time domain simulations were performed.

Figure 1 shows an example of a time domain simulation record. After approximately 125 seconds the water depth may be seen to attain an average value about which the time domain depth record thereafter oscillates.

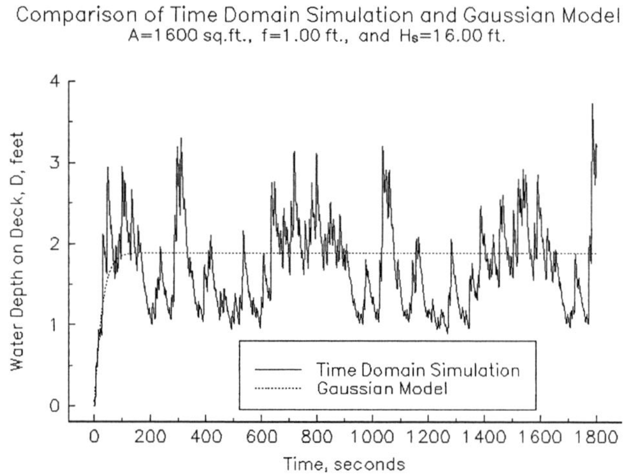

Figure 1: Example of Simulated Time Domain Record of Water Depth on Deck

Expected Build-Up Time

As detailed in SNAME Annex A (1995A) and Hutchison et al. (1995), it is possible to derive a closed form expression for the expected value of the water depth as a function of time, in terms of the flooded deck area, A, the width of the damage opening, W, and the average in-flow rate, $\overline{Q}_{IN}$. Since $\overline{Q}_{IN}$ is a function of residual freeboard, f, significant wave height, H_S, and damage width, W, the expected water depth as a function of time depends on A, W, f and H_S.

The dotted line in Figures 1 and 2 shows the expected build-up process for water on deck (labeled Gaussian model). As illustrated by Figures 1 and 2, the agreement is excellent between the expected trend and the mean trend of the simulated time domain data.

Figures 3 and 4 compare the expected (i.e., average) time required to build up to 99% of the asymptotic average water depth with the build-up time required for the first passage above the asymptotic average water depth, as determined from the sample time domain record. Figure 3 is for a flooded deck area of 1600 square feet, and Figure 4 is for a flooded deck area of 400 square feet. The damage width is constant for all cases at W = 10 feet.

The flooded deck area in Figure 3 is four times that in Figure 4 and consequently the build-up time is longer for the cases shown in Figure 3 than for the cases in Figure 4.

One can see that the build-up time is strongly dependent on the value of the asymptotic average water depth on deck. Excepting those cases where the asymptotic average water depth on deck is small, the build-up time is quite short. The importance of this finding is that all the most hazardous cases are achieved with great rapidity; it is only the (most likely) inconsequential cases, where small average water depths are achieved, that build up slowly.

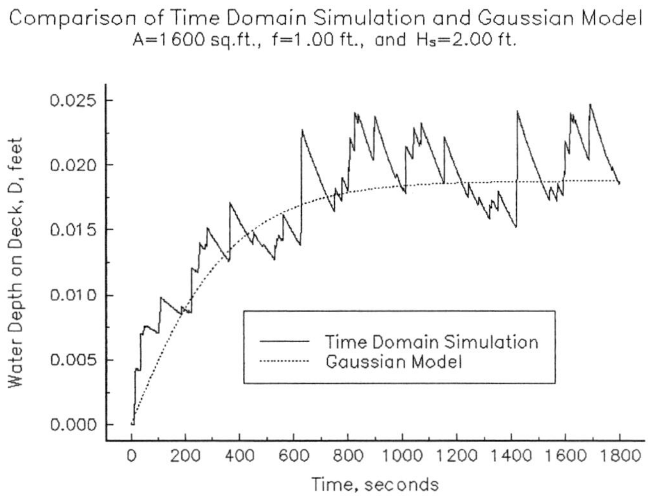

Figure 2: Example of Simulated Time Domain Record of Water Depth on Deck, Showing Comparison with the Expected Build-Up Model

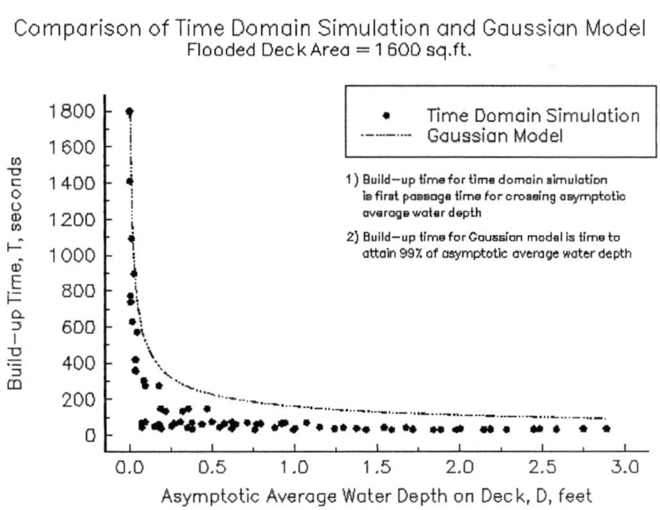

Figure 3: Comparison of Build-Up Times between Time Domain Simulation and Expected Build-Up Model (Labeled Gaussian Model)

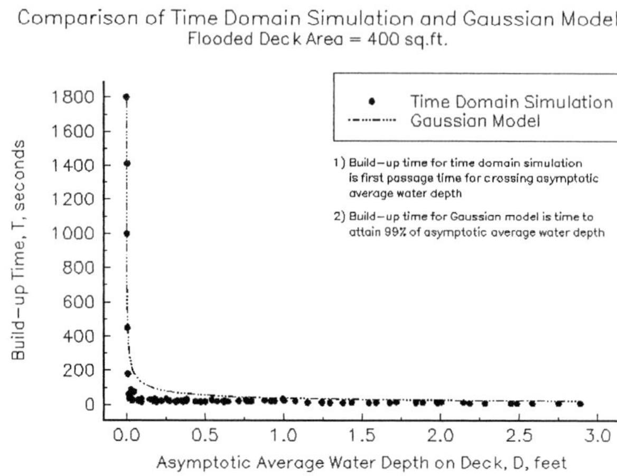

Figure 4: Comparison of Build-Up Times between Time Domain Simulation and Expected Build-Up Model (Labeled Gaussian Model)

Probability Density and Cumulative Probability Distributions

Certain results regarding the stochastic water-on-deck process can be obtained only in the time domain.

Among results that can be obtained from the time domain are sample values for the probability density and cumulative probability distributions for the water depth on deck. Examples of these are shown in Figures 5 and 6.

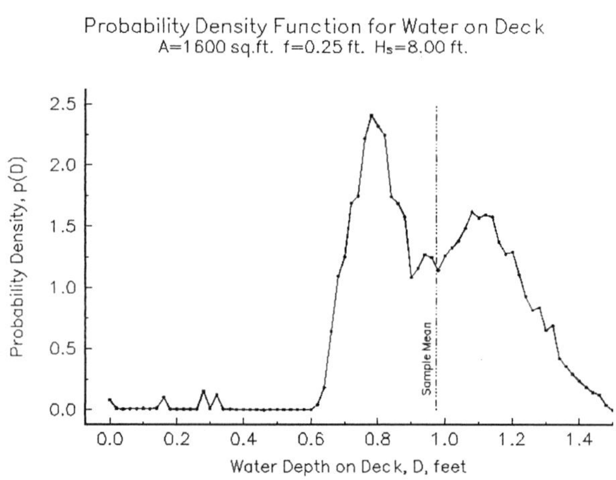

Figure 5: Example of a Sample Probability Density Function for Water Depth on Deck

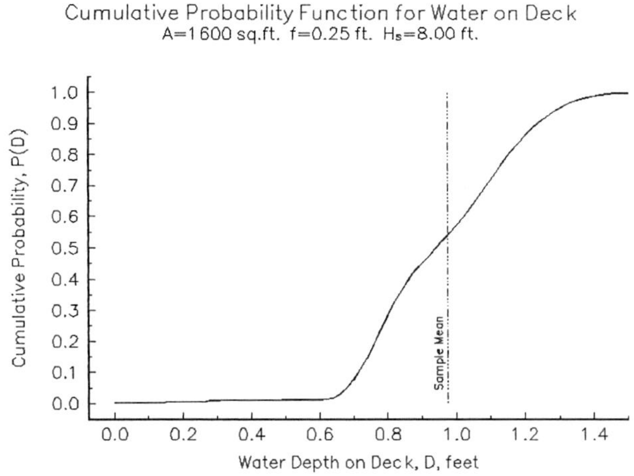

Figure 6: Example of a Sample Cumulative Probability Distribution of Water Depth on Deck

One of the interesting features is the bimodal character of the probability density distribution shown in Figure 5. It was observed in many, though certainly not all, of the cases simulated.

Persistence

Another interesting probability result that may be obtained from the time domain simulations is sample values for the persistence of the water depth process. The persistence measures the average duration of the stochastic water depth process above or

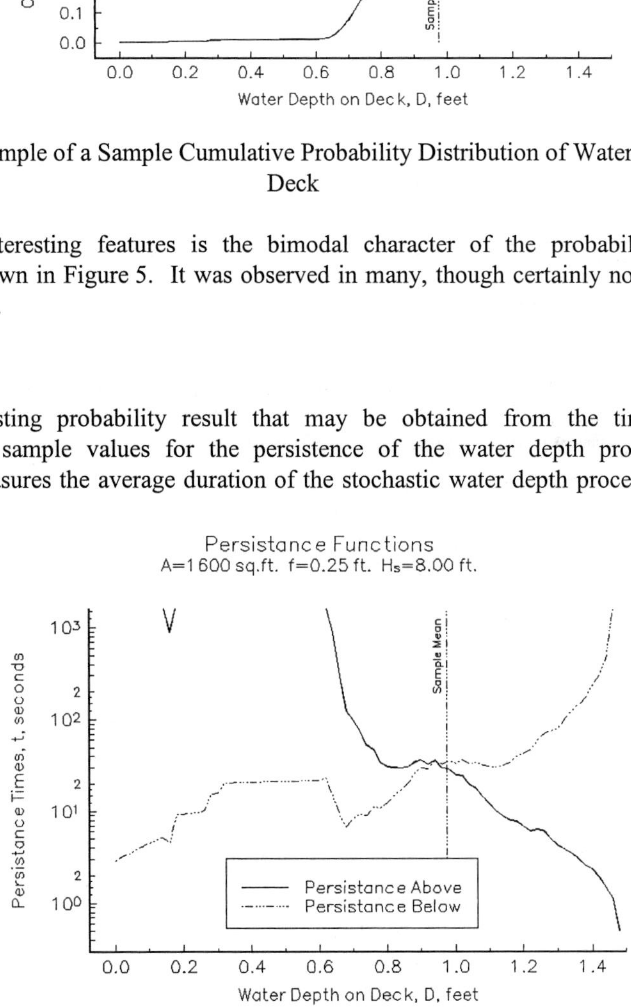

below any specified threshold value. Figure 7 depicts an example of persistence functions sampled in the time domain.

Figure 7: An Example of Sample Persistence Functions for Water Depth on Deck

Figure 7 indicates that the water depth in this case persists at or above the asymptotic average water level (sample mean) for an average duration of about 17.5 seconds, and that it persists below that level for an average of approximately 20 seconds. The average recurrence interval for process upcrossings of the asymptotic average water depth is the sum of the persistences above and below that threshold, or approximately 37.5 seconds.

The water level in this case persists at or above a 1.2 foot depth for approximately 7.0 seconds and below this level for approximately 45 seconds.

A study of the dependency trends of persistence with respect to the independent process parameters such as flooded deck area, freeboard, significant wave height and the width of the assumed damage opening, has not been completed at this time.

PRESSURE HEAD FORMULATION OF WEIR FLOW EQUATION RESULTS

Subsequent to the 28 February 1995 submission of SNAME Annexes A and B (1995A&B) to the IMO Panel of Experts, and prompted by correspondence with Dr. Vassalos of the University of Strathclyde, the SNAME Ad Hoc Panel investigated the application of the pressure head weir flow, Equation 1, throughout. On theoretical grounds the pressure head form of the weir flow equation is regarded as more correct and accurate than the velocity superposition form applied during the first phase of the SNAME analytical studies, but the disadvantage is the loss of separation between in-flow and out-flow processes.

$$dQ = K\,\mathrm{sign}(h_{OUT} - h_{IN})\sqrt{|h_{OUT} - h_{IN}|}\;dA \tag{1}$$

where: K is an empirical weir flow coefficient.

h_{IN} is the instantaneous head measured on the inside of the flux plane at any specified elevation above the deck.

h_{OUT} is the instantaneous head measured on the outside of the flux plane at any specified elevation above the deck.

dA is the differential element of flow area in the flux plane at the specified elevation, $dA = W\,dz$, where W is the width of the damage opening and dz is a differential element of elevation.

dQ is the differential volume flow rate across the differential area dA in the flux plane.

The method of Gaussian integral equations was applied using the pressure head formulation of the weir flow equation and the final results differ by only a small amount from those obtained using the velocity superposition method. Thus, for the purposes of regulation and rule making it may suffice to adopt the velocity superposition method and gain the advantages associated with the separation of in-flow and out-flow processes.

It should also be noted that, for the purposes of scientific investigation and engineering (but possibly not for the purposes of regulation and rule making), the method of Gaussian integral equations may be applied to cases based entirely on the pressure head formulation of the weir flow equation, and including additional outflow devices such as flow biased freeing ports and deck drains.

The fundamental idea behind the analysis that follows is that the <u>average</u> net volume flux is zero once equilibrium has been established between the in-flow and out-flow processes.

Figure 8 shows two curvilinear lines, one marked "Weir Flow Model, Based on Velocity Superposition" and the other marked "Weir Flow Model, Based on Pressure Head." A straight line approximation suggested by the SNAME Ad Hoc RO-RO Safety Panel to the IMO Panel of Experts, and data from the IMD model tests (which will be discussed in a subsequent section of this paper) are also shown.

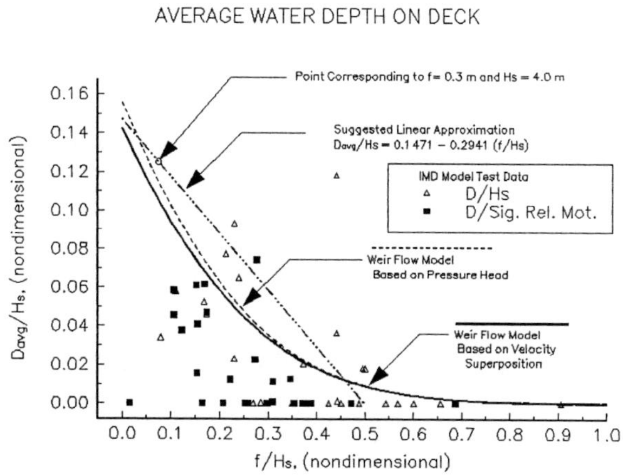

Figure 8: Comparison of Asymptotic Average Water Depth on Deck as Estimated by Pressure Head and Velocity Superposition Forms of the Weir Flow Equation

The curve marked "Weir Flow Model, Based on Pressure Head" was obtained using the Gaussian integral approach by solving the following equation for unknown asymptotic average water depth, D:

$$\overline{Q}_{NET} = 0 = \int_{f+D}^{\infty} N(0,\sigma,\eta)\left\{K\,W\left(\tfrac{2}{3}\right)(\eta - f - D)\sqrt{(\eta - f - D)}\right\} d\eta$$

$$+ \int_{f+D}^{\infty} N(0,\sigma,\eta)\left\{K\,W\,D\sqrt{(\eta - f - D)}\right\} d\eta$$

$$+ \int_{f}^{f+D} N(0,\sigma,\eta)\left\{K\,W\left(\tfrac{2}{3}\right)(f + D - \eta)\sqrt{(f + D - \eta)}\right\} d\eta + \int_{f}^{f+D} N(0,\sigma,\eta)\left\{K\,W(\eta - f)\sqrt{(f + D - \eta)}\right\} d\eta$$

$$+ \int_{-\infty}^{f} N(0,\sigma,\eta)\left\{K\,W\left(\tfrac{2}{3}\right) D\sqrt{D}\right\} d\eta \qquad (2)$$

where:

- K is a dimensional flooding coefficient
- W is the width of the damage opening
- f is the freeboard
- H_S is the significant wave height
- η is the wave elevation

σ is the standard deviation of the wave elevation process, $\sigma = H_S / 4$

$N(0,\sigma,\eta)$ is the Gaussian (normal) probability density function with zero mean and standard deviation, σ

$$N(0,\sigma,\eta) = \frac{e^{-\left\{\eta^2/2\sigma^2\right\}}}{\sigma\sqrt{2\pi}}$$

Note that K W is a common factor, which may be factored out of Equation 2.

The equation for D has been solved using a numerical root-finding procedure. The result is the curve shown in Figure 8 labeled "Weir Flow Model, Based on Pressure Head" and graphed using a short dashed line. As shown in Figure 8, the pressure head equations lead to a slightly greater predicted depth of water on deck at low freeboard values when compared with the corresponding results obtained from the velocity superposition equations, but the difference is not large. At values of f/H_S greater than 0.45 the difference is negligible. The pressure head equations approach more closely the point of interest to certain Nordic parties, corresponding to 0.5 m of water depth in 4.0 m significant waves for a vessel with 0.3 m freeboard.

Overall, there is excellent agreement between the *pressure head* and *velocity superposition* weir flow models. The advantage which the SNAME Ad Hoc Panel finds with the velocity superposition method is the ability to decouple the in-flow and out-flow processes, which greatly facilitates the process of evaluating out-flow credits for freeing ports and deck drains, as was done in SNAME Annex B (1995B), Hutchison (1995), Hutchison et al. (1995) and Hutchison et al. (1996).

COMPARISON WITH PHYSICAL MODEL TEST RESULTS

The findings of this analytical study have been compared with data measured in physical model tests at National Research Council Canada, Institute of Marine Dynamics (IMD) (Pawlowski et al., 1994, and Molyneux, 1995). Those physical model tests have also been presented in SLF39/INF.16.

The model tests at IMD were of a prismatic ship floating in waves with six degrees-of-freedom. The tests therefore do not correspond precisely with the assumptions inherent in the study of a stationary ship. Of the many cases studied in the IMD experiment program, only those cases that did not capsize are a source for data regarding the asymptotic average water volume and depth.

For the stationary assumptions in the simulation, the relative motion is the same as the wave height. The experimental points are plotted in two ways in Figure 8. The first method of plotting the points (corresponding to the open triangle symbols) made use of the significant wave height to normalize both freeboard and average water depth on deck. In the second method of plotting the experimental points (corresponding to the solid square symbols), the freeboard and average water depth are normalized by the significant double amplitude of relative motion (between the deck edge and the local wave surface) measured at the point of damage, instead of the significant wave height. There is slightly less scatter of the experiment results normalized by relative motion, and the theoretical line is more conservative than the observed data when presented on the basis of relative motion.

There are few experimental points shown in Figure 8 at low values of f/H_S, and those few do not approach the theoretical curve. The theoretical curve is for a stationary ship that does not sink, trim, heel or capsize in response to the water burden on deck. The experimental points are for a free floating model, which will in fact sink, trim, heel and sometimes capsize. However, the experimental points in Figure 8 were only obtained from those cases where the model did not capsize and an asymptotic average water accumulation could be determined. Most experimental cases with a free floating ship and very small freeboard (i.e., small f/H_S) ended in capsize, and therefore no asymptotic average water accumulation could be determined. The measured data confirm the predictions that above a certain ratio of freeboard to wave height there is very little water on the deck.

It is interesting to observe that even at relatively low values of f/H_S there are also some cases when there are very low volumes of water on the deck. Although the instrumentation in the model was not designed to measure very low values of water, video records of the experiments confirmed that the volumes of water on the deck in these cases were negligible.

From the video tapes it was seen that below a critical value of wave height a lot of the water was coming onto the deck through the damage in the deck and not through the side. In these cases it was very easy for the water to drain back out through the hole in the deck, without flooding it. The other factor that has to be considered is the relationship between relative motion and roll angle. The flow of water onto the deck did not become significant until the root-mean-square roll angle was greater than approximately 2 degrees. In these cases the majority of the relative motion was coming from heave. The water remained relatively static and easily drained off the deck. For higher roll angles a wave system built up on the deck, which affected the drainage rates. None of these factors is included in the simulations, since the only route for the flood water was through the damage in the side and vessel motions are ignored as are flood water dynamics.

DEPENDENCIES INDICATED BY MATHEMATICAL MODEL

The dependence of the main dependent variables examined in this paper on the independent parameters, is summarized in the following table.

TABLE 1

Dependence of Dependent Variables on Independent Parameters

Dependent Variables		Independent Parameters				
		f	H_S	W	A	C_W
Average In-Flow Rate	$\overline{Q}_{IN}$	yes	yes	yes	no	yes
Average Out-Flow Rate	$\overline{Q}_{OUT}$	yes	yes	yes	no	yes
Asymptotic Average Water Depth	$\overline{D}$	yes	yes	no	no	no
Average Build-Up Time	t	yes	yes	yes	yes	yes

The most important result is that, under the assumptions of this study, the asymptotic average water depth is independent of the width of the assumed damage opening, the flooded deck area and the weir flow coefficient. The only dependencies for the asymptotic average water depth of the stationary ship are freeboard and significant wave height.

ACKNOWLEDGMENTS

The SNAME Ad Hoc RO-RO Safety Panel wishes to acknowledge joint industry financial assistance supporting this research effort from Washington State Ferries, the Alaska Marine Highway System and The Glosten Associates, Inc.

The mathematical theories that were developed were compared with model test data for damaged ships freely floating in waves, which had previously been performed at National Research Council Canada, Institute for Marine Dynamics, under joint sponsorship of the Transportation Development Centre and Canadian Coast Guard, Ship Safety Branch and the Institute for Marine Dynamics.

This synoptic technical paper is substantially based on a paper originally presented at Cybernautics 95, SNAME California Joint Sections Meeting, held aboard the *Queen Mary* at Long Beach, California, 21-22 April 1995 (Hutchison et al., 1995). Co-authors David Molyneux of IMD and Patrick Little of the U.S. Coast Guard contributed to the original source paper.

References

Pawlowski JS, Molyneux D and Cumming D. (1994). Analysis of Experience on RO-RO Damage Stability, IMD TR-1994-27 (protected).

SNAME Ad Hoc RO-RO Safety Panel. (1995A). for the IMO Panel of Experts. "Water Accumulation on the Deck of a Stationary Ship," Annex A to the second position paper.

SNAME Ad Hoc RO-RO Safety Panel. (1995B). for the IMO Panel of Experts. "Freeing Port Effectiveness of Water on Deck," Annex B to the second position paper.

Hutchison BL, Molyneux D and Little P. (1995). Time Domain Simulation and Probability Domain Integrals for Water on Deck Accumulation, *Cybernautics 95*, SNAME California Joint Sections Meeting.

Hutchison BL (1995). Water-on-Deck Accumulation Studies by the SNAME Ad Hoc Ro-Ro Safety Panel, *Workshop on Numerical & Physical Simulation of Ship Capsize in Heavy Seas*, Ross Priory, Loch Lomond, Glasgow, Scotland, 24-25 July.

Molyneux D. (1995). Estimates of Steady Water Depths on the Deck of a Damaged RO-RO Ferry Model, IMD TR-1995-26.

Hutchison BL, Little P, Molyneux D, Noble PG, and Tagg R.D. (1996). Safety Initiatives from the SNAME Ad Hoc Ro-Ro Safety Panel, *Ro-Ro 96 Conference*, Lübeck, Germany, 21-23 May.

D. Vassalos, M. Hamamoto, A. Papanikolaou and D. Molyneux (Editors)

AN EXPERIMENTAL STUDY ON FLOODING INTO THE CAR DECK OF A RORO FERRY THROUGH DAMAGED BOW DOOR

Nobuyuki Shimizu[1], Roby Kambisseri[2] and Yoshiho Ikeda[3]

[1] Imabari Shipbuilding Co., Ltd., Japan
[2] Department of Ship Technology, Cochin University of Science & Technology, Kerala, India
[3] Department of Marine System Engineering, Osaka Prefecture University, Japan

ABSTRACT

The authors have done experiments with a model of a RORO car ferry with opening at the bow, instead of a bow door. The model was run into regular head waves, allowing only heave and pitch motions. The flooding of the car deck increased with the increase of bow opening size, wave height and Froude number, and the peak is around $\lambda/L=1$. The motions changed with the flooding, which ranges from 0 to full volume of the car deck. Theoretical calculations revealed that there is heavy loss of (static) righting moment due to the shift of water on deck. This study suggests that a ship should be slowed down or stopped, on damage, to reduce or avoid further flooding.

KEYWORDS

Stability, Damage, Flood Water, Head Sea, Bow Door, Car Ferry, Car deck, Estonia

INTRODUCTION

Shortly after midnight on Sept. 28, 1994, the 15,566 tonnes, ferry ESTONIA sank in the stormy Baltic sea, during the overnight crossing from Tallin to Stockholm, causing the death of 852 of the 989 persons on board, JAIC of ESTONIA (1995). Estonia was running at a speed of around 14.5 knots during the time of accident, facing wind speeds of 15-20 m/s and significant wave heights of 3.5-4.5 m. It sank about 30 minutes after the leak of water through the sides of the forward ramp was observed on the TV monitor of the engine control room. The bow visor (55 ton) separated from the bow and was lost, and the inner door (forward ramp) was fully opened allowing large amounts of water to enter the car deck resulting in the sinking of the ship.

Few experimental and theoretical studies had been done on the survivability of damaged RORO ships in U.K., MSA-U.K (1994), Vassalos D. and Turan O (1994), Denmark, MSA-U.K (1994) and Canada, John T. Stubbs, Peter Van Diepen, Juan Carreras and Joseph H. Rousseau (1995), for the damage in the midship region. So far, studies on the damage or loss of bow door were not done. The nature of water entering the ship through a bow opening and the phenomenon due to the presence and motion of water on deck is unknown. By the present studies, the authors are trying to understand these problems. They have done experiments with the model of a Car Ferry, with bow openings in calm sea and regular head waves at different advance speeds. The heave and pitch motions and the wave profiles were measured. From these experimental data, the amount of water on deck with respect to time, minimum amount of water on deck which will cause static capsize, the time required for the static capsize were determined.

EXPERIMENT

Model

A 2 m long model of a 1500 GRT Car ferry of length 75 m was used for the experiment. Principal particulars are shown in Table 1. In model, height of car deck from keel is 0.161 m and depth of the car deck is 0.131 m. The top of the car deck is covered by acrylic sheet. The bow of the model was cut open to represent the missing bow door. The experiments were done with two widths of openings of 0.10 m (Full) and 0.05 m (Half). A bodyplan of the model is shown in Figure 1.

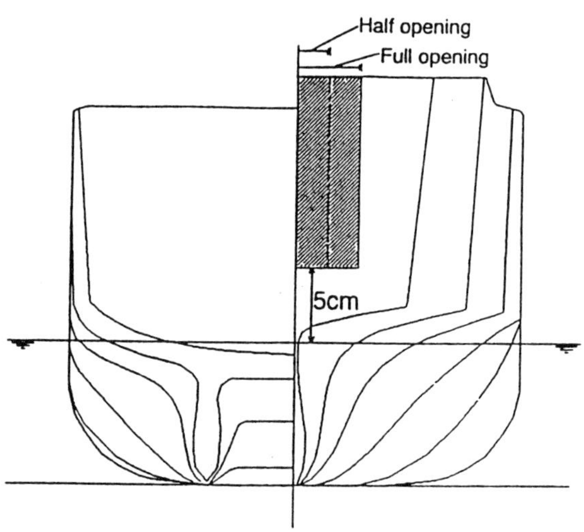

Figure 1: Body plan of the model

TABLE 1
PRINCIPAL PARTICULARS OF SHIP AND MODEL

	Model	Ship
L_{pp} (m)	2.000	75.000
B (m)	0.360	13.500
D (m)	0.308	8.500
d (m)	0.111	4.150
W (ton)	0.046	2409.4
KG (m)	0.149	5.600
Scale	1/37.5	1/1

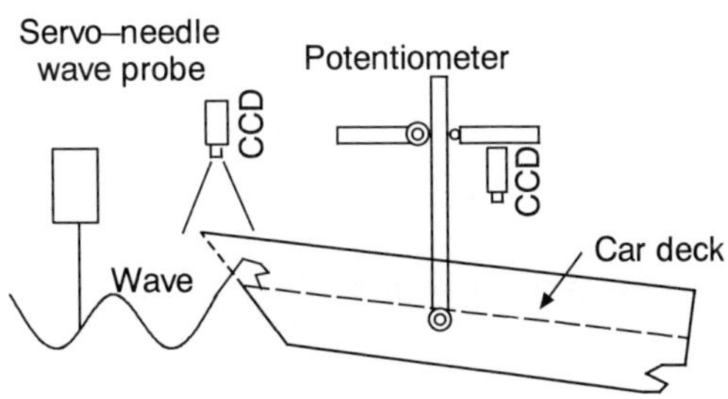

Figure 2: Experimental set-up

Experimental set-up

The model was connected to a Motion Measuring Device, which was fitted on the Towing carriage of a Towing Tank (70 m × 3 m × 1.6 m) in the Osaka Prefecture University. The connections allowed freedom only for the pitch and heave motions. These two motions and the wave profiles were measured using potentiometers and servo-needle wave probe, at definite intervals of time. Also the views of the bow opening and the decks were recorded using two video cameras. Figure 2 shows the experimental set-up.

Experimental condition

The model was run at different Froude numbers with regular head waves of different λ/L (λ is wavelength) and wave heights. To get the mean value of the heave motion data, which means the time-variation of the sinkage of the ship, heave values were measured in calm condition for zero advance speed, before every run.

The ranges of values are as shown in TABLE 2.

TABLE 2
EXPERIMENTAL CONDITION

Fn	0, 0.1, 0.2, 0.3	
λ/L	0.6, 0.8, 1.0, 1.2, 1.4, 1.6, 1.8	
Wave height (m)	0.04, 0.08, 0.10	
Bow opening width (m)	Full = 0.10	Half = 0.05

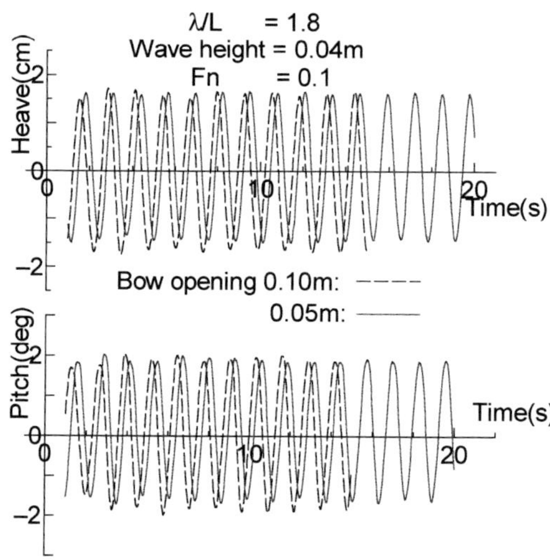

Figure 3: Measured ship motions of a damaged model
- no flooding occurs

Analysis and Results

Heave and pitch traces were plotted for each run. Some examples are shown in Figures 3 to 6. Figure 3 is for a case without any flooding of the car deck. Its heave and pitch motions are almost steady. Figure 4 is for light flooding of the car deck. In this case, the mean heave changes with time due to the flooding of car deck, and the rate for the full opening case is much larger than that for the half opening case.

Figure 5 is for heavy flooding, at high advance speed. The model was rapidly drowned due to flooding in the full opening case. Due to the heavy flooding (in the full opening case) the mean pitch increases with the increase of the mean heave causing the model to nose-dive into the water within a short time span. The mean pitch at 12 sec (when the experiment was aborted) is about 4 degrees by bow. But in the half opening case, the mean heave increases slowly showing only light flooding. The mean pitch by bow which increased initially became zero on further flooding; the flooding at the bow causes the initial bow-trim (reaching up to 0.5 degrees) which is reduced by the spreading of the flood water to the aft region. It should be noted that in some cases non-linear motions are observed as shown in Figure 6. It is left to explain, however, why the non-linear motions are caused. These examples of the experiments demonstrate that the characteristics of the ship motions significantly depend on the advance speed, the wave period, the wave height and the area of the bow opening. Three photographs during the experiment are shown in Figure 7.

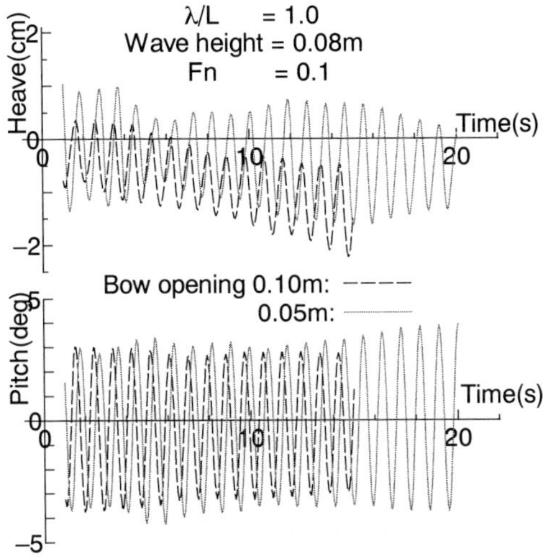

Figure 4: Measured ship motions of a damaged model
- light flooding occurs

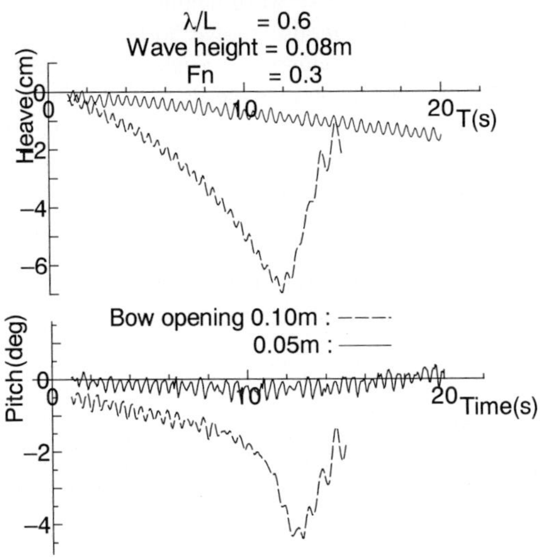

Figure 5: Measured ship motions of a damaged model
- heavy flooding occurs

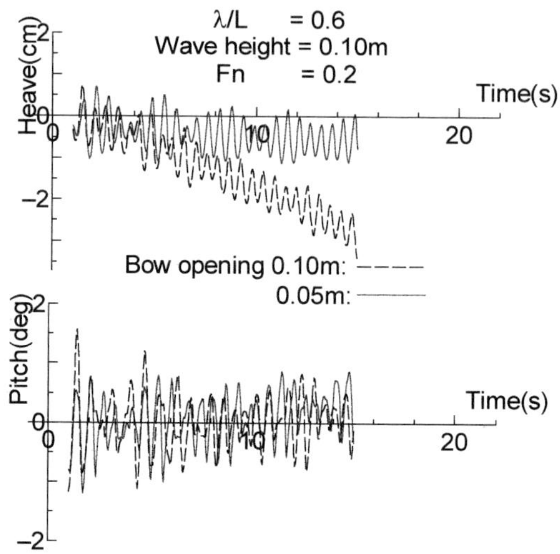

Figure 6: Measured ship motions of a damaged model
- non-linear motions can be seen

Water depth on the car deck at the bow is calculated from the heave, pitch and wave profile data of each run and a sample of it is shown in Figure 8. As shown in that figure maximum values of these depths are found for the first three seconds of the run and also for the whole period of the run including the first three seconds. These two maximum values of depths are plotted in Figure 9. In these figures, relative wave heights about the car deck at bow which is calculated by Ordinary Strip Method, are also shown, to be compared with the experimental results, Tasaki R (1960), Takaishi Y., Ganno M., Yoshino T., Matsumoto N. and Saruta T (1972), Ohkusu M (1994). In these calculations the flooding of the car deck is not taken into account and hence, in all the figures, the measured water levels are more than the calculated values. In most of the cases - included in the figures - there is flooding of the car deck, since the water height from car deck level is positive, which is also confirmed by water on deck calculation using mean heave. In most of the cases, the maximum water level at bow is more for the full width bow opening case than for the half width one. The time at which the maximum values occur varies from the initial stages of the run to its final stages. As advance speed increases, the water level increases. The difference between the measured and the calculated water level increases with advance speed because the calculation method does not take into account advance speed effect, except the frequency change due to speed.

The water on deck at definite time intervals can be calculated using the mean value of the heave motion. Some of the results are shown in Figures 10 and 11. Figure 10 shows the effect of Froude number and Figure 11 shows the effect of λ/L. It can be seen that the water on deck increases with increase of Froude number, increase of wave height and wider bow opening. For the same wave height and Froude number, the maximum amount of water on deck is when λ/L is around 1.0.

Figure 7: Photographs during experiment

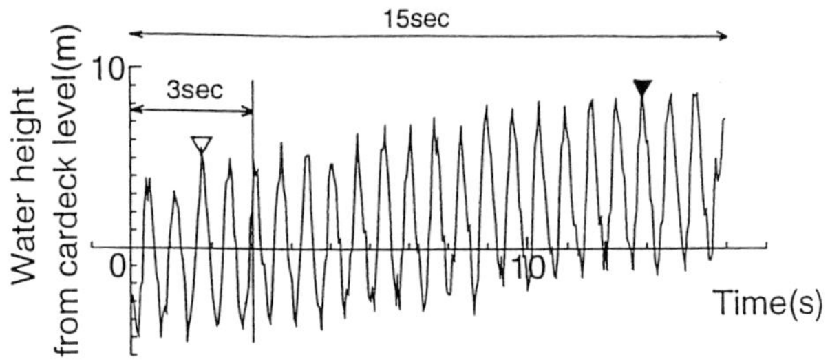

Figure 8: Trace of water level at the bow door

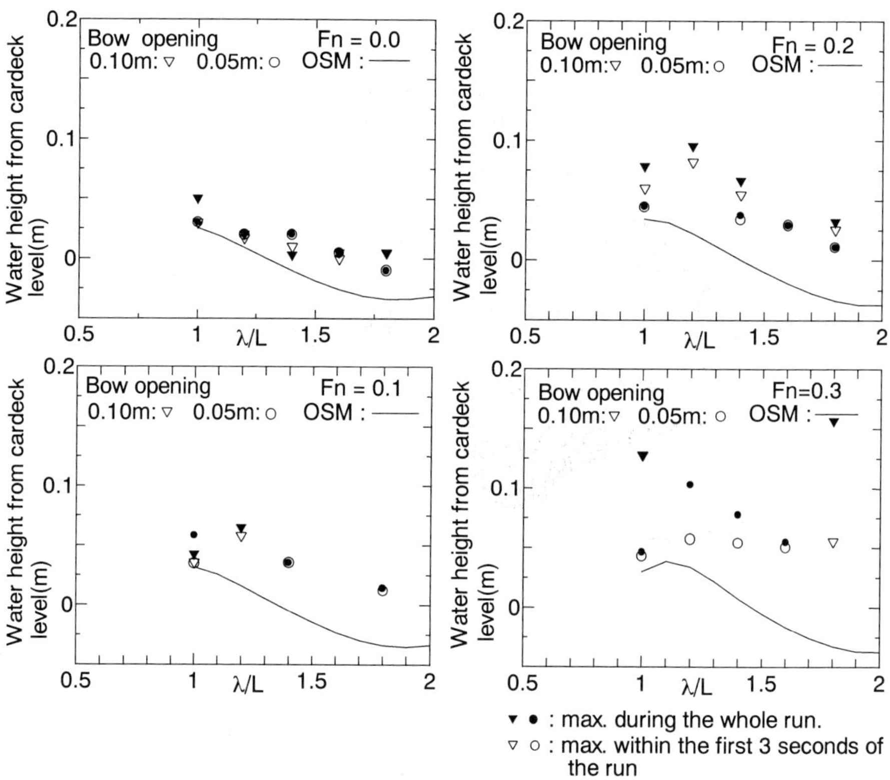

Figure 9: Comparison between experiment and calculated results of water level at bow door (Wave height=0.08m)

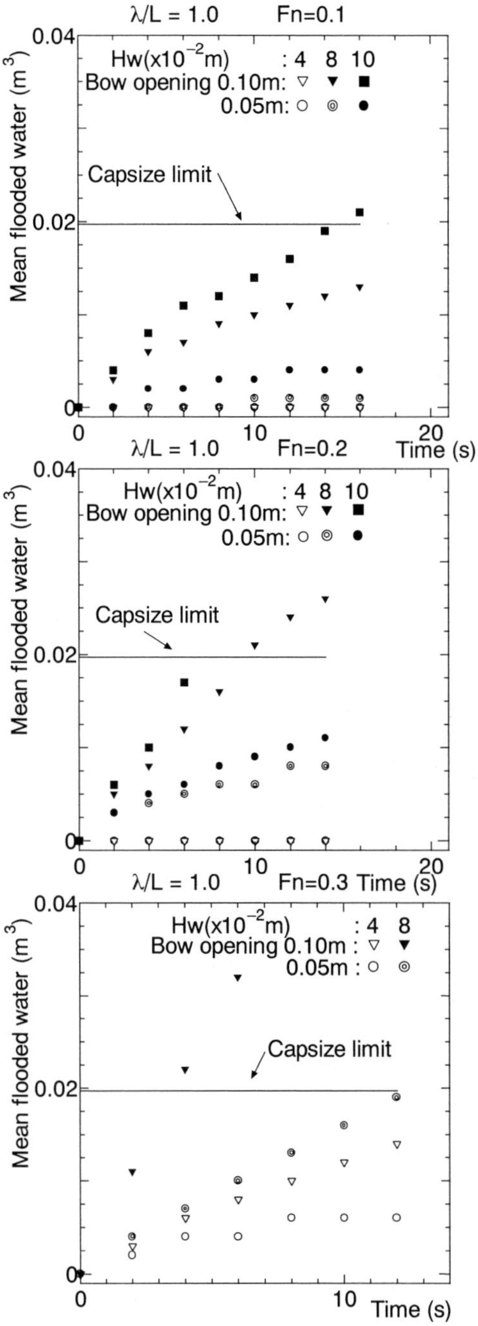

Figure 10: Water on deck - Effect of Froude number

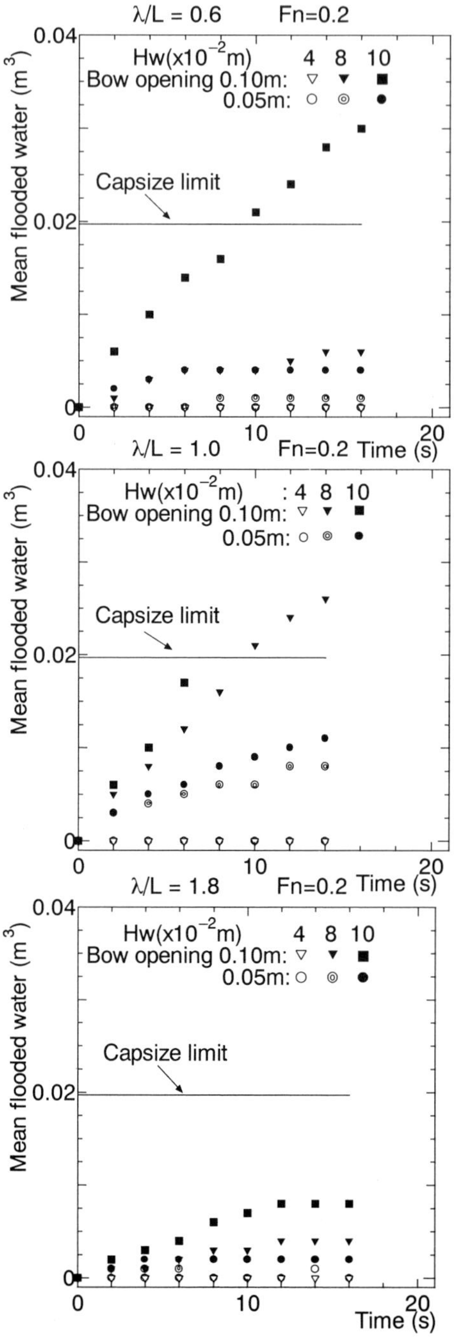

Figure 11: Water on deck - Effect of λ/L

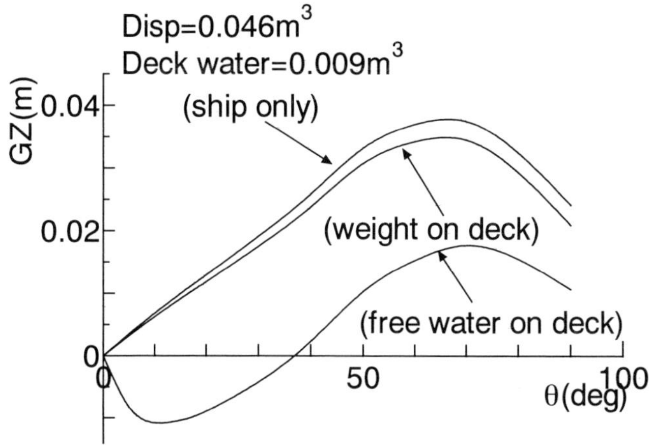

Figure 12: Loss of righting lever due to water on deck

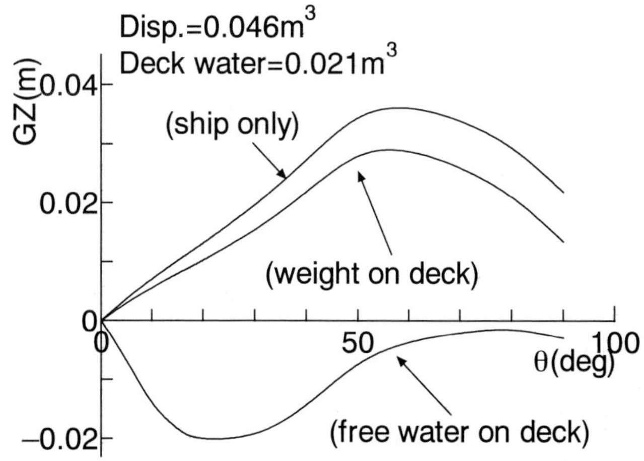

Figure 13: Loss of righting lever due to water on deck (Capsizing case)

The GZ curves (including the effect of deck water) and the static equilibrium angles of heel for the experimental ship were found for different amounts of water on deck. Figure 12 shows that the loss of righting lever due to the effect of deck water is very high. It can be seen in Figure 13 that though the ship capsizes due to the effect of deck water, it is very stable for a fixed weight on deck, weighing the same as the deck water. Figure14 shows the effect of deck water on the equilibrium angle of heel. The minimum amount of water on car deck for static capsize is 1040 m^3 for the ship and 0.0197 m^3 for the model, which is shown in Figures 10 and 11 as a capsize limit. The time for the accumulation of this water on the deck of the model is taken as the capsizing time. Capsizing time and the number of waves encountered before capsize are given in Table 3. At Fn=0.3 and λ/L=1.8, the ship capsizes within just 4 waves.

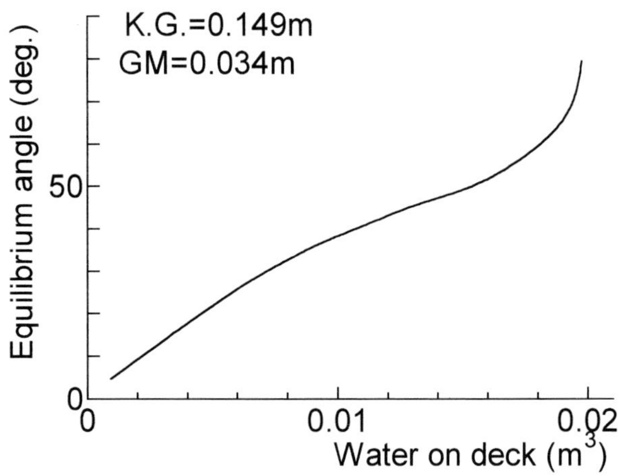

Figure 14: Water on deck and equilibrium angle

The obtained values of the water on deck in the present experiments are for a model that is prevented from heeling or rolling. In the real case, the model or ship will heel due to the movement of water on deck, which will change the area of bow opening exposed to water and the entry of water into the car deck. This may speed up capsizing. This may suggest that an experiment of a ship in three degrees of freedom, heave, pitch and roll should be carried out.

It can be seen that the water that causes capsizing occupies only a small portion of the total volume of the car deck. The unrestricted flow of water into the full-length car deck will surely cause capsizing. However, reducing the speed or stopping the ship can reduce the rate of flow of water into the ship. This may suggest that the navigation of a damaged RORO ship is also very important for survival.

CONCLUSIONS

An experimental study was conducted on a model of a car ferry with openings at the bow, in regular head seas. The amounts of water accumulated on the deck are found from the data. The main observations are the following:

(1) The mean sinkage in the half-opening case is much less than half the sinkage of the full opening case.

(2) For the same Froude number and same wave height, the maximum height of water at the bow opening and the maximum amount of flood water on the car deck occur when λ/L is around 1.0.

(3) When the Froude number is 0.3 and λ/L is 1.8 the model (or the ship) capsizes within just 4 waves.

TABLE 3
STATIC CAPSIZE TIME AND WAVES ENCOUNTERED

Capsize time in sec. (number of waves encountered)

Fn	λ/L	0.6		1.0		1.8	
	Wave height	Opening width					
		Full	Half	Full	Half	Full	Half
0.1	4cm	-	-	-	-	-	-
	8cm	-	-	-	-	-	-
	10cm	-	-	15 (16.6)	-	-	-
0.2	4cm	-	-	×	×	×	×
	8cm	-	-	9 (11.9)	-	-	-
	10cm	9 (16.8)	-	7 (9.3)	-	-	-
0.3	4cm	-	-	-	-	18	-
	8cm	9 (20)	-	3.5 (5.4)	12.5 (19.4)	3.5 (3.6)	×
	10cm	×	×	×	×	×	×

Full = Opening width of 10cm × : No data available
Half = Opening width of 5cm - : No capsize with in 20Sec.

References

JAIC of ESTONIA. (1995), Part-Report, covering Technical issues, by the Joint Accident Investigation Commission of Estonia, Finland and Sweden on the capsizing of MV ESTONIA.

John T. Stubbs, Peter Van Diepen, Juan Carreras and Joseph H. Rousseau (1995). *TP 12310E Flooding Protection of RO-RO Ferries, Phase I, (Vol.1)*, Transports Canada.

MSA-U.K. (1994). Reports on Research into Enhancing the Stability and Survivability Stands of RO-RO Passenger Ferries, *Marine Safety Agency, U.K.*

Takaishi Y., Ganno M., Yoshino T., Matsumoto N. and Saruta T. (1972). On the Relative Wave Elevations at the Ship's Side in Oblique Seas. *Journal of the Society of Naval Architects of Japan* **132**, 147-158. [Japanese]

Tasaki R. (1960). On the Shipping Water in Head Waves. *Journal of the Society of Naval Architects of Japan* **107**, 47-54. [Japanese]

Vassalos D. and Turan O. (1994). Damage Scenario Analysis: A tool for assessing the Damage survivability of passenger ships. *STAB '94*

Contemporary Ideas on Ship Stability
D. Vassalos, M. Hamamoto, A. Papanikolaou and D. Molyneux (Editors)

DAMAGE STABILITY TESTS WITH MODELS OF RO-RO FERRIES-A COST EFFECTIVE METHOD FOR UPGRADING AND DESIGNING RO-RO FERRIES

M. Schindler

Danish Maritime Institute, Lyngby, Denmark.

ABSTRACT

The 'Regional Agreement Concerning Specific Stability Requirements for Ro-Ro Passenger Ships' (the Stockholm Agreement) as an alternative to the deterministic method, allows for an approval based on model tests, provided that such tests are performed in accordance with the specified procedure and the Guidance Notes. DMI's experience on the field of the experimental damage stability, successfully built up since late '80-ies, was an important factor in the development and definition process leading to the final model criteria and the test methods prescribed by The Stockholm Agreement.

KEYWORDS

Model tests, survivability of passenger ro-ro ferries, dynamic damage stability, the Stockholm Agreement, SOLAS damage / side damage, damaged bow port.

INTRODUCTION

Testing of models representing damaged ships exposed to rough seas has become a very important tool for investigation of problems in the field of damage stability. The high degree of complexity caused by strong non-linear dynamic effects related to wave motions, the response of the damaged ship, and water ingress, make these problems difficult to simulate by means of mathematical modelling.

The same dynamic effects mean that physical model testing makes heavy demands on the model construction and the test techniques. DMI has developed criteria for both of these. The model construction methods and the model test technique used will be briefly presented in this article.

DMI has a leading position with regard to damage stability problems, drawing on more than 25 years of experience. In recent years, this position has been considerably strengthened due to the fact, that DMI has taken a very important part in damage stability research and development projects initiated by the Department of Transport, the former UK Maritime and Coastal Agency (MCA), and a Nordic co-operation on Safety of Passenger / Ro-Ro Vessels. This project was followed by a special investigation ordered by the Nordic Shipowners' Association. A large number of new models have been built and tested since then on commercial basis. Most of them with relation to the Stockholm Agreement, but some took part in customers' private research projects.

Two of the 'commercial' models were later rebuilt and adopted for the purpose of the Research Project 423, the joint MCA and DMI investigation on the importance of the transverse radius of gyration, which was performed at DMI. Further more, three of the DMI built models, after the end of the respective projects, were shipped to Marintek (Norway), University of Strathclyde (UK) and BMT (UK), where they were used in different research projects. The models specially designed for these investigations showed a very high degree of reliability, some of them spending several hundreds of hours in water. DMI's test technique, rigorously followed during every single test series, produced reliable results with excellent repeatability.

PROBLEM FORMULATION

As indicated above, the water ingress into a damaged Ro-Ro ferry is a complex process which depends on a wide range of parameters describing the Ro-Ro ferry characteristics, the damage characteristics and the environmental conditions.

In a calm sea, the damaged Ro-Ro ferry, which is designed according to present regulations, will find an equilibrium condition which is characterised by the whole Ro-Ro deck being well above the waterline. Adherence to present regulations will assure that for a given 'standard' damage the Ro-Ro ferry shall survive with a certain margin of safety against capsize. To demonstrate the fulfilment of the safety margin the still water stability calculations shall be made for the Ro-Ro ferry in damaged condition. This calculation includes the effect of water flooding freely onto the Ro-Ro deck at higher angles of heel. As required by the 'Stockholm Agreement', for the Ro-Ro ferries under the SOLAS resolution or equivalent, an assumed amount of water on the Ro-Ro deck has to be included in the calculation.

Exposure of the damaged Ro-Ro ferry to waves, particularly beam seas, may have the effect that water flows in and out of the damaged opening. This water flow is caused by the varying water head inside and outside the Ro-Ro deck, the relative wave elevation, the ship motions and the motion of water trapped on the Ro-Ro deck.

The overall effect of the waves is that a net increase of water on the Ro-Ro deck can occur. An additional amount of water trapped on the Ro-Ro deck will cause an increase in the mean heel of the Ro-Ro ferry. Unless the forces acting on the Ro-Ro ferry are in balance, the ferry will capsize.

Based on experience, the net water ingress on the Ro-Ro deck of a free floating ferry will depend on the following parameters:

- Ro-Ro Ferry Characteristics:
  - Size / dimensions / displacement / freeboard
  - Loading condition (KG, GZ-characteristics)
  - Subdivision below Ro-Ro deck
  - Cross-flooding capability
  - Arrangement on Ro-Ro deck

- Damage Characteristics:
  - Shape
  - Size
  - Location
  - Damaged freeboard

- Environmental Conditions:
  - Sea state (H_s , T_p)
  - Relative wave direction
  - Wind forces and direction

The importance of these parameters for survivability in waves has been confirmed on many occasions using models of different Ro-Ro ferries. This was first identified and systematically investigated by model testing in the course of the entire project financed by the UK Department of Transport, which was initiated after the capsize of 'Herald of Free Enterprise'. More details of Phase 1 of this project can be found in Pucill & Velschou (1990), while the foundlings of Phase 2 are described in Velschou & Schindler (1990) and Velschou & Schindler (1994).

As mentioned above, the amount of water trapped on the Ro-Ro deck is the decisive factor for survivability of a damaged Ro-Ro ferry in waves. This aspect was the main objective of the test programme 'Testing of Large Scale Flooding' performed at DMI, which was part of the entire 'Joint North-West European Research Project' established after the capsize of 'Estonia' in 1994. The new international damage stability requirements with allowance for 'extra' water on deck which are defined in the 'Stockholm Agreement', are mainly derived from the results and conclusions of this project. The results and conclusions of the experimental work performed at

DMI during this project were a decisive factor for formulation of the required amount of 'extra' water on deck, depending on the environmental conditions and the ferry characteristics. This problem is discussed in Damsgaard & Schindler (1996). Further more, the very confident and reliable experimental results achieved in the course of the project paved the way for the model test method, as an alternative to the traditional deterministic method in the process of the approval procedure with regards to damage stability requirements.

The chapter below gives some indication about the DMI standard for the model construction methods and the model test procedures, which satisfy the prescriptions of the Stockholm Agreement and which normally are applied in course of the approval model tests as performed by DMI.

CONSTRUCTION AND MODELLING TECHNIQUE

The GRP model is built over a foam plug. The Napa system hull definition is used to prepare drawings for milling the plug, used for casting the GRP shell of the hull. For this purpose the plug dimensions are reduced by the thickness of the GRP casting (3 mm).

The model hull shell is normally manufactured in 3 mm GRP as are the tank top in damaged compartments, the main deck and the subdivision bulkheads. This method yields the best guarantee for water tightness. The intact compartments within the damaged zone are formed in foam and are normally covered by topcoat. If applicable, sponsons or any other external buoyancy in the intact parts of the model are also made of foam and covered by topcoat. The damaged part of the external buoyancy is cast using the 3 mm GRP. If required by the design concept of the individual ship, the 'double skin' on both sides of the damaged part of the model is preserved, allowing for cross-flooding of the space inside the external buoyancy. Ventilation of all damaged compartments below the main deck is provided.

The important details on and above the main deck are modelled using either 3 mm GRP, foam or transparent 3 mm polycarbonate plates, up to the required depth of the model. The transparent 3 mm polycarbonate plates are used, where inspection for leaking water is required.

Two threaded steel rods are cast in the glass-fibre structure at the centre line of the model to provide lifting points and for mounting a longitudinal steel profile on which ballast is placed to adjust the model centre of gravity, KG. Miscellaneous aluminium profiles are fitted to provide sufficient model stiffness.

Bilge keels and fender lists are included in the model. Other appendages, like the rudders, thruster tunnels, shaft brackets, stabiliser fins and sea chests can be included in the model in a simplified form.

The damage openings in the outer shell and penetrated bulkheads are modelled in accordance with SOLAS, i.e.:

Length (outer shell) : 3 m + 0.03 X Subdivision Length

Shape : Rectangular in sides and isosceles triangles in all decks

Penetration : B/5

The height of the damage opening corresponds to the top of the model.

The correct permeabilities are modelled in all damaged compartments below the main deck by means of dummy blocks (the normal practise), while permeabilities on and above the vehicle deck are not simulated.

Loading of the model is performed in accordance with information provided by the Client, specifying one or two intact loading conditions. The position of the centre of gravity (KG) of the model is adjusted by changing the vertical position of the ballast lift

The Guidance Notes related to the SOLAS/CONF. 4/36 Agreement define 0.25 x L as the maximum radius of the longitudinal gyration, I_{yy}. The same document defines 0.40 x B as the maximum value of the transverse radius of gyration, I_{xx}, to be accepted for the actual ship. Prior to the tests in water, a fully equipped and loaded model is subjected to oscillation tests in air using a special designed cradle for determination of the longitudinal radius of gyration, I_{yy}.

MODEL TEST PROCEDURE

The other hydrostatic and dynamic properties of the model are checked in water. Inclining tests in water with model in intact condition are made in order to check the intact GM. A given weight included in the total displacement of the model with a mass close to the G position of the model is moved horizontally in transverse direction by a known distance. The resulting intact GM is calculated based on the heel angle readings.

Roll decay tests with the model in the intact and the damaged condition are made in order to check the model's roll periods. The model is forced to roll and the radius of transverse gyration corresponding to each condition is calculated using the expression:

$$T = \frac{2\,\pi\,i}{\sqrt{g\,GM}}$$

as given by the Guidance Notes.

The model tests in waves are performed in DMI's towing tank which has a length of 240 m, width of 12 m and depth of 5.4 m. In one end the tank is equipped with a powerful computer controlled wave maker which is able to generate any physically realistic wave spectrum with maximum wave height up to approximately 0.9 m in model scale. A wave absorbing beach is situated at the opposite end of the tank.

Each test starts with the model placed in the tank approximately 20 m from and beam onto the wave maker with the damage side facing the incoming waves. The model is allowed to cross flood. The equilibrium angles are checked prior to each test with an automatic digital level gauge which reads to 0.1° accuracy.

A reference zero reading is taken on all sensors when the model is at rest, and thereafter, the wave maker is started. When, after a few seconds, the waves reach the location of the model, the data recording and video recording are started.

The model is allowed to drift freely, followed by the carriage such that it during the whole test is in approximately the same position relative to the moving wave gauge. A soft line connected to each end of the model at the water line prevents the model from drifting into the tank wall. No intervention is normally required to prevent the model to yaw as the natural yaw angle very rarely exceeds 10°.

As required, the minimum time for each individual test is 30 minutes full-scale. However, the tests continue slightly longer, as the excessive time is used for evaluation of stationarity of the model heel angle at the end of each test.

Instrumentation consists of the instruments monitoring the model motions in waves and the probes measuring the wave height in the tank. The unit containing the roll and pitch gyro is mounted in a watertight compartment below the Ro-Ro deck. Two accelerometers record heave accelerations of the model. Relative water level recorders mounted on the outside of the model determine its attitude relatively to the sea level. The drift speed of the model is determined as the towing carriage speed.

The signals obtained from a stationary wave recorder, placed 20 m from the wave maker, are used for documentation of the wave quality, while the signals obtained from a mobile wave recorder, mounted on the towing tank carriage in line close to the free-drifting model, are used for documentation of the actual wave height.

The data and video recordings are stopped after the end of each test. Immediately after that, the logging programme prints the statistical values Mean, RMS, Max. and Min. of each channel. They are inspected together with the graphic representation of time history of the gyro and of the signals recording carriage speed to confirm the validity of the test.

After completion of each test, the model is towed back to the starting position, water is drained from the deck and the model is inspected for water leakage into closed intact compartments.

As specified by the Stockholm Agreement, the model must survive the tests as described above in at least 5 realisations of each of the two sea states as described below, with the same significant wave height, viz.:

- Short waves corresponding to spectral peak period $T_p = 4\sqrt{H_s}$, using a narrow band Jonswap spectrum with peakedness factor $\gamma = 3.3$.

- Long waves corresponding to spectral peak period $T_p = 6\sqrt{H_s}$ or the roll period of the damaged ferry, whichever is the lower, using a broad-band PM spectrum (Jonswap spectrum with $\gamma = 1$).

This means, that at least ten individual test in waves are required for each of the damage cases investigated.

DMI RECORD

After completion of the main investigation under the ‘Joint North-West European Research Project’, but before the final adoption of new stability requirements formulated in the ‘Stockholm Agreement’ were made, two important investigations were undertaken at DMI, using the same ferry model.

The first one was ordered by a group of the North European Ship Owners to investigate the effectiveness end the effects of some different concepts of ‘not full height’ transverse bulkheads on a number of different arrangements on ro-ro deck. The conclusions were very similar to the findings made during the second phase of the Department of Transport, ‘Ro-Ro Passenger Ferry Safety Studies’, Velschou & Schindler (1990) and Velschou & Schindler (1994) and they confirmed the certain risks related to the concept.

The second one was a direct follow up of the main investigation under the ‘Joint North-West European Research Project’ with the objective to validate the agreed methodology formulated in the ‘Stockholm Agreement’ regarding the requirements for the approval procedures based on the model experiments.

After the final adoption of the model test method as an alternative to the deterministic method, DMI was the first institution to perform model tests which led to the approval of a Ro -Ro ferry in accordance to the Stockholm Agreement. Since then, 14 models representing different Ro-Ro ferries ranging from about 40 m in length to 200 m were built and tested at DMI using the procedures as described above. They were used in more than 20 individual test series related to the ‘Stockholm Agreement’; either as feasibility studies to investigate their ability to satisfy the requirement or as the final approval tests in the process of upgrading to the standards of the ‘Stockholm Agreement’.

The important tool considerably increasing rate of the successful damage stability tests is DMI’s in-house simulation method by means of a numerical analysis of survivability of a given ferry. The numerical simulation method has been based on a large number of model test results and for the first time it was applied successfully to a British ferry prior to performing the model tests. Since then, this simulation method is used frequently at DMI to predict chances to survive prior to the model tests. If required, any modifications to the ferry can be assessed by means of this method, before the final decision for the model test is made.

OTHER INVESTIGATIONS BY MODEL TESTS INVOLVING 'SIDE DAMAGE' CASES

Gradually, several conventional (applying to SOLAS) Ro-Ro ferries have been model tested. Some of them with respect to the SOLAS/CONF. 3/46 (Stockholm) Agreement above and others subjected to intensive government-controlled and commercial research programmes aimed for better understanding of their behaviour. No model of a fast mono-hull Ro-Ro ferry applying to MSC.36(63) HSC-code has until 1996 been investigated seriously by model testing in the damaged condition.

On this background, Fincantieri C.N.I. Yard in Genoa as a private venture, authorised Danish Maritime Institute to undertake model tank testing with one of their fast passenger Ro-Ro ferry designs with the objective of investigating her ability to survive in rough seas in regard to the two worst damage positions.

The entire test programme was divided in two phases;

- Phase One was dedicated to the investigation of survivability of the ferry 'as is', i.e. generally in accordance with prescriptions in the 'Stockholm Agreement'

- The objectives of Phase Two were to find the limits for survivability in terms of three decisive parameters, regardless of the fact that they may lie beyond the limits of the existing design.

These three parameters were:

- Displacement / Damaged Freeboard
- Intact / Damaged GM
- Significant Wave Height

In the most extreme configuration the model was tested at 131.1% of the design displacement and at a freeboard of only 23% of the design damaged freeboard. The actual intact GM corresponded to 67.7% of the design intact GM, while the actual damaged GM corresponded to only 38.5% of the design damaged GM. The model was tested in waves of the significant wave height corresponding to 4 m, i.e., the maximum value prescribed by the 'Stockholm Agreement'.

The following sentence is from the final conclusion, which in its full form is present in Schindler (1997). 'During the entire test programme the model showed an excellent stability and never capsized. As this model was tested in loading and damage conditions well beyond the limits of the present design (regarding decisive parameters for survivability of damaged Ro-Ro ferries), the observed result shows (the actual vessel) to be extremely safe with respect to damage stability. Based on model test results, her survivability is considerably better than indicated by static calculations as prescribed by the 'Regional Agreement on Specific Stability Requirement' (the Stockholm Agreement), which refers to Ro-Ro passenger ferries for which MSC.36(63) HSC-code do not apply.

The second project to be mentioned in this paragraph basically originates from the 'Joint North- West European Research Project'. During the course of the model testing programme of this Project undertaken by DMI, different opinions of the importance of the transverse radius of gyration on survivability of damaged Ro-Ro ferries were presented. Regardless of ferry size, concept, loading condition, or any other factor which might have influence on an individual ferry's non-dimensional transversal radius of gyration, I_{xx}/B, the maximum value of I_{xx} allowed by the Stockholm Agreement for the purpose of the approval tests corresponds to 0.40 x B, as a maximum. The more exact importance of the transverse radius of gyration on survivability of damaged Ro-Ro has never been a subject for closer investigation by model tests.

Identifying the need for more research on this matter, the UK Secretary of State for the Environment, Transport and Regions invited DMI to submit a tender for such service. The objectives of the joint UK MCA and DMI research project was to provide a better indication and improve the understanding of the effects of the transverse radius of gyration on survivability of a damaged Ro-Ro passenger ship by an investigation based on model tests. Two models, which originally were designed to the standards in accordance with the Model Test Method for investigations under the Stockholm Agreement approval tests were used. The modifications necessary for this investigation included redesigning the ballast lifting arrangements of both models to facilitate the required control of the transverse radii of inertia of the models.

For each individual transverse position of ballast (radius of gyration), the survival criteria was the lowest value of GM, at which each model 'just' survived the test in that particular model configuration and significant wave height. This was achieved by demonstrating that the model 'just' capsized during a corresponding test at exactly the same condition, but with the GM reduced by the minimum step length available. The vertical position of the centre of gravity was controlled by adjusting the vertical position of the ballast with the minimum step length corresponded a 10 cm change in GM in ship scale. The available total range of model centre of gravity was sufficient to find the capsize / survive conditions of all tested model configurations in the applied sea states.

The transverse radius of gyration was controlled by a transverse shift of ballast using four predefined positions.

Therefore, the aim of the redesigned lifting arrangement was to ensure an accurate control of model centre of gravity KG and transverse radius of inertia I_{xx} independently of each other, and with absolute repeatability.

Each of the models was tested in waves of two different significant wave heights and with three different arrangement concepts on the main deck:

- Centre casing
- Centre casing and 1 intact transversal bulkhead
- Side casing

One model represented a 160 m long ferry which is a one-compartment ship with an open shaft arrangement, while the second one represented a 140 m long ferry which is a two-compartment ship with a twin skeg arrangement. In general, the tested models represented two different concepts. The most relevant are listed below , but the scale of both models was the same.

Ship 1	Ship 2
One-compartment ship	Two-compartment ship
Regulation: A-265	Regulation: Solas '90
Passed the Stockholm Agr. Tests*)	Did not pass the Stockholm Agr. Tests*)
GM_{intact} during tests: 0.8 m - 1.4 m	GM_{intact} during tests: 1.9 m - 2.8 m

*) The original 'as is' configuration including centre casing configuration.

The report, in MCA (1999) concluded: 'Disregarding these differences, the survivability of both models in the conditions as tested was practically unaffected by the applied range of the transversal radius of gyration which roughly represents shifting a mass corresponding to 50% of the total displacement by a distance of 1/3 the total beam B. Until the reliable statistics of the present Ro-Ro ferries' transversal radius of gyration are available, it is felt that the above range of the transversal radius of gyration covers a sufficiently wide band to either side of 40% of beam to be representative for the majority of the existing Ro-Ro ferry designs.' More details from this project are available in MCA (1999).

Bow Port Investigation - Free Sailing Model Tests with Models of Ro-Ro Passenger Ferries as Part of the Upgrading Process of Existing Ships

In parallel with the services described above, DMI can now offer a new type of investigation related to safety at sea. Modifications of the forward ramp/collision bulkheads on existing Ro-Ro ferries, as prescribed by the regulations related to the 'Stockholm Agreement', is an expensive process, which very often results in a considerable loss of capacity.

For the purpose of the UK-national approval process, DMI has developed a new test method to document the Ro-Ro ferries sea worthiness without need for expensive re-construction of the bow port arrangement. The free-sailing tests in head seas simulating the damaged visor/forward ramp are frequently performed in DMI's towing tank with radio controlled, self-propelled and totally free-sailing models representing passenger Ro-Ro ferries. These tests require the same degree of precision as is required from the damage stability investigations under the Stockholm Agreement. The test method for the purpose of such investigation was agreed upon with the Maritime and Coastguard Agency, UK.

Up to now, models of six UK domestic ferries, originally built and tested for the purpose of side damage investigations under the Stockholm Agreement took part in the 'Bow Port Investigations' as described in this paragraph. The main factor for good results are the freeing ports present on each of the ferries. The tested models showed sufficient stability as they were able to free the vehicle deck of water efficiently through the existing freeing ports in the model sides.

The importance of the freeing ports can be illustrated by the results from a series of free-sailing tests in head seas simulating the damaged visor/forward ramp with a model representing a 'conventional' ferry with a closed deck (without freeing ports). This ferry is about 200 m long with side casings on the main deck and has a very considerable intact freeboard. See Fig.1.

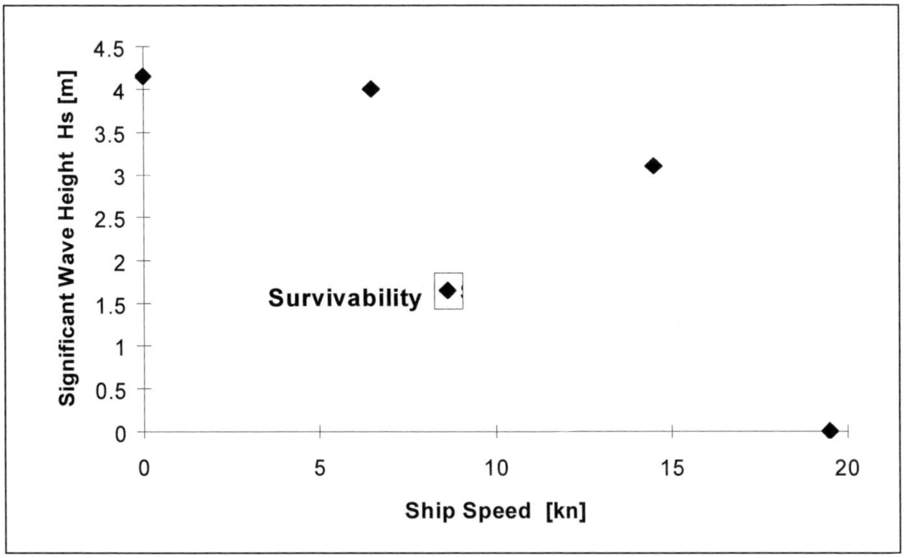

Figure 1 : Survivability in waves at different speeds. No freeing ports.

IMPORTANCE OF THE MODEL TEST METHOD

The public founded and the private research projects described above are of a great importance for better understanding of the mechanisms and the dynamic effects which determine capsize or survival of a damaged ferry in waves. Observations and conclusions made were an important factor for formulation of the existing international damage stability requirements, which apply to the Ro-Ro ferries. The same observations and conclusions are an important tool in hands of designers of new and safer Ro-Ro ferries.

The model test method is now approved and a recognised alternative to the deterministic method in the process of upgrading and approval of the existing Ro-Ro ferries with regard to the required standards. Good results achieved during such investigations have already resulted in considerable money saving for the Ro-Ro ship owners.

References

Damsgaard, A. and Schindler, M. (1996). Model Tests for Determining Water Ingress and Accumulation, *Int. Seminar on The Safety of Passenger Ro-Ro Vessels Presenting the Results of the Northwest European Research & Development Project, The RINA*, London 7 June.

Maritime and Coastguard Agency (1999). Research Project 423 - Transverse Radius of Gyration, Summary Report , UK , 22 June.

Pucill, K.F. and Velschou, S. (1990). Ro-Ro Passenger Ferries Safety Studies - Model Test of Typical Ferry, *Int. Symposium on the Safety of Ro-Ro Passenger Ships, The RINA*, London 26-27 April.

Schindler, M. (1997). Damage Stability Tests of a Model Representing a Fast Ro-Ro Ferry, *Symposium on The Safety of High Speed Craft, The RINA*, London 6 February.

Velschou, S. and Schindler, M. (1994). Ro-Ro Passenger Ferry Damage Stability, *The 12th Int. Conf. on Marine Transport Using Roll-on/Roll-off Methods, RoRo94*, Gothenburg 26 - 28 April.

Velschou, S. and Schindler, M. (1994). Ro-Ro Passenger Ferry Damage Stability Studies - A Continuation of Model Tests of Typical Ferry, *Symposium on Ro-Ro Ship's Survivability, The RINA*, London 25 November.

Contemporary Ideas on Ship Stability
D. Vassalos, M. Hamamoto, A. Papanikolaou and D. Molyneux (Editors)

ABOUT SAFETY ASSESSMENT OF DAMAGED SHIPS

Roby Kambisseri[1] and Yoshiho Ikeda[2]

[1] Department of Ship Technology, Cochin University of Science & Technology, Kerala, India
[2] Department of Marine System Engineering, Osaka Prefecture University, Japan

ABSTRACT

In this paper, a new approach to ensure the after damage survivability of ships is discussed. Severity of damage is measured by the size of damage opening. Required safety depends on the value lost if the ship sinks. A safer ship will be the one that can survive a larger damage opening, anywhere over its hull. In impact damage, size of damage opening will be influenced by the strength of structure at the region of impact. Survivability after damage, in a sea state, is to be assessed by a Capsizing Probability, considering also the effect of water shipped into the damaged region and the fluctuating restoring ability of the ship in waves.

KEYWORDS

Damage stability rules, Safety level, Survivability criteria, Damage opening size, Capsizing probability, Structure, Sinking, Ship

INTRODUCTION

A block diagram of safety level, damage and damage stability is shown in Figure 1, which will give a broad outline of the parameters involved in them, Roby Kambisseri and Yoshiho Ikeda (1996).

Required safety of a ship should be related to the (equivalent - financial - value) loss that can be caused by the sinking of that ship. The more the value involved in a ship, the safer it should be against sinking. In the case of damage stability, a more valuable ship should be able to withstand severe or larger size of damage (opening) on its hull or in other words should be able to withstand larger damages or heavier collision. The size of the opening caused on the hull of a ship will depend on the impact load and the strength of the structure in and around the location of collision. For the same damage opening, size of the space open to (i.e., the damage space) will depend on the subdivision of the ship and the location of damage.

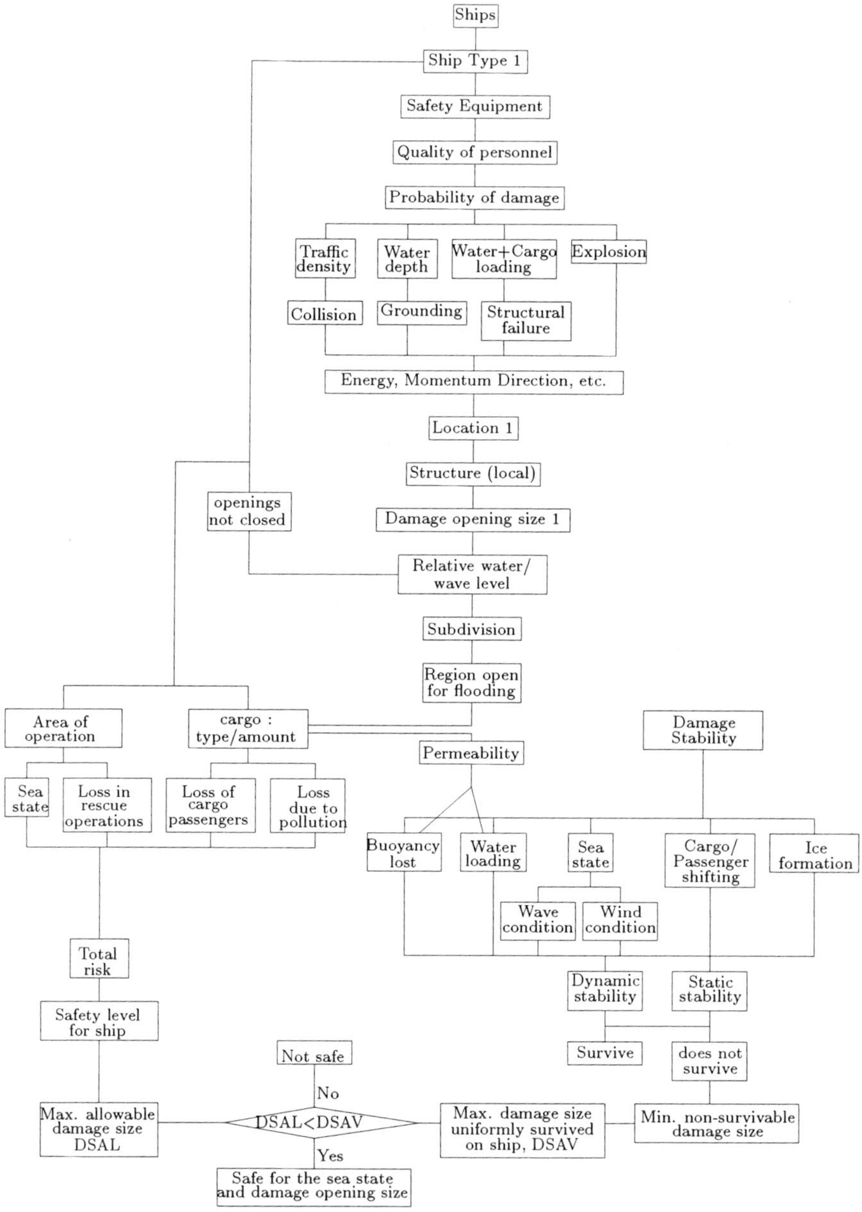

Figure 1: Block diagram of safety level, damage and damage stability

The weakest region, or the smallest damage opening for which a ship becomes unsurvivable, should determine the actual safety or damage surviving ability of that ship. Strength of the structure of a ship can be included if the size of damage opening is determined from impact load. Loss caused by the sinking of a ship or if mapped into a safety scale can be used to determine the size of the damage opening or the damage opening caused by an impact load that should be survived by the ship. The ship should achieve this in the sea states it has to operate. Survivability in a sea state is assessed by a capsizing probability study of the ship taking also into account the righting ability in the sea condition as well as the change in the righting ability due to the presence of water in the damaged region.

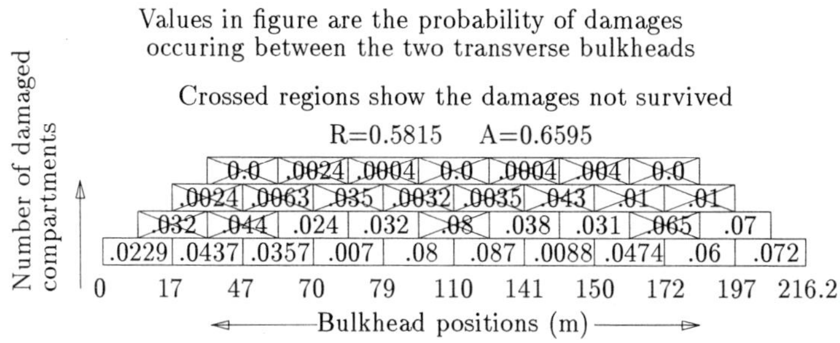

Figure 2: R, p_i, and A for D=16 m

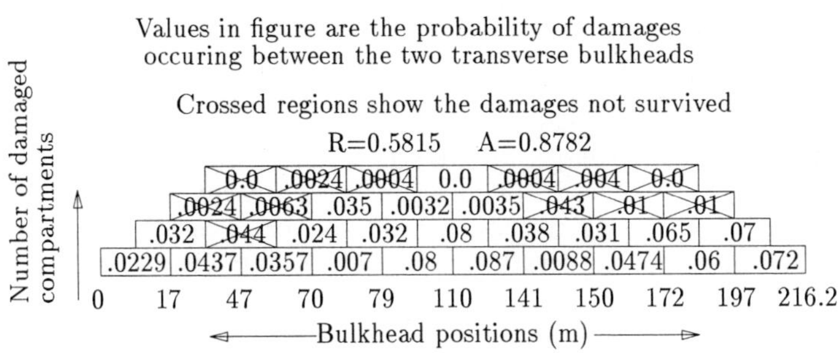

Figure 3: R, p_i, and A for D=18 m

DEFECTS OF PROBABILISTIC RULES

Probabilistic damage stability rules can be found in Soce (1992) and A.265 (1973), and a study on the proposals to modify the probabilistic rules are made by Roby Kambisseri and Yoshiho Ikeda (1995a).

Minor damage

Figures 2 and 3 show the probabilities of collision damages of a ship with different depths, Roby Kambisseri and Yoshiho Ikeda (1996). All non-survivable cases of damages (s_i=0) are shown crossed. Both cases satisfy the present criteria based on probabilistic approach (i.e.,

Attained subdivision index, A is not less than Required subdivision index, R), but can be lost by minor damage/s. The smallest damage that can sink these ships, with different A values, is the same. i.e. A higher A value does not ensure a higher level of safety.

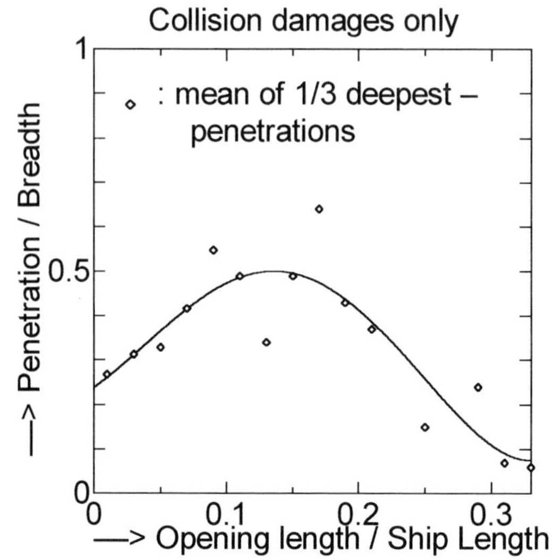

Figure 4: Change of damage penetration with damage opening length

Structural strength

In the present damage stability rules, extent of damage opening is determined on the basis of statistical data of (collision) damage. Real damage opening size, however, depends on many factors like strength of struck and striking ships, mass, velocity and bow shape of striking ship, striking angle and yawing velocity, etc. So, for the same collision, damage opening could vary with the variation in strength of local and global structure of ships.

Damage Penetration

In the probabilistic rules, damage penetration is assumed to increase with the damage length. Variation of damage penetration with the damage opening length is given in Figure 4, Roby Kambisseri and Yoshiho Ikeda (1996). This is drawn using the means of the $(1/3)^{rd}$ deepest penetrations of collision damage data used in the probabilistic rule. Very long damage openings may not be deep and such damages can be survived using longitudinal subdivisions, Roby Kambisseri and Yoshiho Ikeda (1995b). Damage penetration significantly depends on the structural strength of struck ships.

Survivability criteria

Present stability criteria are based on calm water righting ability curve and without considering the additional loading due to free water in damaged spaces, Soce (1992), A.265 (1973).

A new approach

The deficiencies of the present damage stability rules, which are mentioned above, call for new criteria and a method for its application, Roby Kambisseri and Yoshiho Ikeda (1996). In future, developments and new ideas in ship technology should be reflected in the damage stability rules. In the following sections, a concept of a new approach to future damage stability rules will be proposed.

SURVIVABILITY

How to determine whether a ship can survive a sea with a portion of its hull damaged? For this, the remaining intact portion of the hull, the damaged space with or without water which is or not flowing in and out of it and the water (waves) supporting the ship and their contributions are to be included into the intact stability analysis of ships.

A possibility that may not occur to intact ships is that the damaged ships may capsize due to loss of longitudinal stability, Nobuyuki Shimizu, Roby K. and Yoshiho Ikeda (1996).

The contribution from the water in the damaged space may increase or decrease the restoring ability. Studies show that when a ship is flooded, the ship rolls about the heel angle caused by flooding, John T. Stubbs, Peter van Diepen, Juan Carreras and Joseph H. Rousseau (1995). This is because, when the ship heels, the water gets trapped between two slanting planes, without much scope for its flowing and the amount of energy required to move the water over this sloped surface to the other side of the ship, is much greater than the energy contribution of the exciting forces. When the ship rolls, the motion of inside water lags behind the rolling motion and the energy of flow and the impact at the sides of the ship will have a roll damping effect and will be safer for the ship. So, the static effect seems to be critical to the damaged stability of ships. However, further studies are needed to see if there is any adverse dynamic effect.

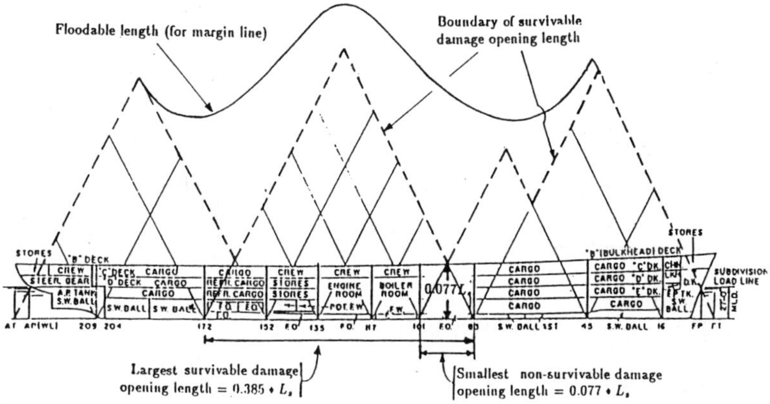

Figure 5: Survivable damage opening length boundary - A

For any sea state a capsizing probability, Umeda N. and Ikeda Y (1994), can be predicted considering all the factors that contribute to the stability of a ship. A required minimum value of capsizing probability can be used to classify the survivability after damages. The capsizing probability will determine whether a ship with a damage is survivable or not.

The variation in the methods employed in the estimation of motion in waves, effect of damage water, etc., may result in different values of capsizing probability for a ship. A rule-software could be used to standardise the calculation of capsizing probability and to ascertain the safety level of ships.

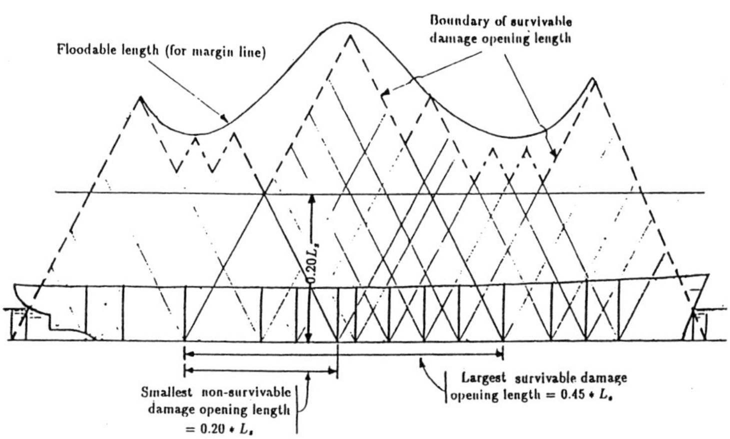

Figure 6: Survivable damage opening length boundary - B

SAFETY PARAMETER

How to grade the after damage safety of ships? If there is more than one design, meeting the same requirements, how to determine which one has the best damage stability? Figures 5 and 6 show two subdivisions based on the same floodable length curve. The first one can survive a largest damage opening of size $0.38L_s$ and the smallest damage opening size it cannot survive is $0.077L_s$. L_s is the maximum length on or below subdivision waterline. The second one can survive a largest damage opening of size $0.45L_s$ and the smallest damage opening size it cannot survive is $0.20L_s$. The second subdivision is safer against any damage caused by damage openings of size up to $0.20L_s$ where as an opening of size $0.077L_s$ can sink the first one. The second subdivision will be safer from the point of view of damage stability, since it will not be sunk even by a collision or impact much heavier than the one which will capsize the first. Safety level of damaged ships can be graded using the minimum damage opening size that the ship cannot survive. i.e., the parameter, which defines damage safety, is the size of the damage opening. Damage opening size depends also on the strength of the structure. To include it, safety parameter is to be changed to the impact load and the ship should survive all the damages caused by the damage openings created by that impact. This will help to identify and strengthen structurally weak locations on a ship.

Algorithm to find minimum non-survivable damage opening size

The maximum survivable damage opening, damages caused by which can be survived anywhere on a ship, can be found by the following method, Roby Kambisseri and Yoshiho Ikeda (1996).

Select the minimum damage opening length ($L_{m,d}$) which can cause a damage as follows:

a. If one compartment damage (no bulkheads damaged), we can assume, $L_{m,d} = 0.1$ m.

b. If two compartment damage (one bulkhead damaged), we can assume, $L_{m,d} = 0.5$ m.

c. If three or more compartment damage (two or more bulkheads damaged), $L_{m,d}$ = distance between foremost and aft most of the bulkheads damaged.

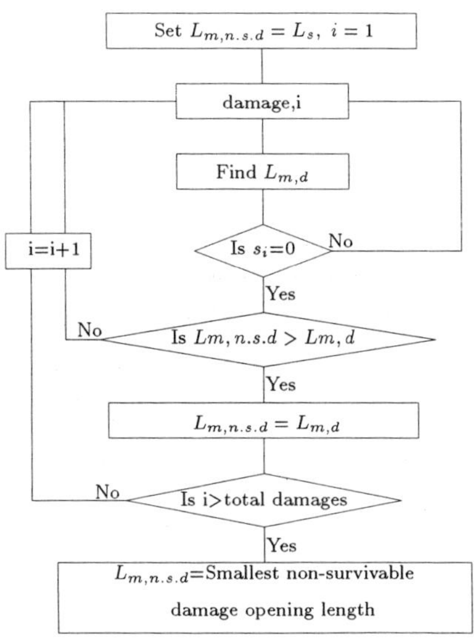

Figure 7: Flow chart to find the smallest non-survivable damage opening length

The flow chart in Figure 7 will give the smallest damage opening length ($L_{m,ns.d}$), some of the damages caused by which cannot be survived by the ship.

If $L_{m,ns.d}$ is 0.1m then the ship is non-survivable for at least one damage in which no bulkhead is involved. If $L_{m,ns.d}$ is 0.5m then the ship can survive all cases where bulkheads are not damaged and is non-survivable at least for a one bulkhead damage. If $L_{m,ns.d} > 0.5$ m then the maximum damage opening length survivable over the hull will be little less than $L_{m,ns.d}$.

FIXING SAFETY LEVEL

It is always difficult to define the safety level required for each transportation system. It may be relative and vary from time to time. The required safety level, however, depends on each ship and has to be measured by the loss generated by accident, including pollution of marine environment in addition to the loss of a ship including its passengers, crew and cargo, loss in rescue operation, etc. Economic aspects of ship's operation also have significant effect on the required safety level. Therefore it depends on ship type, size, capacity, missions, region of operation, etc.

The more valuable or the more dangerous an object is, it is to be given better protection so as to avoid it from loosing or to avoid it causing destruction. The level of protection given should be commensurate with the value of the object or the loss it could cause. The same is also applicable to ships. Level of safety of a ship should depend on the value of the ship including its cargo and passengers and the loss it could cause by way of pollution, etc.

Shipping is a profit-oriented operation. The cost required to prevent the loss of a ship should not be disproportionate with the value loss that the loss of ship will cause. Here comes the conflict between the shipping and regulating agencies. Shipping agencies are eyeing at profit and regulators are aiming at safety. So, it is appropriate to have regulations that will allow safety within the economic feasibility.

A ship should be designed to have enough capsizing probability for all damages caused by damage openings up to a certain size; the size is determined for each ship to match its required safety level. It need not survive larger damages, since the risk involved may not be sufficient enough to bear the cost of ensuring survival of a bigger damage.

Three models are given for determining the damage opening size required to be survived by a ship. Perfect survivability or a certain survival probability is assumed to be guaranteed for all damages caused by the damage openings up to this size; the survival probability decreases with the increasing damage opening size.

Model 1

In this model, Roby Kambisseri and Yoshiho Ikeda (1996), (Persons On board)/$(L\ B\ d)^{1/3}$ is assumed to give the safety level. The following relation maps the safety level and damage opening size.

$$\text{Percentage damage opening size} = K \times (\text{Persons On board})/(L \times B \times d)^{1/3} \qquad \textbf{(1)}$$

Where L is the length, B is the breadth and d is the draft of the ship. K is 0.25 for the data in Table 1, assuming a maximum possible damage opening size of $0.24L_s$, which is the same as that in the present rules. This is a very simple model and with a suitable function for safety level, damage opening sizes and their mappings can be found.

Model 2

In this model, Roby Kambisseri and Yoshiho Ikeda (1996), a safety level is to be fixed using probable loss due to the sinkage of the ship. Using computer simulations of structural damage, damage experiments and damage statistics, the probability for each damage opening size can be fixed. A relation mapping the required safety level or risk involved, to the size of damage opening could be used to ensure a survivability standard varying with safety.

TABLE 1
TRIAL RELATIONS FOR SAFETY LEVEL

P is the number of passengers.

Length (m)	Breadth (m)	D (m)	Passe-ngers	Service Speed (Kn)	P/L	$8P/L^{0.5}$	$0.25P/(LBd)^{1/3}$
166.0	27.2	8.7	341	18.5	2.05	11.47	2.51
211.9	32.2	7.5	2621	21.5	12.37	28.14	17.65
105.0	19.5	4.7	7	17.5	6.67	20.66	8.23
157.7	23.4	5.8	114	22.0	0.72	6.80	1.03
118.5	24.0	7.7	16	24.0	0.14	2.94	0.14
150.0	29.0	6.6	2	20.0	1.33	9.24	1.63
285.0	40.0	8.0	37	22.5	12.98	28.83	20.55
155.8	21.4	5.2	306	20.0	1.96	11.21	2.96
256.0	32.2	12.0	27	23.5	0.11	2.6	0.15
320.0	58.0	20.8	36	15.5	0.11	2.68	0.12
156.0	26.7	9.7	28	20.0	0.18	3.39	0.20
174.3	32.2	11.2	34	14.5	0.2	3.53	0.21
176.0	32.0	11.3	25	14.3	0.14	3.02	0.16
143.0	23.0	6.2	415	21.0	2.90	13.63	3.8
126.8	18.4	3.6	1606	19.5	12.67	28.47	19.75
107.0	19.5	5.3	473	15.2	4.42	16.82	5.31
185.0	30.8	7.5	2866	20.0	15.49	31.49	20.49
320.0	58.0	20.8	4	15.7	0.13	2.83	0.14
177.0	26.3	7.0	1887		10.66	26.12	14.77
183.0	28.0	7.5	2394		13.08	28.94	17.74
190.8	28.0	7.5	2566		13.45	29.34	18.75
224.0	31.5	7.8	3554	22.3	15.87	31.87	23.41
177.0	26.3	6.8	1396		7.89	22.47	11.03
190.8	28.0	7.5	1896		9.94	25.22	13.85
171.6	32.0	6.8	3313	21.5	19.31	35.15	24.78
170.0	27.8	6.0	14	21.0	8.24	22.96	11.48
163.1	26.8	6.7	1669	23.0	10.24	25.6	13.54
103.0	18.9	5.0	138	17.5	13.40	29.28	16.168
170.0	25.5	6.6	711	22.85	4.18	16.36	5.81
103.8	19.2	4.7	75	18.0	7.23	21.50	8.9
101.8	20.7	5.3	417	16.0	4.1	16.19	4.66

A graph similar to Figure 8, showing the relationship between the safety level and the damage opening length, can be used to find the damage opening length required to be survived by a ship. The initial part of the graph is the probability distribution of (collision) damage - maximum probable damage opening length of which is around $0.35L_s$ - which is modified and extended so that the damage opening length is the length of the ship when the safety level is the maximum. This graph and the ranges of both its axes could be different for different types of ships and could vary even between different ships of a type. This model has some flexibility in accommodating the changes in the concept of safety, changes in the structural strengths of ships and the changes in mapping between safety level and damage opening. A relation to determine the safety level and the mapping should be fixed only after careful study.

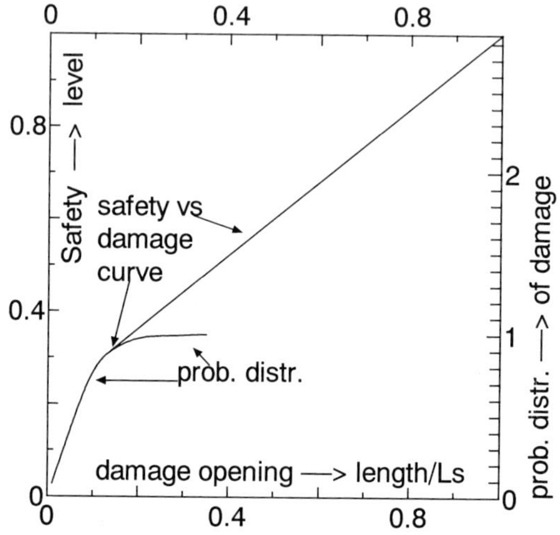

Figure 8: Required survivable damage opening length (model 2)

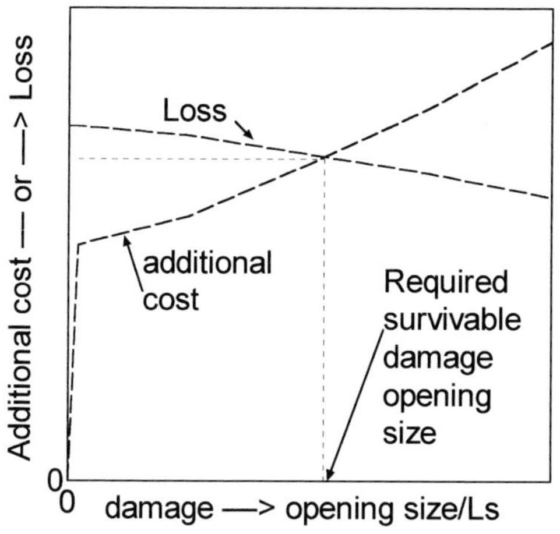

Figure 9: Required survivable damage opening size (model 3)

TABLE 2
SALIENT FEATURES OF PRESENT DAMAGE STABILITY RULES AND NEW PROPOSAL

Sl. no.	Features compared	Deterministic Rule	Probabilistic Rule	Present Conceptual Proposal
1	Safety parameter	Factor of subdivision, F	Required Sub-division Index, R	Survivable damage opening size
2	Safety parameter depends on	Length of ship, No. of passengers	Length of ship, No. of passengers, Life boat capacity	Probable loss or Risk due to capsize of ship
3	Damage space considered	All damages whose length not greater than Floodable length	All damages caused by damage openings of up to $48/L_s$ but ≤ 0.24	All damages caused by damage openings of a certain size
4	Damage opening size based on	Floodable length (in effect)	Collision damage statistics	Probable loss due to capsize of the ship
5	Damage Survivability	All damages should meet the after damage stability requirements	Only a percentage of the above damages need be survived	All the above damages to meet the survival criteria
6	Ship Structure	Not considered	Not considered	Survivable damage opening size to depend on the structure
7	Defect of the Approach	For the same damage opening, survivability decreases with increase of required safety	Minor damage non-survivability could be present	?
8	Survivability Criteria	Based on calm water GZ curve	Based on calm water GZ curve	Based on restoring ability at sea and Capsizing Probability
9	Loading due to flood water	Not considered	Not considered	To be included
10	Survivable Sea State	Not specified	Not specified	Sea state to be specified

Model 3

This is a complex model. In this the damage opening size that a ship is required to survive can be found using the relation shown in Figure 9. This figure could be made for each ship. The additional cost is the cost required to make a ship, which cannot survive any damage, to one, which can survive all damages caused by a damage opening size. Loss is the loss or a proportion of the loss involved in the loss of the ship due to a damage caused by the damage opening size. It must be taken into account that the probability of collision and capsize due to that is very small.

Comparison of Rules and New Concept

The salient features of the present rules and the new approach are given in Table 2 for comparison and better understanding.

CONCLUSIONS

The stability criteria and the method to vary safety level of ships applied in the present damage stability rules are not sufficient for ensuring safety of damaged ships as well as to grade the relative safety of ships. On the basis of recent studies a new way of arriving at realistic survivability criteria is proposed in this paper, which is used to estimate a capsizing probability. Safety level of a ship is connected to the value loss that can occur with the sinking of the ship. Size of a damage opening is identified as the parameter that should define the safety level. All the damages caused by openings up to this size are to be survived by a ship. Three models to determine the damage opening size that should be survived by a ship are also shown in this paper.

References

A.265. (1973). IMO Resolution A.265 (VIII). *International Maritime Organisation.*

John T. Stubbs, Peter van Diepen, Juan Carreras and Joseph H. Rousseau. (1995). TP12310E Flooding Protection of RO-RO Ferries, Phase I. *Transportation Development Centre, Policy and Co-ordination, Transport Canada* **1.**

Nobuyuki Shimizu, Roby K. and Yoshiho Ikeda. (1996). An Experimental Study on Flooding into the Car Deck of a RORO Ferry through Damaged Bow Door. *Journal of the Kansai Society of Naval Architects, Japan* **225.**

Roby Kambisseri and Yoshiho Ikeda. (1995a). A Comparative Study of the Probabilistic Damage Stability Rules and Proposals. *Contemporary Ideas on Ship Stability, Univ. of Strathclyde.*

Roby Kambisseri and Yoshiho Ikeda. (1995b). Compartmentation - Best Guide for Damage Stability or what is Wrong with the Floodable Length Approach of Subdivision. *Contemporary Ideas on Ship Stability, Univ. of Strathclyde.*

Roby Kambisseri and Yoshiho Ikeda. (1996). A New Approach to Damage Stability Rule (1st Report) - Discussion on the Present Rules and the Concept of the New Approach. *Journal of the Kansai Society of Naval Architects, Japan.* **226.**

Soce. (1992). SOLAS consolidated edition.

Umeda N. and Ikeda Y. (1994). Rational Examination of Stability Criteria in the Light of Capsizing Probability. *STAB'94* **2.**

D. Vassalos, M. Hamamoto, A. Papanikolaou and D. Molyneux (Editors)

SURVIVABILITY OF DAMAGED RO-RO PASSENGER VESSELS

Bor-Chau Chang[1] and Peter Blume[2]

[1]TU Hamburg-Harburg,
Lämmersieth 90, D-22305 Hamburg, Germany; email: bschang@ms26.hinet.net
[2]Hamburg Ship Model Basin (HSVA),
Bramfelder Strasse 164, D-22305 Hamburg, Germany

ABSTRACT

The survivability of damaged ro-ro passenger vessels is investigated by both model tests and theoretical motion simulations in irregular seaways. The simulation combines nonlinear equations for roll and surge motions with a linear treatment of heave, pitch, sway, and yaw, using the strip method. The volume of water in each damaged compartment is corrected at each time step. Special emphasis was placed on simulating realistically the motion of water on deck. A series of model tests regarding the survivability of damaged ships was carried out at HSVA in accordance with the "Model Test Method" required by IMO. The simulation method was comprehensively validated by comparisons with model tests. Based on the results of model tests and motion simulations, fundamental relationships between capsizing safety of damaged ro-ro ships and form parameters of the ship, locations of the damage and vehicle deck subdivisions can been established.

KEYWORDS

Survival, damage case, ro-ro ship, simulation, model test, seakeeping

INTRODUCTION

Concerns for the safety and vulnerability of ro-ro ships have been expressed constantly in the past. After several disasters with ro-ro passenger vessels, there has been strong pressure for increasing the safety level of this ship type, resulting in a new SOLAS-Resolution, SOLAS(1974), and requirements accepted at the Stockholm Conference (known as the *Stockholm Regional Agreement*), IMO (1996). The latter demands that a vessel satisfies SOLAS'90 criteria in the presence of a given height of water on the vehicle deck. An alternative has been allowed also: to perform model tests to detect whether the damaged ship will capsize in certain seaways. The experiments have to be performed in accordance with the *Model Test Method* required by IMO. Existing ro-ro passenger ships shall comply with the provisions of the new agreement from a date between April 1, 1997 and October 1, 2002, depending on the present standards of safety of the ships.

Ships which fail to satisfy the Stockholm Regional Agreement have to be modified for further operating. To this end, above all the vehicle deck arrangement will be considered; however, the effect of various deck arrangements is still uncertain. Therefore the present research is intended to show such influences in model experiments and to validate a motion simulation method for testing the survival conditions of damaged ships in waves.

METHOD OF SIMULATION

In the following, the method is only sketched because all details are explained in Chang (1999), Kröger (1986a,b) and Petey (1986). The ship is considered as a six-degree-of-freedom system travelling at a given mean angle relative to the dominant direction of a stationary seaway. The seaway is simulated as a superposition of a large number of component waves having random frequency, direction and phase angle. The random quantities are computed from a given sea spectrum. During the simulation the chosen mean ship speed and mean wave encounter angle remain constant, whereas the instantaneous ship speed and heading are influenced by the ships motions which are simulated in all six degrees of freedom. For the heave, pitch, sway and yaw motions, the method uses response amplitude operators determined by means of the strip method, whereas the roll and surge motions of the ship are simulated, using nonlinear motion equations coupled with the other four degrees of freedom. Thus the four first mentioned motions have been treated linearly, including hydrostatic and hydrodynamic forces. Both the wave exciting moment and the roll moment induced by the sway and yaw motions of the ship are determined by response amplitude operators. The following nonlinear motion equation is used for the determination of rolling:

$$\ddot{\varphi} = \{ M_d - m(g-\ddot{\zeta})\, h_s - I_{xz}[(\ddot{\theta} + \theta\dot{\varphi}^2)\sin\varphi - (\ddot{\phi} + \phi\dot{\varphi}^2)\cos\varphi] + M_{wind} + M_{cd} + M_{sy} + M_{waves} \} / \{ I_{xx} - I_{xz}(\phi\sin\varphi + \theta\cos\varphi) \} \tag{1}$$

where a dot designates time derivatives,

φ, θ, ϕ = roll, pitch and yaw angle
m = mass of ship including the water on the vehicle deck and compartments
$g, \ddot{\zeta}$ = gravitational acceleration and heaving acceleration at the c.o.g.
h_s = righting lever in an "effective" longitudinal wave
M_d = damping moment ($= -d_L\ \dot{\varphi} - d_Q\ \dot{\varphi}\,|\,\dot{\varphi}\,|$)
M_{wind} = moment due to wind
M_{cd} = moment due to water motion on the vehicle deck and in compartments
M_{sy} = moment due to sway and yaw motions, using response amplitude operators determined by means of the strip method
M_{waves} = exciting moment due to waves for the non-oscillating ship, using response amplitude operators determined by means of strip method
I_{xx} = moment of inertia about longitudinal axis through the centre of gravity of the ship, including added inertia due to water on the vehicle deck, in flooded compartments and due to the outside water
I_{xz} = product of inertia relating to the centre of gravity of the ship

For computing righting levers h_s due to the ship's heel in a hydrostatic pressure field under the wavy water surface, Grim's effective wave concept, Grim (1961), in the form modified by

Söding (1982) is used. The height Z of the wavy water surface along the vertical plane through the longitudinal ship axis is approximated in the form

$$Z(x,t) = a(t) + b(t)x + c(t)\cos(2\pi x/\lambda_E) \qquad (2)$$

in the region of ship length using the method of least squares of errors. The length between perpendiculars L_{pp} is used as length of the effective wave λ_E. Grim showed that the response amplitude operators between regular waves and the quantities $a(t)$, $b(t)$, $c(t)$ in (2) can be computed easily. Using these transfer functions together with the heave and pitch transfer functions, the mean ship immersion, its trim and the effective regular wave height are computed for every time step during the simulation. The righting arm is interpolated from tables, computed before starting the simulation, depending on these three quantities and the heel angle.

In the simulations, time is advanced in small increments. The rate of inflow and outflow of water through any opening is estimated from the motion of the internal and the external water surface relative to the openings at each time step, Söding (1982). The openings can be located at the shell of a ship or at internal subdivisions between compartments; they may be intended as openings, or they may be produced by a damage, e.g. due to a collision. The variations of the mass, the centre of gravity and moment of inertia of the ship due to the inflow and outflow are considered by varying the above quantities. The forces and moments due to the interior fluid motion in partly flooded rooms and on the vehicle deck are also determined (M_{cd}) and added to the other moments due to wave excitation, wind etc. Two different methods of computing the internal water flow are used, depending on the height of the water. For low fill depth compared to the tank width, the velocity vector of fluid particles is almost parallel to the tank bottom. The velocity component perpendicular to the tank bottom is neglected, and the depth-averaged water velocity is computed from the so-called shallow-water equations in two dimensions for an accelerated reference system, Petey (1986). Glimm's method (a random choice method), Petey (1986) and Dillingham (1981), is used to obtain the solution, because this method is capable of dealing with the frequently occurring cases of hydraulic jumps and of a partially dry bottom. However, the shallow-water equations give correct results only for small heel (about $< 25°$). Therefore, if the average water fill depths are larger than about 15% of the tank width, or if the heel angle exceeds 25°, the deep water method is used: The free surface of the liquid is assumed oblique but plane, since the greatest natural period of the fluid oscillation is much smaller than the dominant period of the ship motions in this case. A single degree of freedom equation of motion of the liquid free surface is derived from Lagrange's equation, Söding (1982) and Petey (1988), which can be solved in the time domain using the familiar Runge-Kutta integration scheme. The change between these two methods is made automatically during the simulation according to the actual situations of the deck or compartments in question.

COMPARISON WITH MODEL EXPERIMENTS

For three ships, and with different subdivisions on the ro-ro deck, damage positions, trims and drafts, the limit between safe and unsafe damaged metacentric heights is determined by testing the survivability with different KG values. The limit was determined both by simulation and model tests. The simulation method is validated by comparing both results.

Ships and Investigated Conditions

For the investigations three existing ro-ro passenger vessels, that is ship A, B and C, were chosen. They are considered to be typical representatives of the kind used in the Baltic sea, Blume & Chang (1998). Each ship was investigated with three different subdivisions on the ro-ro deck. Version A is the original design. Version E is a subdivision by two full transverse bulkheads. Version F is a subdivision by longitudinal bulkheads which are at distance larger than $B/5$ from the ship's sides. The damage opening had a length of $3m+0.03L_s$ and ranged vertically from the bottom up to 16 *m* above the base. The depth of penetration was $B/5$.

Ship A

The main dimensions of ship A are:

Length between perpendiculars L_{pp}	144.00 *m*
Subdivision length L_s	153.00 *m*
Breadth B	29.00 *m*
Height to vehicle deck at midship	8.10 *m*
Draught	6.40/6.20/5.90 *m*
Trim	0.0 *m*
Investigated damage cases (Figures 1 and 2)	D9+10 and D11+12

Figures 1 and 2 show that compartments below the vehicle deck were flooded through an opening on port side. Figures 3a to 3c show the three variations of subdivisions on the ro-ro deck for ship A. The statical equilibrium floating positions of the damaged ship are shown in Table 1 for each draft and damage case, Blume & Chang (1998).

TABLE 1: THE STATICAL EQUILIBRIUM FLOATING POSITIONS OF THE DAMAGED SHIP A

Damage case		D11+12			D9+10		
Draft of intact ship	[*m*]	6.40	6.20	5.90	6.40	6.20	5.90
Damaged:							
Mean draught	[*m*]	7.28	7.06	6.73	7.24	7.01	6.67
Trim*	[*m*]	−2.08	−2.06	−2.06	0.18	0.17	0.13
Heel angle**	[°]	−0.30	−0.30	−0.40	−1.70	−2.20	−2.90
Residual freeboard (SOLAS)	[*m*]	0.11	0.32	0.63	0.38	0.50	0.66
Metacentric height GM_D	[*m*]	4.11	3.61	3.13	3.63	3.04	2.38

* "+" means forward trim

** "+" means heeling to starboard

Ship B

The main dimensions of ship B are:

Length between perpendiculars L_{pp}	165.00 *m*
Subdivision length L_s	171.00 *m*
Breadth B	26.00 *m*
Height to vehicle deck at midship	8.00 *m*
Draught	5.75 *m*

Trim	−1.0/0.0/+1.0 *m*
Investigated damage cases (Figures 4 and 5)	D7+8 and D11+12

Figures 4 and 5 show that compartments below the vehicle deck were flooded through an opening on port side as for ship A. The three variations of vehicle deck subdivisions are shown in Figures 6a to 6c. The statical flooding results in a small heel to port side, see Table 2.

TABLE 2: THE STATICAL EQUILIBRIUM FLOATING POSITIONS OF THE DAMAGED SHIP B

Damage case		D11+12		D7+8	
Draft of intact ship	[*m*]	5.75	5.75	5.75	5.75
Trim of intact ship	[*m*]	0.0	−1.00	0.0	+1.00
Damaged:					
Mean draught	[*m*]	6.47	6.50	6.07	6.06
Trim	[*m*]	−1.93	−2.98	0.41	1.38
Heel angle	[°]	−0.20	−0.20	−2.10	−3.70
Residual freeboard (SOLAS)	[*m*]	0.46	−0.13	1.34	0.66
GM_D	[*m*]	2.34	2.04	2.31	1.40

Ship C

The main dimensions of ship C are:

Length between perpendiculars L_{pp}	166.00 *m*
Subdivision length L_s	170.80 *m*
Breadth B	27.20 *m*
Height to vehicle deck at midship	8.70 *m*
Draught	6.00 *m*
Trim	−1.07/0.0/+1.04/+1.08 *m*
Investigated damage cases (Figures 7 and 8)	D6+7 and D10+11

Three vehicle deck subdivisions as shown in Figures 9a to 9c were investigated. The statical flooding results in a small heel to port side or to starboard, see Table 3.

TABLE 3: THE STATICAL EQUILIBRIUM FLOATING POSITIONS OF THE DAMAGED SHIP C

Damage case		D10+11		D6+7		
Draft of intact ship	[*m*]	6.00	6.00	6.00	6.00	6.00
Trim of intact ship	[*m*]	0.0	+1.04	0.0	−1.07	+1.08
Damaged:						
Mean draught	[*m*]	6.42	6.46	6.42	6.36	6.47
Trim	[*m*]	1.27	2.30	−0.94	−1.93	0.05
Heel angle	[°]	−7.90	−8.50	6.40	6.20	6.60
Residual freeboard (SOLAS)	[*m*]	0.08	−0.37	0.26	−0.16	0.61
GM_D	[*m*]	2.16	1.98	2.49	2.54	2.45

Model Tests and Simulations

A series of model tests regarding the survivability of the damaged ships were carried out at the Hamburg Ship Model Basin (HSVA) in accordance with the "Model Test Method" required by IMO. Each model was built in a scale 1:30 and fitted with two propellers, two rudders and bilge keels. During the tests the model was exposed to long-crested irregular waves in the "dead ship" condition. This means: the model was drifting freely without speed ahead more or less parallel to the wave crests, with the opening on the side facing the wave. The measurements were started when the sea state was fully developed at the location of the model, and were generally continued for a time corresponding to 33 minutes in full scale.

The sea states were defined by (a) a JONSWAP spectrum with peak enhancement factor $\gamma=3.3$ having a modal period T_p= 8 s, and (b) a Pierson-Moskowitz spectrum with modal period 12 s. The significant wave height $H_{1/3}$ was 4 *m*. For each case different *KG* values were tested in order to determine the limit between safe and unsafe conditions with respect to capsizing. For each sea state repeated runs with different seaway realizations were performed. For details about the model tests see Blume & Chang (1998).

Motion simulations were performed for the same conditions as used in the model tests; however, the duration of each simulation was 50 minutes. The time step of the simulation was 0.2 s for the ship motion and 0.02 s for the motion of the water on deck. The seaway was simulated as a superposition of 50 regular waves. The effect of wind was neglected as in model tests. The statical equilibrium floating position of the damaged ship was used as initial condition in all simulations, as also in the model tests.

Comparisons and Discussion of Results

Table 4 shows the results of both model tests and motion simulations only for an example case, by giving the number N of accomplished model test runs or simulations and the number N_c of runs in which the ship capsized. For the complete results of all investigated cases see Chang (1999).

In order to determine the limiting GM_D between capsize and survival accurately, additional simulations were carried out with GM_D being changed in steps of 5 cm. The results are also included as in Table 4. The ship is considered to survive under that condition if she never capsizes in eight simulations with different realizations of the same seaway. Tables 5 to 7 show the limiting GM_D values for the three ships based on the results of the simulations and model tests. To derive the limiting GM_D value from the experiments without additional tests, the residual area E_R of the statical righting lever curve in still water of the respective damage case beyond the maximum 5-minutes-average of the measured heeling angles was used to represent the survivability limits: the limiting GM_D is achieved where the function $E_R(GM_D)$, linearly extrapolated from the measurements without a capsize, crosses the line $E_R=0$. Figure 10 illustrates the process for an example case, Blume & Chang (1998).

Comparing the results, generally a good correlation is found between model tests and simulations. Substantial differences between the results of simulations and model tests are shown only in a few cases: For ship A, sea state 4*m*/8*s*, with the subdivision Version F, draft T=6.2 *m*; for ship B, damage case D7+8, trim dT=0 *m*, with the subdivision Version A; for ship C, damage case D10+11, trim dT=0 m, with subdivision Version A.

In view of the very complicated water and ship motions and the small residual righting levers (compared to the ship dimensions and to her motion amplitudes) which determine whether a condition is safe or unsafe, the close coincidence between experimental and simulated limiting GM_D values is surprisingly good. A one-degree change in the equilibrium heel angle changes the results more than the difference between tests and simulations (Table 5)! The simulation method was comprehensively validated by comparisons with model experiments.

The effects of additional vehicle deck subdivisions (Version E and F) on damage survivability can be noticed by examining the results presented in Tables 5 to 7 as follows: In general a subdivision by transverse bulkheads is quite effective whereas a subdivision by longitudinal bulkheads hardly changes the limiting GM_D. But the effects of additional vehicle deck subdivisions strongly also depend on other factors, such as the length of the subdivided room, the damage location, and the arrangement of the trunks on the vehicle deck.

CONCLUSIONS

1. Unreasonably large metacentric height is required for the ship not built according to SOLAS 90 standard (ship A) in its original state to avoid capsizing.
2. Generally additional vehicle deck subdivisions increase the capsize resistance. Transverse bulkheads are found to be a better alternative than longitudinal subdivisions with respect to survivability.
3. Our simulation model is capable of predicting, with good engineering accuracy, the limit of damaged metacentric heights between safe and unsafe with respect to capsizing. It shall be used for systematic parameter investigations, to identify those parameters which are most significant for the survivability of damaged ro-ro vessels.

REFERENCES

Blume, P. and Chang, B.C. (1998), *Überlebensfähigkeit von RoRo-Fähren im Leckfall*, Bericht Nr. 1623, Hamburg Ship Model Basin (HSVA), August 1998

Chang, B.C. (1999), *On the Survivability of Damaged Ro-Ro Vessels Using a Simulation Method*, Bericht Nr. 597, Arbeitsbereiche Schiffbau der Technische Univ. Hamburg-Harburg, Hamburg

Dillingham, J. (1981), Motion Studies of a Vessel with Water on Deck, *Marine Technology, Vol.* **18** No. 1, 1981

Grim, O. (1961), Beitrag zu dem Problem der Sicherheit des Schiffes im Seegang, *Schiff und Hafen*, No. **6**

IMO (1996), Regional agreement concerning specific stability requirements for ro-ro passenger ships, *IMO circ. Letter No.* 1891 dated 29th April 1996

Kröger, P. (1986a), Rollsimulation von Schiffen im Seegang, *Schiffstechnik* **33**, p.187

Kröger, P. (1986b), Ship Motion Calculation in a Seaway by means of a Combination of Strip Theory with Simulation, *3rd International Conference on Stability of Ships and Ocean Vehicles (STAB)*, Gdansk

Petey, F. (1986), Numerical Calculation of Forces and Moments due to Fluid Motion in Tanks and Damaged Compartments, *STAB 86 Proceedings*

Petey. F. (1988), *Ermittlung der Kentersicherheit lecker Schiffe im Seegang aus Bewegungssimulationen*, Bericht Nr. 487, Institut für Schiffbau, Hamburg

Söding, H. (1982), *Leckstabilität im Seegang*, Bericht Nr. 429, Institut für Schiffbau, Hamburg

SOLAS (1974), *International Convention for the Safety of Life at Sea (SOLAS) 1974, including amendments*

TABLE 4: NUMBER N OF MODEL TESTS OR SIMULATIONS AND NUMBER OF THEM ENDING WITH A CAPSIZE N_C; SHIP A, DAMAGE CASE D11+12, TRIM OF INTACT SHIP dT=0.0 M, WITHOUT SUBDIVISION ON VEHICLE DECK (VERSION A)

Draft of intact ship	Damaged ship	Sea state 4*m*/8*s*				Sea state 4*m*/12*s*			
		Test		Simulation		Test		Simulation	
T [m]	GM_D [m]	N	N_c	N	N_c	N	N_c	N	N_c
		without subdivision on vehicle deck (Version A)							
5.9	**4.45**	2	-	4	-	2	-	4	-
	3.24	2	-	4	-	2	-	4	-
	2.85	5	-	8	-	5	-	8	-
	2.60	-	-	8	-	-	-	-	-
	2.55	-	-	8	2	-	-	-	-
	2.49	5	1	8	4	5	-	8	-
	2.45	-	-	-	-	-	-	8	-
	2.40	-	-	-	-	-	-	8	3
6.2	**3.76**	2	-	4	-	-	-	4	-
	3.65	-	-	8	-	-	-	-	-
	3.60	-	-	4	1	-	-	-	-
	3.55	-	-	4	2	-	-	-	-
	3.45	-	-	4	4	-	-	-	-
	3.33	2	1	4	4	-	-	8	-
	3.30	-	-	-	-	-	-	4	3
	2.48	1	1	4	4	-	-	4	4
		without subdivision on vehicle deck (Version A) with −1° added heeling angle							
5.9	**3.24**	2	-	4	-	2	-	4	-
	2.95	-	-	8	-	-	-	-	-
	2.90	-	-	8	2	-	-	-	-
	2.85	2	1	4	1	2	-	8	-
	2.80	-	-	-	-	-	-	8	-
	2.75	-	-	-	-	-	-	8	1

TABLE 5: THE LIMITING GM_D VALUES BETWEEN CAPSIZE AND SURVIVAL BASED ON MODEL TESTS AND SIMULATIONS **FOR SHIP A;** TRIM OF INTACT SHIP dT=0.0 *M*

Draft of intact ship *T*	Sea state 4*m*/8*s* D 11+12 Test GM_D	Sea state 4*m*/8*s* D 11+12 Simulation GM_D	Sea state 4*m*/8*s* D 9+10 Test GM_D	Sea state 4*m*/8*s* D 9+10 Simulation GM_D	Sea state 4*m*/12*s* D 11+12 Test GM_D	Sea state 4*m*/12*s* D 11+12 Simulation GM_D	Sea state 4*m*/12*s* D 9+10 Test GM_D	Sea state 4*m*/12*s* D 9+10 Simulation GM_D
	without subdivision on vehicle deck (Version A)							
5.9 m	2.54 m	2.60 m	3.28 m	3.25 m	2.32 m	2.45m	-	3.25 m
6.2 m	3.55 m	3.65 m	4.09 m	4.14 m	-	3.33 m	-	4.14 m
6.4 m	-	-	4.30 m	4.50 m	-	-	-	4.45 m
	without subdivision on vehicle deck (Version A) with −1° added heeling angle							
5.9 m	3.20 m	2.95 m	-	-	2.76 m	2.80 m	-	-
	with longitudinal bulkheads on vehicle deck (Version F)							
5.9 m	1.53 m	1.67 m	1.90 m	1.69 m	-	1.62 m	-	1.64 m
6.2 m	1.82 m	2.22 m	2.36 m	1.90 m	-	2.17 m	-	1.90 m
6.4 m	-	-	2.36 m	2.00 m	-	-	-	2.00 m
	with two full transverse bulkheads on vehicle deck (Version E)							
5.9 m	1.38 m	1.34 m	1.78 m	1.91 m	-	1.29 m	-	1.91 m
6.2 m	1.51 m	1.57 m	2.18 m	2.28 m	-	1.57 m	-	2.28 m
6.4 m	-	-	2.42 m	2.46 m	-	-	-	2.46 m

TABLE 6: THE LIMITING GM_D VALUES BETWEEN CAPSIZE AND SURVIVAL BASED ON MODEL TESTS AND SIMULATIONS **FOR SHIP B**; DRAFT OF INTACT SHIP T=5.75 *M*

Trim of intact ship dT	Sea state 4*m*/8*s* D 11+12 Test GM_D	Sea state 4*m*/8*s* D 11+12 Simulation GM_D	Sea state 4*m*/8*s* D 7+8 Test GM_D	Sea state 4*m*/8*s* D 7+8 Simulation GM_D
	without subdivision on vehicle deck (Version A)			
0. m	2.30 m	2.44 m	1.13 m	1.67 m
−1.0 m (Trim by stern)	2.73 m	2.77 m	-	-
+1.0 m (Trim by bow)	-	-	1.50 m	1.69 m
	with longitudinal bulkheads on vehicle deck (Version F)			
0. m	1.19 m	1.44 m	1.60 m	1.68 m
−1.0 m	1.54 m	1.67 m	-	-
+1.0 m	-	-	1.60 m	1.68 m
	with two full transverse bulkheads on vehicle deck (Version E)			
0. m	1.11 m	1.07 m	1.21 m	1.19 m
−1.0 m	0.78 m	0.78 m	-	-
+1.0 m	-	-	0.96 m	1.11 m

TABLE 7: THE LIMITING GM_D VALUES BETWEEN CAPSIZE AND SURVIVAL BASED ON MODEL TESTS AND SIMULATIONS **FOR SHIP C**; DRAFT OF INTACT SHIP T=6.0 M

Trim of intact ship	Sea state 4*m*/8*s*			
	D 10+11		D 6+7	
	Test	Simulation	Test	Simulation
dT	GM_D	GM_D	GM_D	GM_D
	without subdivision on vehicle deck (Version A)			
0. m	1.02 m	1.48 m	1.94 m	1.95 m
+1.04 m (Trim by bow)	0.71 m	0.75 m	-	-
+1.08 m	-	-	1.92 m	1.63 m
−1.07 m (Trim by stern)	-	-	2.10 m	2.15 m
	with longitudinal bulkheads on vehicle deck (Version F)			
0. m	1.20 m	1.42 m	1.83 m	2.01 m
+1.04 m	1.11 m	1.09 m	-	-
−1.07 m	-	-	2.00 m	2.24 m
	with two full transverse bulkheads on vehicle deck (Version E)			
0. m	1.07 m	1.00 m	1.24 m	1.38 m
+1.04 m	0.54 m	0.51 m	-	-
−1.07 m	-	-	1.29 m	1.40 m

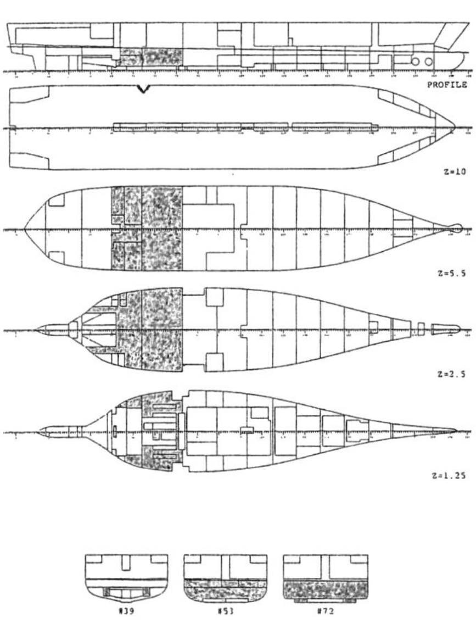

Figure 1: Damage case D11+12 of ship A

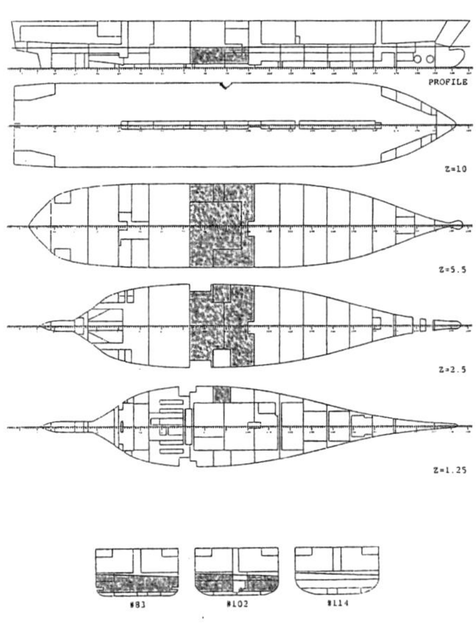

Figure 2: Damage case D9+10 of ship A

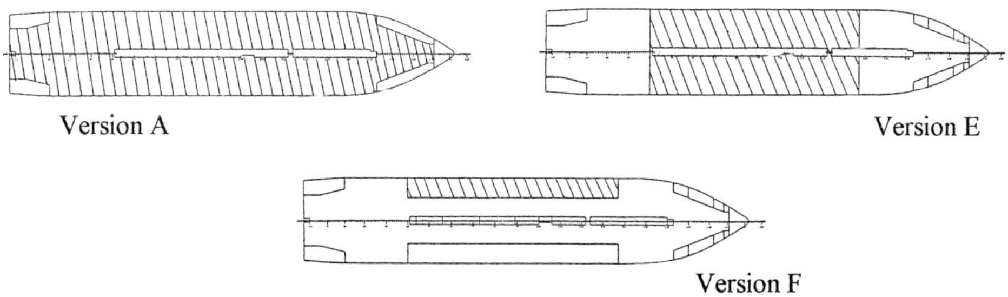

Figure 3: Variations of vehicle deck subdivision for ship A

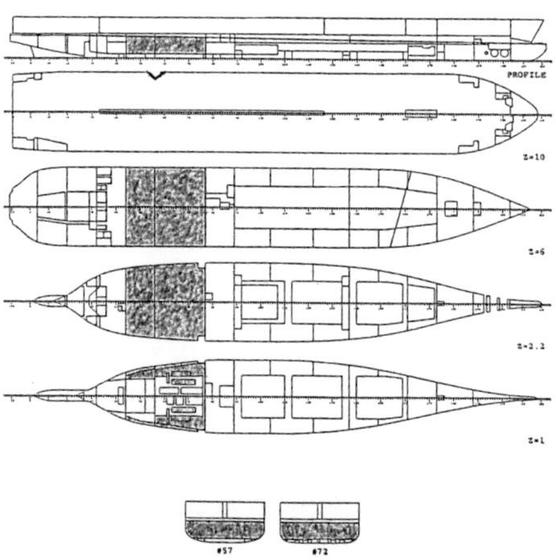

Figure 4: Damage case D11+12 of ship B

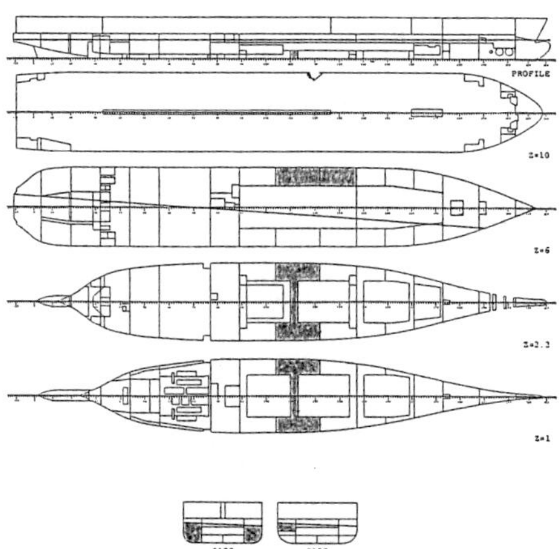

Figure 5: Damage case D7+8 of ship B

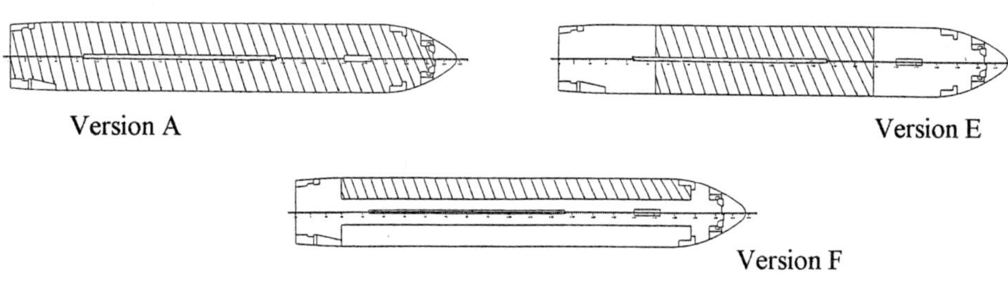

Figure 6: Variations of vehicle deck subdivision for ship B

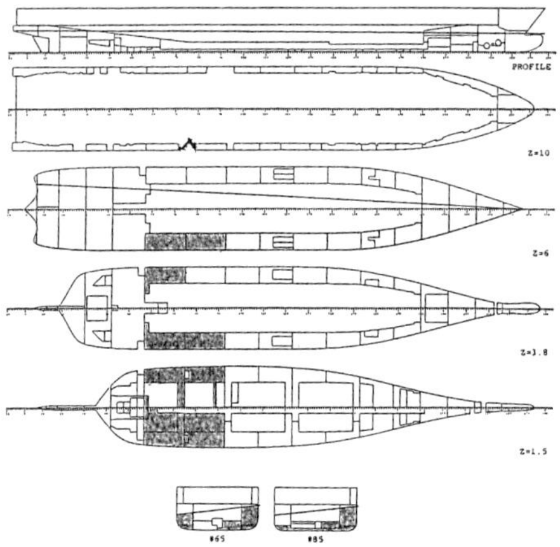

Figure 7: Damage case D6+7 of ship C

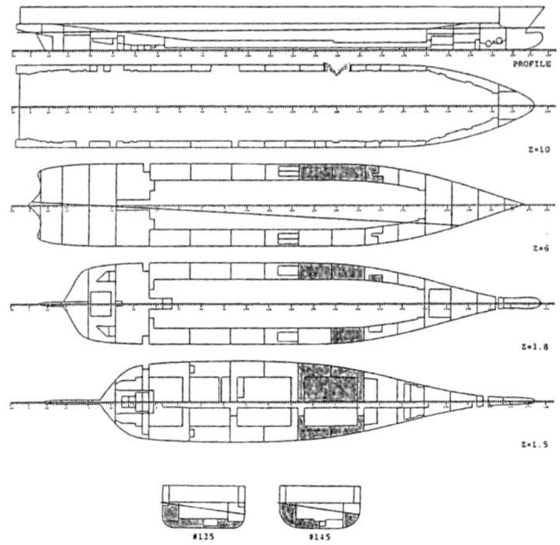

Figure 8: Damage case D10+11 of ship C

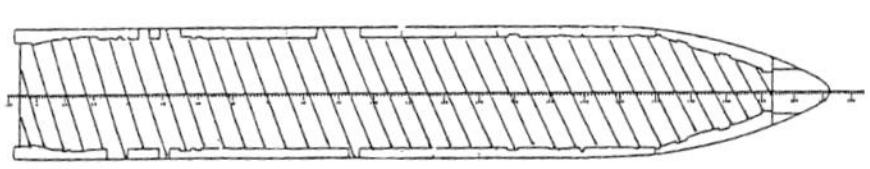

Version A

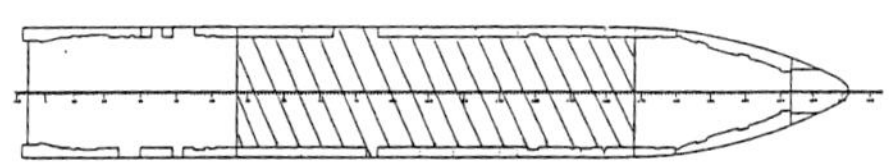

Version E

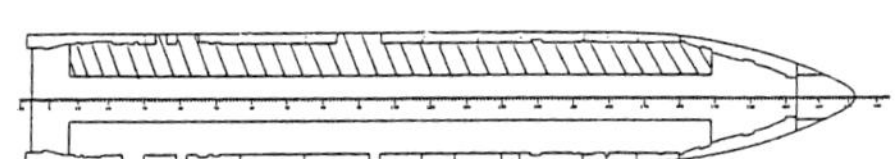

Version F

Figure 9: Variations of vehicle deck subdivision for ship C

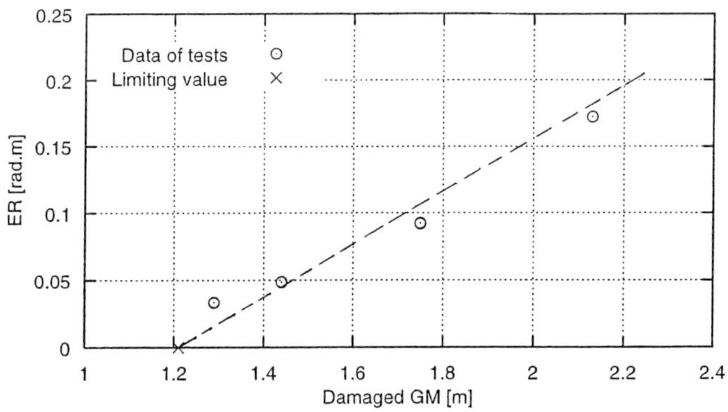

Figure 10: Derivation of the limiting GM_D value from the experiments

D. Vassalos, M. Hamamoto, A. Papanikolaou and D. Molyneux (Editors)

DYNAMICS OF A SHIP WITH PARTIALLY FLOODED COMPARTMENT

Jan O. de Kat

Maritime Research Institute Netherlands (MARIN)

ABSTRACT

This paper presents a brief outline of a six degree-of-freedom, nonlinear time domain model that is capable of simulating the large amplitude motion response of an intact or damaged ship in waves and wind. The part of the mathematical model related to the damage fluid effects is discussed in some detail; the present model neglects sloshing. Experiments were conducted with a tanker model in low steepness beam waves and with different amounts of fluid inside the ship. Roll decay tests in calm water provide useful material for validation of the simulation model. Furthermore, predicted and measured heave and roll responses in waves are compared and discrepancies in roll due to sloshing are highlighted.

KEYWORDS

Ship dynamics, flooding, sloshing, model tests, simulations.

INTRODUCTION

To describe the complete behaviour of an intact ship in waves and wind is not a trivial undertaking; yet the presence of fluid inside a ship may add another dimension of complexity. The dynamics of a ship with internal fluid has been investigated extensively in the past in the context of roll-stabilizing tanks, water trapped on deck, dock ships, LNG carriers and oil tankers, among others. A ship in a damaged condition presents some similarities within this context, but the degree of complexity increases as we consider water ingress through a damage opening, possibly with progressive flooding through and between compartments. The purpose of this paper is to revisit the elementary case of a ship with a rectangular compartment, which is not connected to the ocean and which can be partially flooded with a fixed amount of water.

Recent research on this topic has been reported by Armenio et al. (1996) and Francescutto and Contento (1994), who describe a time domain simulation model that couples the roll motion of a ship to the fluid dynamics inside a tank, where the internal fluid problem is dealt with by solving the Navier-Stokes equations numerically. The ship roll motion is described by a single degree of freedom model; the approach has been applied to a fishing vessel. Petey (1986) also describes a simulation model that can solve the coupled roll and fluid motions in the time domain. Rakhmanin and Zhivitsa (1994) derive equations for the coupled sway and roll motions for a ship with a partially filled compartment, and they illustrate that the fluid being a non-solid should result in the adjustment of, for example, the moment of inertia of the vessel-liquid combination. Rakhmanin (1962) provides detailed expressions that determine the extent to which the moment of inertia of a "solidified" fluid is reduced by considering it as a liquid.

This paper starts with a brief description of the mathematical model for a ship with one or more damage openings. Rather than embarking with the more general damage scenario involving water ingress and progressive flooding, we first consider the case of a ship with a fixed amount of fluid inside an isolated compartment. To this end experiments have been carried out with a tanker model in beam waves. A number of the test conditions have been simulated and comparisons between simulations and model tests are discussed. The paper concludes with observations and recommendations.

THEORY

The mathematical model described here has its roots in intact stability: It was developed to simulate the large amplitude motions of a steered ship in severe seas and wind. The model consists of a nonlinear strip theory approach, where linear and nonlinear potential flow forces are added to maneuvering and viscous drag forces. It is capable of predicting seakeeping and maneuvering behaviour in moderate to severe seas, as well as extreme motions related to surfriding, broaching and capsizing (see De Kat, 1994, and De Kat et al., 1994). The mathematical model has been extended to account for the effects of floodwater inside the ship, with the possibility of modeling water ingress and outflow through openings between compartments.

The derivation of the equations of motions is based on the conservation of linear and angular momentum. These are given in principle in the inertial (earth-fixed) reference system, defined by the system of axes (x^e,y^e,z^e). As the exciting forces are more easily described in the ship-fixed coordinate system, the Euler method is applied for deriving the equations of motion in terms of a rotating, ship-fixed coordinate system. Here the fluid inside the ship is considered in a dynamics sense as a free particle with concentrated mass; using this approach classical rigid body dynamics can be used to derive the equations of motion. A similar methodology has been applied by Letizia and Vassalos (1995).

For a damaged ship the equations of motion are expressed with respect to the center of gravity of the intact ship, G. The origin of the local, ship-fixed coordinate system with axes (x,y,z) is located at G. Although in terms of local coordinates G is independent of time, the center of gravity of the overall damaged ship will be time dependent, which introduces a number of cross-coupling terms in the equations of motion.

The center of gravity of the flood water is located by the vector **r** with respect to G. The location vector $\mathbf{r} = (x_f, y_f, z_f)$ and its derivatives (velocities and accelerations) are defined in the local ship-fixed coordinate system *Gxyz* with respect to G.1

The combined set of equations for the conservation of linear and angular momentum gives the equations of motion for the ship system with six degrees of freedom. Here it is convenient to consider the full 6 x 6 mass matrices, as is done for the intact case. All terms that contain linear or rotational ship accelerations are transferred to the left hand side (LHS) of the equations of motion. Using the generalized mass matrices, the equations of motion in the ship-fixed coordinate system are written as follows:

$$\left([\mathbf{M_0}]+[\mathbf{a}(\infty)]+[\mathbf{M_f}]\right).\ddot{\mathbf{x}}_G = \begin{pmatrix} \Sigma F_x \\ \Sigma F_y \\ \Sigma F_z \\ \Sigma M_x \\ \Sigma M_y \\ \Sigma M_z \end{pmatrix} - \begin{pmatrix} (m_0+m_f).(w_G q - v_G r) \\ (m_0+m_f).(u_G r - w_G p) \\ (m_0+m_f).(v_G p - u_G q) \\ (I_{zz,0} - I_{yy,0})qr \\ (I_{xx,0} - I_{zz,0})pr \\ (I_{yy,0} - I_{xx,0})pq \end{pmatrix} - [\dot{\mathbf{M}}_\mathbf{f}].\dot{\mathbf{x}}_G + \text{additional terms} \qquad (1)$$

The matrix $[\mathbf{M}_0]$ is the generalized mass matrix of the intact ship, $[\mathbf{a}]$ is the added mass matrix that is part of the linear radiation forces (the convolution integrals are part of the terms in the RHS). m_0 is the time-independent mass of the intact (dry) ship, m_f is the mass of the flood water inside the compartment. $[\mathbf{M}_f]$ is the matrix containing all ship-acceleration related inertia terms associated with the flood water in a damaged compartment and it is given by:

$$[\mathbf{M_f}] = \begin{bmatrix} m_f & 0 & 0 & 0 & m_f.z_f & -m_f.y_f \\ 0 & m_f & 0 & -m_f.z_f & 0 & m_f.x_f \\ 0 & 0 & m_f & m_f.y_f & -m_f.x_f & 0 \\ 0 & -m_f.z_f & m_f.y_f & I_{f,xx}+m_f.r_x^2 & -m_f.x_f y_f & -m_f.x_f z_f \\ m_f.z_f & 0 & -m_f.x_f & -m_f.x_f y_f & I_{f,yy}+m_f.r_y^2 & -m_f.y_f z_f \\ -m_f.y_f & m_f.x_f & 0 & -m_f.x_f z_f & -m_f.y_f z_f & I_{f,zz}+m_f.r_z^2 \end{bmatrix} \qquad (2)$$

The summation signs in the RHS represent the sum of all external force contributions, which result from:

- Froude-Krylov force
- Wave radiation (convolution integrals)
- Diffraction
- Viscous and maneuvering forces
- Propeller thrust and hull resistance
- Rudder and appendages
- Wind
- Internal fluid

The "additional terms" in the RHS of the equations of motion stem from cross products, which appear when expressing the conservation of momentum in a ship-fixed coordinate system, and from the motion of the fluid relative to the ship. For example, the conservation of linear momentum in the ship-fixed system is given by the following equation:

$$\mathbf{F} = (m_0 + m_f).(\dot{\mathbf{v}}_G + \omega \otimes \mathbf{v}_G) + m_f.(\dot{\omega} \otimes \mathbf{r} + \dot{\mathbf{v}}_f + \omega \otimes (\omega \otimes \mathbf{r}) + 2.\omega \otimes \mathbf{v}_f) + \dot{m}_f.(\dot{\mathbf{v}}_G + \dot{\mathbf{v}}_f + \omega \otimes \mathbf{r}) \tag{3}$$

where the vectors $\mathbf{v}_G = (u_G, v_G, w_G)$ and □ = (p, q, r) represent the linear and angular velocity vector in the ship-fixed coordinate system, respectively; $\mathbf{v}_f$ is the velocity vector of the center of gravity of the flood water expressed in the ship-fixed reference system All terms resulting from the above expression are retained and, except for the ship acceleration terms, they are put in the RHS of the equations of motion. The conservation of angular momentum can be derived in a similar fashion.

Forces Associated with Internal Fluid

Taking a quasi-static approach, the damage fluid causes an additional vertical force to act on the ship. In the inertial (earth-fixed) reference frame, the force vector is the following:

$$\mathbf{F}_f^{\,e} = \begin{pmatrix} 0 \\ 0 \\ m_f . g \end{pmatrix} \tag{4}$$

The above vector can be expressed in terms of the ship-fixed coordinate system by the following transformation:

$$\mathbf{F}_{dam} = [\mathbf{T}_{3f}]\mathbf{F}_f^{\,e} \tag{5}$$

where the matrix $[\mathbf{T}_{3f}]$ is the transformation matrix containing the Euler angles. The moment vector about G (in the ship-fixed system) caused by the flood water is given by:

$$\mathbf{M}_{dam} = \mathbf{r} \otimes \mathbf{F}_{dam} \tag{6}$$

An important assumption made here is that the water level remains horizontal at all times, which implies that sloshing effects are neglected. Possible consequences of this assumption are investigated below. Another critical assumption is the homogeneous distribution of permeability inside any compartment.

EXPERIMENTS

Model tests have been carried out at MARIN with a 200 kDWT tanker at scale 1:82.5 in a basin measuring 200 m x 15 m x 1 m. The particulars of the tanker are given in table 1 and in figure 1. The tanker was equipped with a compartment symmetrically positioned amidships; the length of the compartment is 82.50 m and the width is 31.76 m. The tanker was tested with the compartment positioned at two levels with respect to the baseline: (1) at a height of 5.20 m above the keel and (2) at 16.50 m above the keel. Different fill levels were achieved by adding color-dyed water to the compartment, which was closed off with a perspex lid to prevent fluid from spilling out in the event of sloshing. This paper discusses results pertaining only to the case with the compartment located closest to the keel.

TABLE 1
PARTICULARS OF TANKER

Length, Lpp (m)	310.20
Beam, B (m)	47.20
Draft, T (m)	16.00
Depth, D (m)	26.07
Displaced weight (tf)	202,600
Center of gravity above baseline, KG (m)	10.0
Metacentric height, GM (m)	9.50
Natural roll period - intact (s)	10

Tests comprised roll decay tests and tests in regular beam waves with periods ranging from 8 to 12 s. The wave height (crest-trough) was between 2 m and 3 m. The model was kept in position by means of a soft-spring mooring arrangement as shown in Fig. 1, at zero forward speed Fluid levels in the compartment ranged from zero ("intact" ship) to 16 m fill depth, which corresponds to a maximum fill ratio h/w = 0.5. Fig. 2 shows the natural period of the fluid in transverse direction as a function of fill ratio; the period for the first fundamental sloshing mode is given by:

$$T_f = 2\sqrt{\frac{\pi.w}{g.\tanh(\frac{\pi.h}{w})}} \tag{7}$$

For shallow fill depths (h/w < 0.15) the period can be based on the critical shallow water wave speed:

$$T_f = \frac{2w}{\sqrt{gh}} \tag{8}$$

RESULTS AND DISCUSSION

Heave and roll motion responses are the primary items of interest in this study. Results for heave are presented in terms of Response Amplitude Operators (RAOs), given by the heave amplitude per unit wave amplitude. Roll decay tests provide useful information on the natural roll period and damping in calm water. Roll RAOs are obtained by the ratio of the roll amplitude per unit wave amplitude. These results are presented for the following cases:

1. Intact ship
2. Fluid level: h = 16 m (h/w = 0.5; volume = 41927 m^3) – no sloshing
3. Fluid level: h = 1 m (h/w = 0.03; volume = 2620 m^3) – subresonant condition
4. Fluid level: h = 4 m (h/w = 0.13; volume = 10482 m^3) – resonant sloshing

In case 2, with h = 16 m, the natural sloshing period lies well below the range of wave periods tested. Hence, for the conditions tested the fluid level remains horizontal and the simulations should be able to represent the behaviour quite well.

Case 3 presents a very shallow fill depth; here the sloshing period lies well above the wave periods tested. However, the fluid level does not stay horizontal in those test conditions: the fluid motion is characterized by hydraulic bores -- at any one time three bores could be running across the tank simultaneously. One might be tempted to believe that in spite of the fluid dynamics the motion of the ship may not be affected significantly, as the mass of fluid is small compared with the mass of the ship.

Case 4 is a true sloshing condition: the sloshing period lies within the range of wave periods and is approximately equal to the natural roll period of the ship. In the model tests the water would splash forcefully against the compartment's lid for all wave periods tested. Here we would expect large deviations between simulations and measurements.

An overview of the figures showing comparisons between tests and simulations is given in table 2.

TABLE 2.
OVERVIEW OF FIGURES WITH EXPERIMENTAL AND SIMULATION RESULTS

	Heave RAO	Roll decay	Roll RAO
Intact ship (no fluid)	Fig. 3	Fig. 7	Fig. 11
1 m fill height	Fig. 4	Fig. 8	Fig. 12
4 m fill height	Fig. 5	Fig. 9	Fig. 13
16 m fill height	Fig. 6	Fig. 10	Fig. 14

Heave in beam waves

Figures 3, 4, 5 and 6 show that the heave response in beam waves is predicted consistently well for all fluid fill levels considered. The characteristics of the heave response change gradually with the amount of fluid in the tank; heave tends to decrease with increasing fluid levels, but not at all exciting frequencies.

Roll decay

For the intact case, the predicted roll decay follows the measured roll decay very closely (see figure 7) – both the roll period and decay are very close. This suggests that viscous roll damping is modeled adequately in these conditions, which then leads to the assumption that hull roll damping is modeled properly also in the cases with internal fluid present. For the largest fill depth considered here, with fill ratio h/w = 0.5 and fluid weight being around 20% of the displacement of the intact ship, the predicted roll decay is quite similar to the measured results (see figure 10). Here the natural period of the tank is much shorter than the roll period of the ship - 5 s versus 10 s, and no sloshing occurs (the fluid surface stays horizontal). It is interesting to note the slightly higher damped behaviour in the simulation model than in the case of the model test. The natural period of the ship with fluid is fractionally longer than the intact ship – the increase in natural period is about 1 second.

Figures 8 shows that even the presence of a relatively small amount of fluid (h/w = 0.03 with fluid weight being approximately equal to 1% of the displacement of the ship) inside the tank can have a significant influence on the roll decay. The measured roll decay shows a monotonically, albeit not continuous, decrease in successive roll amplitudes, whereas the simulated roll decay underpredicts the roll damping. During the first roll period the simulated and measured periods are approximately equal; subsequently, however, the predicted roll period stays constant, while the measured period does not remain the same (it decreases from around 11 to 10 seconds).

The discrepancies in roll become even more significant for the 4 m fill level case. The measured roll motion decays rapidly during the first two cycles, but then shows a beat-type of behaviour. The sloshing of the fluid inside the ship keeps it from attaining static equilibrium for some time. The natural period is the same as the predicted period during the first cycle (around 11 s), but then increases and increases again. Obviously a different mathematical model would be required to simulate properly the observed dynamics.

Roll response in beam waves

For the ship with no internal fluid, the RAO for roll obtained by simulation is reasonably close to the RAO derived from the model tests (see figure 11). The correspondence in the resonant peak is good, while the roll is overpredicted at the highest wave frequencies; the same applies to the case with the largest fill depth (see figure 14). In the latter case the fluid level remains horizontal at all times and the roll moment caused by the fluid is in phase with the roll motion.

For the case of shallowest fill depth tested (h/w = 0.03) the measured and predicted roll response agree quite well for frequencies higher than the roll resonant frequency, see figure 12. At the resonant roll frequency the measured roll response is lower than the simulation results.

The largest discrepancies between simulation and model test occur for the true sloshing case, where the fill height is 4 m, see figure 13. Here the fluid acts as an active damper over the whole range of wave frequencies tested, rather like a well tuned anti-roll tank. The roll moment exerted by the fluid on the ship is out of phase with the roll motion and therefore provides an effective source of roll damping.

CONCLUSIONS

This paper presents a six degree-of-freedom mathematical model that is capable of simulating the dynamics of a damaged (and intact) ship. To validate the simulated dynamic response of a damaged ship in a stepwise fashion, model tests have concentrated initially on a vessel with a floodable compartment without any openings to the sea. For this purpose a tanker model with different fluid fill levels was tested in low steepness beam waves. Predicted heave motions in beam seas compare very well with measurements. Roll motions, however, show less consistent agreement: in non-resonant conditions the roll motion amplitudes agree fairly well with measured values, but in (fluid) resonant conditions the simulation model overpredicts the roll response significantly.

References

Armenio, V., La Rocca, M. and Francescutto, A. (1996). On the Roll Motion of a Ship with Partially Filled Unbaffled Tanks, *International Journal of Offshore and Polar Engineering*, Part 1 & 2, Paper no. IJOPE-JC-154.

De Kat, J.O. (1994). Irregular Waves and Their Influence on Extreme Ship Motions, *Proceedings of the Symposium on Naval Hydrodynamics*, Santa Barbara, Aug., pp. 39-58.

De Kat, J.O., et al. (1994). Intact Ship Survivability: New Criteria from a Research and Navy Perspective, *Proceedings of the STAB '94 Symposium*, Melbourne (FL), Nov.

Francescutto, A. and Contento, G. (1994). An Experimental Study of the Coupling Between Roll Motion and Sloshing in a Compartment, *Proceedings of the Fourth (1994) International Offshore and Polar Engineering Conference*, Vol. III, Osaka, April, pp. 283-291.

Letizia, L. and Vassalos, D. (1995). Formulation of a Non-Linear Mathematical Model for a Damaged Ship Subjected to Flooding, *Proceedings of the Sevastianov Symposium*, Kaliningrad, May.

Petey, F. (1986). Numerical Calculation of Forces and Moments Due to Fluid Motions in Tanks and Damaged Compartments, *Proceedings of the Third International Conference on Stability of Ships and Ocean Vehicles STAB '86*, Gdansk, Sept., pp. 77-82.

Rakhmanin, N.N. (1962). The Experimental Study of Dynamic Properties of the Ship with Partially Flooded Compartments, Troodi of CNII of A.N. Krylov, Vyp. 191, Sudpromgiz, (in Russian).

Rakhmanin, N.N. and Zhivitsa, S. (1994). Prediction of Motion of Ships with Flooded Compartments in a Seaway, *Proceedings of the STAB '94 Symposium*, Melbourne (FL), Nov.

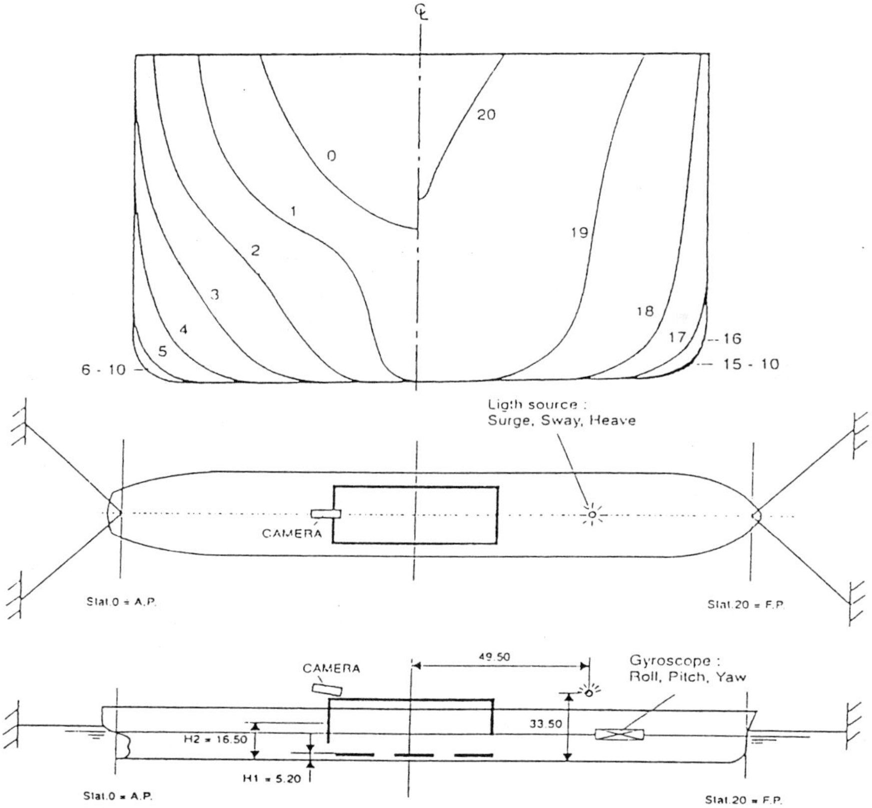

Figure 1: Particulars of tanker and test setup

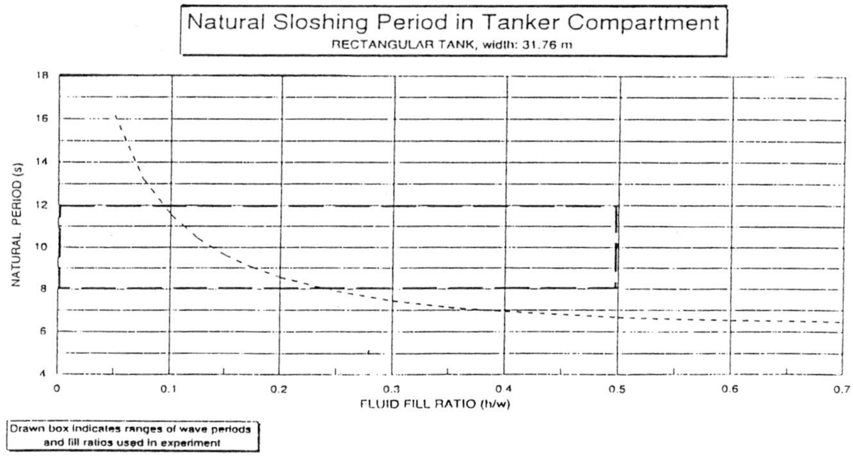

Figure 2: Sloshing period (fundamental mode) of fluid inside rectangular tank as a function of fill ratio

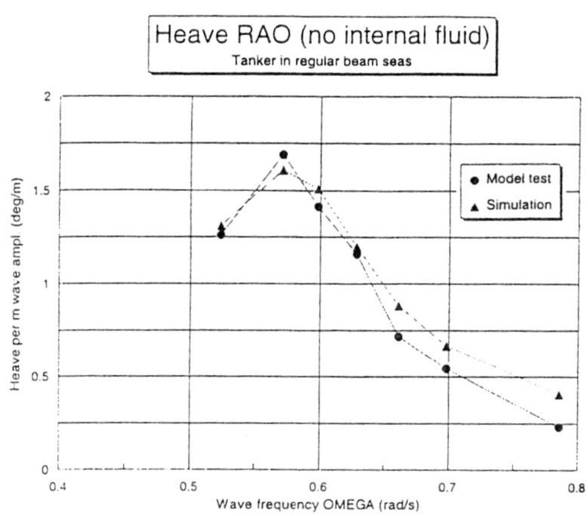

Figure 3: Response Amplitude Operator for heave in beam waves (ship without internal fluid)

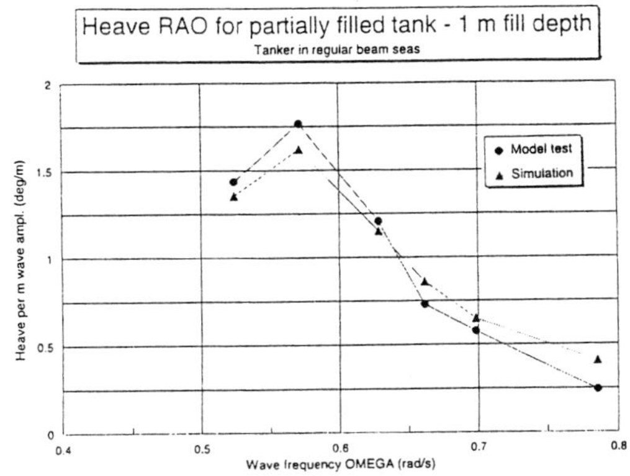

Figure 4: Response Amplitude Operator for heave in beam waves (fluid level inside compartment: 1 m)

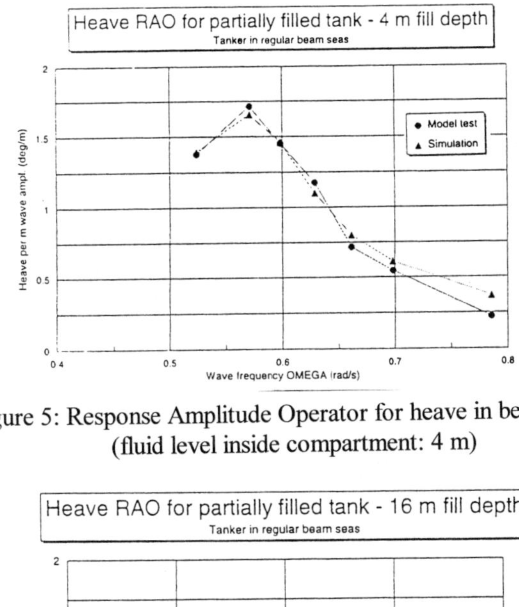

Figure 5: Response Amplitude Operator for heave in beam waves (fluid level inside compartment: 4 m)

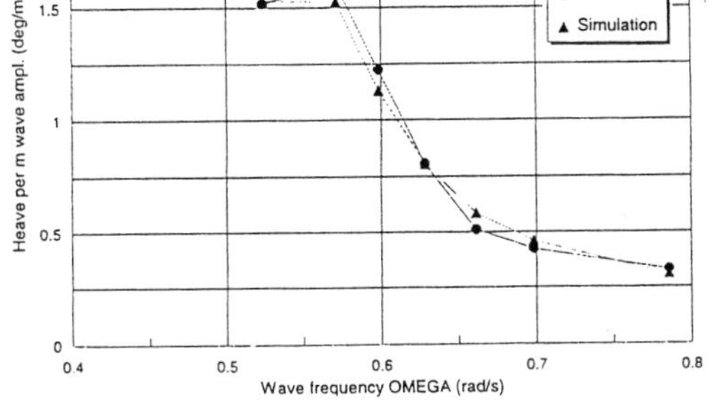

Figure 6: Response Amplitude Operator for heave in beam waves (fluid level inside compartment: 16 m)

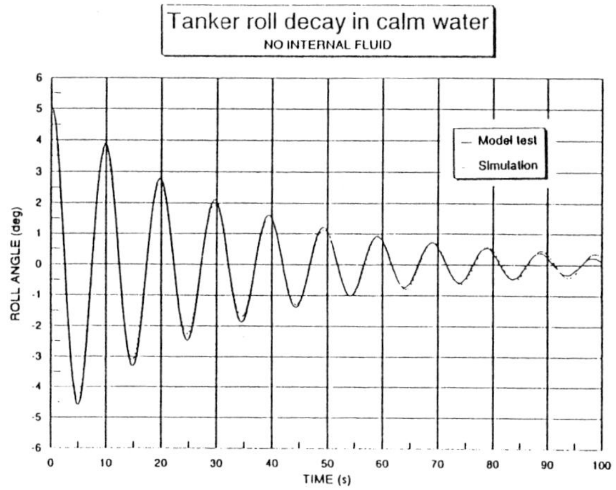

Figure 7: Roll decay for ship without internal fluid

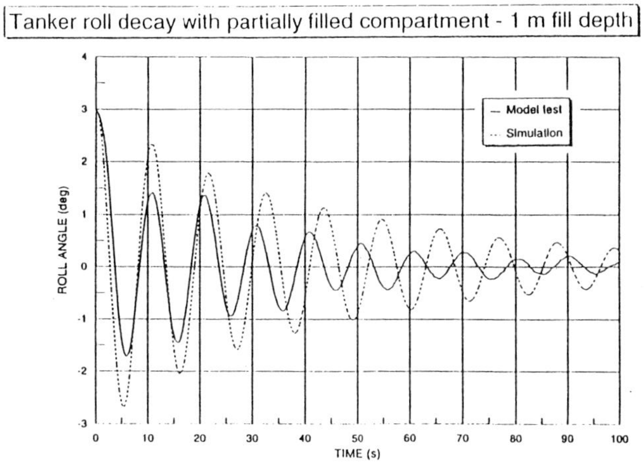

Figure 8: Roll decay for ship with 1 m fluid level

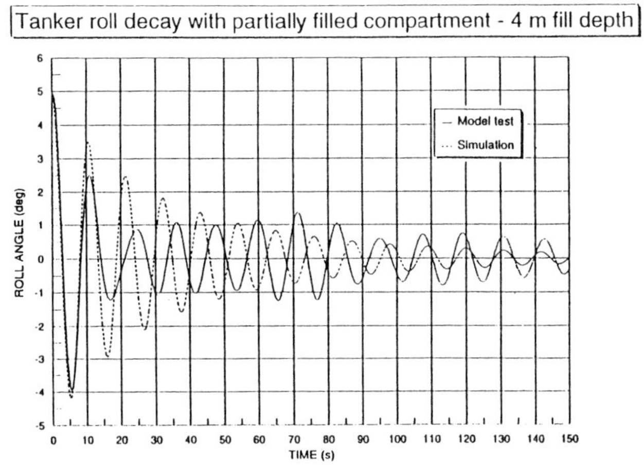

Figure 9: Roll decay for ship with 4 m fluid level

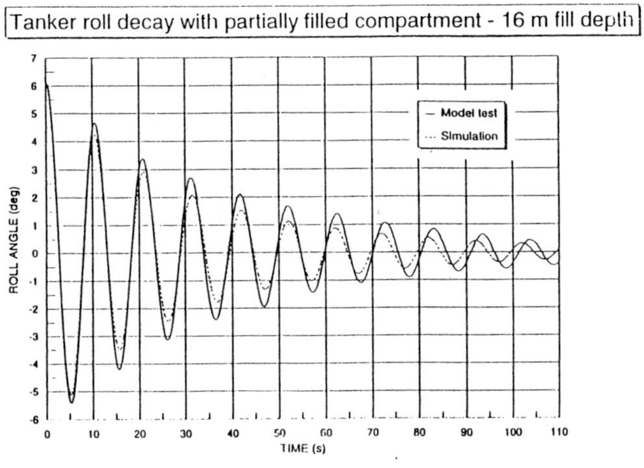

Figure 10: Roll decay for ship with 16 m fluid level

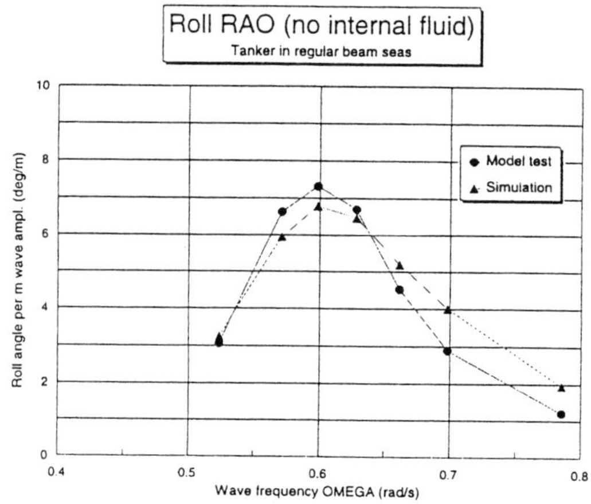

Figure 11: Response Amplitude Operator for roll in beam waves (ship without internal fluid)

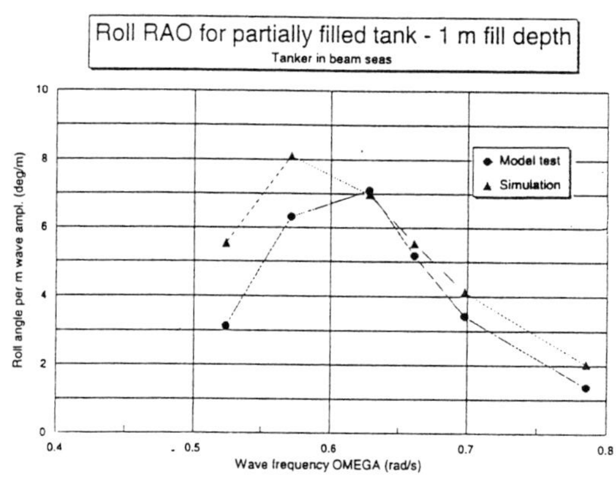

Figure 12: Response Amplitude Operator for roll in beam waves (fluid level inside compartment: 1 m)

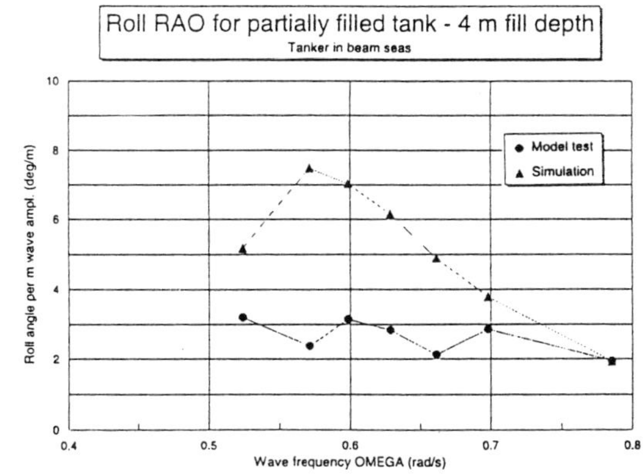

Figure 13: Response Amplitude Operator for roll in beam waves (fluid level inside compartment: 4 m)

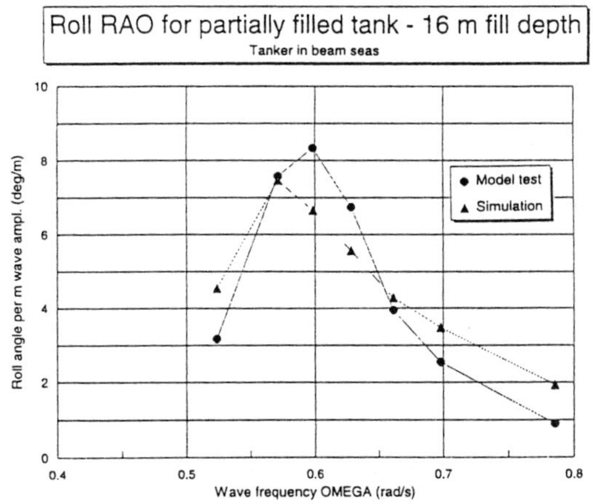

Figure 14: Response Amplitude Operator for roll in beam waves (fluid level inside compartment: 16 m)

Contemporary Ideas on Ship Stability
D. Vassalos, M. Hamamoto, A. Papanikolaou and D. Molyneux (Editors)

RO-RO PASSENGER VESSELS SURVIVABILITY- A STUDY OF THREE DIFFERENT HULL FORMS CONSIDERING DIFFERENT RO-RO-DECK SUBDIVISIONS

Anneliese E. Jost, Germanischer Lloyd and Dr. Peter Blume, HSVA

ABSTRACT

Based on the newly defined Stockholm Agreement's water on deck criteria model tests where carried out aiming at a better understanding of the critical water height on the ro-ro deck.

For this purpose three differently aged vessel types with respectively different operating freeboards have been included in this study. The results achieved are shown and discussed.

KEYWORDS

RoRo Passenger Ship Safety, Damage Stability, Stockholm Agreement, Model Tests, Capsizing Safety Margin

1. BACKGROUND

Legislative Activities have finally calmed down that had been pushed forward by the tragic losses of ro-ro passenger vessels the most recent of which had been the Estonia. The drastic measures newly required are being put into place now. Once they will be completely in place, time will proof their effectiveness.

Legislative tools dedicated to ro-ro passenger vessels cover other than a set of 1995 SOLAS Amendments, IMO Resolutions requiring the upgrading of shell doors scantlings and their securing as well as a Regional Agreement on a specific damage stability standard applicable to ships travelling in the North and Baltic Seas.

With regard to damage survivability it should be noted that the amended SOLAS [1] requirements and the Regional agreement [2] enforce an increase in standard that will become mandatory to both new and existing passenger vessels. One of the newly installed criteria is a survivability standard that requires to show sufficient residual stability even with accumulating water on the damaged ro-ro deck. The respective legislation does allow to chose showing compliance with this standard either by calculation or by model tests. In both cases the regional agreement does give detailed procedures which are to be complied with.
The scope of this research project (sponsored by BMBF[1]) was to find a better understanding of the critical amount of water on deck relative to the existing trading practice and eventual upgrading of the ship types. Thus three existing hull forms that had been built in accordance with different subdivision and survivability standards due to their individual keel laying and or reconstruction dates were chosen as sample ships. It was anticipated that some kind of subdivision of the ro-ro deck would become necessary during the course of the proposed upgrading and the effect of such subdivision on the survivability of these hull forms was to be studied.

2. HYDROSTATIC EVALUATION OF THE SAMPLE SHIPS

Today's subdivision standard is based on the SOLAS 1974 requirement that has been considerably amended after 1988. Up to 1988 for any passenger ship compliance had to proven with existing intact stability criteria the requirement of floodable lengths and certain damage stability criteria. Usually the damage stability requirements would be the crucial criteria and the designs freeboard. Both the intact stability criteria and the floodable length requirement have not been reconsidered while the damage stability criteria had been amended by a requirement of a residual stability standard. As usual this new standard became applicable to new ships and showed a remarkable influence on the designs.
Ro-ro passenger ferries operating in the Baltic and North sea areas are required to comply with a survivability standard that includes the flooding of the ro-ro (bulkhead) deck. The requirement is that acceptable residual stability standards are to be maintained even with a certain flooding height $(h_W \leq 0.5m)$. Compliance with this standard can either shown by model testing or calculation.
Technically, two different survivability standards for new and existing ro-ro passenger ferries can not be justified. Consequently the amended standard became mandatory also to existing vessels.

The vessels chosen for this project are considered to be typical representatives of the kind used in this area:

1. Ship A
   This ship was built in accordance with SOLAS 74/83 standard. Her subdivision below the bulkhead deck consists of a relative high number of transverse bulkheads combined with some longitudinal subdivision. The operating freeboard is minimised by making use of

[1] German Ministry for Research and Technology

cross flooding that decrease or eliminate heeling angles in damaged condition. In this case the criteria of non-submerging the margin was the governing criteria as residual stability standards had not been imposed.

2. Ship B
   Although this ship had been built prior to ship A, with regard to survivability standards she presents a later design standard since she was completely re-built when entering a different service after the 1988 SOLAS amendments came into force. With this conversion in accordance with SOLAS 74/88 she features a very similar subdivision below the bulkhead deck, the operating freeboard was remarkably increased due to the requirement of a minimum heeling lever curve in damaged condition.

3. Ship C
   This vessel was only designed when the new standard had already been established. Her subdivision below the bulkhead deck is governed by longitudinal bulkheads located at B/5 measured from the shell. Not all of these are equipped with cross flooding ducts. Respectively heeling angles in the damaged water lines are not eliminated although they are legally limited.

The main dimensions of the vessels and some general remarks regarding the damage stability calculations are shown next.

Main Dimensions:	**Ship A**	**Ship B**	**Ship C**
length $\mathbf{L_S}$ [m]	151.46	177.21	179.5
beam **B** [m]	29.00	26.00	27.2
height **D** [m]	8.10 /16.20	8.0 /	8.7 /
draught **t** [m]	6.40/ 6.20 / 5.90	5.75	6.0
trim [m]	0.0	-1.0 /0.0 / +1.0	-1.0 /0.0 / +1.0
modelled damaged Compartments	9/10 and 11/12	7/8 and 11/12	6/7 and 10/11

All three vessels are narrow and evenly subdivided below the bulkhead deck [figures 1]. All three ships had in their original configuration non-watertight divided ro-ro decks.
Each one was subjected to a SOLAS 90 damage stability calculation in order to establish a governing double compartment that was taken into account regarding the "water on deck" criteria.
The Stockholm agreement requires that unless the governing criteria is located within 10% of the mid-length a second double compartment located within this margin has to be demonstrated. Taking this into account the models of all three vessels were built to cover two SOLAS damages over the length. Where the crucial condition turned out to be within the a.m. mid length part the second damage was chosen to be near the vessels shoulder.
For vessels B and C, where the operating freeboard was considered rather large, it was understood that further increase of freeboard would only effect the survivability to become

more favourable by allowing for less GM_O values. In those cases the introduction of trim was considered to be of more interest.
Generally, for all three vessels three different ro-ro deck versions were investigated. Starting point was

Version A the open deck configuration.

In the course of this project. By calculation a variety of different deck subdivision alternatives were considered. Within the scope of this project only two further versions could be expanded on in model testing:

Version E2 with 2 full height bulkheads in transverse ship direction and

Version F with side casings.

In terms of compliance with "SOLAS 90" standard, in the damage stability investigation the side casing version was the version that seemed to best cope with the criteria.
Summarising, the scope the research project covered variations of damage location, subdivision on the ro-ro deck and GM as well as draughts and trims respectively as shown below.

	Ship A			Ship B			Ship C		
	Draught variations			Trim variations			Trim variations		
	6.4 m	6.2 m	5.9 m	-1.0 m	0.0 m	+1.0 m	-1.0 m	0.0 m	+1.0 m
GM variations	X	X	X	X	X	X	X	X	X
Sea State realisations	X	X	X	X	X	X	X	X	X

3. MODEL TESTING

Within the model tests the model is subjected to a long-crested irregular seaway. The damaged model is free to drift and is placed in beam seas with the damage hole facing the oncoming waves.
The survival criteria in these model tests are such that at least five experiments for each peak period need to be carried out. The test duration shall be such that a stationary state has been reached and a minimum of 30 min. in full scale time are proven. The model is considered to survive when the angles of roll do not reach more than 30° against a vertical axis occurring more frequently than 20% of the rolling cycles or the steady heeling angle becomes greater than 20°.
To cover the scope of these tests the model has to be such that the hull is thin enough in the damaged areas and the main design features such as watertight bulkheads, air escapes permeabilities and etc. above and below the bulkhead deck can be modelled to represent the real situation.

Each of the models used in this project was built accordingly in GRP and the interior for two damage locations was mostly built of ply wood in suitable size. These two damage locations were chosen in accordance with the SOLAS damage stability as described above. The damage holes were opened successively. Each one sized in accordance with the SOLAS damage and a penetration depth of B/5. Both of the damage locations per ship were subjected to draught or trim variations and to address the critical wave height the GM_L was continuously decreased in order to find the boarder line between safe and capsizing.
This borderline is of course subject to uncertainties that are usually involved in methods that include statistics. The results cover for each tested GM_L the observation survived/ not survived. This statement of result does not cover the residual margin that might exist versus an exact limiting condition on the edge of capsizing. For this reason it would be necessary to evaluate many more runs in testing such capsizing margins to finally achieve the answer to a question of critical "water on deck"-height.

4. COMPUTER SIMULATIONS

The scope of this investigation covered also a computer simulation of the relative motions of the damaged ro-ro passenger vessels in the seaway. The software utilised in this respect is based on methodologies developed at the Institut für Schiffbau, Hamburg. (Kröger [3] and Petey [4]) which was expanded and adapted to such model tests by Chang [5]
As an example table 4 and 5 cover the results of simulations for ship A, tables 6 and 7 ship B and tables 8 and 9 ship C respectively. The simulations confirm the outcome of the model testing.
In order to get an improved understanding of the boarder line between safe and unsafe in this context the simulations covered the same scope of GM variations, only in much smaller steps.

5. CONCLUSIONS

The realisation of the seaway and the accumulated water on deck is shown in figures2 and 3. Due to the a.m. uncertainties the measured accumulated water heights cannot be considered representing the critical water height on deck. The scheme of water height measurements are shown in figures 3a.
Figure 2 shows a typical time history of a capsizing. Obviously, the identification of the "critical" water height on deck is judgmental.
In order to produce a better understanding of such critical height of water and in conjunction with earlier survivability test made at the HSVA model basin, it is proposed that the residual stability lever curve is employed to help judge the residual stability margin. It is proposed that the residual area of the heeling lever curve of the respective damage case beyond the measured heeling and rolling angle is used to represent survivability borders. (figure 4)
With regard to improving safety of such existing vessels, and in order to limit the necessary increase in GM to a practicable margin further subdivision versions on the ro-ro deck were investigated. As described above three versions per ship were tested with the scope of the tested configurations shown in Table 1.

TABLE 1
TEST MATRIX AND SURVIVABILITY FOR SHIP A

Initial draught	Original configuration Version A		With additional bulkheads Version E2		With side casings Version F	
	L 11/12 GM_L	L9/10 GM_L	L 11/12 GM_L	L9/10 GM_L	L 11/12 GM_L	L9/10 GM_L
6.4 m	-	4.2 m	-	2.3 m	-	2.4 m
6.2 m	3.5 m	4.2 m	1.3 m	2.0 m	1.8 m	2.4 m
5.9 m	2.6 m	3.2 m	1.4 m	1.7 m	1.6 m	1.9 m

The conclusion to be drawn from such result can be summarised as follows:

- the increase in freeboard effects a drastic decrease in GM requirement of the damage case.
- For this vessel not built in accordance with "SOLAS 90" standard the requirement in GM in the original configuration becomes impracticably high.
- Any subdivision introduced on the ro-ro deck significantly influences the survivability of the vessel in damaged condition.
- Other than the calculation the test show better results by the introduction of the proposed two full height bulkheads.

A graphical evaluation of the maximum allowable KG-values are shown in figure 5, 6 and 7.

In case of ship B and ship C the freeboard effect was not considered of the same predominance. A result of the trim variations, however is that obviously small trim angles do have some effect on the GM-requirement as shown in Tables 2 and 3.

TABLE 2
TEST MATRIX AND SURVIVABILITY FOR SHIP B

Initial trim	Original configuration Version A		With additional bulkheads Version E2		With side casings Version F	
	L 11/12	L 7/8	L 11/12	L 7/8	L 11/12	L 7/8
	GM_L	GM_L	GM_L	GM_L	GM_L	GM_L
-1.0 m	2.73 m	-	0.78 m	-	1.54 m	-
0.0 m	2.30 m	1.13 m	1.11 m	1.21 m	1.19 m	1.29 m
1.0 m	-	1.37 m	-	0.96 m	-	1.60 m

TABLE 3
TEST MATRIX AND SURVIVABILITY FOR SHIP C

Initial trim	Original configuration Version A		With additional bulkheads Version E2		With side casings Version F	
	L 10/11	L 6/7	L 10/11	L 6/7	L 10/11	L 6/7
	GM_L	GM_L	GM_L	GM_L	GM_L	GM_L
1.06 m	0.71 m	1.92 m	0.54 m	-	1.11 m	-
0.0 m	1.02 m	1.94 m	1.07 m	1. 24 m	1.20 m	1.83 m
-1.07 m	-	2.10 m	-	1.29 m	-	2.00 m

An overview of the survivability results obtained by numerical simulation is shown in Tables 4 to 9.

The ship motion, the relative motion between the damage opening and the water surface and the water height on deck were measured during the tests. The most important information is the roll-response, which indicates whether or not the vessel is regarded as capsizing or not. Figures 3a show as a section of the measured time and water height records such a typical capsizing. Encountering a group of high waves the vessel is forced to heel to a rather large angle by the load introduced by the collected water on deck. In the following calmer period the vessel recovers very slowly to a more upright position. Another group of larger waves approaching again increases the heeling continuously to lead to the eventually observed capsize. Provided the initial GM is slightly larger the vessel shows better recovering times before encountering the next group of high waves.

Once the critical KG/GM values had been generated as describing the limit between safe and unsafe, the residual stability parameters were calculated. Accordingly the respective maximum lever arms and stability ranges were derived.

The results indicate clearly that the "open deck" Version iterates a larger requirement of residual stability. Both of the subdivided Versions (side casings and two transverse bulkheads), however, showed requirements in a similar order.

Based on the known dependency of the stability parameters on B/T and the observation that the deck area involved (i.e. length and beam of the wetted ro-ro cargo area) and trim have large impact it is proposed to use these geometric features to describe the residual stability characteristic.

$$GZ^{*} = GZ_{MAX} * \frac{T}{B} * \frac{1}{(1.368 - \exp(-(A_{RO}/bL)^{2})) * (0.2 + \exp(-(57.3 * \Delta T/L)^{2}))}$$

$$F_b^{*} = (F_b + B/2 * \sin\Phi_{Stat}) / \varsigma_{\omega sig}$$

For Version A all three hulls are shown in figure 4a versus the freeboard.

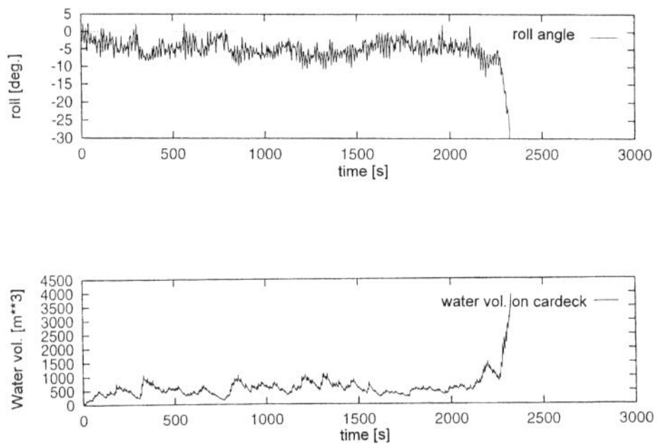

Figure 2: Typical Capsize Scenario

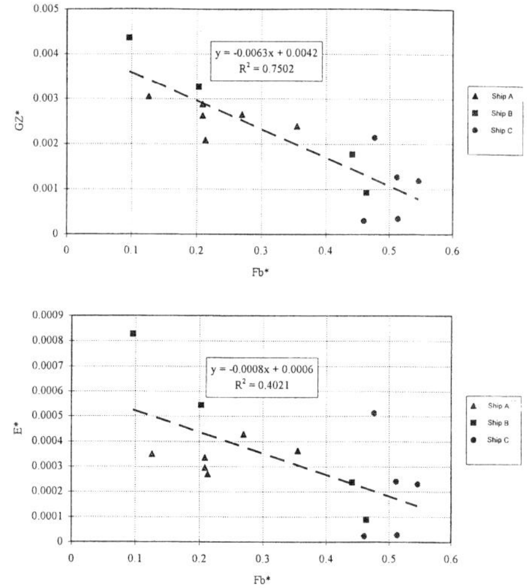

Figure 4a: Standardised maximum lever arm and area curves demonstrated as a function of freeboard

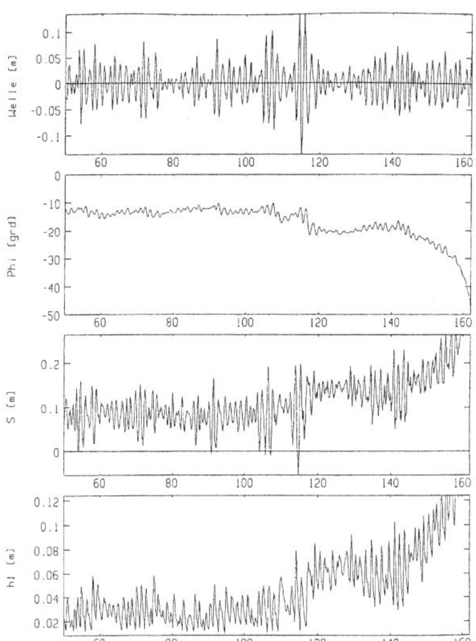

Fig 3a.1: Section of time records measured for a capsizing situation

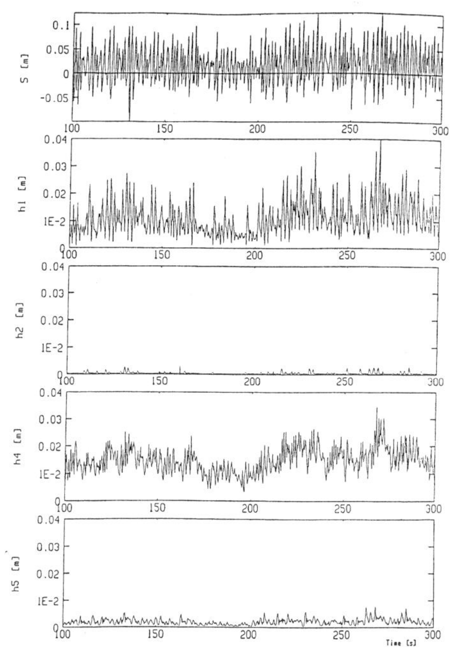

Fig 3a.2: Section of time records of water height

Table 4: Number of tests resp. simulation runs N and number of observed capsizing incidents N_C
Ship A: damage case 9/10: no trim

initial draught T [m]	damage GM_L	Test N	Test N_C	Simulation N	Simulation N_C	Test N	Test N_C	Simulation N	Simulation N_C
	Open deck				(Version A)				
5.9	**3.83**	2	-	2	-	-	-	4	-
	3.45	2	-	2	-	-	-	4	-
	3.25	2	-	8	-	-	-	8	-
	3.22	-	-	4	1	-	-	4	1
	3.17	-	-	8	1	-	-	4	2
	3.12	-	-	4	1	-	-	8	4
	3.06	1	1	8	1	-	-	8	5
	3.02	-	-	4	2	-	-	-	-
6.2	**4.44**	2	-	2	-	-	-	4	-
	4.14	-	-	8	-	-	-	8	-
	4.09	2	1	8	2	-	-	4	2
	4.05	-	-	8	3	-	-	8	2
	4.00	-	-	4	1	-	-	4	2
	3.95	1	1	4	1	-	-	4	2
	3.73	1	1	6	4	-	-	4	4
6.4	**4.50**	-	-	8	-	-	-	-	-
	4.45	-	-	8	1	-	-	8	-
	4.40	-	-	-	-	-	-	4	4
	4.31	2	-	4	4	-	-	4	4
	4.10	1	1	4	4	-	-	4	4
	3.60	1	1	4	4	-	-	4	4
	3.03	1	1	4	4	-	-	4	4
	Side casings				(Version F)				
5.9	**2.52**	2	-	4	-	-	-	4	-
	2.17	2	-	4	-	-	-	4	-
	1.76	2	1	4	-	-	-	8	-
	1.69	-	-	8	-	-	-	8	-
	1.64	-	-	4	2	-	-	8	-
	1.59	-	-	-	-	-	-	8	1
	1.44	-	-	4	2	-	-	-	-
6.2	**2.69**	2	-	4	-	-	-	4	-
	2.36	2	1	4	-	-	-	4	-
	2.20	-	-	8	-	-	-	4	-
	2.10	-	-	4	-	-	-	4	-
	2.00	-	-	8	-	-	-	8	-
	1.90	-	-	8	-	-	-	8	-
	1.85	-	-	8	1	-	-	8	1
6.4	**2.82**	2	-	4	-	-	-	4	-
	2.62	2	-	4	-	-	-	4	-
	2.40	2	-	4	-	-	-	4	-
	2.20	-	-	8	-	-	-	8	-
	2.10	-	-	8	-	-	-	8	-
	2.00	-	-	8	-	-	-	8	-
	1.95	-	-	8	1	-	-	8	1
	With transverse bulkheads				(Version E)				
5.9	**2.06**	2	-	4	-	-	-	4	-
	1.91	-	-	8	-	-	-	8	-
	1.86	-	-	8	2	-	-	4	2
	1.81	-	-	4	3	-	-	4	3
	1.74	2	-	8	3	-	-	4	4
	1.70	-	-	4	3	-	-	4	4
	1.65	-	-	4	2	-	-	-	-
	1.55	1	1	4	1	-	-	4	4
	1.45	-	-	4	4	-	-	-	-
6.2	**2.70**	2	-	4	-	-	-	4	-
	2.37	2	-	4	-	-	-	4	-
	2.28	-	-	8	-	-	-	8	-
	2.23	-	-	4	2	-	-	4	1
	2.13	-	-	4	4*	-	-	4	4
	2.04	2	-	4	4	-	-	4	4
	1.90	1	1	4	4	-	-	4	4
6.4	**2.91**	2	-	4	-	-	-	4	-
	2.57	2	-	8	-	-	-	4	-
	2.46	-	-	8	-	-	-	8	-
	2.40	-	-	8	3	-	-	8	1
	2.34	2	-	4	4	-	-	4	4
	2.10	1	1	4	4	-	-	4	4

Table 5: Number of tests resp. simulation runs N and number of observed capsizing incidents N_C
Ship A: damage case 11/12: no trim

initial draught T [m]	damage GM_L	Test N	Test N_C	Simulation N	Simulation N_C	Test N	Test N_C	Simulation N	Simulation N_C
	Open deck				(Version A)				
5.9	**4.45**	2	-	4	-	2	-	4	-
	3.24	2	-	4	-	2	-	4	-
	2.85	5	-	8	-	5	-	8	-
	2.60	-	-	8	-	-	-	-	-
	2.55	-	-	8	2	-	-	-	-
	2.49	5	1	8	4	5	-	8	-
	2.45	-	-	-	-	-	-	8	-
	2.40	-	-	-	-	-	-	8	3
6.2	**3.76**	2	-	4	-	-	-	4	-
	3.65	-	-	8	-	-	-	-	-
	3.60	-	-	4	1	-	-	-	-
	3.55	-	-	4	2	-	-	-	-
	3.45	-	-	4	4	-	-	-	-
	3.33	2	1	4	4	-	-	8	-
	3.30	-	-	-	-	-	-	4	3
	2.48	1	1	4	4	-	-	4	4
	Open deck 1° additional heeling towards the waves				(Version A)				
5.9	**3.24**	2	-	4	-	2	-	4	-
	2.95	-	-	8	-	-	-	-	-
	2.90	-	-	8	2	-	-	-	-
	2.85	2	1	4	1	2	-	8	-
	2.80	-	-	-	-	-	-	8	-
	2.75	-	-	-	-	-	-	8	1
	Side casings				(Version F)				
5.9	**1.93**	2	-	4	-	-	-	4	-
	1.67	-	-	8	-	-	-	-	-
	1.62	2	-	4	3	-	-	8	-
	1.57	-	-	-	-	-	-	8	2
	1.48	2	1	4	3	-	-	4	2
6.2	2.42	-	-	4	-	-	-	-	-
	2.22	-	-	8	-	-	-	-	-
	2.17	-	-	4	3	-	-	8	-
	2.12	2	-	4	4	-	-	4	2
	1.83	2	-	4	4	-	-	4	4
	1.40	-	-	4	4	-	-	-	-
	With transverse bulkheads				(Version E)				
5.9	**2.32**	2	-	4	-	-	-	4	-
	1.94	2	-	4	-	-	-	4	-
	1.56	2	-	4	-	-	-	4	-
	1.45	2	-	4	-	-	-	4	-
	1.39	-	-	8	-	-	-	-	-
	1.34	-	-	8	-	-	-	8	-
	1.29	-	-	8	1	-	-	8	-
	1.24	-	-	-	-	-	-	4	2
6.2	**1.66**	2	-	4	-	-	-	4	-
	1.57	-	-	8	-	-	-	8	-
	1.52	-	-	8	2	-	-	4	2
	1.47	-	-	4	3	-	-	4	3
	1.37	2	-	4	3	-	-	4	4
	1.09	1	1	4	4	-	-	4	4

Table 6: Number of tests resp. simulation runs N and number of observed capsizing incidents N_C
Ship B: damage case 11/12: initial draught t = 5,75m

initial trim d T [m]	damage GM_L	Test N	Test N_C	Simulation N	Simulation N_C
	Open deck			(Version A)	
0.	2.44	-	-	8	-
	2.39	-	-	8	2
	2.34	5	-	8	3
	2.26	2	2	8	8
	2.19	1	1	8	7
-1.0	3.02	2	-	8	-
	2.87	2	-	8	-
	2.77	-	-	8	-
	2.74	-	-	8	2
	2.71	2	1	8	3
	Side casings			(Version F)	
0.	1.94	2	-	8	-
	1.64	2	-	8	-
	1.44	-	-	8	-
	1.39	-	-	8	5
	1.34	2	-	8	5
	1.16	1	1	8	8
-1.0	2.47	2	-	8	-
	2.24	2	-	8	-
	2.02	2	-	8	-
	1.79	2	-	8	-
	1.67	-	-	8	-
	1.62	-	-	8	4
	1.57	2	-	8	5
	1.30	1	1	8	8
	With transverse bulkheads			(Version E)	
0.	1.77	3	-	8	-
	1.47	2	-	8	-
	1.17	2	-	8	-
	1.07	-	-	8	-
	1.02	1	1	8	1
-1.0	1.38	2	-	8	-
	1.16	2	-	8	-
	0.93	2	-	8	-
	0.83	-	-	8	-
	0.78	-	-	8	-
	0.73	2	-	8	4
	0.65	1	-	8	5

Table 7: Number of tests resp. simulation runs N and number of observed capsizing incidents N_C
Ship B: damage case 7/8: initial draught t = 5,75m

initial trim d T [m]	damage GM_L	Test N	Test N_C	Simulation N	Simulation N_C
	Open deck			(Version A)	
0.	2.31	5	-	8	-
	1.72	2	-	8	-
	1.67	-	-	8	-
	1.62	-	-	8	8
	1.44	2	-	8	7
	1.38	2	-	8	8
+1.0	2.10	2	-	8	-
	1.79	2	-	8	-
	1.74	-	-	8	-
	1.69	-	-	8	-
	1.64	-	-	8	5
	1.49	1	1	8	8
	Side casings			(Version F)	
0.	2.27	2	-	8	-
	1.96	2	-	8	-
	1.77	2	-	8	-
	1.68	-	-	8	-
	1.63	-	-	8	2
	1.58	2	1	8	4
+1.0	2.27	2	-	8	-
	1.89	2	-	8	-
	1.68	-	-	8	-
	1.63	-	-	8	2
	1.58	1	1	8	7
	With transverse bulkheads			(Version E)	
0.	2.13	2	-	8	-
	1.75	2	-	8	-
	1.44	2	-	8	-
	1.29	2	-	8	-
	1.19	-	-	8	-
	1.14	-	-	8	5
+1.0	1.55	2	-	8	-
	1.31	2	-	8	-
	1.11	-	-	8	-
	1.06	-	-	8	6
	1.01	2	-	8	8

Table 8: Number of tests resp. simulation runs N and number of observed capsizing incidents N_C
Ship C: damage case 10/11: initial draught t = 6,0m

initial trim d T [m]	damage GM_L	Test N	Test N_C	Simulation N	Simulation N_C
	Open deck			(Version A)	
0.	**2.16**	5	-	8	-
	1.62	2	-	8	-
	1.48	-	-	8	-
	1.43	-	-	8	2
	1.33	-	-	8	3
	1.23	-	-	8	2
	1.08	2	-	8	2
	0.3	2	1	8	5
+1.04	**1.53**	2	-	8	-
	1.12	2	-	8	-
	0.75	-	-	8	-
	0.70	-	-	8	2
	0.11	2	-	8	4
	Side casings			(Version F)	
0.	**1.90**	2	-	8	-
	1.42	2	-	8	-
	1.37	-	-	8	1
	1.27	-	-	8	7
	1.17	2	2	8	7
+1.04	**1.78**	2	-	8	-
	1.54	2	-	8	-
	1.14	2	-	8	-
	1.09	-	-	8	-
	1.04	-	-	8	5
	With transverse bulkheads			(Version E)	
0.	**2.03**	2	-	8	-
	1.74	2	-	8	-
	1.35	2	-	8	-
	1.25	-	-	8	-
	1.15	-	-	8	-
	1.05	-	-	8	-
	1.00	-	-	8	-
	0.95	-	-	8	1
+1.04	**1.68**	2	-	8	-
	1.31	2	-	8	-
	0.56	2	-	8	-
	0.51	-	-	8	-
	0.46	-	-	8	3

Table 9: Number of tests resp. simulation runs N and number of observed capsizing incidents N_C
Ship C: damage case 6/7: initial draught t = 6,0m

initial trim d T [m]	damage GM_L	Test N	Test N_C	Simulation N	Simulation N_C
	Open deck			(Version A)	
0.	**2.49**	5	-	8	-
	2.05	2	-	8	-
	1.95	-	-	8	-
	1.90	-	-	8	2
	1.77	1	1	8	2
+1.08	**2.33**	2	-	8	-
	2.07	2	-	8	-
	1.90	1	1	8	-
	1.78	-	-	8	-
	1.68	-	-	8	-
	1.63	-	-	8	-
	1.58	-	-	8	1
-1.07	**2.47**	2	-	8	-
	2.24	2	-	8	-
	2.15	-	-	8	-
	2.08	1	1	8	2
	Side casings			(Version F)	
0.	**2.31**	2	-	8	-
	2.12	2	-	8	-
	2.01	-	-	8	-
	1.96	2	-	8	2
	1.85	2	1	8	3
-1.07	**2.29**	1	-	8	-
	2.19	-	-	8	1
	2.14	-	-	8	3
	2.09	2	-	8	5
	1.95	1	1	8	6
	With transverse bulkheads			(Version E)	
0.	**2.00**	2	-	8	-
	1.82	2	-	8	-
	1.63	2	-	8	-
	1.43	-	-	8	-
	1.38	-	-	8	-
	1.33	-	-	8	1
-1.07	**2.20**	2	-	8	-
	2.00	2	-	8	-
	1.83	2	-	8	-
	1.65	2	-	8	-
	1.45	-	-	8	-
	1.40	-	-	8	-
	1.35	-	-	8	3

SIMULATION OF LARGE AMPLITUDE SHIP MOTIONS AND OF CAPSIZING IN HIGH SEAS

Apostolos D. Papanikolaou[1], Dimitrios A. Spanos[2] and Georgios Zaraphonitis[3]

[1]Professor, Head of Ship Design Laboratory, National Tech. Univ. of Athens
[2] Dr.-Eng. Cand., Ship Design Laboratory, National Tech. Univ. of Athens
[3] Ass. Professor, Ship Design Laboratory, National Tech. Univ. of Athens

ABSTRACT

The formulation of a mathematical model for the simulation of large amplitude ship motions and capsize of a damaged ship at zero forward speed in waves is presented. A numerical solution (algorithm), developed for the purpose of systematic evaluation of the so obtained theoretical model, is outlined and results from an application of the method to a Ro-Ro passenger ship are presented and discussed.

KEYWORDS

Large amplitude, ship motions, simulation, capsize, non-linearities, flood water dynamics, memory effects.

INTRODUCTION

In view of recent regulatory developments of the International Maritime Organisation (*ref: IMO-SOLAS 95, Reg. 14, Stockholm Regional agreement*) to allow the physical modelling of the damage stability of Ro-Ro passenger ships in waves as an alternative to the so-called 'water on deck' regulatory concept, it becomes evident that the availability of proper computer algorithms, allowing the simulation of capsize of a damaged ship in waves is of great importance. This is justified by the fact, that such a procedure allows the necessary flexibility and efficiency to address systematically alternative design measures as well to improve the survivability of the ship and her compliance with pertinent SOLAS regulations, of quite complex nature.

Based on earlier related work by Zaraphonitis and Papanikolaou (1993), a six degrees of freedom mathematical model of ship motions in waves, at zero forward speed, has been formulated and solved numerically in the time domain, allowing the simulation of ship motions and the prediction of capsizing under specific environmental conditions. A similar general theoretical approach has been presented earlier by Letizia and Vassalos (1995).

The non-linear equations of ship motions, accounting for the effect of flooding, have been exactly formulated based on large amplitude rigid body dynamics. However, in order to simplify their solution, the mass of the flood water is assumed to be concentrated at the centre of ship's volume occupied by the fluid. A semi-empirical water ingress/outflow model accounting for the damage opening and the effective pressure head is used for the modelling of the water flow into and out the damaged ship compartments. Radiation and wave diffraction forces are calculated from hydrodynamic coefficients evaluated by a 3D frequency domain diffraction panel code, applying the impulse response function concept. Froude-Krylov (undisturbed incident wave) and hydrostatic forces are calculated by direct pressure integration over the instantaneous wetted ship surface.

An advanced numerical integration method, based on the extrapolation technique, is used for the numerical integration of the formulated 6 DOF non-linear differential equations of motion. The employed method proved to be very fast and accurate. A DEC 3000 Alpha workstation was used for the development and systematic numerical evaluation of the developed algorithm.

MATHEMATICAL MODEL

Co-ordinate Systems

Four co-ordinate systems are used to express the equations of motion. Let $OXYZ$ be an inertial co-ordinate system, with OZ vertical and positive upwards and $Gx'y'z'$ a body-fixed co-ordinate system with G located at the centre of gravity of the intact ship. We introduce also a co-ordinate system $OX'Y'Z'$ which is always parallel to the body-fixed co-ordinate system and $Gxyz$ which is always parallel to the inertial co-ordinate system. When the ship is at rest, point O coincides with G and all co-ordinate systems coincide with each other. The instantaneous position of the ship is uniquely defined by the position vector $\vec{X}_G$ of point G with respect to the inertial co-ordinate system and the three Euler angles (θ roll, φ pitch and ψ yaw).

Let P be a point in space and $\vec{X}$ its position vector with respect to the inertial co-ordinate system. Let $\vec{X}'$, $\vec{x}$ and $\vec{x}'$ be the position vectors of point P with respect to systems $OX'Y'Z'$, $Gxyz$ and $Gx'y'z'$ respectively. These vectors can be transformed to each other using a co-ordinate transformation matrix **R**:

$$\vec{X} = R\vec{X}' = \vec{X}_G + \vec{x} = \vec{X}_G + \mathbf{R}\vec{x}' \tag{1}$$

The full mathematical expression for the co-ordinate transformation matrix **R** is given in appendix A.

In the following, all vectors or matrices expressed with respect to $OX'Y'Z'$ or $Gx'y'z'$ will be marked with an $(')$, while one or two dots over a variable, or function, denotes first or second time derivative.

Equations of Motion

We consider the complete dynamic system consisting of the intact ship and the flood water. In order to simplify the derivation and solution of the motion equations of the above dynamic system, the mass of the flood water is assumed to be concentrated at its centre of gravity ('lump mass concept'). This is a rather reasonable assumption, since as already discussed in previous work, e.g. D. Vassalos (1994), the effect of sloshing is expected to be weak. Sloshing can induce considerable dynamic effects when the excitation frequency is close to the natural frequency of the flood water and the ship's mass small. However, the possibility of significant sloshing effects is herein rather small, because the roll natural frequency of Ro-Ro ferries is usually very low, especially in damage condition, and the absolute ship's size comparably large.

According to Newton's second law:

$$\vec{F} = \frac{d}{dt}\iiint_V \rho\dot{\vec{X}}dV \tag{2}$$

For the angular motion the expression is:

$$\vec{M} = \frac{d}{dt}\iiint_V \rho\vec{X} \times \dot{\vec{X}}dV \tag{3}$$

where $\vec{F}$ and $\vec{M}_C$ are the sum of all external forces and moments (about C) applied to the dynamic system characterized by a mass density distribution ρ on the entire volume V, consisting of the intact ship and the flood water mass, expressed in the inertial co-ordinate system.

Let W be the centre of gravity of the flood water and $\vec{X}_W$ its position vector, expressed with respect to the inertial co-ordinate system. Equation 2. takes the form:

$$\vec{F} = m_S\ddot{\vec{X}}_G + m_W\ddot{\vec{X}}_W + \dot{m}_W\dot{\vec{X}}_W \tag{4}$$

where m_S is the mass of the intact ship and m_W is the mass of the flood water.

It can be easily proved that:

$$\iiint_V \rho\vec{X} \times \dot{\vec{X}}dV = m_S\vec{X}_G \times \dot{\vec{X}}_G + \mathbf{R}(\mathbf{I}\vec{\omega}') + m_W\left(\vec{X}_G + \vec{x}_W\right)\times\left(\dot{\vec{X}}_G + \dot{\vec{x}}_W\right) \tag{5}$$

where $\vec{x}_W$ is the position vector of *W*, expressed in *Gxyz*. Let $\vec{M}_G$ be the sum of all external moments about point *G*, also expressed in *Gxyz*.

$$\vec{M}_G = \vec{M}_C - \vec{X}_G \times \vec{F} \tag{6}$$

Introducing Eqn. 5. and Eqn. 6. in Eqn. 3. and after some manipulations the equation of angular motion takes the following form:

$$\vec{M}_G = \mathbf{R}\left(\mathbf{I}\dot{\vec{\omega}}' + \vec{\omega}' \times \left(\mathbf{I}\vec{\omega}'\right)\right) + m_W \vec{x}_W \times \left(\ddot{\vec{X}}_G + \ddot{\vec{x}}_W\right) + \dot{m}_W \vec{x}_W \times \left(\dot{\vec{X}}_G + \dot{\vec{x}}_W\right) \tag{7}$$

where **I** the intact ship inertia matrix and $\vec{\omega}'$ the angular velocity vector expressed in the 0body-fixed co-ordinate system. The details of the derivation of the above equations can be found in Zaraphonitis (1997). In appendix B, Eqn. 4. and Eqn. 7. are transformed in a more suitable form for the employed numerical integration scheme.

Exciting Forces

The external forces and moments consist of the following parts:

Froude-Krylov and hydrostatic forces and moments are calculated by direct numerical integration of the incident wave and hydrostatic pressure respectively over the instantaneous wetted surface. Integration is extended up to the instantaneous free surface, taking into account the ship's motions and the free surface elevation due to the incident wave. The distortion of the free surface due to the diffraction of the incident wave system and due to radiation is at this stage omitted.

Neglecting the nonlinearity of the radiation problem and following the so-called Impulse Response Function concept, (see, Cummins 1962 and De Kat 1988), radiation forces in the time domain can be calculated from the added mass and damping coefficients, as following:

$$F'_{R,i}(t) = -A_{ij}(\infty)\ddot{X}'_{Gj} - \int_0^\infty K_{ij}(\tau)\dot{X}'_{Gj}(t-\tau)d\tau,\ \mathrm{i,j}=1,\ldots,6 \tag{8}$$

and

$$K_{ij}(\tau) = \frac{2}{\pi}\int_0^\infty B_{ij}(\sigma)\cos(\sigma\tau)d\sigma \tag{9}$$

where $A_{ij}(\infty)$ is the infinite-frequency added mass coefficient and $B_{ij}(\sigma)$ is the frequency-dependent damping coefficient of the ship, calculated herein by the 3D frequency domain panel program NEWDRIFT (see, Papanikolaou 1988), while t and σ are the time and wave frequency variables, respectively. The integral of Eqn. 9. for the Kernel functions $K_{ij}(\tau)$ is calculated numerically using Filon's method. Due to the fast decay of the Kernel functions, the integration of the convolution integral in Eqn. 8. is truncated at an appropriate upper limit.

A quadratic roll damping model is used to account for viscous effects.

Diffraction forces and moments are approximated by the linear superposition of the elementary diffraction forces associated with each of the component waves composing the encountered wave train:

$$F'_{Di}(t) = \mathrm{Re}\sum_{n=1}^{N} F_i^D(\sigma_n \beta_t)\zeta_n(X_G, Y_G, t),\ \mathrm{i=1,\ldots,6} \tag{10}$$

where $F_i^D(\sigma_n, \beta_t)$ is the frequency-depended diffraction coefficient of mode i, $\zeta_n(X_G,Y_G,t)$ is the instantaneous wave elevation at point G of wave component n and β_t is the relative wave heading (β_t=β-ψ), while β is the wave heading in the inertial co-ordinate system

Water Ingress Model

The rate of inflow or outflow of flooding water $\dot{m}_W$ is calculated by integration of the elementary flow rate over the surface A of the opening:

$$\dot{m}_W = \iint_A dQ \tag{11}$$

dQ is expressed by a semi-empirical formula, Hutchinson (1995):

$$dQ = -gKsign(H_{out} - H_{in})\sqrt{|H_{out} - H_{in}|}dA \tag{12}$$

where K is an empirical weir flow coefficient, H_{out} is the height of the external free surface and H_{in} is the height of the internal free surface.

Natural Seaway Modelling

Two different approaches were used for the modelling (realisation) of the incident wave spectrum by a finite number of harmonic waves. According to the first approach, we introduce a lower and an upper limit for the wave frequency σ_{min} and σ_{max}. The continuous incident wave spectrum is discretised by a number of N harmonic wave components with frequency:

$$\sigma_n = \sigma_{\min} + n\Delta\sigma \tag{13}$$

and amplitude:

$$\mathrm{a}_n = \sqrt{2S(\sigma_n)\Delta\sigma} \tag{14}$$

where:

$$\Delta\sigma = \frac{\sigma_{\max} - \sigma_{\min}}{N} \tag{15}$$

Following the second approach, the area under the incident wave spectrum curve between σ_{min} and σ_{max} is subdivided in N parts of equal area *ds*. The incident wave spectrum is decomposed in *N* harmonic wave components of equal amplitude $a = \sqrt{2ds}$ and frequency corresponding to the centre of the nth elementary part. In both cases, the phase angles of the regular waves are randomly distributed in the range 0 to 2π.

The mean wave energy of the discretised wave systems resulting from the above two approaches, obviously equals the wave energy of the incident irregular seaway. In figures 2 and 3 a JONSWAP spectrum with significant wave height H_s=4.0m and peak period T_p=8sec is presented, along with its discretization calculated by the above described two approaches (amplitudes and corresponding frequencies of the individual harmonic waves).

Since in the first approach the frequencies σ_i of the harmonic wave components are equally spaced, the resulting wave system is periodic by $2\pi/\Delta\sigma$. In order to simulate a genuine irregular seaway for a sufficiently long time by this first approach, the number N of the harmonic wave components should be very large. Therefore, the second approach is considered to give a more efficient representation of a proper incident irregular seaway realisation and is adopted in the following (see, Spanos (1997)).

NUMERICAL SOLUTION

The resulting system of differential (actually: *integro*-differential) non-linear equations is integrated numerically in the time domain, using an advanced integration method based on the extrapolation scheme described by Stoer and Bulirsch (1980). This method proved to be very fast and accurate, especially for this specific type of problems. The basic aspects of the adopted numerical procedure for the calculation of the exciting forces and for the implementation of the simulation scheme are presented in more detail in D. Spanos et al. (1997).

At the development stage, the computer program was running on a DEC-ALPHA 3000 workstation computer. Simulation time was about 15 times slower than real time for the case of one compartment flooding and for an incident wave train consisting of 20 wave components. In the meantime, almost real time simulations could be achieved by a more modern but still quite small workstation of DEC ALPHA PW 433 MHZ type. The prospects, the software to run efficiently on a PC Pentium III computer, are standing excellent.

DISCUSSION OF RESULTS

Simulation records for the motion of an existing Ro-Ro vessel, in service between the Greek mainland and the Aegean Sea islands, are presented. The main characteristics of the Test Ship are presented in Table 1.

In Figure 1 the discretisation of the vessel by 2x177 surface panels is presented. The incident wave is herein described by a JONSWAP spectrum with $H_s = 4.0\,m$ significant wave height and $T_p = 8\,sec$ peak period (Figs. 2, 3). The incident wave heading β is equal to 90° (beam waves). The realisation of the assumed wave spectrum was achieved by use of 20 elementary waves.

Two types of ship motion simulations are presented in the following. At first the ship is considered intact, floating freely at the free surface at zero forward speed and excited by the transverse irregular seas. In the second case, one ship compartment of 66 m in length, located 3 m ahead of the intact ship's LCG and extending from side to side and from the main (car) to the upper deck, is considered flooded. The flood water mass is set equal to 10% of the intact ship's displacement and is kept constant throughout the simulation (no water inflow or outflow is herein considered).

In Fig. 6 the simulation records for both studied cases (intact and flooded hull) are presented, assuming the same wave excitation. In the first row of two graph sets, the wave elevation at the ship's centre of gravity is presented, followed by the results for the heave, roll and pitch motion. Note that capsizing of the flooded ship herein occurs when the coupled heave-pitch and roll motions tend to come into resonance. In Figures 4 and 5 the same results for the roll motion are presented in the common form of the well-established phase diagrams.

CONCLUSIONS

A mathematical model and the corresponding numerical solution procedure for the simulation of large amplitude motions and capsize of a damaged ship is presented, followed by numerical results from the application of the method to a typical Greek Ro-Ro vessel. Further work is currently underway towards the refinement of the mathematical model and the computer algorithm in order to increase the accuracy and speed of the algorithm. In the near future, a series of systematic experiments in the NTUA towing tank is scheduled in order to experimentally fully validate the accuracy of the method.

After all, it is the opinion of the authors that the presented simulation model will be a valuable tool in the process of designing a Ro-Ro vessel, enabling the designer to analyse the impact on damage stability of different design solutions and to maximise the survivability of the vessel, before proceeding to expensive and tiresome experimental investigations.

ACKNOWLEDGEMENTS

The authors wish to acknowledge the support to the present research by the Greek Secretariat General for Research and Technology (code ΠΕΝΕΔ 1995).

The study has been also supported through technical information provided by the Greek Shipowners Association for Passenger Ships, the Union of Greek Coastal Passenger Shipowners and the Hellenic Chamber of Shipping.

References

Cummins W. E. (1962). The impulse response function and ship motions. *Schiffstechnik.* **9:47**, 101-109.

De Kat J. O. (1988). *Large amplitude ship motions and capsizing in severe sea conditions.* Ph.D. Dissertation. Dept. of Naval Architecture and Offshore Engineering. University of California. Berkeley.

Hutchinson L. (1995). Water on-deck accumulation studies by the SNAME ad hoc Ro-Ro safety panel. *Proc. of Workshop on Numerical & Physical Simulation of Ship Capsize in Heavy Seas*, Ross Priory, Glasgow.

Letizia L. and Vassalos D. (1995). Formulation of a non-linear mathematical model for a damaged ship subject to flooding. *Proc. of the Sevastianov Symposium,* Kaliningrad.

Papanikolaou A. D. (1988). NEWDRIFT: The six DOF three-dimensional diffraction theory program of NTUA-SDL for the calculation of motions and loads of arbitrarily shaped bodies in regular waves. *Internal Report*. NTUA-SDL.

Spanos D. (1997). Theoretical-numerical modelling of large amplitude ship motions and of capsizing in heavy seas. *Dr. Eng. Thesis*. Dep. of Naval Architecture. NTUA. in progress.

Zaraphonitis G. , Papanikolaou A. D. and Spanos D., (1997). On a 3-D mathematical model of the damage stability of ships in waves. *Proc. of the 6th Int. Conf. on Stability of Ships and Ocean Vehicles.* Varna.

Stoer B. and Bulirsch R. (1980). *Introduction to numerical analysis.* Springer-Verlag. New York.

Vassalos D. (1994). A realistic approach to assessing the damage survivability of passenger ships. *Trans. SNAME.* **102**, 367-394.

Zaraphonitis G. (1997). Formulation of the equations of motion for a damaged ship in waves. *Internal Report.* Ship Design Laboratory. NTUA.

Zaraphonitis G. and Papanikolaou A. D. (1993). Second order theory and calculations of motion and loads of arbitrarily shaped 3D bodies in waves. *Journal Marine Structures.* **6**.

APPENDIX A

Coordinate Systems Transformation

When the ship is at rest, point O coincides with G and all co-ordinate systems coincide with each other. When the ship is moving, the position and the orientation of the body-fixed co-ordinate system with respect to the inertial one is uniquely defined by the position vector $\vec{X}_G$ and the set of the three so-called Euler angles: roll (θ), pitch (φ) and yaw (ψ).

To obtain the body-fixed co-ordinate system from the inertial one, the later is supposed to be translated to *Gxyz* and then rotated by an angle ψ about the yaw axis, then by an angle φ about the new pitch axis and finally by an angle θ about the new roll axis.

The transformation matrix between the inertial and the body-fixed coordinate system is given by:

$$\mathbf{R} = \begin{bmatrix} \cos\varphi\cos\omega & \sin\theta\sin\varphi\cos\psi - \cos\theta\sin\psi & \cos\theta\sin\varphi\cos\psi + \sin\theta\sin\psi \\ \cos\varphi\sin\psi & \sin\theta\sin\varphi\sin\psi + \cos\theta\cos\psi & \cos\theta\sin\varphi\sin\psi - \sin\theta\cos\psi \\ -\sin\varphi & \sin\theta\cos\varphi & \cos\theta\cos\varphi \end{bmatrix} \tag{16}$$

Let $\vec{\omega}$ and $\vec{\omega}'$ be the angular velocity vector expressed with respect to the inertial and the body-fixed coordinate systems respectively:

$$\vec{\omega} = \mathbf{R}\vec{\omega}' \tag{17}$$

It can be proved, see Zaraphonitis (1997), that:

$$\vec{\omega}' = \mathbf{B}\begin{bmatrix}\dot{\theta} & \dot{\varphi} & \dot{\psi}\end{bmatrix}^{\mathrm{T}} \tag{18}$$

where:

$$\mathbf{B} = \begin{bmatrix} 1 & 0 & -\sin\varphi \\ 0 & \cos\theta & \sin\theta\cos\varphi \\ 0 & -\sin\theta & \cos\theta\cos\varphi \end{bmatrix} \tag{19}$$

APPENDIX B

Equations of Motion

In order to proceed with the numerical integration of the equations of motion, Eqn. 4. and Eqn. 7. must be transformed into a more appropriate form.

From Eqn. 4. we obtain:

$$\left(m_S + m_W\right)\ddot{\vec{X}}_G = \vec{F} - m_W\ddot{\vec{x}}_W - \dot{m}_W\left(\dot{\vec{X}}_G + \dot{\vec{x}}_W\right) \tag{20}$$

From Eqn. 7. it can be proved that:

$$\vec{M}_G = \mathbf{R}\left(\mathbf{I}\dot{\vec{\omega}}' + \vec{\omega}' \times (\mathbf{I}\vec{\omega}')\right) + \frac{m_W}{m_S + m_W}\vec{x}_W \times \vec{F} + \\ + \frac{m_S m_W}{m_S + m_W}\vec{x}_W \times \ddot{\vec{x}}_W + \frac{m_S \dot{m}_W}{m_S + m_W}\vec{x}_W \times \left(\dot{\vec{X}}_G + \dot{\vec{x}}_W\right) \tag{21}$$

and from Eqn. 18.:

$$\dot{\vec{\omega}}' = \mathbf{B}\begin{bmatrix}\ddot{\theta} \\ \ddot{\varphi} \\ \ddot{\psi}\end{bmatrix} + \dot{\mathbf{B}}\begin{bmatrix}\dot{\theta} \\ \dot{\varphi} \\ \dot{\psi}\end{bmatrix} \tag{22}$$

Inserting Eqn. 22. in Eqn. 21. and after some manipulation we derive:

$$\mathbf{I}\cdot\mathbf{B}\begin{bmatrix}\ddot{\theta} \\ \ddot{\varphi} \\ \ddot{\psi}\end{bmatrix} = \vec{M}'_G - \vec{\omega}' \times (\mathbf{I}\vec{\omega}') - \mathbf{I}\begin{bmatrix} -\dot{\varphi}\dot{\omega}\cos\varphi \\ -\dot{\theta}\dot{\varphi}\sin\theta + \dot{\theta}\dot{\psi}\cos\theta\cos\varphi - \dot{\varphi}\dot{\psi}\sin\theta\sin\varphi \\ -\dot{\theta}\dot{\varphi}\cos\theta - \dot{\theta}\dot{\psi}\sin\theta\cos\varphi - \dot{\varphi}\dot{\psi}\cos\theta\sin\varphi \end{bmatrix} - \\ - \frac{m_W}{m_S + m_W}\vec{x}'_W \times \vec{F}' - \frac{m_S m_W}{m_S + m_W}\mathbf{R}^{\mathrm{T}}\left(\vec{x}_W \times \ddot{\vec{x}}_W\right) - \frac{m_S \dot{m}_W}{m_S + m_W}\mathbf{R}^{\mathrm{T}}\left(\vec{x}_W \times \left(\dot{\vec{X}}_G + \dot{\vec{x}}_W\right)\right) \tag{23}$$

Let $\vec{\xi}$ be a 13-dimensional vector, with:

$$[\xi_1 \quad \xi_2 \quad \xi_3]^{\mathrm{T}} = \vec{X}_G \tag{24a}$$

$$[\xi_4 \quad \xi_5 \quad \xi_6]^{\mathrm{T}} = [\theta \quad \varphi \quad \psi]^{\mathrm{T}} \tag{24b}$$

$$\xi_{\mathrm{i}} = \dot{\xi}_{\mathrm{i}-6}, \ \mathrm{i}=7,\ldots,12 \tag{24c}$$

$$\xi_{13} = m_W \tag{24d}$$

The equations of motion (Eqn. 20. and Eqn. 23.) can take the form:

$$\dot{\xi}_{\mathrm{i}} = \xi_{\mathrm{i}+6}, \ \mathrm{i}=1, \ldots, 6 \tag{25a}$$

$$\dot{\xi}_i = f(\xi_1,\ldots,\xi_{13},t), \ \mathrm{i}=7, \ldots, 12 \tag{25b}$$

Finally, from Eqn. 11. and Eqn. 12. we can derive:

$$\dot{\xi}_{13} = f(\xi_1,\ldots,\xi_{13},t) \tag{26}$$

TABLE 1
TEST SHIP MAIN CHARACTERISTICS

L_{BP}	142.00 M
B	22.80 M
T	6.40 M
$D_{MAIN\ DECK}$	8.00 M
$D_{UPPERDECK}$	12.90 M
LIGHT SHIP	7884 T
DISPLACEMENT	11354 T
KG	9.874 M

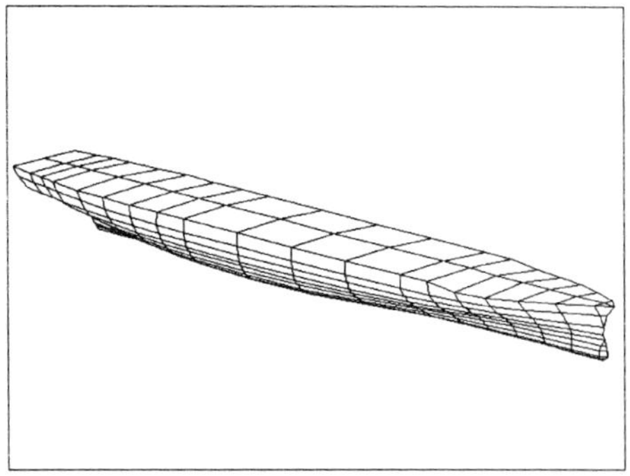

Figure 1. Ship discretisation 2x177 panels.

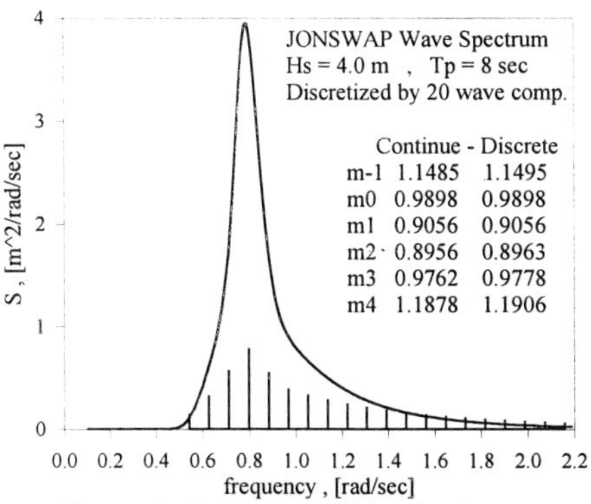

Figure 2. Spectrum discretisation – 1st approach.

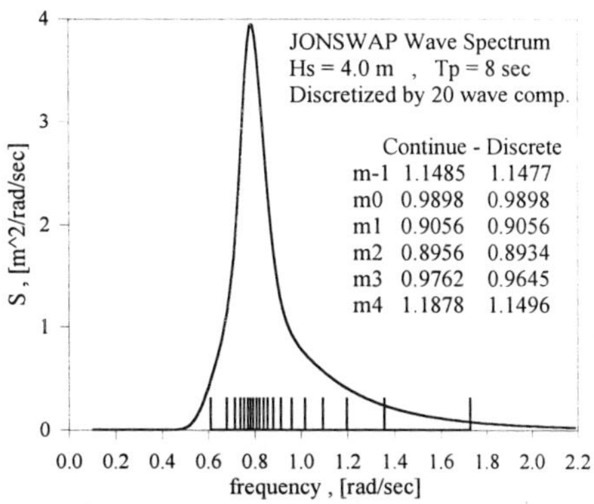

Figure 3. Spectrum discretisation – 2nd approach.

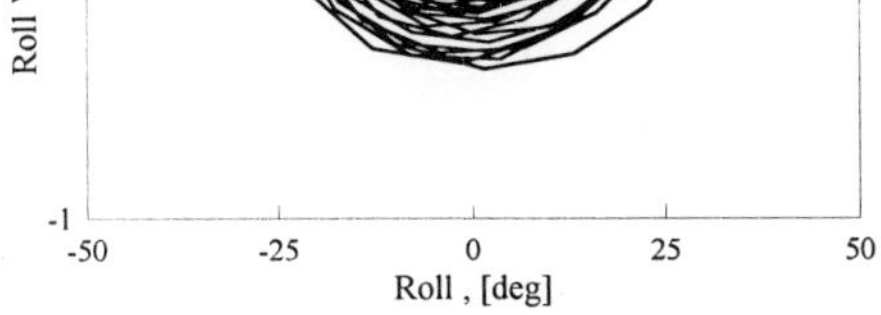

Figure 4. Roll phase diagram, intact case.

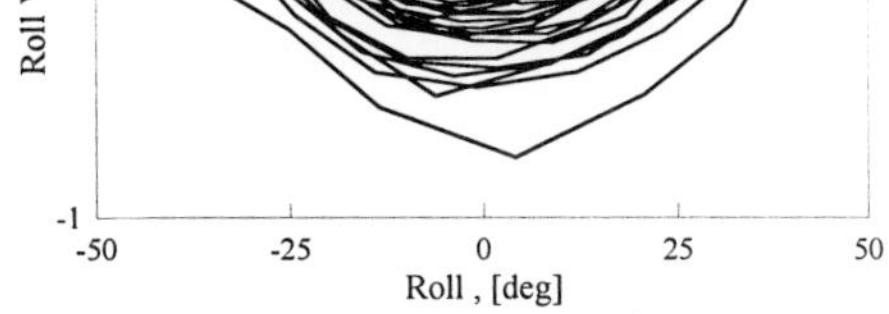

Figure 5. Roll phase diagram, flooded case.

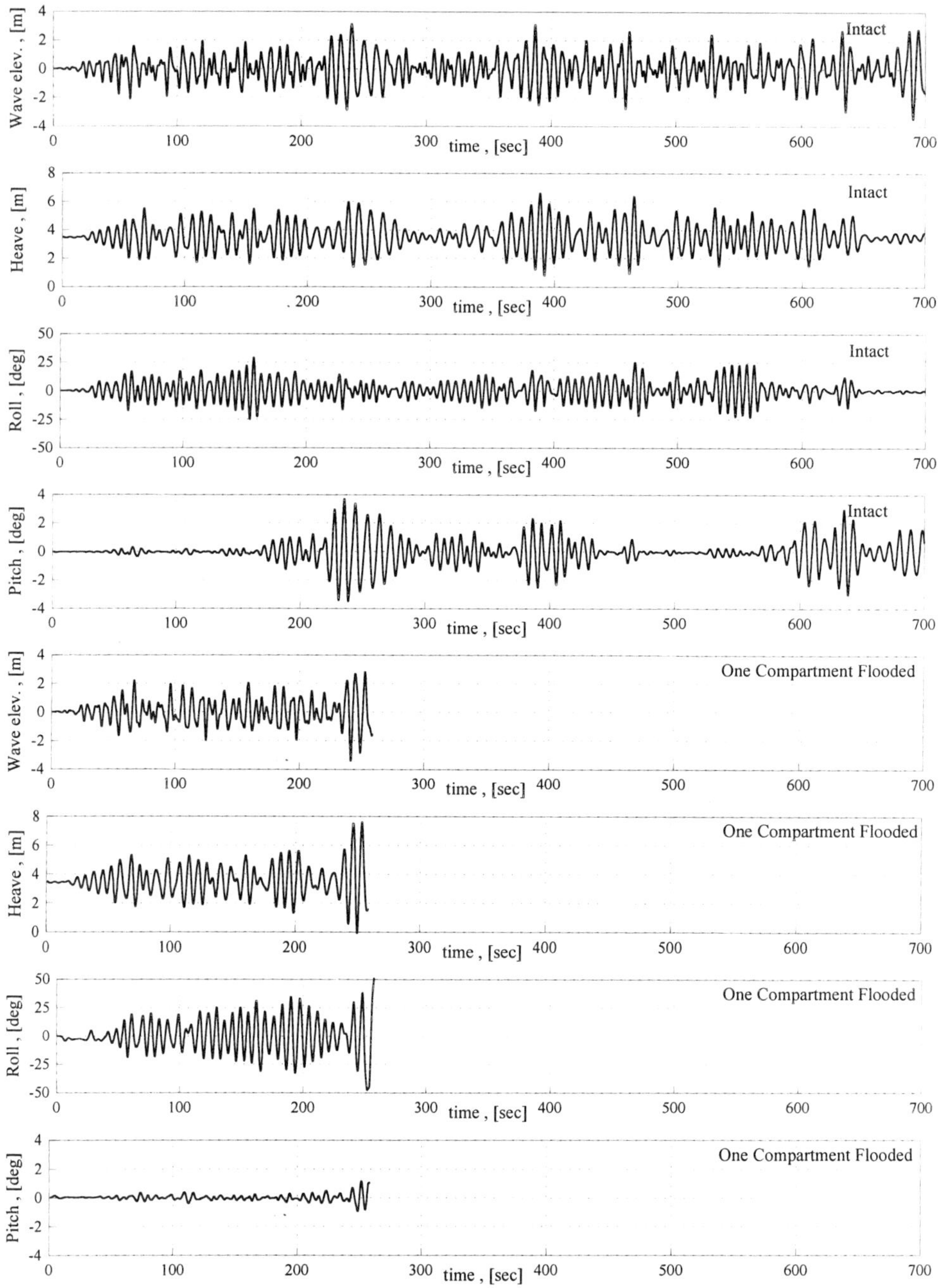

Fig. 6. Simulated records. Intact case - Flooded case

Contemporary Ideas on Ship Stability
D. Vassalos, M. Hamamoto, A. Papanikolaou and D. Molyneux (Editors)

ON THE CRITICAL SIGNIFICANT WAVE HEIGHT FOR CAPSIZING OF A DAMAGED RO-RO PASSENGER SHIP

T. Haraguchi[1], S. Ishida[1] and S. Murashige[2]

[1]Ship Research Institute, Ministry of Transport
6-38-1, Shinkawa, Mitaka, Tokyo 181-0004, Japan
[2]Department of Mathematical Engineering, The University of Tokyo, Japan
7-3-1, Hongo, Bunkyo, Tokyo 113-8656, Japan

ABSTRACT

Since the accident of Estonia, studies have been continued on the stability of RO-RO passenger ships in damaged condition in waves. Because the phenomena are complicated and affected by many factors, studies should be conducted in various conditions. In this study experiments in beam waves were carried out, having the characteristics of Japanese ships and waves around Japan in mind. Discussions were mainly focused on the relation between the height of water on deck and the critical significant wave height for capsizing, and on the effect of the peak period of wave spectrum. The main conclusions are as follows,

(1) When a ship has no initial heel angle capsizing does not occur, but an initial heel angle of as small as 2 degrees put the ship in dangerous condition for capsizing.

(2) The critical significant wave height for capsizing is affected by the peak period of the wave spectrum. The longer the peak period is, the higher the critical significant wave height becomes.

(3) The relation between the critical height of water on deck and the critical significant wave height proposed by UK gives a good estimation in short waves. However, thinking of the waves in various locations in the world, an equation applicable to a wider range of peak period is desired.

KEYWORDS

Damaged stability, capsizing, RO-RO passenger ships, height of water on deck, critical wave height, peak period of wave spectrum, initial heel angle.

INTRODUCTION

After the capsizing accident of Estonia occurred in the Baltic Sea in 1994, the International Maritime Organization (IMO) reviewed the measures to enhance the safety of RO-RO passenger ships and amended the International Convention for the Safety of Life at Sea (SOLAS) in November 1995. Subsequent to this amendment, a proposal, SLF 40/4/5(1996), was submitted by the United Kingdom to Sub-committee on Stability and Load Lines and on Fishing Vessels Safety (SLF40) of IMO, which explicitly includes the flooded free water effect on deck to the stability. From this background, much study has been continued on how to prevent sea water from accumulating on RO-RO deck, Rousseau (1997), and on stability performance when flooding into the deck has occurred, Vassalos & Jasionowski et al. (1998), Vassalos & Conception & Letizia (1997), Vassalos & Pawlowski & Turan (1996), Shimizu & Roby & Ikeda (1996), Hamano & Roby & Ikeda (1997), Ishida & Murashige & Watanabe et al. (1996), Ishida & Murashige (1997).

In the experiment of this study, a model of a typical RO-RO passenger ship in Japan was used because Japanese ships tend to have a different proportion from Northwest European ships and because oceanographic phenomena are a little different from Northwest Europe. In this paper at first, the effect of initial heel angle was mentioned. Subsequently, the relation between the critical height of water on deck and the critical significant wave height for capsizing was discussed. It is because the validation of the important proposal by UK,

$$hc = 0.085\ Hsc^{1.3} \tag{1}$$

to Japanese ships is necessary. This equation shows that a ship should have the stability performance to bear the water height on deck of hc in the critical significant wave height, Hsc. At the same time, equation (1) can be explained that the critical significant wave height should cause the amount of water on deck corresponding to hc. The experimental results are shown on the effects of initial heel and of peak period of wave spectrum to the equation (1).

EXPERIMENT

Model Ship and Experimental Conditions

The model ship is a 1/48.6 scale model of a typical and oceangoing RO-RO passenger ship in Japan. The characteristics of the ship are larger L/B and smaller B/d ratios than those of Northwest European ships. The principal particulars are shown in Table 1. Figure 1 shows the damaged opening, flooded compartments (the central part, colored dark) and the position of 7 water level gauges on the vehicle deck. For realistic modeling

TABLE 1
PRINCIPAL PARTICULARS IN INTACT AND DAMAGED CONDITIONS

(scale ratio : 1/48.6)

	Ship		Model	
	Intact	Damaged	Intact	Damaged
Lpp(m)	170.00		3.500	
Bmld(m)	25.00		0.515	
Dmld(m)	9.50		0.196	
Mean Draft(m)	6.60	8.2	0.136	0.17
Trim(m)	0.00	-1.3	0.000	-0.0259
Condition	Full Load Departure Condition			
Δ(ton)	15020		0.128	
KG_0(m)	10.86		0.224	
G_0M(m)	1.41	2.8	0.029	0.057
Fbd[1)](mid)(m)	2.90	1.3	0.060	0.027
Fbd[2)](mid)(m)	2.20	0.6	0.045	0.012
Tr(sec)	17.90	13.4	2.570	1.93

1) Designed Deck Height, 2) Lower Deck Height

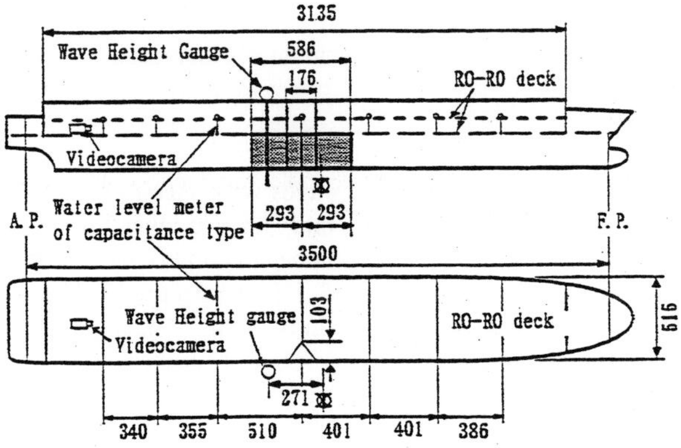

Figure 1: Damaged RO-RO Ship Model

of the shell plating in the damaged compartment as well as the vehicle deck, the model ship was made of FRP. In accordance with SOLAS'90, two compartments are damaged with an opening in the central part of the ship and are designed to make flooding symmetrical. Two vehicle deck heights were tested, one originally designed (hereafter called the designed deck height) and the other lowered by 0.7m (hereafter called the lower deck height) including the ceiling. GM values in both conditions were kept the same. In addition to the condition of no initial heel angle, the conditions with initial heel angles of 2 degrees and 4 degrees to weather (damaged) side were tested, assuming cargo shifting or asymmetrical flooding. Only the condition of the lower deck height with 2 degrees of initial heel angle just does not comply with SOLAS'90 because of insufficient range of positive GZ value. The other conditions sufficiently satisfy SOLAS'90.

Measuring System

The experimental apparatus is shown in Figure 2. The model ship was placed with the damage opening facing the oncoming waves. Swaying, heaving and rolling motions were set free but yawing is loosely restricted by means of a string. The measuring time was set to be 30 minutes in the real ship scale. The carriage followed the model ship to let it drift freely, however, when the drifting speed is too large, drifting was a little controlled by the string.

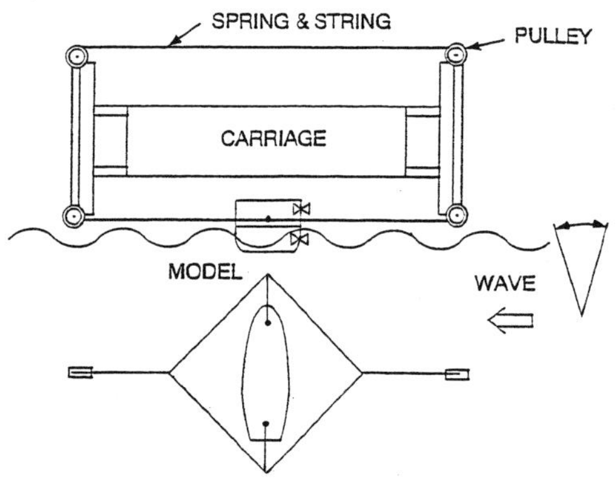

Figure 2 : Experimental Apparatus

Incident waves

All experiments were carried out in irregular waves with spectra of JONSWAP type. As for the wave period, 13.7 sec., 11.6 sec., 9.5 sec. and 7.4 sec. were used (listed in order of the discrepancy from the natural rolling period in damaged condition). The maximum ratios of wave height to wavelength are 1/25, 1/15, 1/12 and 1/10 respectively. These wave heights are the highest ones that the wave maker can generate in the tank. The significant wave heights used in the experiment are shown in Table 2.

TABLE 2
SIGNIFICANT WAVE HEIGHT, MODEL CONDITION AND OCCURRENCE OF CAPSIZE

Deck Height	Initial Heel Angle (deg)	Peak Period of Wave Spectrum (sec)			
		13.65	11.55	9.45	7.35
Designed Deck Height	0°	10.55O	12.58O	10.53O	6.37O
	2°	10.55×	9.44×	6.32×	5.10×
	2°	9.76O	8.58×	5.74O	4.50×
	2°	8.79O	7.55O	5.05O	3.82O
	4°	7.53×	7.55×	5.74×	3.82×
	4°	7.13×	6.29×	5.49×	3.47×
	4°	6.59O	5.39O	5.05O	3.06O
Lower Deck Height	0°	10.55O	12.58O	10.53O	6.37O
	0°	8.79O	9.44O	8.42O	5.10O
	2°	5.27×	4.72×	3.16×	2.55×
	2°	4.79×	4.19×	2.81×	2.18×
	2°	4.63×	4.02×	2.69×	1.91O
	2°	4.39O	3.78O	2.53O	1.53O

Numbers in the table are significant wave height (m)
O:Non-capsize ×:Capsize

EFFECT OF INITIAL HEEL

The Case without Initial Heel

In Table 2 marks O represent non-capsize and marks × represent capsize in 90 degrees. As shown in this table, the ship only capsized with initial heel and she did not in case of no initial heel. In the case of no initial heel, the ship heeled to lee side in both heights of the vehicle deck. Consequently the deck edge of the damaged side (weather side) became higher than that before flooding. Thereafter flooding into the vehicle deck stopped and that avoided capsizing.
This fact agrees with the former experimental results of the authors, Ishida & Murashige & Watanabe et al. (1996).

Figure 3 shows GZ curves for the designed deck height and the lower deck height. Plus angle represents heel to damage side (weather side). The stability performance of intact side is far superior to that of damage side. Therefore, it is natural that the ship did not capsize when she heeled to lee side. It is concluded that heeling to lee side is very safe when there is no opening in the lee side because it stops further flooding, and because the righting moment itself is large in that direction. According to the wave statistics, Watanabe & Tomita & Tanizawa (1992), 1/20 as the ratio of wave height to wavelength is the highest in the sea areas around Japan. Moreover, the ratios of 1/20 or over were used in this experiment, excluding the wave with the peak period of 13.7 sec. Therefore, it is concluded that this ship at the GM value in the experiment is hard to capsize without initial heel in the sea area around Japan. A circular letter issued by IMO, IMO Circular letter No1891 (1996), includes a provision that a ship shall be thought to have capsized when its steady heel angle exceeds 20 degrees. In this experiment steady heel angles were less than 20 degrees except for only one case, hence this ship is supposed to be hard to capsize in terms of the steady heel angle, too.

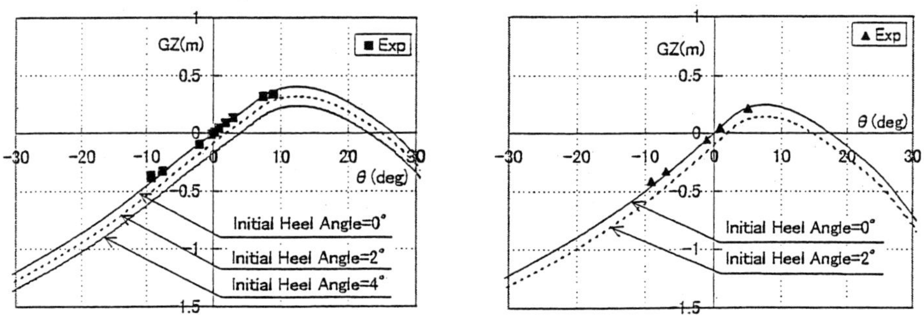

(a) Designed Deck Height (b) Lower Deck Height

Figure 3 : GZ Curves

The Case with Initial Heel

On the other hand, when the model ship had an initial heel to weather side, the mean heeling direction was always weather side and she capsized in many cases. An initial heel to weather side lowered the deck edge, enhanced flooding into the vehicle deck, increased steady heel to weather side and again lowered the deck edge. This chain led to capsizing. This is the same phenomenon in the case of a ship with a center casing, Ishida & Murashige & Watanabe et al. (1996), except that the stability performance reduced by the initial heel. This fact indicates that if the water flooded into the vehicle deck accumulates on the damage side, it will increase the risk of capsizing. Table 2 shows that the wave height at which the ship capsized became lower with increasing initial heel angle and with descending deck height. This is because the stability performance deteriorates with increasing initial heel and with descending deck height, as shown in Figure3.

The ship capsized at about only 2m of the significant wave height in the condition of 2 degrees of initial heel and the lower deck height, in which the stability is slightly in short of the requirements of SOLAS'90. So, it can be said that capsizing is unavoidable in this condition. However even if the deck height was raised to the designed value from this condition, in which SOLAS'90 is satisfied, capsizing took place with the significant wave height of 3.5m. It can be concluded that the risk of capsizing is very high when a ship has an initial heel to the damage side (weather side).

CRITICAL HEIGHT OF WATER ON DECK AND CRITICAL SIGNIFICANT WAVE HEIGHT

Definitions of Critical Height of Water on Deck and Critical Significant Wave Height

The height of water on deck from the outer mean sea surface is an important factor for stability in waves. The value is not zero in general even if time average for a certain period is made. Direct measurement of this quantity is very difficult, so usually it is evaluated statically as a function of the amount of water on deck and the heeling angle in calm water.

Examples of time histories of the amount of water on deck are shown in Figure 4. For evaluating the height of water on deck, the time-averaged volume for a certain period in steady condition was used. In non-capsized case the period was selected at the last stage of experiment, and in capsized case the period was just before the capsizing motion. As for the heeling angle, time average in the same period was used.

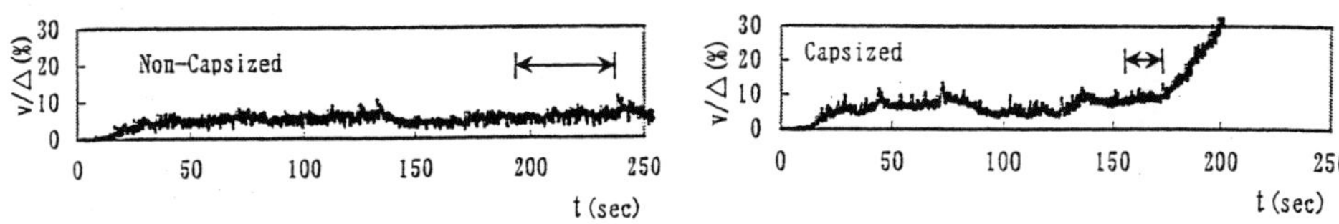

Figure 4 : Time History of Amount of Water on Car Deck(v : Amount of Water on Deck, Δ : Displacement)

When the ship capsized the critical height of water on deck (hcritical) was defined as the average of the heights of water on deck in two experiments, one capsized the other not, in the same conditions except for the significant wave height. Similarly, the critical significant wave height (Hscritical) and the critical angle for capsizing (θ critical) were defined as the averages in the two experiments. Hereafter, affixed "critical" represents the critical values obtained from the experiment with this manner.

On the other hand, another critical height of water on deck from mean sea surface should be defined (hc), which is calculated without considering waves explicitly. In calculating hc, the heeling angle is fixed to a critical value (θ c), in which the GZ curve has the maximum value in damage side. The amount of water on deck is decided to make GZ zero at the angle of θ c. Hereafter, affixed letter "c" represents the values obtained by this method. In addition, the critical significant wave height (Hsc) is defined by equation (1).

Critical Height of Water on Deck

Figure 5 shows the relationship between hcritical and Hscritical with black marks. For comparison some steady conditions of non-capsized cases (Figure 4) are also shown with empty marks. The horizontal three lines show hc's, which were calculated from GZ curves as mentioned. The solid curve represents equation (1). The results of four peak periods of wave spectrum are included.

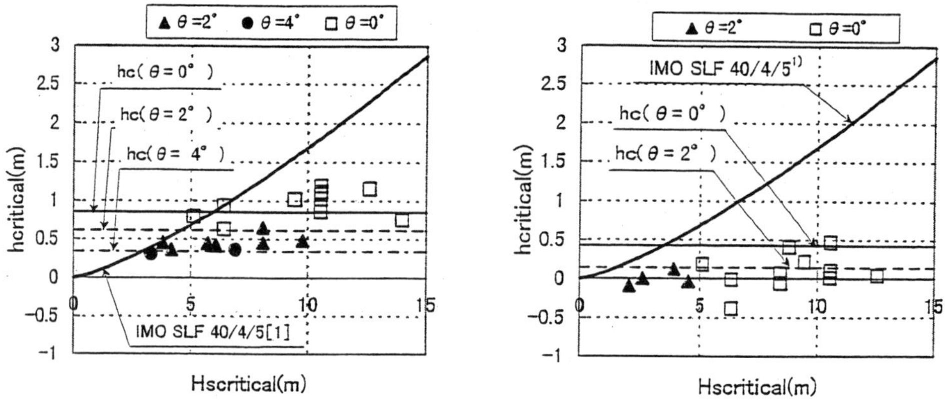

(a) Designed Deck Height (b) Lower Deck Height

Figure 5 : Critical Height of Water on Car Deck(θ : Initial Heel Angle)

Figure 5(a) for the designed deck height indicates that hcritical (black marks) is equivalent to or somewhat smaller than hc (horizontal lines) and that hcritical remains almost constant or goes up slightly for the significant wave height. Figure 6 shows the ratio of hcritical to hc. The ratios are ranging between 0.5 and 1.0, and capsizing especially took place at the smaller value than hc when the initial heel angle is 2 degrees. As will be mentioned later, the ratio of the critical heel angle (θ critical/ θ c) is greater than 1. The height of water on deck has a tendency to decrease when the heel angle increases, as shown in Figure 7. Consequently, it is supposed that if θ critical/ θ c approaches 1, hcritical/hc would approach 1. It is concluded that the critical height of water on deck can be roughly estimated by hc in various ship and wave conditions.

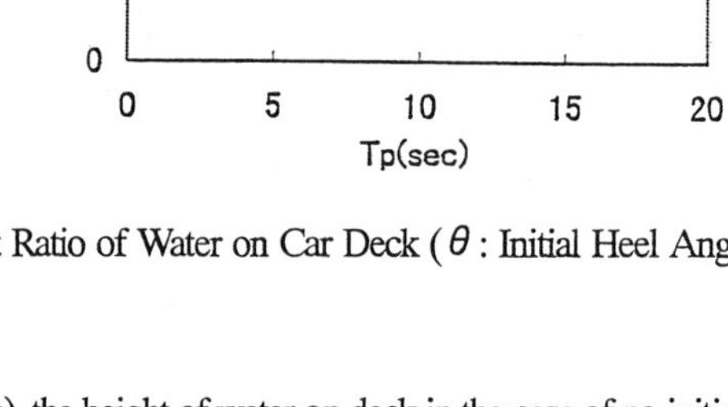

Figure 6 : Critical Height Ratio of Water on Car Deck (θ : Initial Heel Angle, Tp : Peak Period of Wave Spectrum)

Additionally in Figure 5(a), the height of water on deck in the case of no initial heel (empty marks) shows a similar tendency, while some of them are greater than hc. This is because, being different from the case of initial heel, the ship heels to lee side (intact side), hence it has a stability great enough to bear large amount of flood water.

Meanwhile, in Figure 5(b) for the lower deck height, the tendency of hcritical is similar to that in Figure 5(a). It might seem to be strange that some hcritical values are negative. It means flooding might be still continuing at the moment of capsizing. However the discrepancies between hcritical and hc are almost the same as Figure 5(a). It is thought that the main cause is the dynamic effect and that hcritical might depend somewhat on the time histories of wave elevation. Anyway, the stability of this condition is very small as shown Figure 3(b), so the ship capsized with a small amount of water on deck.

In the case of no initial heel (no capsize) in Figure 5(b), the ship heeled to lee side as was the case with the designed deck height. The height of water on deck was lower than the one in Figure 5(a) and sometimes lower than the mean sea surface. This can be explained by the reasons that the ship was easy to heel because of the small stability as shown in Figure 3(b) and that the height of water on deck is sensitive to heel angle as shown in Figure 7. It should be noted that the smaller amount of water on deck makes the height more sensitive to heel angle and that the amount was in fact small because water ingress stopped after she heeled to lee side in a short time.

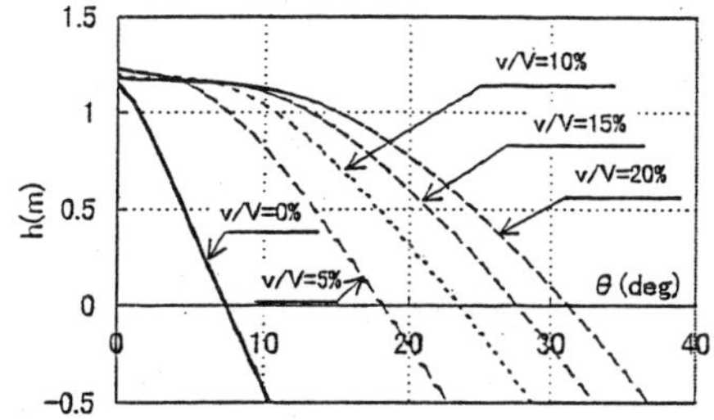

Figure 7 : Relation between Height of Water on Car Deck and Heel Angle (v : Volume of Water on Deck, V : Volume of Car Deck)

Effects of Peak Period of Wave Spectrum on Critical Significant Wave Height

Figure 8 shows the comparisons of the results in Figure 5(b) and equation (1) for each peak period of the wave spectrum. These figures prove that when the peak period is 7 sec. Hscritical is smallest and is in good agreement with equation (1). However with the increase of peak period of wave spectrum Hscritical increases. Figure 9 shows the ratio of Hscritical to Hsc for the designed deck height. It can be seen that the ratio is increasing as the peak period increases, being in agreement with the tendency in Table 2. When the peak period is 13 sec. Hscritical is greater than Hsc by a factor of two.

As shown in Figure 5(b), the variation of Hscritical for the lower deck height is smaller than the designed deck height. But when the peak period is 13 sec., Hscritical is greater than Hsc by three times. This fact indicates that the ship can survive in higher waves than equation (1) in longer waves and that the critical value depends on wave steepness.

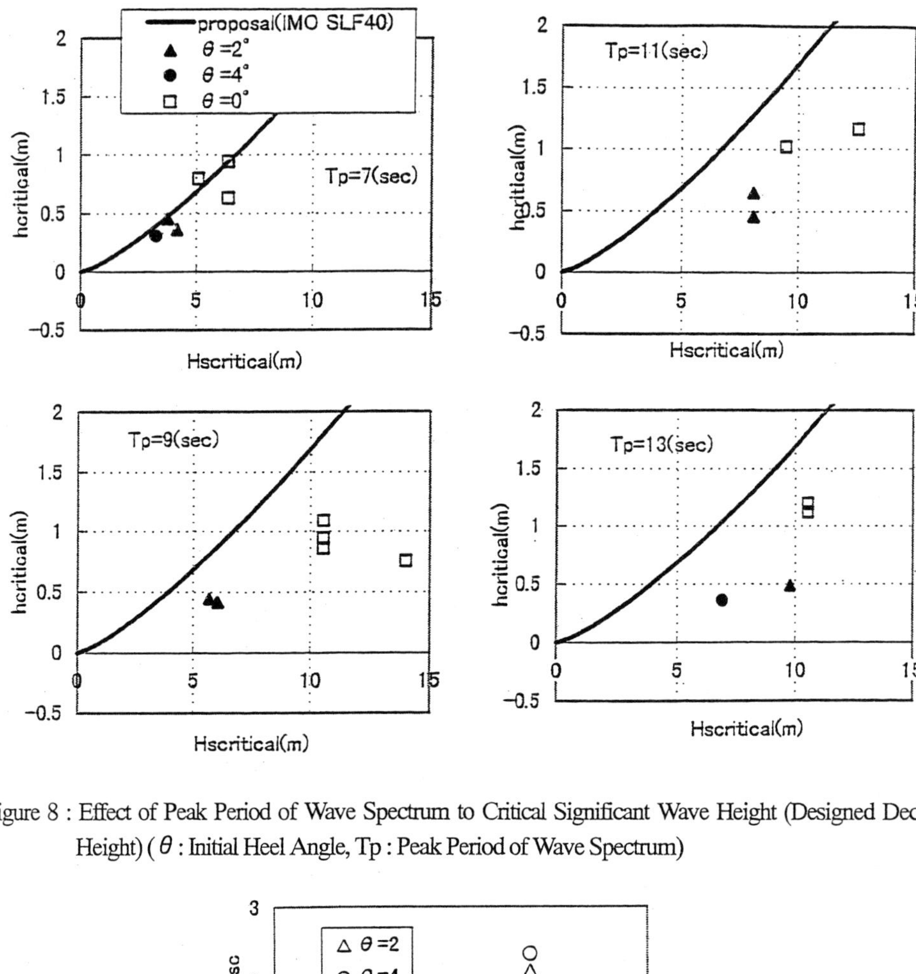

Figure 8 : Effect of Peak Period of Wave Spectrum to Critical Significant Wave Height (Designed Deck Height) (θ : Initial Heel Angle, Tp : Peak Period of Wave Spectrum)

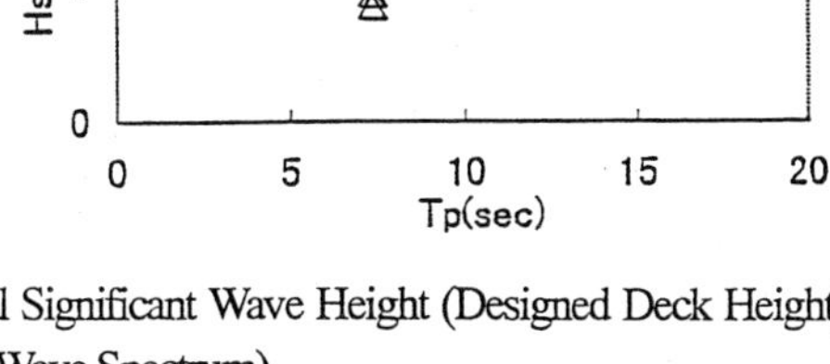

Figure 9 : Ratio of Critical Significant Wave Height (Designed Deck Height) (θ : Initial Heel Angle, Tp : Peak Period of Wave Spectrum)

COMPARISON BETWEEN PROPOSAL BY UK AND EXPERIMENTAL RESULTS

As discussed with Figure 5 and Figure 8, hcritical is small compared with equation (1) when the peak period of wave spectrum is long. Therefore, the equation (1) gives rather large amount of accumulated water, i.e. the required stability performance is higher than the necessary in reality. From the standpoint of wave height, Hscritical is 2 or 3 times as large as Hsc, i.e. the critical wave height might be underestimated in some peak period of the wave spectrum. In other words the dependence of the critical wave height on the peak period of wave spectrum is not reflected to equation (1).

Equation (1) was obtained from many simulations and experiments, conducted in the same spectrum type as this experiment, JONSWAP type. But the peak period was from 4 sec. to 9 sec. and the significant wave height was from 1m to 8m, which was decided considering the crowded sea areas around Europe. As a result, it is convinced that the equation almost agrees with the results of this experiment up to 9 sec. of the peak period. Conversely, it might be natural that the equation is not applicable to peak periods greater than 9 sec.

In order to apply the equation to various sea areas in the world, it should be considered that the critical significant wave height varies by the peak period. According to the database on waves statistics in the sea area around Japan, Watanabe & Tomita & Tanizawa (1992), the frequency of occurrence of waves whose period is 9 sec. or over has exceeded a negligible level especially in the side of Pacific Ocean. So an equation applicable to a wider range of peak periods is desired.

CONCLUSIONS

A capsizing experiment was carried out in beam seas with a JONSWAP spectrum, using a model of a typical oceangoing RO-RO passenger ship in Japan. The main conclusions are as follows,

(1) The ship, which satisfies SOLAS'90, does not capsize in the condition of no initial heel even in high waves, which are rarely seen in wave statistics around Japan. However she often capsizes with a little initial heel to damage side. The critical significant wave height for capsizing is lower than that in which the ship without initial heel survives.

(2) The critical significant wave height for capsizing is affected by the peak period of the irregular wave spectrum. The longer the peak period is, the higher the critical significant wave height becomes.

(3) The relation between the critical height of water on deck and the critical significant wave height proposed by UK gives a good estimation as long as the peak period of the wave spectrum is short. However, considering the waves around Japan and other areas in the world, the relation should include the effect of peak periods.

ACKNOWLEDGEMENT

This study was conducted as a part of RR71 project by Ship Research Association. The authors would like to be grateful for the support from the chairman of RR71 committee, Professor

Fujino from Tokyo University, and from the leader of its working group for damaged stability, Professor Ikeda from Osaka Prefecture University, as well as from the other committee members.

References

Hamano T. & Roby K. & Ikeda Y. (1997). A New Approach to Damage Stability Rule (2nd report)-Experiments to Identify the Motion Characteristics of Damaged Ships, *Journal of the Kansai Society of Naval Architects* 228 (in Japanese).

IMO Circular letter No1891 (1996). AGREEMENT CONCERNING SPECIFIC STABILITY REQUIREMENTS FOR RO-RO PASSENGER SHIPS UNDERTAKING REGULAR SCHEDULED INTERNATIONAL VOYAGES BETWEEN OR TO OR FROM DESIGNATED PORTS IN NORTH WEST EUROPE AND THE BALTIC SEA, ANNEX1 "Significant wave heights", Appendix MODEL TEST METHOD.

Ishida S. & Murashige S. & Watanabe I. et al. (1996). Study on Damage Stability with Water on Deck of a RO-RO Passenger Ship in Waves, *Journal of the Society of Naval Architects of Japan* 179 (in Japanese).

Ishida S. & Murashige S. (1997). STABILITY OF A RO-RO PASSENGER SHIP WITH A DAMAGE OPENING IN BEAM SEAS, *Proceeding of STAB'97*.

SLF 40/4/5 (1996). HARMONIZATION OF DAMAGE STABILITY PROVISION IN IMO INSTRUMENTS, A Proposal on New Damage Stability Framework for RO-RO Vessels based upon Joint North West European R & D Project "Safety of Passenger/RO-RO Vessels" Submitted by Denmark, Finland, Norway, Sweden and the United Kingdom.

Rousseau J.H. (1997). Flooding Protection of RO-RO Ferries Phase Ⅱ.

Shimizu N. & Roby K. & Ikeda Y. (1996). An Experimental Study on Flooding into the Car Deck of a RORO Ferry through Damaged Bow Door, *Journal of the Kansai Society of Naval Architects* 225.

Vassalos D. & Jasionowski A. et al. (1998). Time-Based Survival Criteria for RO-RO Vessels, *The Royal Institution of Naval Architects spring meetings*.

Vassalos D. & Conception G. & l.Letizia (1997). Modeling the Accumulation of Water on the Vehicle Deck of a Damaged RO-RO Vessel, *Third International Workshop on Theoretical Advances in Ship Stability and Practical Impact*.

Vassalos D. & Pawlowski M. & Turan O. (1996). Joint North West European Project, Safety of Passenger RO-RO Vessels - Task 5, A Theoretical Investigation on the Capsizal Resistance of Passenger RO-RO Vessels and Proposal of Survival Criteria, University of Strathclyde Marine Technology Centre.

Watanabe I. & Tomita H. & Tanizawa K. (1992). Winds and Waves of the North Pacific Ocean (1974～1988), *Papers of Ship Research Institute*, Supplement No14.

Contemporary Ideas on Ship Stability
D. Vassalos, M. Hamamoto, A. Papanikolaou and D. Molyneux (Editors)

EXPLORATION OF THE APPLICABILITY OF THE STATIC EQUIVALENCE METHOD USING EXPERIMENTAL DATA

Andrew Kendrick[1], David Molyneux[2], Andre Taschereau[3], and Tom Peirce[4]

[1]Fleet Technology, Ltd;
[2]Institute for Marine Dynamics
[3] Transportation Development Centre
[4]Operational Dynamics

ABSTRACT

The Static Equivalency Method (SEM) developed by the Strathclyde University research team (see Vassalos et al., 1997) offers a simple predictor of capsize for a wide range of damaged ship conditions. Data generated by Transport Canada's experiment program of capsize investigations was reanalyzed to investigate the validity of the predictor equations, and their range of applicability. Excellent agreement was found for cases in which the dynamic roll component of ship behaviour is small, and where internal sloshing is limited, in particular for ships with centreline casings. The SEM capsize predictor was also applied successfully to cases with freeing ports, though more work is needed to define water build-up under these conditions. Some limitations of the model predictions were highlighted, and potential sources of error identified. Results of the project were used to provide recommendations as to how the SEM could be applied as a component of deterministic or fully probabilistic damaged stability criteria.

KEYWORDS

RO-RO Ferries, Capsize Prevention, Waves, Flooding, Static Equivalent Method

INTRODUCTION

The extensive international research which followed the ferry catastrophes of the 1980's and early 1990's led to the development of numerous proposals for ways in which safety could be enhanced in future operations. These included recommendations for new stability criteria, for design features and operational procedures. Under considerable time pressure, IMO introduced amendments to the SOLAS Convention in 1990, and then declined to make further global modifications in 1995, leading to a number of Northern European nations adopting by regional agreement of their own, more stringent, requirements.

As is too often the case, the regulatory decisions were made before the researchers tasked with clarifying the issues had managed to fully disseminate, assimilate, and evaluate each other's results. In some cases, the fact that policy decisions were taken removed the support for continuing with promising research which had only yielded preliminary or inconclusive results. However, the Canadian government decided to persevere with its own program to try and provide explanations for some puzzling phenomena and indicate the most promising directions for future safety initiatives.

BACKGROUND

Transport Canada Research Program

Transport Canada initiated a multi-phase program of research in 1993 entitled, "Flooding Protection of RO-RO Ferries". The objective of the program was to examine the survivability of monohull RO-RO ferries, fitted with freeing ports, under various conditions of ferry loading, residual freeboard after collision damage amidships and the prevailing sea state.

Phase I of the program, which used a highly simplified model of a large Ro-Ro vessel was completed in March 1995 and a set of reports, with supporting data which confirmed the benefits of freeing ports, were provided to IMO, Transport Canada (1995i), Stubbs et al (1996). Using the findings of Phase I and other available publications on RO-RO ferry capsize, Phase II investigated the relationships which describe the capsize phenomenon, using a more ship-shaped model of a smaller ferry. The results of this work are summarized, Transport Canada (1995ii), Transport Canada (1997) and Molyneux et al (1997) and were also provided in full to IMO. It noted that parameters other than those contained in the SOLAS 90 criteria appeared to provide better insight into safety than the SOLAS standards themselves.

Although Canada had decided to remain with the standard SOLAS 90 approach for the time being, it was recognized that both this and the SOLAS '90+50' option are imperfect predictors of safety. There was thus a continuing desire to develop a better understanding of the mechanisms causing capsizes involving water on deck, which are quite different from those which govern intact capsize under more extreme wave climates.

The Static Equivalency Method

The Static Equivalency Method (SEM) is based on a number of insights and hypotheses, some of which are common to other investigations of capsize, including the earlier phases of the Transport Canada project. Vassalos et al (1997) provides more details of its workings.

It is presumed that it is the accumulation of water on the vehicle deck that causes the ship to capsize. The required capsize volume (or weight) of water on deck is assumed to be that which would cause the ship to loll to its angle of maximum GZ, θ_{GZm} , in the flooded condition. Any additional heel with this volume on deck, or any additional volume at the same heel angle, will create a larger overturning moment. This will be resisted by a smaller restoring moment. Thus, the ship will inevitably capsize.

The depth of water on deck at the critical condition corresponds to an elevation, *h*, above the mean external sea level, and this elevation can, in turn, be correlated with the significant wave height, H_S , through an empirical equation:

$$h = 0.085 H_S^{1.3} \qquad (1)$$

Kinetic wave energy is, in effect, transformed into potential energy.

It is assumed that the process is quasi-static, as the time frames associated with capsize are significantly longer than the wave or ship roll periods. Based on this assumption, dynamic effects do need to be accounted for either in the stability calculation approach or in the correlation of water elevation and wave height.

It is worth noting that the relationship given above has been modified, Vassalos et al (1998), to include residual freeboard.

The new version is:

$$h = 0.088 H_S^{(0.97+0.46F)} \qquad (2)$$

where F is the residual freeboard.

The stability calculations needed to predict capsize water volume can be undertaken in several ways which should yield essentially identical results. None of these are 'standard' routines for commercial stability analysis software packages, but reprocessing of their normal output data allows the important quantities to be calculated with modest effort.

Project Objectives

The basic Static Equivalency Method, as outlined above, offers an appealingly simple means to predict capsize. However, the published descriptions of the method left a number of concerns, including:

i) the influence of relative motions on the accuracy of the results;
ii) the influence (if any) of sea spectrum;
iii) the influence of ship size and configuration;
iv) the ability of the method to represent adequately centre and side casing influences;
v) the potential for treating freeing port effects in the method.

As all of these variables had been explored (to varying degrees) in the Canadian experimental program. It was therefore hoped that the experimental data could be used, first, to check the basic validity of the SEM predictions, and then to examine some or all of these presumed second-order effects.

ANALYSES

The value of θ_{GZm} was first found from analysis of the damaged hulls from Phases I and II of the project. The "equilibrium" weight/volume of water on the car deck was then established for this (imposed) heel angle. Sinkage, deck edge immersion, and internal water level were supplementary outputs.

The data sets obtained during the experiment phases of the project were reviewed to identify conditions expected to be most relevant to the testing of the static equivalency hypotheses. This was done both qualitatively (through the characteristics of the data traces) and quantitatively (through comparisons with the numerical analyses). Initially the selection of specific experiments was undertaken independently by the authors, searching on a variety of criteria. The resulting lists were then collated to produce an agreed set, which was confined to cases which the SEM has been developed to handle, i.e., fully-enclosed car decks, rather than those with bulwarks or open ends.

It was not considered necessary to reprocess many non-capsize runs, as the previous data analysis was expected to have provided representative mean values for the volume, heel, and motion parameters under safe conditions. However, some 'marginally safe' runs were re-examined, particularly where closely related capsize runs showed unexpected characteristics. Rapid capsizes were also excluded from reprocessing, due to the difficulty of identifying any specific critical point in the process.

Basic Capsize Prediction

As explained above, the Static Equivalency Method predicts wave height to cause a capsize by relating this to the buildup of water on deck, and assuming that the required volume on deck corresponds to the predictions of static stability calculations. Thus, the initial verification of the SEM considered its ability to predict the critical volume of water on deck and the corresponding wave height to cause a capsize. Secondly, heel angles and relative motions just prior to capsize were compared.

Comparisons of predicted and actual volumes of water in capsize conditions for the Phase II model is shown in Figure 1. (Phase I results were similar.) As can be seen, the critical volume data shows some scatter about the expected lines, with a tendency to under predict the volumes at the higher values. Neither the scatter nor this under prediction was unexpected. The capsize process is itself random in nature Vassalos et al (1997), and any model tests have some lack of precision. This work also observed an under prediction of survivability under conditions where the damaged ship has high residual stability. Comparisons of measured and predicted volumes are an indication of the validity of the general methodology, but the most important question is, obviously, whether the method can accurately predict the sea conditions under which capsize can be expected to occur.

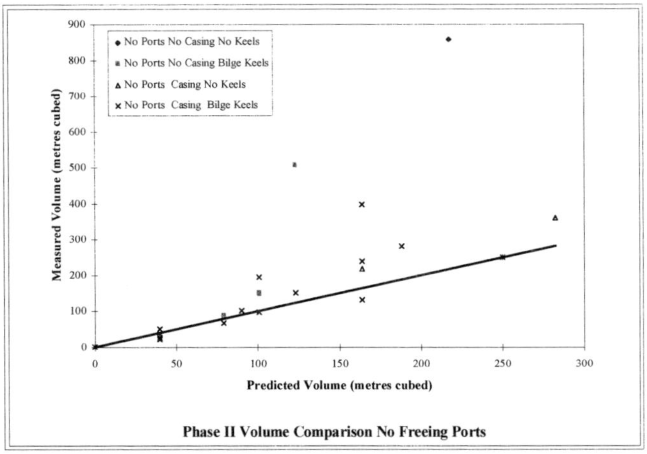

Figure 1: Phase II Volume Comparisons

The predicted and measured capsize wave heights are compared in Figures 2 and 3. In the experimental program, the significant wave heights were pre-determined and spaced sufficiently to obtain differences that were more than those obtained due to random nature of wave height. This provided two data points for each condition, the maximum significant wave height the model survived and the minimum significant wave height to cause a capsize. These data are shown in Figures 2 and 3 by diamonds and squares respectively. In cases where no capsize was observed during experiments, the measured data is shown with an arrow, to indicate that the capsize would most likely occur in higher waves. There were also cases where no survival was observed, and these are also marked with arrows pointing to lower wave heights.

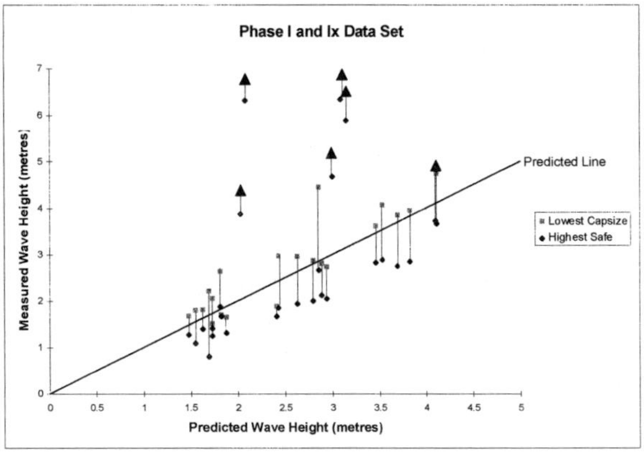

Figure 2: Predicted and Measured Capsize Wave Heights, all Phase I data

As can be seen, almost all the measurements bracket the predicted value for both models tested. This excellent correspondence between prediction and measurement is the key result of the project, as it demonstrates the SEM's ability to account for a range of variables in a single, simple approach to capsize prediction. The prediction capability holds good for both the large, simplified, and small 'realistic' models, answering another of the important questions regarding the method.

The data sets plotted here cover only cases with a centre casing; the set without casing is discussed later in the paper.

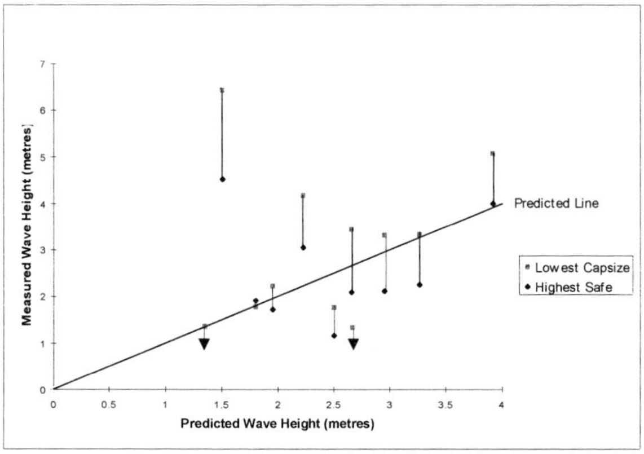

Figure 3: Predicted and Measured Capsize Wave Heights, Phase II

Effects of Supplementary Factors

Two of the factors within the SEM that the authors felt required further investigation were the regression equation linking internal water elevation and wave height, and the physical meaning of the relationship. The capsize prediction capability appeared to provide a reasonable validation of the equation, but closer examination of the significant wave height and relative motion data indicated that the relationship was more complex than implied by the SEM's basic formula. Understanding this mechanism is important for applying the SEM to deigns outside the data set used to develop the method.

Figure 4 relates the Phase I motion data at the damage opening to the wave height, and shows that there is not a constant relationship between the two. Very similar results were observed in the original research that led to the development of the SEM. The resulting equation derived during the simulations for the SEM was

$$Hsr = 3.185\ Hs^{-0.69} \qquad (3)$$

The values derived from the analysis of the experiment data were

$$Hsr = 3.125\ Hs^{-0.54} \qquad (4)$$

However, changing the formulation of the SEM to include the more complex relationship between Hsr and Hs does not significantly improve the accuracy of the predicted significant waveheight to cause a capsize. The factors modifying the influence of wave height thus remain somewhat unclear, and more extensive simulations and analyses of the phenomenon are likely to be needed to gain further insights. It is possible that the heave and roll components of the relative motion need to be considered separately, and there is a certain amount of evidence from the test series that changes in roll amplitude have relatively little effect on performance, as will be seen from discussions below.

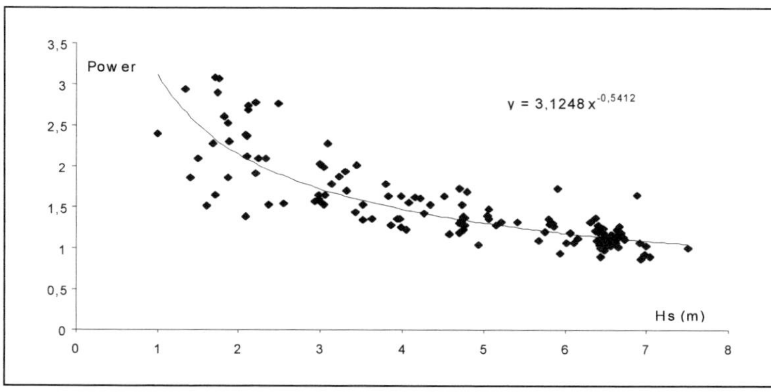

Figure 4: Regression on power p between significant height of relative motion $H_{sr} = H_s^{\,p}$ and significant wave height H_s for all data from Phase I. Average value of $p = 1.50$.

Freeing Ports

A considerable amount of effort was used in the earlier phases of the program to investigate freeing port effectiveness in preventing capsize. Several different configurations were used, including both flapped and permanently open ports. Compared to a fully enclosed deck, the first arrangement allows more outflow for the same inflow, while the second produces both more inflow and more outflow. Flapped ports will thus generally give greater safety than permanent ports, although there are doubts about how reliable most designs would prove in actual service.

Experiments with open freeing ports were reanalyzed, with the main focus on flapped ports. The SEM was not originally intended to account for either option, though it appeared probable that it could be modified relatively easily to investigate flapped ports. Comparing the results with those for the same basic ship conditions, as shown (for example) in Figure 5, the following observations were made:

a) the volumes of water associated with capsize are essentially the same as those for the basic condition (and show even less scatter from the predicted line);

b) the wave heights at capsize for flapped freeing ports are much higher, while those for the permanent openings are more ambiguous.

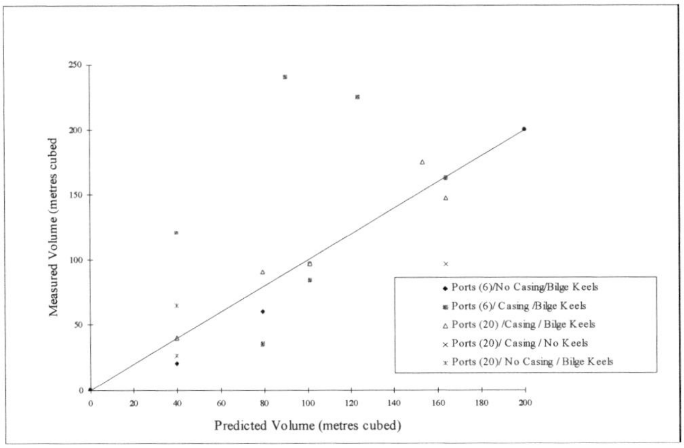

Figure 5: Phase II Volume with Ports

Unfortunately, although the first observation suggests that the basic SEM should remain applicable, insufficient time and resources were available within the project to develop a revised formulation for the wave height/water elevation relationship to account for the ports. This could probably be based on the types of inflow/outflow balancing originally proposed by Hutchison et al (1995) but taking better account of the responses of the ship and the port flow characteristics.

Other variables investigated in the Phase I and II projects which were expected to have some influence on the capsize performance included casing location, presence/absence of bilge keels, and sea spectrum. Points associated with these conditions are identified in Figure 1 and help illustrate the results discussed below.

Sea Spectrum

A very limited number of runs were made with a spectrum other than JONSWAP, and only the most tentative of conclusions can be drawn from the data. In the two conditions tested with an ITTC spectrum, the model capsized at a higher wave height than with JONSWAP. Unfortunately, no otherwise identical conditions were tested with two spectra of equivalent significant waveheights, but it appears from the results of the non-capsize runs that the volumes of water which built up were less for any given wave height when the ITTC spectrum was used. This was an expected result, as the energy distribution of the two spectra at a given wave height differs significantly, with JONSWAP's being higher at frequencies which produce relative motions of the ship.

As one of the underlying hypotheses of the SEM is that wave energy outside the ship transforms into potential energy raising the internal water level, the regression formula defining this would also be expected to change. However, there is insufficient data to attempt to construct a new relationship at this point. The JONSWAP spectrum is representative of coastal conditions, where most collision damage is likely to occur. Therefore the relationship used in the basic SEM, which is conservative, is considered to be appropriate for most applications.

Bilge Keels/Roll Motion

The experiments in Phase II treated bilge keels fitted as the standard condition, but included a small number of runs without keels. There does not appear to be anything in the data to suggest that the build-up of water between the with and without bilge keel conditions followed different relationships, although there was some difference in relative motions due to the increase in roll motion, when the bilge keels were removed. It appears from results of these and other analyses that the SEM is relatively insensitive to the roll component of motion, as is real risk of capsize.

Casing Influences

The SEM predicts differences in the behaviour of ships with side, centre, or no casings, based on the differences in damaged hydrostatics. The test programs did not consider side casings, but devoted considerable attention to the influence of centre casing versus no casing. In general, the Canadian work indicated that no casing conditions had more survivability than conditions where the casing was present, but there were no obvious ways of quantifying the expected degree of performance improvement.

Figure 1 shows some of the differences between casing and no casing results on the standard SEM volume plots. These appear to show that there is relatively little difference between the two configurations when critical volumes are small, but much greater divergences for larger volumes when dynamic effects become significant. The magnitudes of divergence found in the experimental program for "no casings" cases were significantly larger than from previous numerical simulations Vassalos et al (1997). The reason for this is unclear, though it may be that the treatment of internal waves in the simulation needs further refinement. This "sloshing" is a complex phenomenon, as discussed by Huang and Hsiung (1997).

When dynamic effects become important, it is logical that they will be more beneficial when the flow of water across the deck is unobstructed than when the casing retains it on one side. It is questionable as to whether this type of behaviour would actually have practical meaning, as the flow of water in a real damage event is likely to be restricted by vehicles on deck, etc. This would make it more likely that the response can be treated as quasi-static, and the potential for dynamic enhancements of stability can thus be substantially discounted. In other words, it may be non-conservative to explore ways of accounting for improvements when the centre casing is not present. However, the effect certainly warrants further exploration from a theoretical standpoint.

CONCLUSIONS

The results of the two experimental phases of this program compared very well with the predictions of the Static Equivalency Method for most, though not all, of the conditions investigated. This ability of the SEM to work well over a range of ship forms and conditions means that it can provide a superior correlation with ship survivability over the current SOLAS criteria, and over any of the other simplified methods which have been published to date.

The SEM has some shortcomings, the most significant of which are summarized below. None is considered to invalidate the overall conclusions reached above, but all warrant additional investigation to enhance the current version of the method.

The method does not take full account of dynamic effects, which appear to be of increasing importance when capsizing a ship with good inherent damaged stability and when no casing or other obstructions are present to restrict the flow of water across the deck. The method errors on the side of conservatism, and so the consequences may be acceptable from a regulatory or initial design standpoint. Since the detailed numerical simulations of capsize on which the simplified SEM is based do appear to track all model test data, either model tests or simulations could be used by designers and owners to justify a relaxation in the criteria where appropriate.

The SEM is based on the relationship between static head, h, and relative motion, Hsr. Relative motion was first defined as a function of waveheight and later as a function of waveheight and residual freeboard. We should note that these two equations give different probabilities of survival. The first equation will give a 50% probability, whilst the second predicts the ship will survive for one hour. Based on the results of model experiments and simulations, it appears that relative motion is a more complex function than waveheight alone. Preliminary investigation of the data in this paper does not show a better fit to the more complex equation than to the simple one, and so was omitted from this presentation.

In order to develop the SEM further, it may be necessary to get a better definition of the relationship between h and Hs. Physical model data alone is not sufficient for modifying this relationship with confidence. The combination of numerical simulation and physical model experiments for validation seems to be a more logical way to proceed.

The original data on the SEM does not quantify the amount of scatter in the predicted boundary between survival and capsize. This is a concern for its use in a deterministic analysis of stability, where it is important that the criteria be set near the upper bound of potential capsize behaviour. The modified version is based on a pre-determined probability of survival in a given time, which partially addresses this concern.

The effectiveness of freeing ports cannot yet be quantified using the SEM, although a promising line of approach to this has been identified. This could be carried forwards analytically, though it is probable that additional numerical simulations would also be required to bring the work to a conclusion.

ACKNOWLEDGEMENTS

The authors would like to thank Transport Canada, Ship Safety Branch and Transportation Development Centre for sponsoring this work, along with the Institute for Marine Dynamics and the Canadian Ferry Operators Association, who provided additional support.

References

Huang, Z.J. and Hsiung, C.C. (1997). Experimental Study on Wave Motion Inside a Damaged Ship Compartment. *Ocean Engineering International* **1:2**, 70-73.

Hutchison, B.L., Molyneux, D. and Little, P.(1995). Time Domain Simulation and Probability Domain Integrals for Water on Deck Accumulation. *CyberNautics 95, SNAME California Joint Sections Meeting,* Long Beach, California, 21-22 April 1995.

Molyneux, W.D. et al (1997). Model Experiments to Determine the Survivability Limits of RO-RO Ferries. *Trans. SNAME, Annual Meeting Technical Sessions,* Ottawa, Canada, 16-18 October 1997, *Technical and Research Session*, p. 11.1-11.17.

Stubbs, J.T. et al (1996). Flooding Protection of RO-RO Ferries. *Spring Meetings, RINA, RINA Transactions* **138: Part B**, 103-116.

Transport Canada (1995i). Flooding Protection of RO-RO Ferries, Phase I. *TP12310E* **1&2**.

Transport Canada (1995ii). Flooding Protection of RO-RO Ferries, Phase I extension. *TP12581E* **1&2.**

Transport Canada (1997). Flooding Protection of RO-RO Ferries, Phase II. *TP12991E.*

Vassalos, D. et al (1997). Dynamic Stability Assessment of Damaged Passenger/RO-RO Ships and Proposal of Rational Stability Criteria. *Marine Technology and SNAME News* **34:4**, 241-267.

Vassalos, D. et al (1998). Time Based Survival Criteria for RO-RO Vessels. *Spring Meetings, RINA* **141.**

D. Vassalos, M. Hamamoto, A. Papanikolaou and D. Molyneux (Editors)

MODELLING THE ACCUMULATION OF WATER ON THE VEHICLE DECK OF A DAMAGED RO-RO VESSEL AND PROPOSAL OF SURVIVAL CRITERIA

D. Vassalos, L. Letizia and O. Turan

The Ship Stability Research Centre (SSRC), Department of Ship and Marine Technology
University of Strathclyde, Glasgow, UK

ABSTRACT

Recent research at the University of Strathclyde culminated in the development of a numerical procedure for assessing the damage survivability of damaged Ro-Ro vessels and, using this as a basis, new survival criteria have been proposed and submitted to IMO for consideration by the international shipping community. This paper presents the results of a fundamental study aimed at enhancing the insight into one of the most dominant parameters affecting the survival of a Ro-Ro vessel, the water accumulation on the vehicle deck. The investigation represents an attempt to identify the most important contributing factors to the flooding process by performing a series of experiments using a scaled model of a typical Ro-Ro vessel. The matrix considered involves a range of ship design and environmental parameters in a number of simplified damage scenarios, starting with a fixed model and progressively introducing more degrees of freedom, building up to six degrees of freedom and to more realistic damages in a way that allows for isolation of individual contributions to the water accumulation on the Ro-Ro deck. The results of the experiments are presented and discussed, leading to recommendations for characterising the flooding process for general assessment of damage survivability and to testing the validity of the proposed survival criteria.

KEYWORDS

Damaged Ro-Ro vessels, floodwater accumulation, damage scenarios, experimental testing, Static Equivalent Method (SEM), probability of survival, survival criteria

INTRODUCTION

The limited understanding of the complex dynamic behaviour of a damaged vessel and the progression of flood water through the ship in a random sea state has, to date, resulted in approaches for assessing the damage survivability of ships that rely mainly on hydrostatic properties with potentially serious consequences concerning the loss of life and property whilst endangering the environment. The tragic accidents of the *Herald of Free Enterprise* and more recently of *Estonia* were the strongest indicators yet of the magnitude of the problem at hand, particularly when water enters the deck of ships with large undivided spaces, such as Ro-Ro vessels. The ship loss could be catastrophic as a result of rapid capsize, rendering evacuation of passengers and crew impractical, with disastrous (unacceptable) consequences. Concerted action to address the water-on-deck problem in the wake of these led to the proposal of new stability requirements, known as the *Stockholm Regional Agreement*, or more commonly as *SOLAS '90+50*, pertaining to compliance of existing Ro-Ro vessels with SOLAS '90 requirements whilst accounting for the presence of a maximum 0.5 m height of water on the vehicle deck. In view of the uncertainties in the current state of knowledge concerning the ability of a vessel to survive damage in a given sea state, an alternative route has been allowed which provides a non-prescriptive way of ensuring compliance and one hopes enhanced survivability, namely the "*Equivalence*" route, by performing model experiments in accordance with the requirements of the SOLAS regulation II-1/8. In response to these developments, the shipping industry, slowly but steadily, appears to be favouring the model experiments route, implicitly demonstrating mistrust towards deterministic regulations which, admittedly, lack solid foundations. An attractive alternative route to tackling the water-on-deck problem in a way that allows for a systematic identification of the most cost-effective and survivability-effective solutions has been introduced by the Ship Stability Research Centre (SSRC) at the University of Strathclyde, by making use of a mathematical/numerical model, developed and validated since 1997, describing the dynamic behaviour of a damaged ship in seaway whilst subjected to progressive flooding. This model was made the basis during the Joint North West European Project (JNWEP) for formulating and proposing rational survival criteria to deal with water on deck as part of the probabilistic procedure for assessing damage stability, Vassalos et al (1996). A relevant paper was submitted to IMO and is currently being considered by the working group on harmonisation of probabilistic standards. In the developed mathematical model and the ensuing criteria the process of water accumulation on the Ro-Ro deck as well as the actual amount of water are dominant features. In this respect, an acceptably accurate model of water ingress/egress is a prerequisite to undertaking any investigations on damage survivability.

Deriving from the above, this paper attempts to elucidate some of the basic characteristics of the flooding process considering a typical Ro-Ro vessel, by presenting and discussing results from an extensive experimental programme aimed at enhancing understanding and insight of this complex phenomenon and of producing corroborative evidence to support the proposal of the Joint North West European Project (JNWEP) concerning probability of survival with water on deck. Firstly, a brief introduction is given on the latter and of the mathematical modelling of damaged vessel dynamics and of water ingress/egress available at SSRC.

PROBABILITY OF SURVIVAL WITH WATER ON DECK

The new damage stability framework proposed by the JNWEP is based on the probabilistic concept of survival. This means that the standard of survivability is expressed in terms of the probability that the vessel will survive given a damage with water ingress has taken place. The total probability of survival depends on two factors: the probability that a compartment is being flooded and the probability that the vessel will survive flooding of that compartment.

The concept itself is simple, but it takes a great deal of effort to establish correct formulation of these two factors, particularly when it involves large scale flooding of extensive undivided deck spaces such as the vehicle deck in Ro-Ro ferries. Concerning the latter and taking into account that there are many effects causing a vessel to capsize, the probability of survival can also be divided in two different factors: the probability to survive pure loss of stability, heeling moments, cargo shift, angle of heel and progressive flooding and the probability to survive water on deck as result of wave action. The calculation of this last factor, referred to as survival factor with water on deck, s_w, is based on a concept whereby the critical wave height at which the vessel will capsize is found, and s_w will simply be the probability that this wave height is not exceeded. In this respect, the critical task has been to formulate a connection between the critical sea state and parameters which can be readily calculated without resorting to costly, time consuming numerical simulations or physical model experiments. For this reason all model tests and simulations have been directed towards this goal, and the result is presented in Vassalos et al (1996). A key observation from this work is that vessel capsizal occurs close to the angle where the righting moment has its maximum, i.e. ϕ_{max}, calculated traditionally by using the constant displacement method and allowing for free-flooding of the vehicle deck when the deck edge is submerged. This fact, coupled with observations from physical model experiments and the experience amassed from studying large numbers of numerical tests led to the development of a "*Static Equivalent Method (SEM)*" which allows for the calculation of the critical amount of water on deck from static stability calculations. To this end, a flooding scenario is considered in which the ship is damaged only below the vehicle deck but with a certain amount of water on the (undamaged) deck inside the upper (intact) part of the ship. The critical amount of water on deck is then determined by the amount causing the ship to assume an angle of loll (angle of equilibrium) that equals the angle ϕ_{max}.

Based on this, the volume of water on deck causing the vessel to assume an angle of loll (angle of equilibrium) that equals the angle ϕ_{max}, was compared with the critical volume of water at the instant of capsize and a good correlation was found. The scenario described above and depicted in Figure 1, is believed to represent closely observations of the flooding process near the capsize boundary or when a stationary (steady) state is reached with the water on deck elevated at an average height, *h*, above the mean water plane, as a result of the wave action and vessel motions. It was subsequently shown that this height is a unique measure of ship survival in damaged condition - the higher the water elevation the higher the sea state needed to elevate the water to this level and the higher the capsizal resistance of the ship - that could be applied universally to all the arrangements studied, involving ship size and shape, subdivision arrangements and loading conditions. It follows, that the relationship between h and Hs will also be unique for a given ship, thus allowing the survivability of the vessel to be expressed as a function of the critical significant wave height as denoted below:

$$h_{crit} = f(Hs) = 0.085\,(Hs_{crit})^{1.3} \qquad (1)$$

Where, h_{crit} = the difference between the inner and outer waterline at the instant of capsize

Hs_{crit} = the critical significant wave height

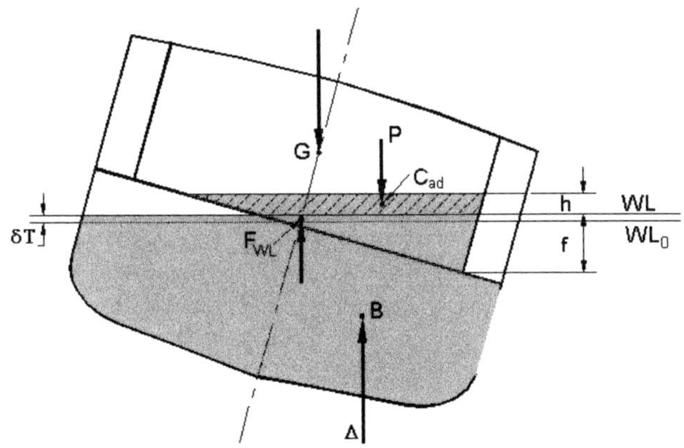

Figure 1: Stability of a Damaged Ship with Water Accumulated on Deck
(*Static Equivalent Method*)

Some additional data and a re-examination of the data presented in Vassalos et al (1996) showed that the effect of damaged freeboard had to be taken into consideration. This, in turn, led to the proposal of the following relationship, Vassalos et al (1997):

$$h_{crit} = f(Hs, F) = 0.088\ (Hs_{crit})^{0.97+0.46F} \qquad (2)$$

At the time, it was conjectured that Hs is raised into some power other than unity as it was the significant wave height measured relative to the deck of the ship at the location of the damage opening in a way that accounts for the vessel motion (i.e., the significant relative wave height - incident) that was the parameter of interest rather than the nominal significant wave height.

MATHEMATICAL MODELLING

The mathematical/numerical models currently available at the University of Strathclyde represent the state-of-the-art, with effort expended towards the development of two models, one following a bottom-up approach (UK Ro-Ro Research Model) and the other a top-down approach (On-going Research Model) described in Vassalos (The Water on Deck Problem of Damaged Ro-Ro Ferries). They have been under development since 1977 and subjected to vigorous validation/calibration during both the UK Ro-Ro Research and the JNWEP. As explained in the above paper, consideration of generalised models necessitates developments to be directed towards three areas:

- Damaged Vessel Dynamics: A non-linear six-degrees-of-freedom seakeeping model that allows the vessel to drift as well as changes in its mass, centre of mass and mean attitude relative to the mean waterplane with time; the same dependence of environmental excitation and hydrodynamic reaction forces on the changing underwater volume of the vessel must also be catered for.
- Water Ingress/Egress: A water ingress/egress model that allows for multiple-compartment flooding in the presence of oscillatory flows and at times of shear flows in extreme wave

conditions. The process of water accumulation on the Ro-Ro deck as well as the actual amount of water at any one instant is crucially important.

- Floodwater/Vessel Interaction: It is very important to emphasise, and it will be demonstrated in this paper, that a study of damage survivability involves two distinct but intrinsically interrelated and highly interacting processes, namely ship motion and flooding. The vessel motion influences considerably and directly the flooding process and conversely, flooding affects both the vessel motion and her attitude. It is essential, therefore, to take both phenomena into consideration when studying the evolution of either. However, the non-stationarity in the vessel motion introduced by the water accumulation coupled with the intermittence of the flooding process itself and the severe non-linearities in the ensuing dynamic system, demand great care in dealing with the many issues of this complex problem. Modelling can be attempted at various levels, e.g.:

  ⇒ ***Level 1:*** *Instantaneous free surface of floodwater parallel to the mean waterplane with floodwater moving in phase with the ship roll motion. This is the simplest to implement* (UK Ro-Ro Research Model).

  ⇒ ***Level 2:*** *Instantaneous free surface of floodwater parallel to the mean waterplane but floodwater moving independently of the ship. Such an approach requires knowledge of transfer moment amplitude and phase lag of floodwater motion w.r.t. ship roll motion. This, in turn, would involve either building up a comprehensive database through a systematic series of model experiments or though computations using CFD techniques.*

  ⇒ ***Level 3:*** *In case the dynamic behaviour of floodwater is considerable and could prove to be dominating or heavily influencing the vessel behaviour, the dynamic system of vessel-floodwater must be treated as two separate worlds interacting. Floodwater sloshing models that allow for random inflow and outflow through multiple openings would necessitate the use of sophisticated viscous flow models using CFD techniques.*

Despite considerable efforts in all three areas over the past 12 years at the Ship Stability Research Centre (SSRC) of the University of Strathclyde the use of a numerical "tool" in its complete form for parametric investigations and routine design applications appears to outside reach at present. Attempting to pave the way forward, this paper focus on the modelling of the process of flooding and its interaction with vessel motion.

Modelling the Water Ingress

Flooding of a damaged compartment is a highly complicated problem to tackle and in need of strong assumptions if easy-to-use equations are to be derived. A theoretical approach to the problem applying the concept of velocity potential would meet a great deal of difficulties, not least the fact that a velocity potential for the water sloshing inside the compartment will need to model outflow, as well as that governing the dynamics of the water flowing in. However, neither of these is readily available, unless employing CFD techniques. Clearly, once this route is abandoned for its evident impracticability, other effects would have to be overlooked or treated empirically. One of these is, for instance, the diffraction of the incident wave by the ship and how this essentially changes the fluid flow characteristics depending on the angle of encounter. Experimental investigations showed, in fact, that the *shielding* due to the presence of a ship, when the damage is on the leeside, has a strong influence on the rate of flooding. On the basis of these considerations, any effort to use the undisturbed wave potential only to describe the flow pattern of the water going through a damage opening will have to be considered partial and incomplete. Having discarded every expectation of using potential theory in approaching this problem, the first, obvious attempt must address hydraulic models with a view to adapting these to the specific case investigated. A systematic search has revealed that in the case of water flooding through the damaged hull of a ship, the most suitable flow configurations with a well established theoretical background appear to be water flow through submerged orifices and channels fed by reservoirs through uncontrollable inlets. The formulations regarding these applications assume steady flow conditions and calm water surface. Furthermore they rely on the application of Bernoulli's equation (and the continuity equation in the case of open channels), corrected by an empirical coefficient to take account of all energy losses. The general form of the equations used to evaluate the volume flow rate is given by:

Flow through a submerged orifice

$$Q = KA\sqrt{2g(h_1 - h_2)} \tag{3}$$

Where, Q is the volume flow rate, A is the area of the opening and the term in the square root represents the theoretical velocity of the water jet.

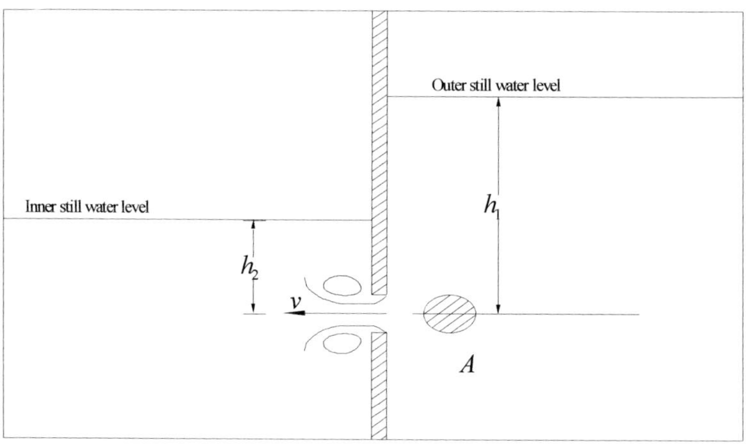

Figure 2: Flow through a Submerged Orifice

Flow from a reservoir to an open channel

$$Q = Kbh\sqrt{2g(H - h)} \tag{4}$$

Where, b is the channel breadth.

As mentioned above, equations (3) and (4) can be found by applying Bernoulli between an upstream section where the water can be considered still and thus the pressure hydrostatic, and a second section at which the mean velocity is calculated. The total pressure head is considered to remain theoretically constant for both sections. Frictional and other losses are taken care of, by introducing the coefficient K.

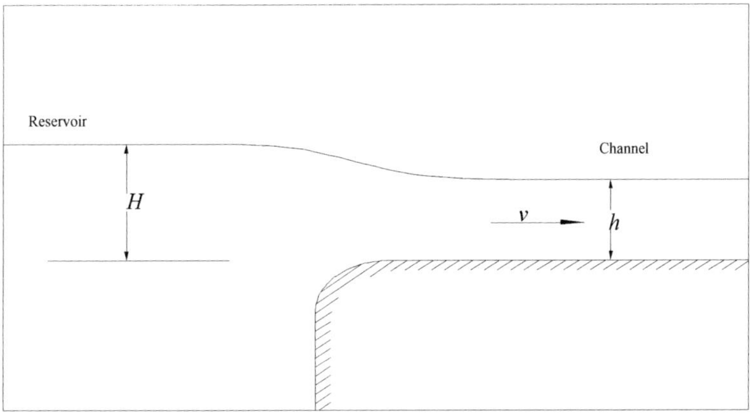

Figure 3: Flow from a Reservoir to an Open Channel

On the basis of the foregoing, a simplified picture of flooding of ship compartments can be considered by generalising the theory used for steady flow cases to a fully dynamic one, on the basis of the assumption that a simplified technique can be acceptably accurate if supported by experimental evidence. Considering, for example, damage below the bulkhead deck, a flooding scenario is depicted by the simplified picture shown in Figure 4, with the two water levels inside and outside the compartment assumed still and horizontal. The resemblance to the static cases illustrated earlier is evident. If Bernoulli's equation is applied at sections A and B, considering the total pressure head remained constant and the velocity zero in the reservoir, the flow rate through the horizontal layer around P can be determined as explained in Vassalos (The Water on Deck Problem of Damaged Ro-Ro Ferries) by:

$$dQ = K.\mathrm{sign}(h_{out} - h_{in}).\sqrt{2g|h_{out} - h_{in}|}.dA \qquad (5)$$

The rate of flooding and hence the time history of the floodwater can be found by integrating dQ over the damage opening height, i.e.

$$Q(t) = K\int_{Damage} \mathrm{sign}(h_{out} - h_{in}).\sqrt{2g|h_{out} - h_{in}|}.dA \qquad (6)$$

Considering that $|h_{out} - h_{in}|$ represents the instantaneous downflooding distance, which is relatively easy to compute, the whole problem of progressive flooding reduces to the evaluation of the coefficient K that can be done experimentally.

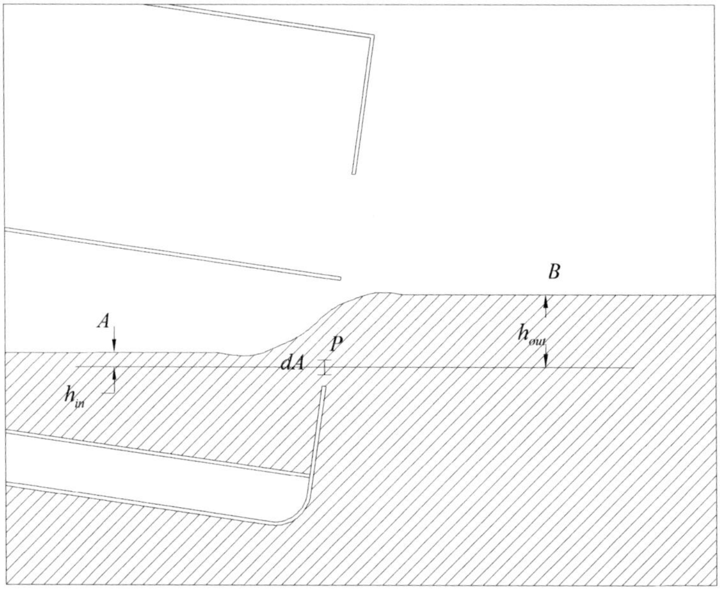

Figure 4: Water Ingress/Egress Main Parameters

In the general case, K can be assumed to be an empirical function of relevant parameters to suit a more realistic case in which incident, diffracted and radiating waves as well as sloshing of the floodwater are present whilst accounting for location, extent and shape of the damage opening and of the internal arrangement of the compartment.

Using a simplified ship model, restrained in beam waves, some preliminary experimental research was undertaken as part of a final year project to check the validity of the foregoing assumptions, concerning the modelling of water ingress/egress, Letizia (1996). A finding of particular importance concerned the dependence of K on the relative water elevation on the two sides of the damage with the mean value reducing substantially as soon as water was present on both sides of the damage opening. More specifically, the mean value for the flooding coefficient when pure inflow occurs was found to be 1.2 whilst with water present on both sides it reduced to 0.7. These results refer to the case when the damage opening faces the oncoming waves. With the damage opening away from the oncoming waves, the corresponding values of K were found to be approximately the same and close to 1.0. Using these values of K, it is evident from Figure 5 that the hydraulic model adopted describes with very well the flooding process, following the trend of the flooding rate with sufficient accuracy.

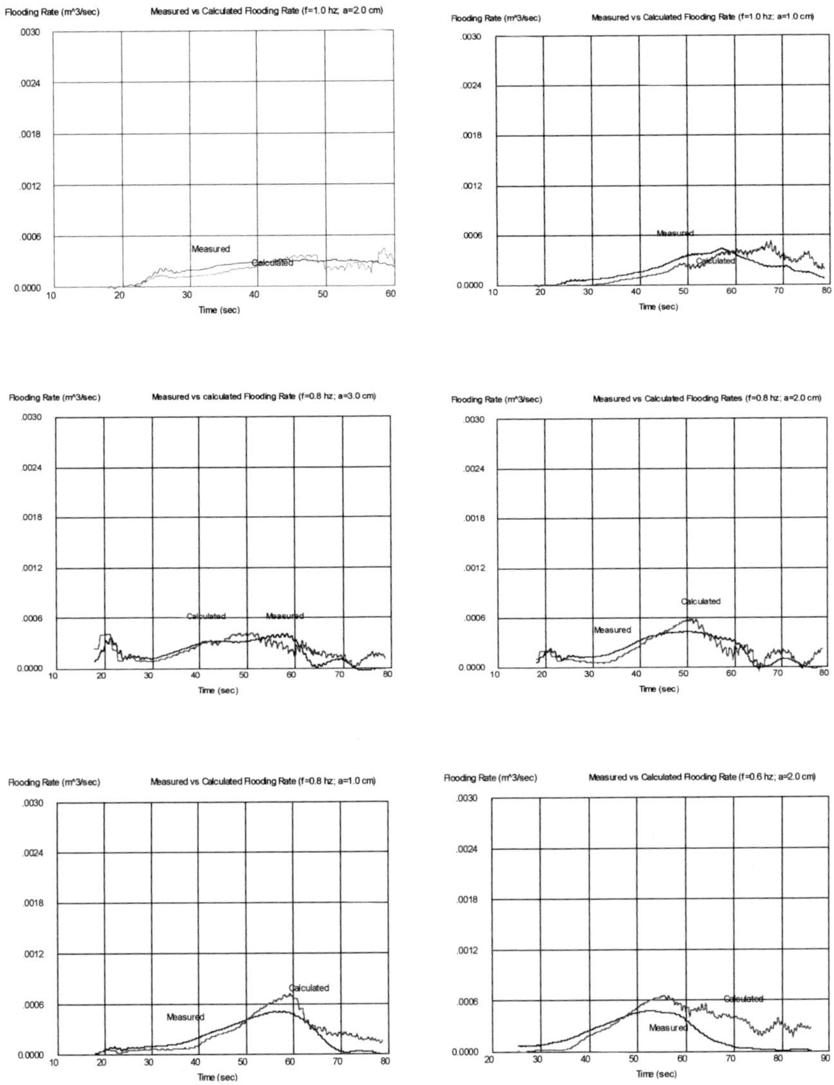

Figure 5: Experimental Flooding Curves

The experimental programme presented in the following provides corroborative evidence for these early findings and is a continuation of the validation/calibration work that took place during the UK Ro-Ro research and the JNWEP, as described in Vassalos et al (1996).

EXPERIMENTAL PROGRAMME

Damage Scenarios and Test Conditions

To foster a better understanding of the water accumulation on Ro-Ro vehicle decks, a series of experiments has been planned and is taking place at the University of Strathclyde. In this paper the results of the first three series of tests are presented, concerning mean asymptotic height of floodwater on the vehicle deck *h*, relative incident wave height, η_{RI} and flooding coefficient K. The additional parameters in the experimental matrix comprise wave and damage opening characteristics, freeboard and KG involving the model in a number of degrees-of-freedom ranging from zero to six. The St Nicholas model was used in these experiments, Vassalos (The Water on Deck Problem of Damaged Ro-Ro Ferries) with the particulars shown in Table 1. There are two compartments open to the sea: the first is above the vehicle deck and extends from 22.25m to 97.25m from the aft perpendicular and the second between the double bottom and the vehicle deck with a length that was adjusted by inserting blocks of foam to allow for changes in the damaged freeboard as shown in Table 2. Additional values for damaged freeboard could be obtained similarly.

TABLE 1
MAIN PARTICULARS OF THE RO-RO VESSEL USED IN THE EXPERIMENTAL INVESTIGATION

L_{BP}	(length between perpendiculars)	=	131.0 m
B	(breadth)	=	26.0m
T	(design draught)	=	6.10m
D	(depth to uppermost continuous deck)	=	18.8m
D_{bd}	(depth to bulkhead deck)	=	7.8m
D_{db}	(depth to double bottom)	=	1.6m
Δ	(displacement)	=	12200 tonnes
C_b	(block coefficient)	=	0.582

TABLE 2
FLOODED COMPARTMENT LENGTHS

Length of Compartment (metres)	Distance from A.P. (positive forward) (metres)		Damaged Freeboard (metres)
	Aft End	Forward End	
13.3	52.6	65.9	1.08
22.8	47.15	69.95	0.6
28.7	43.66	72.36	0.2

Series 1

The parameters investigated in the first series of the experimental programme refer to 1.08m damaged freeboard and cover a range of regular and irregular waves as shown in Tables 3 and 4 and a number of damage scenarios as illustrated in Figure 6. These scenarios attempt to isolate the various contributions to the flooding process, as explained next:

Scenario 1, describes damage to the vehicle deck only

Scenario 2, describes damage to compartments both above and below the bulkhead deck but without penetration

Scenario 3, is the same as scenario 2 but with penetration, allowing water to flow between the two compartments

All three scenarios were tested with a fixed and a heaving model to allow for an evaluation of the contribution of heave motion to the flooding process and of the interaction between the two processes. The damage opening in these scenarios was modelled according to a 100% SOLAS, i.e., damage length at the waterline equal to ($0.03L_s$+ 3.0) metres or 11 metres, whichever is less, using a trapezoidal opening with sides at 15^0 to the vertical.

Scenarios 4 to 6, refer to different shapes of the damage opening, all satisfying SOLAS requirements, with scenario 6 in particular attempting to duplicate the side damage suffered by the European Gateway, following collision with Speedlink Vanguard in 1982.

TABLE 3
REGULAR WAVE PARAMETERS

Wave Height H (metres)	=	2.0, 3.0, 4.0, 5.0
Wave Period T (seconds)	=	16.2, 8.1, 6.5, 5.4, 4.0

TABLE 4
IRREGULAR WAVE PARAMETERS (JONSWAP SPECTRUM)

Significant Wave Height Hs (metres)	Zero Crossing Period T_0 (seconds)
2.0	4.42
3.0	5.42
4.0	6.25
5.0	6.99

Presentation of Results and Discussion

The results derived from testing in regular waves are shown in Figures 7 (heaving model) and 8 (fixed model). With reference to *h*, the following observations can be made:

- *h* is clearly a function of both T and H. The variation of *h*, however, becomes more pronounced with the heaving model. This latter phenomenon is a direct result of the tuning/de-tuning between wave and heave motion. It would be reasonable, for example, to expect a reduction in flooding when the model moves in phase with the waves, whereas the reverse will be true when the model heaves in anti-phase with the waves. In addition, heave motion affects directly the flooding process both in terms of influencing the relative water elevation directly as well as causing water to be sucked into the damaged compartments as the model heaves downwards whilst giving rise to something of a "pumping" action, causing water to flood the vehicle deck through the triangular opening between the two compartments.
- The shape of the damage opening has a clear influence on *h*.

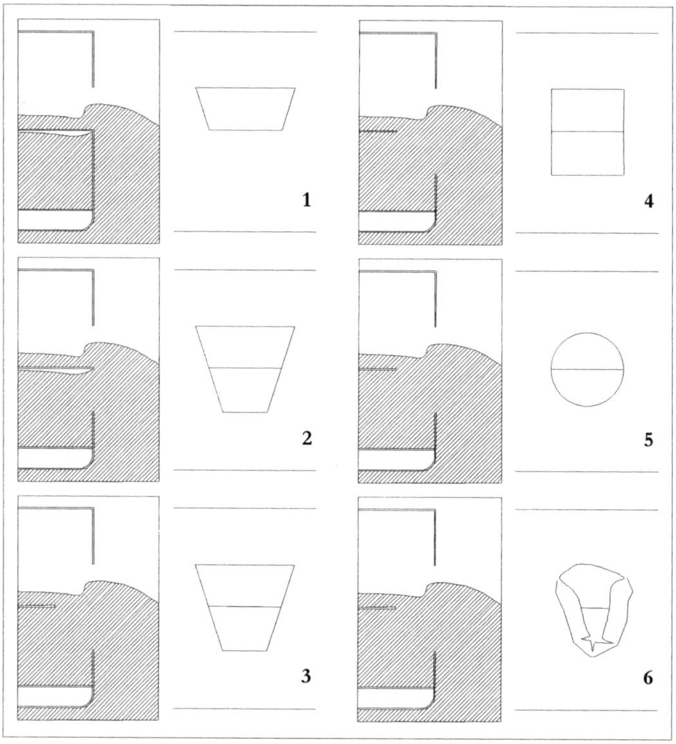

Figure 6: Damage Scenarios Considered in Series 1

It is worth noting that the water ingress and egress are fundamentally different flows. Water egress has more of a constant nature, depending mainly on *h*, whilst water ingress is intrinsically intermittent, especially for larger freeboards. As the flooding process evolves a steady (or stationary) state may be reached when the integral of the sum of inflow and outflow functions oscillates around zero, i.e. when, on the average, inflow and outflow neutralise each other. This leads to larger *h* values with increasing wave heights and decreasing wave periods, as a larger number of higher crests would reach the deck level. It has been observed though, that steep waves tend to diffract and break against the model side, thus giving rise to highly distorted crests, which carry considerably less water. This is the reason why there seems to be an optimal wave frequency for flooding even in the case of the fixed model. As a direct consequence of the observations noted above, two important conclusions can be drawn. Firstly, the vessel motion influences considerably and directly the flooding process and conversely, flooding affects both the vessel motion and her attitude. It is essential, therefore, to take both phenomena into consideration when studying the evolution of either. Secondly, since flooding and ship motion are two compounded processes, testing in regular waves will be unsuitable as it can lead to deceptive results. This second conclusion can be better appreciated by considering that for long wave periods the vessel will be heaving in phase with the waves, hence the distance between sea water and vehicle deck will be nearly constant and flooding will be impaired. Furthermore, a comparison of the average height of water accumulated on deck between Figures 7, 8 (regular waves) and Figure 9 (irregular waves) clearly reveals that *h* is generally smaller in the presence of irregular waves, even though similarity in trends is evident. Finally, it may be observed that the amount of water accumulated on deck in scenarios 1 and 2 is generally greater than the corresponding amount in scenario 3. This difference is attributed to the downflooding of water through the triangular penetration notch, from the vehicle deck to the lower compartment.

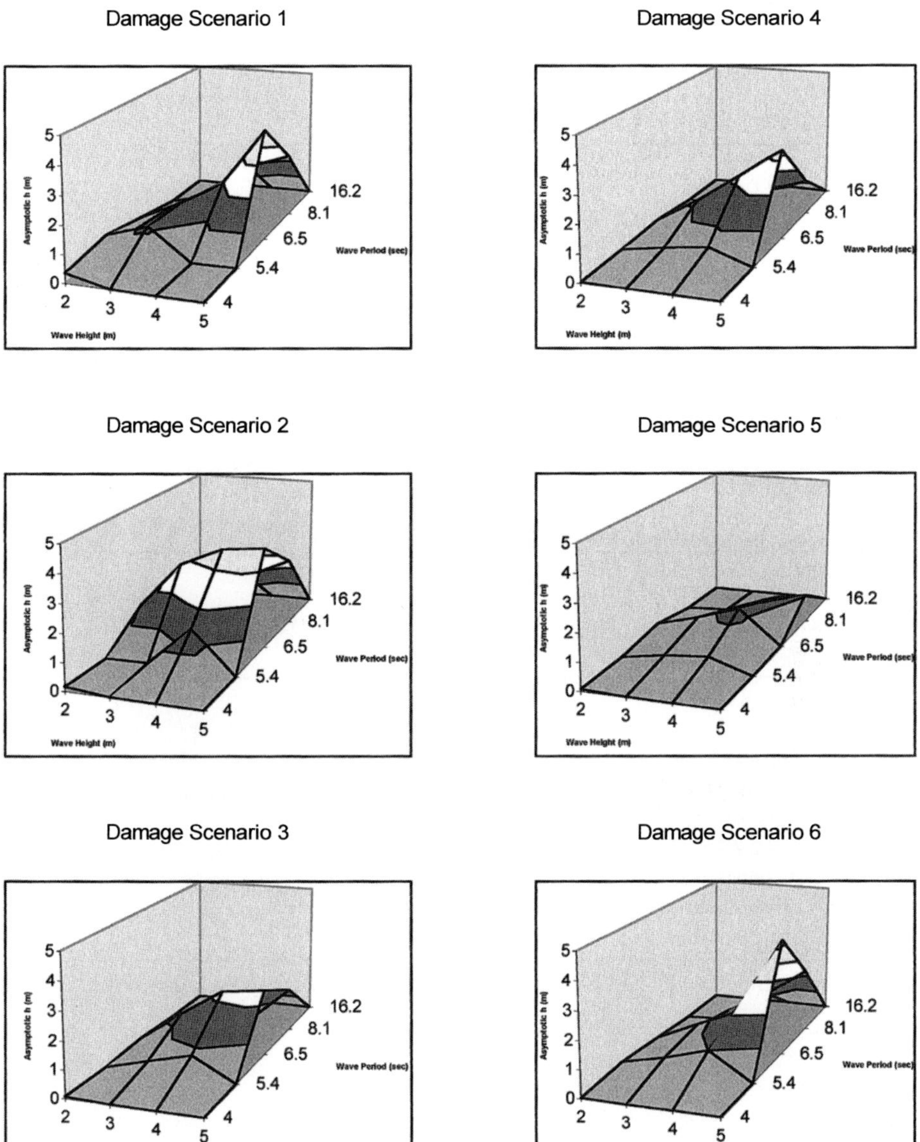

Figure 7: Asymptotic Height of Water on the Vehicle in Regular Waves (Heaving Model)

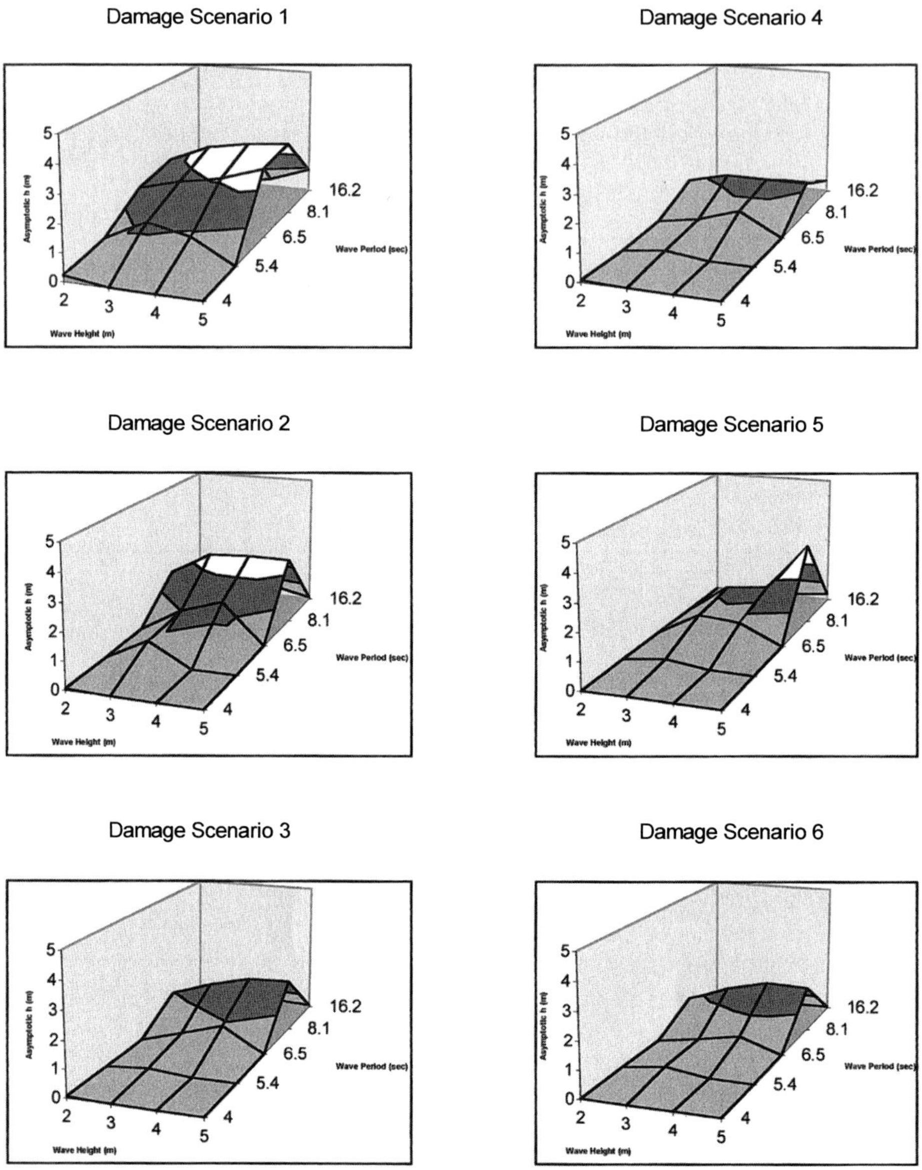

Figure 8: Asymptotic Height of Water on the Vehicle in Regular Waves (Fixed Model)

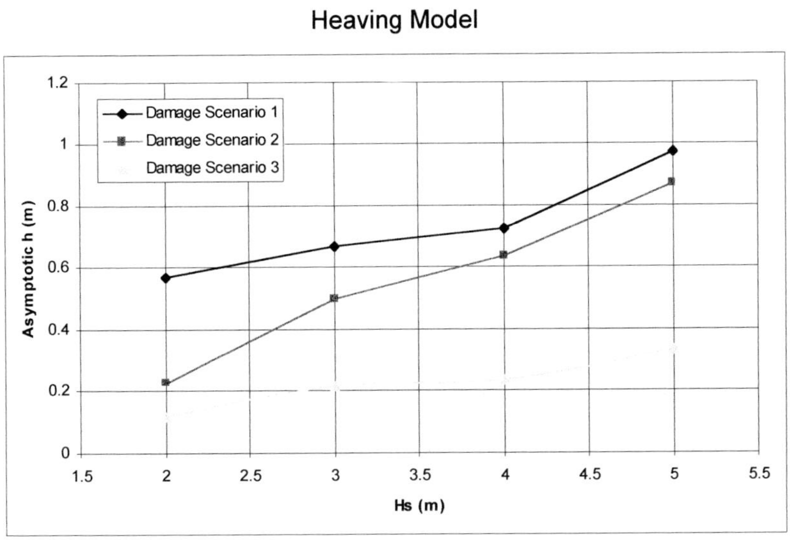

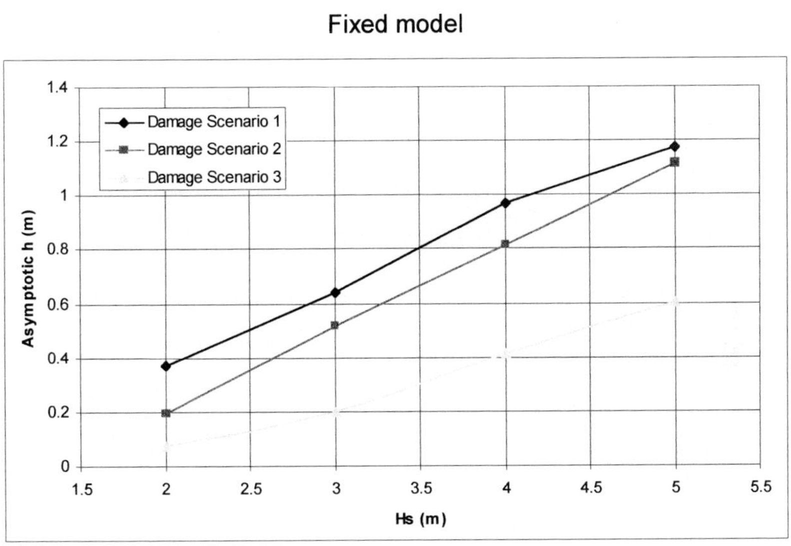

Figure 9: Asymptotic Height of Water on the Vehicle in Irregular Waves

Series 2

In the second series of the experimental programme all three damaged freeboards of Table 2 have been considered but tests were undertaken only in irregular waves as specified in Table 4 and for damage scenario 3. Furthermore, in addition to the fixed and rolling model cases of series 1, the effect of roll motion alone as well as of combined heave and roll motions on the flooding process were also investigated in series 2. Three different loading conditions were also tested, corresponding to KG values of 9, 10, and 11 metres.

Presentation of Results and Discussion

Significant Relative Wave Height

The validity of the conjecture that $Hs^{1.3}$ or $H_S^{0.97+0.46F}$ may represent the significant relative incident wave height (η_{RI}) as explained earlier, has been presently investigated and the results are presented in Figure 10 for varying degrees of freedom. The results clearly show that with the exception of heave motion only a close correlation exists between η_{RI} and Hs. Varying the freeboard, however, appears to result in loss of correlation with $H_S^{0.97+0.46F}$ as shown in Figure 11. This leads to the conclusion that in this case the closer the exponent is to 1.3 the higher the correlation.

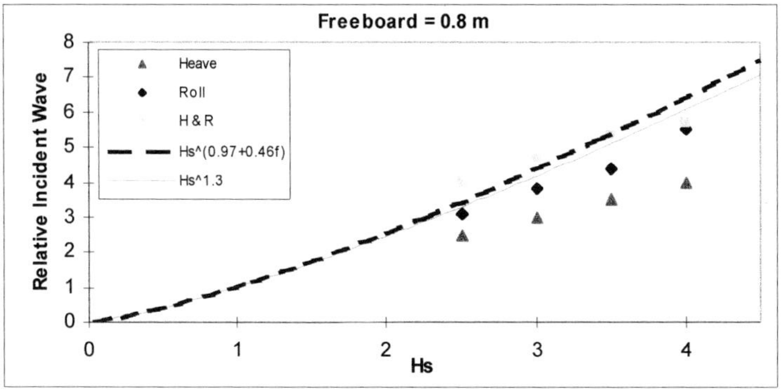

Figure 10: Effect of Motions on Relative Incident Wave Height

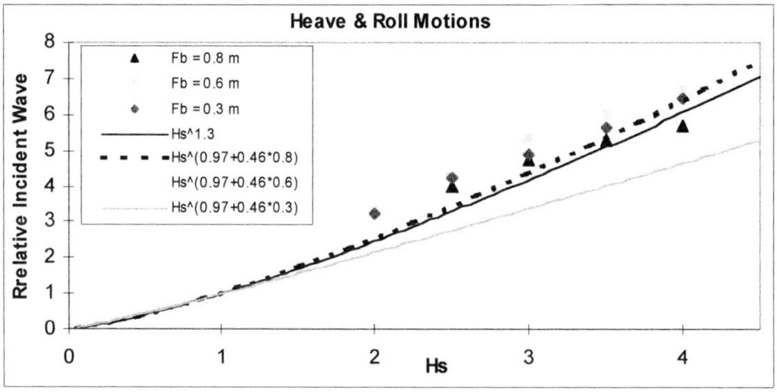

Figure 11: Effect of Freeboard on Relative Incident Wave height

An investigation on the influence of KG in affecting the relative incident wave height revealed that only in the case where the model was allowed to heave and roll, the effect was noticeable.

Volume of Water on Deck

Considering this parameter, it is noteworthy that increasing the degrees of freedom results in lesser amount of water accumulated on deck as shown in Figure 12. This leads to the conclusion that restraining the model in any way will alter the process of flooding. This will be further examined in the planned Series 3 of the experiments where the model will be free to move in all degrees of freedom. The effect of freeboard on water accumulation was also investigated and the results are presented in Figure 13. As expected, increasing the freeboard results in reduced water ingress.

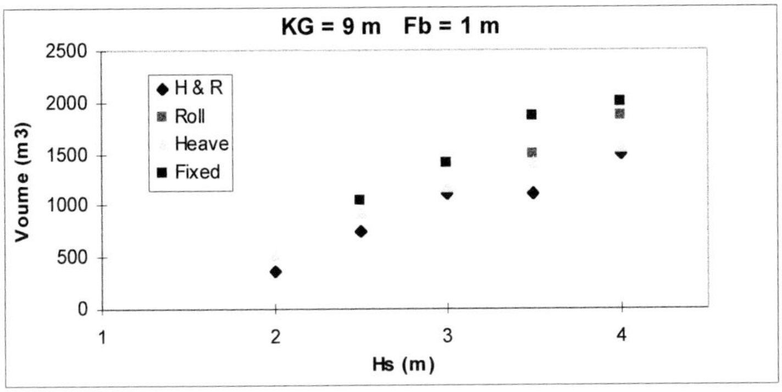

Figure 12: Effect of D.O.F on Water Accumulation

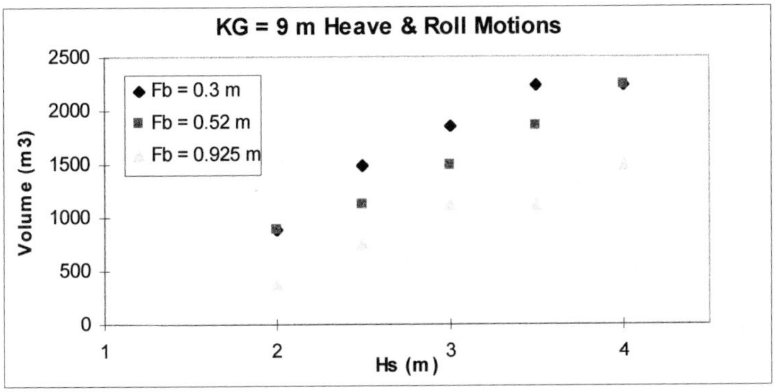

Figure 13: Effect of Freeboard on Water Accumulation

In addition, results appear to indicate water accumulation on deck is at minimum with the stiffest model. Even though a loss of this trend occurs when heave motion is present, this can be explained by observing that roll motion tends to reduce when heave is allowed.

Height of Water on Deck (h)

As no capsize took place in Series 2, the values of h measured should be smaller than that corresponding to h_{crit}, at the point of no return. In spite of this fact, the results obtained show a significant higher h. However, as explained above, the difference between predicted and measured values decreases with increasing number degrees of freedom. In this respect, the measured h when the model is free to move in heave and roll is in reasonable agreement with that predicted by using equation (1). This is shown in Figure 14. Taking this argument still further, it may be claimed that if the ship is also allowed to freely drift, the agreement will be even closer. In Figure 15, on the other hand, the predicted h using equation (2) is compared with the measured values for a range of freeboards. In what appears to be contradicting the trend shown in Figure 13, *h* does not depend critically on freeboard. This seemingly strange finding can be explained by observing that a small amount of water on deck and correspondingly a larger freeboard would imply small deck submergence. Therefore, since the value of *h* is measured from the mean waterplane it might still be possible to have a high *h* value even though the volume of water accumulated on deck is small.

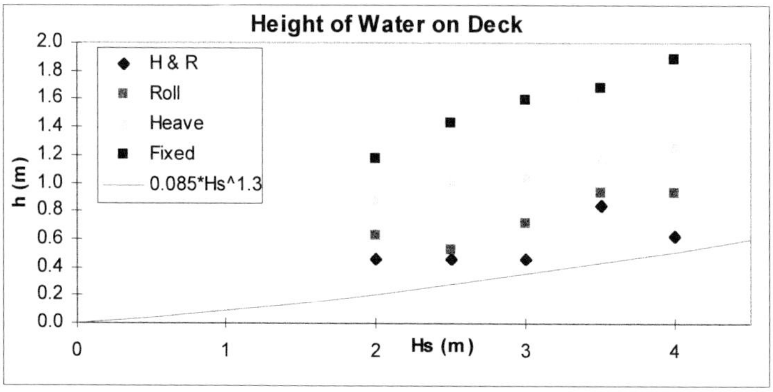

Figure 14: Effect of D.O.F on Height of Water on Deck

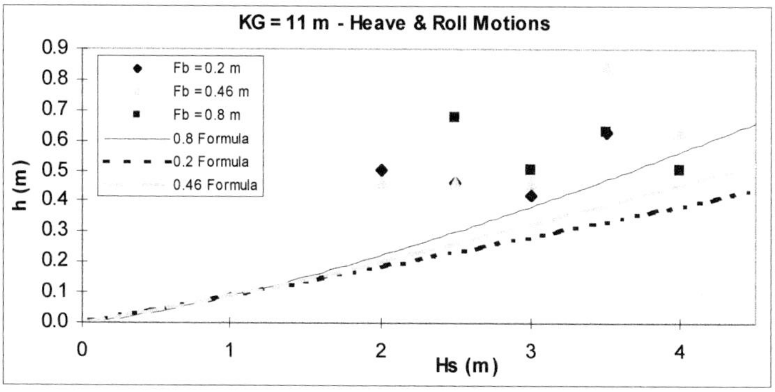

Figure 15: Effect of Freeboard on Height of Water on Deck

Flooding Coefficient

Figure 16 shows the time history of the measured volume of water on deck as well as the estimated volume rates based on incident and diffracted wave fields at the damage opening with K assumed to be equal to 1.0. Using this information the actual K values can be estimated and these are shown in Figure 17. K_u in this figure refers to unidirectional flow and K_b to bi-directional. Through examination of these records it became evident that the value of K changed substantially as soon as water was present on both sides of the damage opening, following the development of a stationary heel. Pursuing this further it was possible to determine average values of these coefficients for a range of freeboards and sea states and these are shown in Figures 18 and 19. This is much in agreement with earlier findings, with the mean value of K_u estimated to be 1.1 and of K_b 0.7. It should be noted that the K values shown in these figures have been calculated on the basis of the estimated diffracted wave field at the damage opening, although similar values of K have been obtained using simply incident waves. Only when the model was restrained to move in roll motion, the use of incident waves in calculating K produced unreliable results.

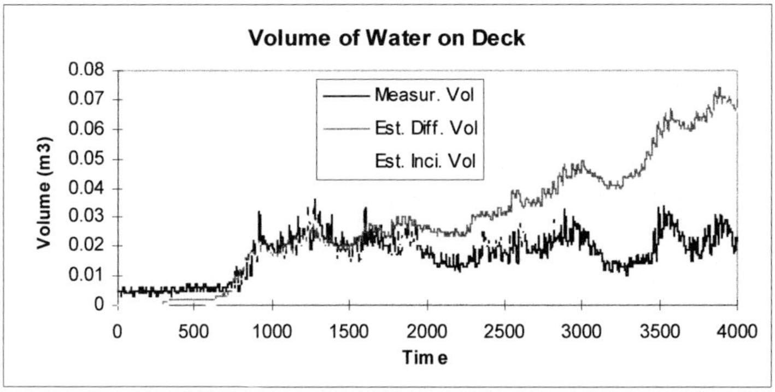

Figure 16: Time Realisation of Water Accumulation on Deck

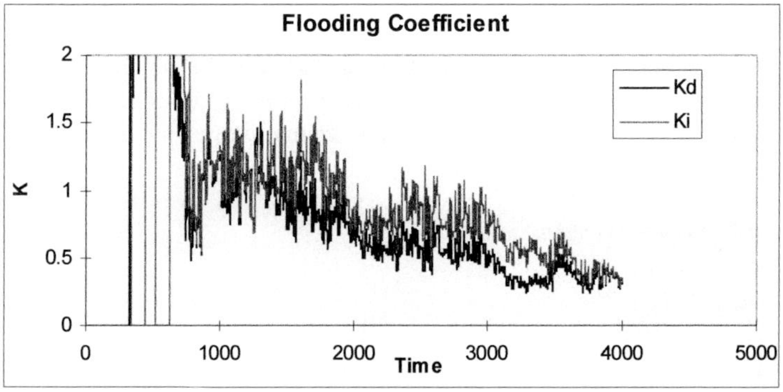

Figure 17: Time Evolution of Flooding Coefficient

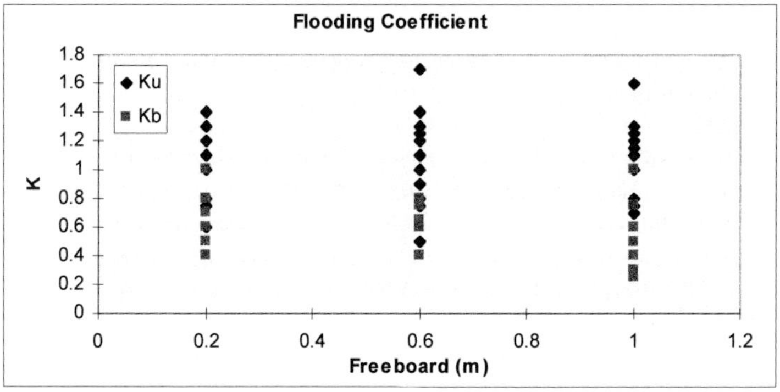

Figure 18: Effect of Freefoard on Flooding Coefficient

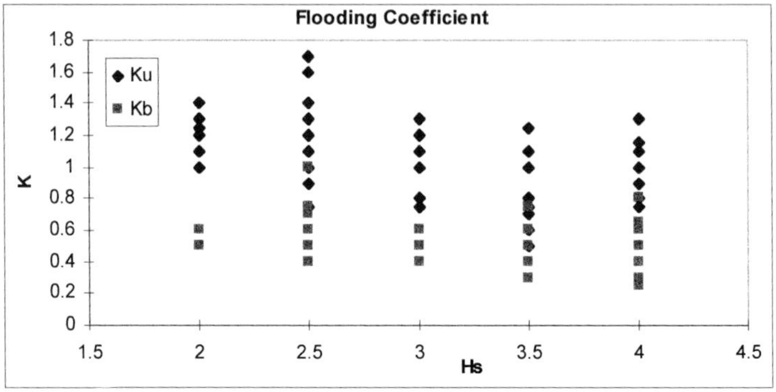

Figure 19: Effect of Seastate on Flooding Coefficient

Series 3

In the third series of the experimental programme all three damaged freeboards of Table 2 have been considered but tests were undertaken only in irregular waves as specified in Table 4 and for damage scenario 3. Furthermore, in addition to the fixed, heaving only, rolling only and heaving and rolling modes of motion of series 1 and 2, water accumulation with the model freely drifting and capable of responding in six degrees-of-freedom is investigated in these tests. Three different loading conditions were also tested, corresponding to KG values of 9, 10, and 11 metres.

Presentation of Results and Discussion

Volume of Water on Deck

Considering this parameter, it is noteworthy that the trend observed in the previous two series, i.e., increasing the degrees of freedom results in lesser amount of water accumulated on deck is also clearly demonstrated in these results, as shown in Figure 20. Moreover, the influence of sway in affecting water accumulation deserves special attention.

Height of Water on Deck (h)

The difference between predicted and measured values concerning h was shown to decrease with increasing number degrees of freedom. In this respect, the measured h when the model is free to move in all degrees of freedom appears to be in reasonable agreement with that predicted by using equation (1). This is shown in Figure 21, with the influence of sway being again noteworthy. In Figure 22, on the other hand, the predicted h using equation (2) is compared with the measured values for two freeboards. Again results do not appear to depend critically on freeboard in a way that clear trends emerge. More emphasis on clarifying this will be placed in the forthcoming series of this on-going experimental programme.

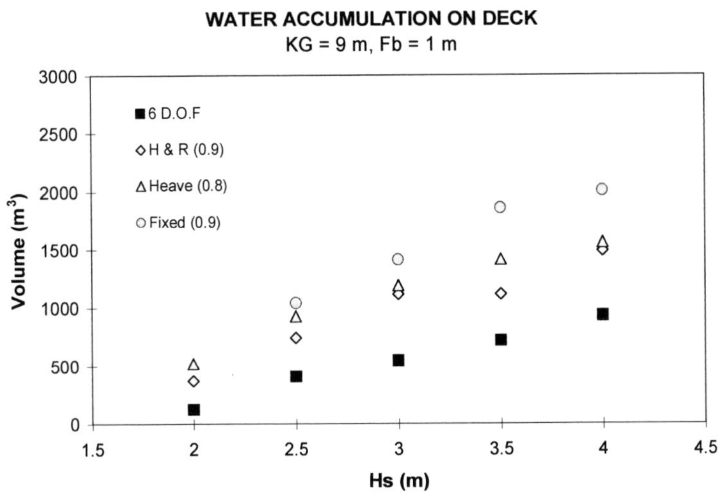

Figure 20: Effect of D.O.F. on Water Accumulation

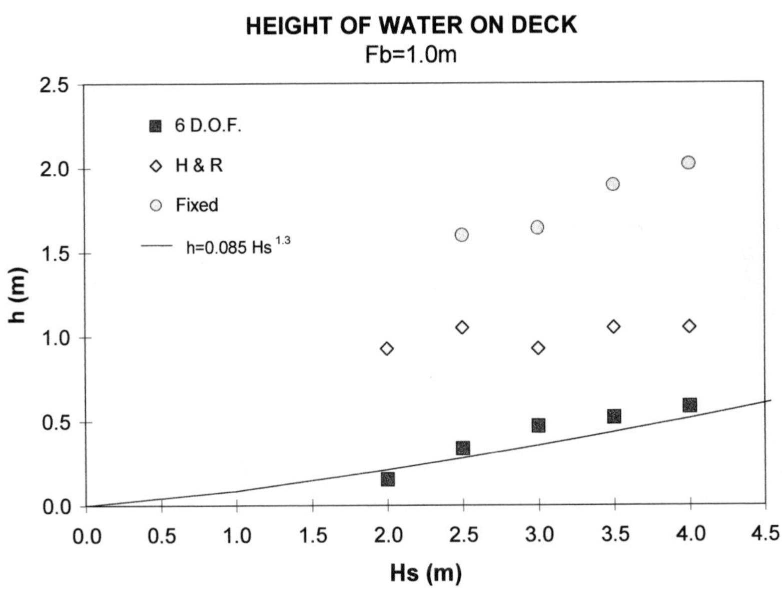

Figure 21: Effect of D.O.F. on Height of Water on Deck

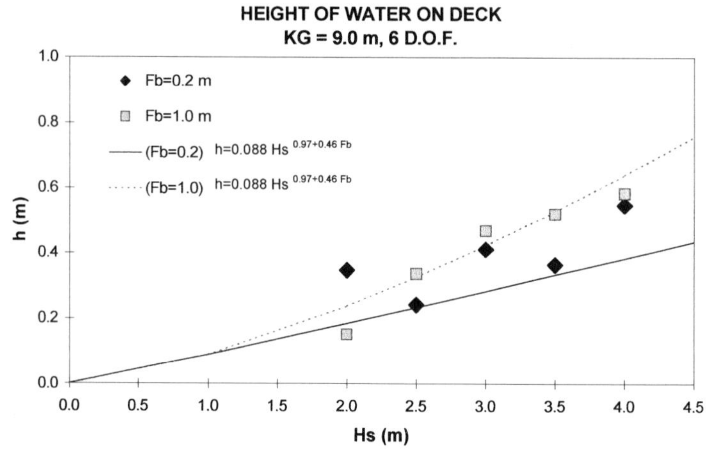

Figure 22: Effect of Freeboard on Height of Water on Deck

CONCLUDING REMARKS

The evidence presented in this paper offers important clues concerning specific contributions to the water accumulation on the Ro-Ro vehicle deck and the characterisation of the flooding process. It also demonstrates that there are good reasons to solicit due consideration of the shape and configuration of the damage openings in determining the amount of water accumulating on vehicle decks and discourage apparently innocuous simplifications often adopted to tackle the problem assessing the damage survivability of this type of vessels. Considering, however, that modelling a damage opening in a way that reproduces reality exactly will not be possible and, in fact, not relevant, it will be particularly helpful for all concerned to appreciate that all that is necessary to progress further in this field is the definition of a generalised damage opening, based on acceptable statistical data that is universally accepted and used in testing for damage survivability in both physical and numerical model tests. It is also very important to emphasise that the study of damage survivability involves two distinct but intrinsically interrelated and highly interacting processes, namely, ship motion and flooding. The non-stationarity in the vessel motion introduced by the water accumulation coupled with the intermittence of the flooding process itself and the severe non-linearities in the ensuing dynamic system, demand great care in dealing with the many issues of this complex problem.

In addition to the above, based on the results derived in series 2 and 3, the following points are noteworthy:

- The values of the flooding coefficient K observed in previous experiments are reconfirmed here. These relate to the value of 1.1 for unidirectional and 0.7 for bi-directional flows, respectively.

- Restraining a model changes the flooding process appreciably and is not recommended, particularly when model testing is used to assess damage survivability.
- The influence of heave motion in affecting water accumulation on deck was shown to be more important than that of roll motion but less so than sway motion.
- The conjecture that $Hs^{1.3}$ represents a wave height modified by the ship motion has been shown to be reasonable.
- Corroborative evidence has now been produced in support of the proposed relationship between the sea state and the height of water on deck and of the probabilistic framework for assessing the damage survivability of passenger/Ro-Ro vessels, proposed by the JNWEP.

References

Vassalos, D., Pawlowski, M. and Turan, O. (1996). A Theoretical Investigation on the Capsizal Resistance of Passenger/Ro-Ro Vessels and Proposal of Survival Criteria. *Final Report, Task 5, The Joint North West European R&D Project,* March.

Vassalos, D, Jasionowski, A and Dodworth, K. (1997). Time-based Survival Criteria for Damaged Ro-Ro Vessels. *Final Report, MCA,* March 1997, 92pp.

Letizia, L. (1996). Damage Survivability of Passenger Ships in a Seaway. *Ph.D. Thesis, Department of Ship and Marine Technology, University of Strathclyde*, November.

3. Special Problems of Ship Stability

Contemporary Ideas on Ship Stability
D. Vassalos, M. Hamamoto, A. Papanikolaou and D. Molyneux (Editors)

DAMAGE STABILITY WITH WATER ON DECK OF A RO-RO PASSENGER SHIP IN WAVES

S. Ishida[1], S. Murashige[1], I. Watanabe[1],
Y. Ogawa[1], T. Fujiwara[1]

[1] Ship Research Institute, Ministry of Transport,
6-38-1, Shinkawa, Mitaka, Tokyo 181-0004, Japan

ABSTRACT

An experiment on the stability of a RO-RO passenger ship with side damage was conducted in beam seas. Capsize only occurred with small GM values, which did not satisfy SOLAS Regulation. In non-capsize conditions the ship attained a stationary condition, with constant mean values of heel angle ϕ and water volume on deck. The effect of experimental parameters on these values and the capsize conditions were discussed. The effect of resonance of roll motion was also investigated. The mean height of water on deck above the calm sea surface H_d, which almost had a certain positive value in various conditions, was proposed as an index for the stationary condition. It was concluded that the possibility of capsize can be evaluated without knowing the exact value of H_d, using an equilibrium curve which is calculated and plotted on H_d- ϕ diagram.

KEYWORDS

Damage stability, capsize, waves, RO-RO ship, vehicle deck, GM, resonance

INTRODUCTION

The safety standard of RO-RO passenger vessels was deliberated at IMO from 1994 to 1995 in order to prevent capsizing disasters like the one of Estonia in 1994. One of the authors, I. Watanabe, joined the discussion in the expert panel. The main topic of it was the stability standard because RO-RO passenger vessels have wide non-separated car decks. Once free flooded water is piled up on them, the large heel moment could be the cause of the capsize because of this feature.

There were several papers published on this problem like Bird & Browne (1973), Velschou & Schindler (1994), Dand (1994) and Vassalos (1994). However flooding, accumulation of flooded water, and the interaction of ship motion and flooded water are so complicated that much more study is necessary for clarifying these phenomena. So we conducted an experiment using a RO-RO passenger ship with a side damage in beam seas in order to contribute to the discussion in IMO.

EXPERIMENT

Table 1 and Figure 1 show the model ship and the damage opening. The damage reaches two compartments and follows the SOLAS Regulation 8.4. A vertically movable vehicle deck is provided. There is some space between the vehicle deck and the hull, so this model simulates a ship with side casing. It should be noted that GM_d (GM in damaged condition) is far larger than the intact value because of the flare. Four wave gauges at the center of the damage and six water level meters are installed on the deck.

The experiment was conducted in irregular waves with JONSWAP spectrum and duration time of 30 minutes in ship scale. The basic significant wave height ($H_{1/3}$) was 4.0m and peak period (T_p) was 8.0sec., but varied keeping the condition of $T_p[sec] = 4\sqrt{H_{1/3}[m]}$. The test was also carried out in some regular waves. The main parameters were GM_d and wave height. Moreover the effect of center casing (CC), height of vehicle deck and initial heel was investigated in some conditions. The damaged side of the ship was kept to weather side.

Table 1
Principal Particulars

	Ship		Model	
	Intact	Damaged	Intact	Damaged
Lpp (m)	101.0		4.3	
B (m)	16.0		0.681	
D (m)	5.7		0.236	
d (m)	4.37	5.22	0.186	0.222
W	3821 t		272.7 kg	
KG (m)	5.87		0.25	
$GM_{(d)}$(m)	1.62	3.12	0.069	0.133
Tr (sec.)	9.40	8.43	1.94	1.74
f (m)	1.17	0.33	0.050	0.014
H (m)	4.84		0.206	

f : freeboard, H : RO-RO deck height

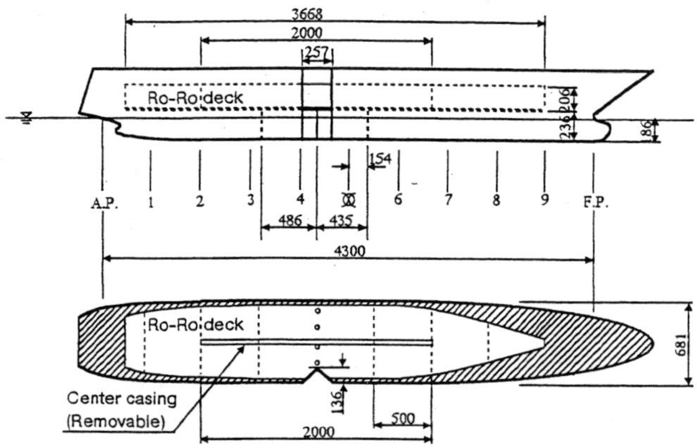

Figure 1 : Model and Damage Opening
(Broken lines and circles in the lower figure show water level meters and wave height gauges, respectively.)

Measured and calculated *GZ* curves plus water volume on deck *w* are shown in Figure 2. Calculated *GZ* curves with constant *w* are also shown in Figure 3. $w/W=10\%$ is equivalent to the flooded water height of 39cm in upright condition. SOLAS Regulation 8.2.3 was satisfied except the condition of $GM_d=1.27m$.

We should keep it in mind that the ship does not roll along the curves in Figure 2 because the area of the damage opening over the vehicle deck is small compared to the whole volume of the deck. This fact means the flooding velocity is not high enough to make *w* equal to the one drawn in Figure 2. In a few cycles of rolling motion we should assume that *w* is almost constant and that the ship rolls along the stability curves shown in Figure 3.

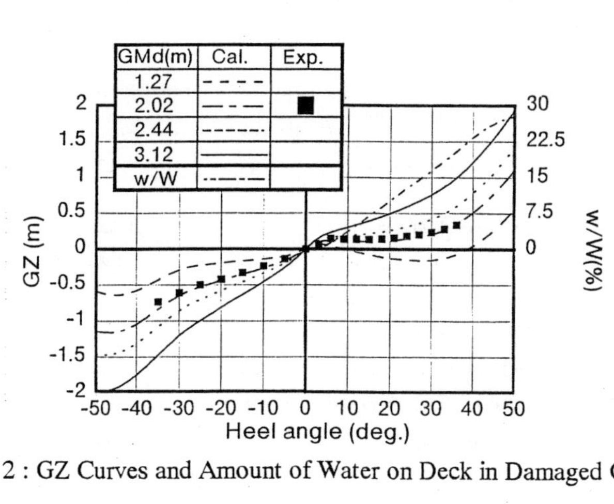

Figure 2 : GZ Curves and Amount of Water on Deck in Damaged Condition

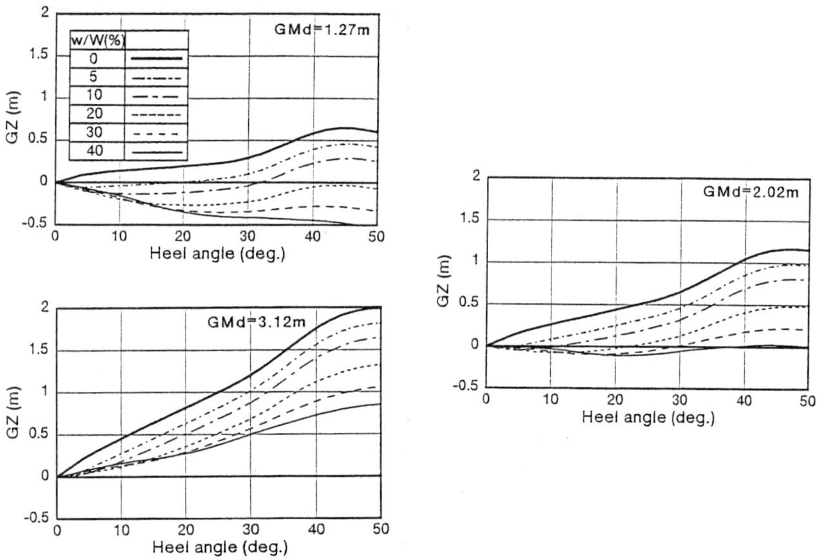

Figure 3 : GZ Curves with Constant Amount of Water on Deck

RESULT IN IRREGULAR WAVES

Figure 4 shows a time history example of roll angle ϕ and w. The model ship with CC heeled to weather side in almost all conditions like this figure, however without CC the direction was always lee side. At the last stage of experiment ϕ and w had constant mean values, ϕ_0 and w_0 respectively, in the stationary condition. The effect of experimental parameters on ϕ_0, w_0 and water ingress rate at the beginning of experiment v was investigated as follows (see Figures 5 and 6).

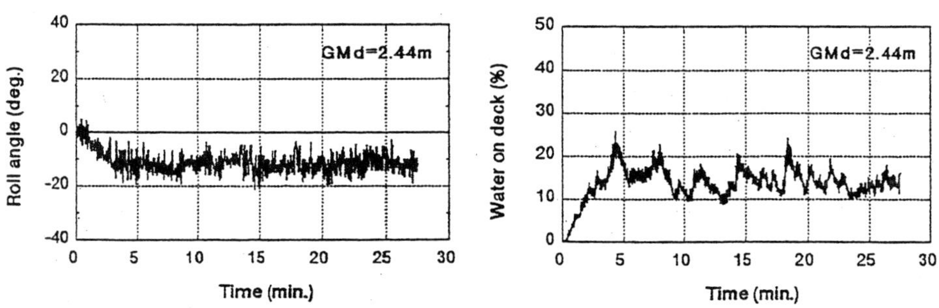

Figure 4 : Time History of Roll Angle and Amount of Water on Deck (With Center Casing)

(1) Effect of GM_d

In Figure 5 (without CC), when GM_d gets smaller ϕ_0 becomes greater, but w_0 and v become smaller because small GM_d leads to a large heel angle to lee side in a short time after flooding, which places the damage opening higher up the sea surface. The effect of wave height is small. The reason seems to be the constant wave slope.

(2) Effect of Center Casing

In Figure 6 (with *CC*) the tendency of ϕ_0 versus GM_d is the same as Figure 5, but to the opposite direction (to weather side) because the flooded water stays mainly in the weather side

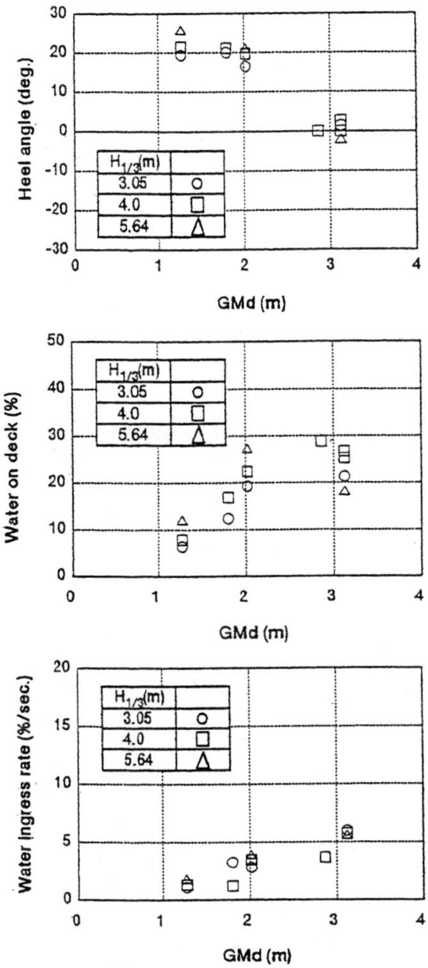

Figure 5 : Experimental Results in Irregular Waves (No Center Casing)

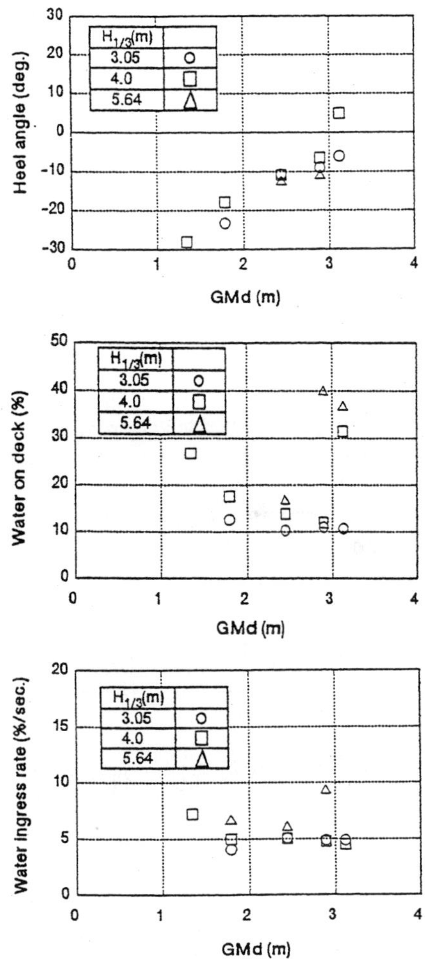

Figure 6 : Experimental Results in Irregular Waves (With Center Casing)

compartment of the deck. The variation of w_0 is not so clear as in Figure 5 because the movement of flooded water between the two compartments is complicated. But in general, the tendency of w_0 is opposite to Figure 5 because heeling to weather side (lowering the damage opening) keeps the water flowing in and out. The almost constant value of v can also be explained by this. It can be seen that with the standard GM_d (3.12m) this ship keeps almost upright condition even if w/W becomes as much as 40% and that it capsizes with the smallest GM_d.

(3) Effect of Initial Heel
When the ship has an initial heel angle of 4 degrees by a shift of weight to weather side, the time histories of ϕ and w are similar to the one with *CC* (Figure 4) because heeling direction is the same. At the case of the smallest GM_d (1.27m) she capsized in about three minutes in model scale. So it can be concluded that heeling to weather side because of *CC* and/or cargo shift leads to a disastrous situation.

EFFECT OF ROLL RESONANCE

The results in regular waves with constant (wave height)/(wave length) ratio of 1/25 are shown in Figure 7. The abscissa ω_i / ω_r is a tuning factor, where ω_i is the frequency of incident wave and ω_r is the natural rolling frequency in damaged condition (but the vehicle deck is undamaged).

It can be seen that not only rolling amplitude and relative water amplitude but also water on deck have peaks around tuning factor = 1. The peaks are notable with *CC* because the motion of the flooded water is reduced by the presence of the *CC*. Moreover the mean heel angle shows some change around the same frequency. This result suggests that irregular waves for the stability test should include significant wave component of the roll resonance frequency and that the interaction of ship motion and water on deck should not be ignored.

Looking through the figures of roll amplitude in Figure 7, the peak frequency ω_{MAX} seems to shift to low frequency side with *CC* and to high frequency side without *CC* in some cases. The factors of shifting ω_{MAX}, characteristic to the damaged RO-RO passenger ships, are listed below.

(1) Large damping (damping in damaged condition is 5 times as large as intact condition according to the free roll test)
(2) Static effect of the water on deck (free water effect, change of moment of inertia, sinkage of the ship)
(3) Dynamic effect of the water on deck.

All the effects of (1), (2) and the damping effect of (3) make ω_{MAX} shift to the low frequency side. But a calculation, modeling the ship and water on deck like a double pendulum, shows an opposite result, Murashige (1996). So, the shift of ω_{MAX} to high frequency side in Figure 7(a) is supposed to be caused by the coupling effect of the ship and water on deck.

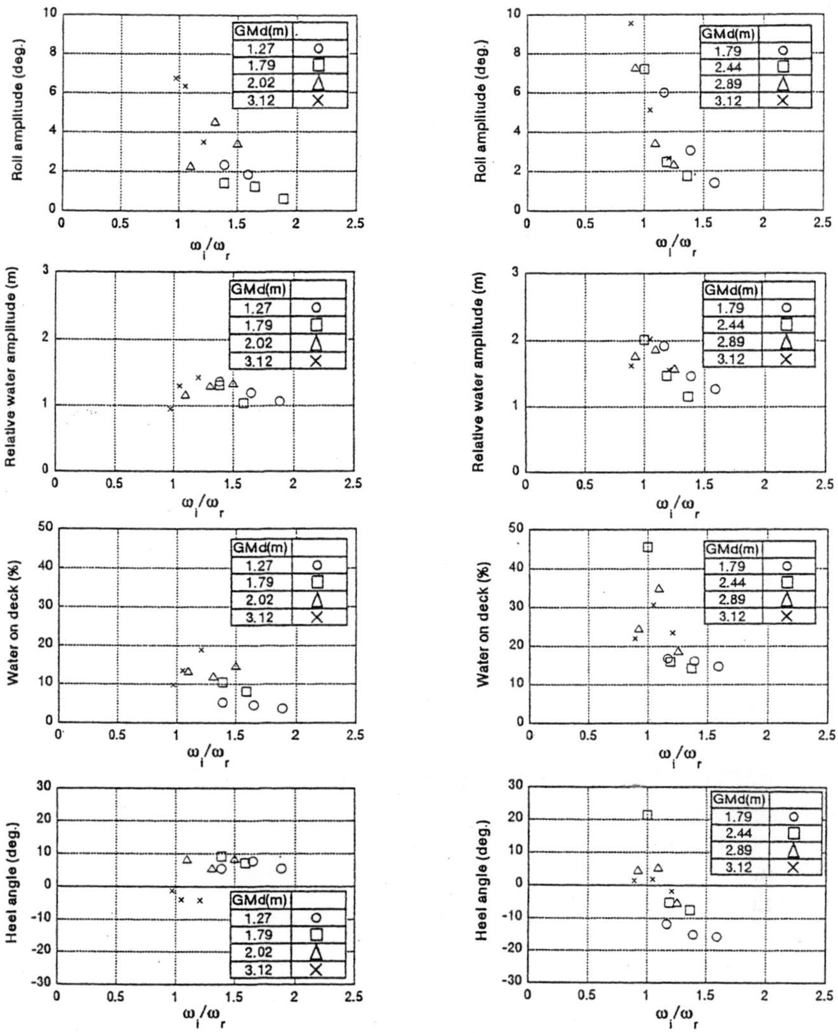

(a) No center casing (b) With center casing

Figure 7 : Frequency Responses in Regular Waves

KEY FACTOR FOR THE BALANCE IN STATIONARY CONDITION

As mentioned in the previous section the water ingress velocity from the damage opening is limited, so in a few cycles of rolling motion the damaged ship moves almost along one of the stability curves shown in Figure 3, with a constant volume of water on deck *w*. When *w* increases by flooding it transfers to another stability curve with less stability. At last when the rolling energy overtakes the dynamic stability the ship will capsize, but if this transference stops in a stationary condition under *a certain balance* she will survive. This *balance* will be discussed below.

The mean heel angle ϕ_0 and the mean water volume on deck w_0 in the stationary condition are shown in Figure 8. When the model capsized the values just before capsize are plotted. The solid lines show the equilibrium angles calculated from the stability curves in Figure 3 for a given w/W. But if the stability curve is almost parallel to the abscissa near the equilibrium point and if some moment like a wind moment works, the equilibrium angle can easily change from the exact one. So quasi-equilibrium angles, the crossing points of the stability curves with the lines of $GZ = \pm 0.0624m$ (2% of GM_d of the standard condition), are also calculated and drawn by broken lines in Figure 8. The zone between these two broken lines will be called a equilibrium zone hereafter.

It can be seen from Figure 8 that the non-capsized experimental results are in or near the equilibrium zones and that capsized results are away from the zone with surplus water on deck. According to Figure 3 GZ is always negative with these GM_d and w/W values, so it is a natural result that she capsized.

It is necessary to know another key factor or key quantity which decides the balance in stationary conditions. As the key quantity we propose H_d, the mean height of the water surface on deck from the calm sea surface. If H_d has a large positive value at a moment the water on

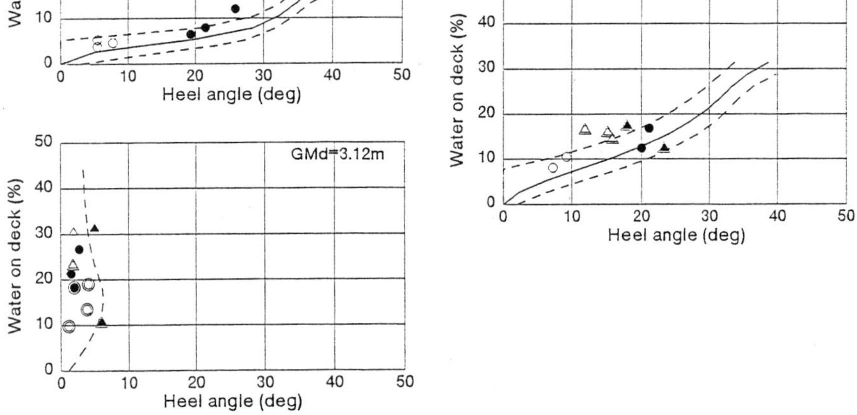

Figure 8 : Amount of Water on deck and Heel Angle in Stationary Conditions

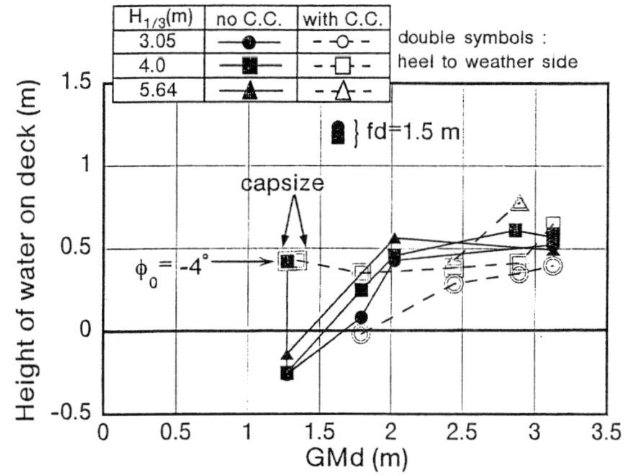

Figure 9 : Height of Water on Deck in Stationary Conditions

deck will flow out and vice versa, so H_d must be within a certain range. Figure 9 shows experimental results of H_d vs. GM_d. It can be seen that H_d has a value between -0.26m ~ 0.78m in ship scale when the freeboard height in damaged condition f_d is standard (0.33m).

It should be necessary to investigate the variation of H_d as a function of ϕ and w before proceeding with the discussion on the experimental results. The calculated H_d in calm water is shown in Figure 10. As long as ϕ is small the water surface on deck is a little higher than the freeboard (0.33m) and is almost constant regardless of w because the flooded water spreads over the whole deck. On the other hand when ϕ_0 becomes large H_d tends to vary widely according to w because flooded water concentrates at the corner of the deck. So, when H_d keeps a certain positive value in waves after the ship heels, that inevitably leads to a large value of w and to a less stable condition.

Returning back to Figure 9, the black and single marks show low H_d values when GM_d is small because the ship heels to lee side and the flooding stops in a short time. As long as wave height is small H_d is also small like the circles in Figure 9. But in other cases in which water is still

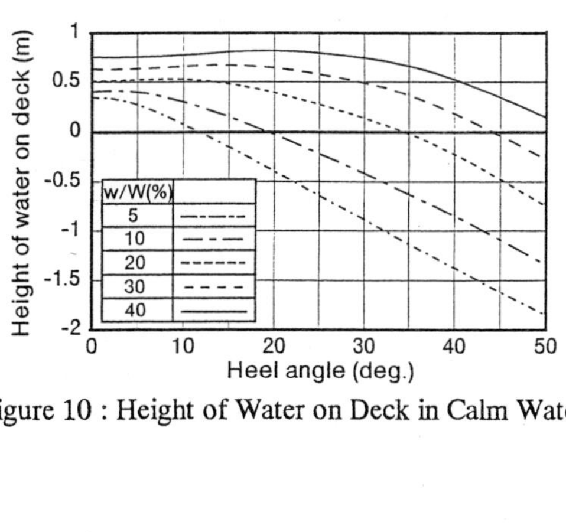

Figure 10 : Height of Water on Deck in Calm Water

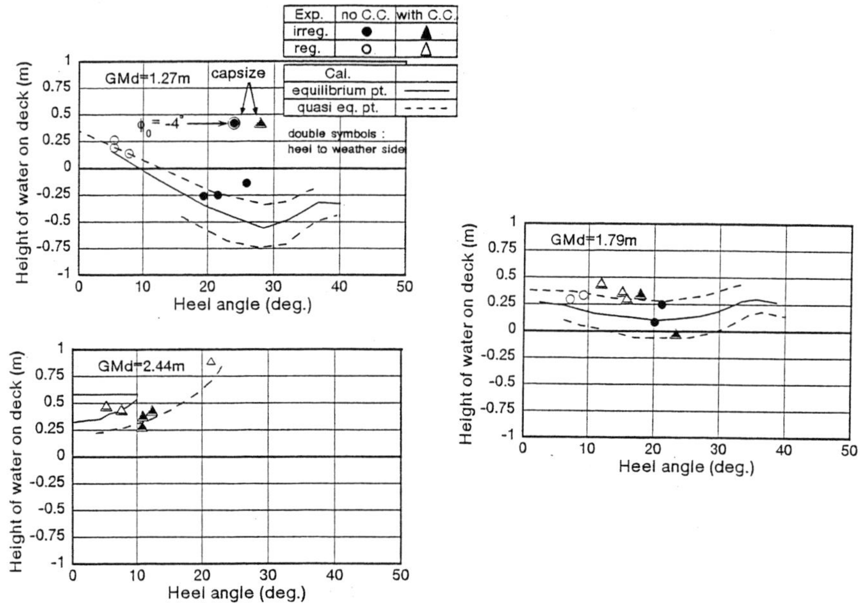

Figure 11 : Height of Water on Deck and Heel Angle in Stationary Conditions

coming in and out of the deck, H_d keeps high values in a small range between 0.4 and 0.8m, regardless of wave height. So if GM_d is very small and the ship heels to weather side by the effect of CC or cargo shift in high waves, the ship will capsize by much flood water on deck. It should be noted that the stability curves of the damaged condition in Figure 2 is calculated under the assumption that $H_d = 0$. This ordinary calculation underestimates the amount of water on deck and the loss of stability in waves.

RISK ESTIMATION FOR CAPSIZE

Figure 11 represents Figure 8 with the y-co-ordinate changed from water on deck to H_d. It can

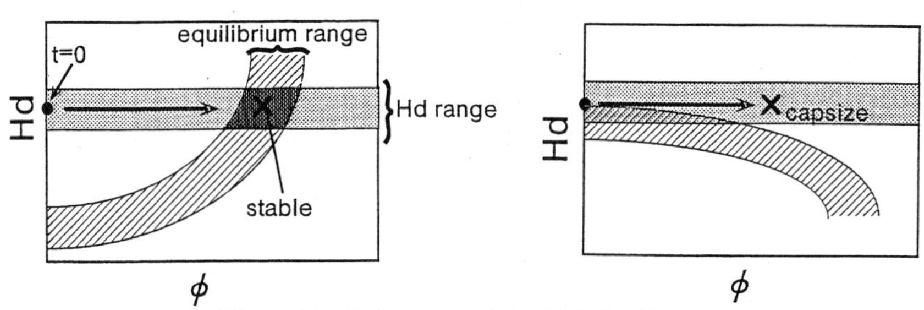

Figure 12 : Judgement of the Risk for Capsize

be seen that when GM_d becomes greater the tendency of the equilibrium zones changes from right side down to right side up. From Figure 11 and the sketch in Figure 12, we can understand the process of the change of ship condition in waves and can estimate the risk for capsizing.

When the first flooding occurs the ship condition locates on the point of $(\phi, H_d) = (\phi_i, f_d)$, where ϕ_i is the initial heel angle. If the ship heels H_d becomes small, but that encourages flooding. So the chain of "flooding", "increase of H_d", "heel (flow out of some water at the same time)", "decrease of H_d" and "flooding" will be repeated. In the dangerous condition, i.e. the wave is high and the ship has a *CC* or $\phi_i < 0$ (weather side), the point in the figures moves almost horizontally to the right, keeping H_d constant and increasing *w*. Finally if the point comes across the equilibrium zone like the figure of $GM_d = 2.44m$ in Figure 11, the movement of the point stops and the ship starts to roll around the stable condition. But if the zone is right side down like the figure of $GM_d = 1.27m$, there is nothing to stop the movement of the point and the ship will capsize with much water on deck.

So the risk for capsizing can be roughly estimated from the tendency of the equilibrium zones without knowing the exact value of H_d. In order to make the ship capsize-resistant it seems crucial to make the *GZ* at big heel angles large enough to make the zone right side up. It can be concluded for this ship that the minimum required GM_d for safety is 1.79m.

CONCLUSIONS

1) The effect of GM_d (*GM* in damaged condition) etc. on the mean heel angle and the mean water on deck in the stationary condition, ϕ_0 and w_0 respectively, was investigated. When GM_d gets larger w_0 also becomes larger, but the ship is stable with smaller ϕ_0 value. The tested ship did not capsize as long as SOLAS Regulation is satisfied.
2) When the ship heels to weather side she becomes unstable with much water on deck if GM_d is small. So cargo shift or existence of center casing might lead to a dangerous situation.
3) The test result in regular waves show that not only ship motion but ϕ_0 and w_0 have some peaks near the resonant frequency of rolling, so the waves for stability test should include that wave component.
4) The water on deck keeps a higher mean surface than sea surface as long as the damage opening of the deck is not made high by heeling to lee side. This difference of water surface, H_d, has a value in a small range in various conditions.
5) In a dangerous condition, i.e. the wave is not so low and the ship has a *CC* or cargo shift, the ship transfers to less stable condition, repeating the chain of "flooding", "increase of H_d", "heel", "decrease of H_d" and "flooding". Finally if the rolling energy overtake the dynamic stability she capsizes.
6) By calculating the equilibrium zone and drawing it on H_d-ϕ diagram the risk of capsize can be roughly estimated without knowing the exact value of H_d.

References

Bird H. and Browne R.P. (1973). Damage Stability Model Experiments. *Trans. of R.I.N.A.*

Dand I.W. (1988). Hydrodynamic Aspects of the Sinking of the Ferry 'Herald of Free Enterprise'. *Trans. of R.I.N.A.*

Dand I.W. (1994). Factors Affecting the Capsize of Damaged RO-RO Vessels in Waves. *Symp. on RO-RO Ships' Survivability.*

IMO RORO/ISWG/1/3/5 (1995). Some Results of Model Test

The Joint Accident Investigation Commission of Estonia, Finland and Sweden (1995). Part-Report Covering Technical Issues on the Capsizing on 28 September 1994 in the Baltic Sea of the RO-RO Passenger Vessel MV ESTONIA.

Murashige S., Ishida S., Watanabe I. and Ogawa Y. (1995). A Model Experiment for a Relation between Flooding of a Ro-Ro Deck and Ambient Waves. *66th General Meeting of Ship Research Institute.* (in Japanese)

Murashige S., Aihara K. and Yamada T. (1996). Nonlinear Roll Motion of a Ship with Water-on-Deck in Regular Waves. *The Second Workshop on Stability and Operational Safety of Ships.*

Nimura T., Ishida S. and Watanabe I. (1994). On the Effect of Hull Forms and Other Factors on the Capsizing of Sailing Yachts. *Journal of the Soc. of Naval Architects of Japan.* **175.** (in Japanese)

Shimizu N., Roby K. and Ikeda Y. (1996). An Experimental Study on Flooding into the Car Deck of a RORO Ferry through Damaged Bow Door. *Journal of the Kansai Society of Naval Architects.* **225**,

Vassalos D. (1994). Capsizal Resistance Prediction of a Damaged Ship in a Random Sea. *Symp. on RO-RO Ships' Survivability.*

Velschou S. and Schindler M. (1994). RO-RO Passenger Ferry Damage Stability Studies - A Continuation of Model Tests for a Typical Ferry. *Symp. on RO-RO Ships' Survivability.*

Watanabe I. (1995). Disaster of Ro-Ro Passenger Ship "Estonia" and Safety Measure in IMO. *66th General Meeting of Ship Research Institute.* (in Japanese)

Contemporary Ideas on Ship Stability
D. Vassalos, M. Hamamoto, A. Papanikolaou and D. Molyneux (Editors)

A STUDY ON CAPSIZING PHENOMENA OF A SHIP IN WAVES

S. Y. Hong[1], C. G. Kang[1] and S.W. Hong[1]

[1]Korea Research Institute of Ships and Ocean Engineering
P.O. Box 23, Yusong, Taejon, 305-600, Korea

ABSTRACT

After the disastrous accident of the Seohae Ferry in Korea's West Sea in 1993, a numerical and experimental study on the capsizing of the ferry was carried out to investigate the cause of the accident. Righting moments for various loading conditions were calculated both in still water and in waves. Capsizing simulations were performed with hydrodynamic coefficients obtained from the strip method (Salvesen, Tuck and Faltinsen, 1970) and nonlinear restoring moments considering relative motion. Runge-Kutta fourth-order method was employed in time integration for the simulation. It was found that deterioration of designed righting moment causes the ship to capsize even in not so extremely severe wave conditions. Nonlinear effects such as hydrodynamic forces due to change of attitude of a ship, effects of green water and freeing ports which are not considered in the calculation were also investigated through model tests.

KEYWORDS

Capsizing, Heeling Angle, Righting Arm, Capsizing Simulation, Model Test, Captive Tests

INTRODUCTION

Capsizing of a ship in waves may be the last disaster that she could experience through her life. That kind of accident usually results in tremendous loss and casualties. Therefore it is important to examine the mechanism of capsizing in waves for the prevention of the accident. Energetic research activities have been carried out by Hamamoto and others(1992, 1994), Kan and Taguchi(1992) on this hot issue during past years.

There was a disastrous ship accident on October 10, 1993 in Korea's West Sea. A passenger ship named Seohae Ferry, capsized and submerged in a moment under estimated 2m wave condition. Usually this wave condition is not regarded as a sufficiently severe one to cause a ship to capsize, but after the investigation it turned out to have been sufficiently severe condition for the ship in this particular accident.

As a result of this accident, of the 362 persons onboard (passengers and crew) more than 290 people died or disappeared. This accident typically showed the importance of stability of a ship in waves. To investigate the cause of this accident within a short time, and due to a lack of witnesses and evidences, we approached with a conventional method under limited assumed ship conditions.

In this paper, numerical and experimental stability analysis is performed for the passenger ship Seohae Ferry (Lee, et al., 1994). Restoring moment is calculated both in calm water and in waves for various loading conditions. Capsizing simulations are performed to examine the possible cases of capsizing. Experiments were also carried out to measure restoring moments in waves and to investigate nonlinear hydrodynamic effects which were not included in the simulation model.

THEORETICAL ANALYSIS

Quasi-static analysis of righting moment

Generally, righting moment of a ship in waves has its minimum value when the wave crest is at the middle of the ship, while maximum value is obtained when the wave trough approaches the same location of the ship. This can be explained by the geometrical property of ships; it has noticeable change of sectional shapes at the stem and stern according to the change of waterline, while it does not at the middle of the body in the same situation. The righting moment of a ship was calculated by the following procedures (Lee, et al., 1994);

1) Search equilibrium waterline for a given heeling angle.
2) Search center of buoyancy of the submerged part.
3) Calculate moments due to buoyancy and gravity.
4) Finally restoring moment is obtained by subtracting the gravity effect from the moment due to buoyancy.

Main particulars of the Seohae ferry are, length between perpendiculars=33m, breadth=6.2m, displacement=220ton, draft at FP and AP are 1.615m and 2.603m, respectively. Designed GM was 0.5m.

Capsizing Simulation

Transverse righting moment($GZ \cdot W$) is one of the most important factors which have influence on the capsizing of ships in waves. W is the displacement of the ship and GZ is the righting arm. For large rolling motion, righting arm is governed by nonlinear effect as illustrated in Figure 1, while GZ is usually approximated by the linear function of rolling angle in the linearized ship motion equation. In the present capsizing simulation model, all the hydrodynamic forces except the lateral righting moment term were obtained from the STF(Salvesen, Tuck and Faltinsen, 1970) strip method. Lateral righting moment is calculated as accurately as possible by considering relative motions between ship and incident waves at each station. Time-domain simulation of rolling motion was performed using the Runge-Kutta fourth-order time integration scheme with linearized hydrodynamic forces and nonlinear righting moment. The equation of motion can be expressed as the following coupled equations.

$$
\begin{aligned}
[M]\{\ddot{\vec{x}}\} &= \{\vec{F}_{dyn} + \vec{F}_{sta}\} \\
\vec{F}_{dyn} &= \vec{F}_{excit.} + \vec{F}_{hdyn} \\
\vec{F}_{sta} &= \vec{C}(\vec{x}(t))
\end{aligned}
\qquad (1)
$$

$\vec{x}$ is the motion vector, $\vec{F}_{dyn}$ is the hydrodynamic external force which is composed of the wave exciting force($\vec{F}_{excit}$) and the hydrodynamic reaction force vector($\vec{F}_{hdyn}$). $\vec{F}_{sta}$ was calculated by considering the relative wave which is a function of motion and the incident wave.

MODEL EXPERIMENTS

Loss of stability of a ship is regarded as one of the major causes of capsizing in waves. Hence, many simulation models have been developed to investigate the role of nonlinearity of righting moment in waves, and the focus is paid on the accurate estimation of righting moment in waves. Experimental study can give more information when a theoretical approach may miss real physical phenomena such as green water on the deck, the effect of green water flow, the effect of freeing port, nonlinear hydrodynamic forces due to change of attitude of a ship and so forth. A series of model experiments were performed at KRISO towing tank of which the length is 200m, breadth 16m and depth 7m. A flap-type wave maker installed at the end of the tank generated target waves. The experimental conditions were as follows:

- scale ratio : 1/12
- heading : following sea
- ship type : ferry
- ship speed : 10 knots
- draft : #106 condition(departure condition)
- wave height : 0, 1, and 2m
- wave length : 35m
- measured items : heeling angle, sinkage and trim for freely floating model test
heeling angle, rolling moment for captive model test

Two methods are devised to get a more realistic relation between righting moment and heeling angle in waves. The first is a freely floating model method in which more accurate value of heeling angle and sinkage can be obtained for a given heeling

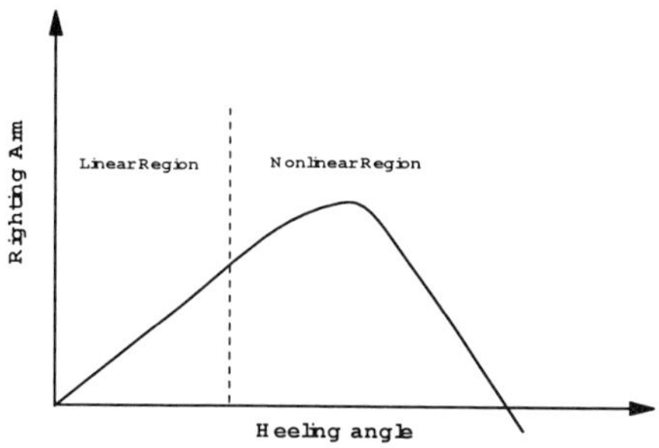

Figure 1: Righting arm vs. heeling angle

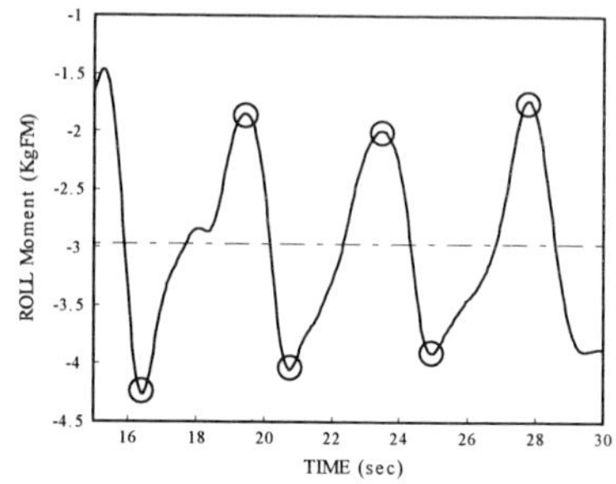

Figure 2: Measured Restoring Moment in Following Waves

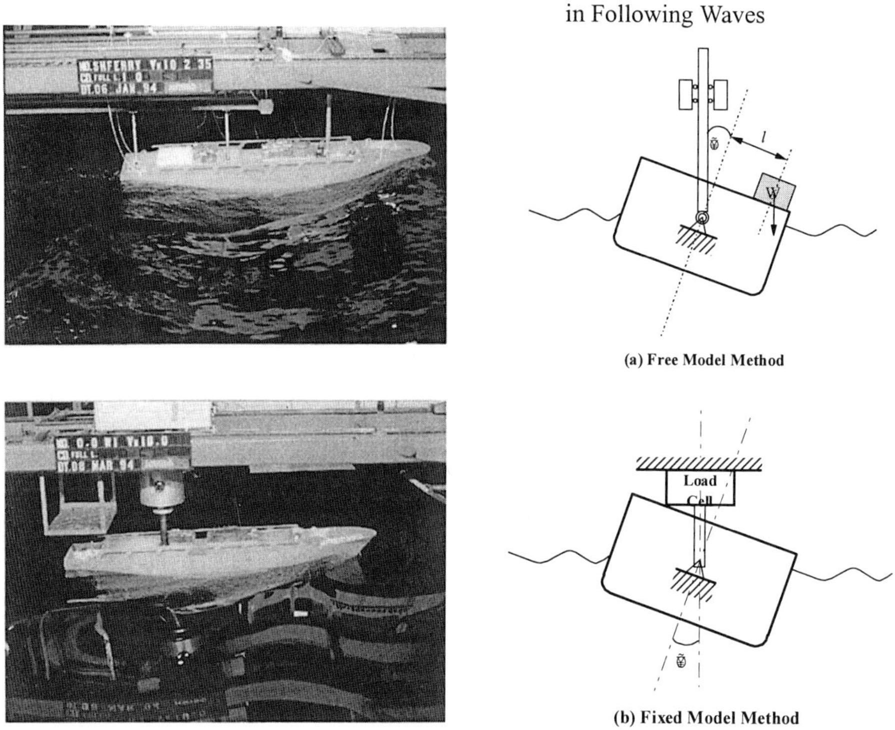

Figure 3: Test setup for measurement of heeling angle and restoring moment

moment. The other was a captive model method where the change of righting moment can be measured for a given heeling angle in waves. The former method is applied first to investigate the effects of Kelvin wave induced by ship's advance as well as to find out the realistic attitude of the ship for a given heeling moment. Restoring moment in waves is essentially nonlinear as shown in Figure 2. However, measured restoring moment signal shows a periodic pattern and it has almost repeated amplitudes. Since the peak value is important in a stability sense, the analysis is made on the basis of peak to peak value approach. Figure 3 shows the test scene and schematic diagrams for both of the experimental methods used in this study.

RESULTS AND DISCUSSIONS

Calculation of righting moment in waves

In order to investigate as many capsizing conditions as possible, restoring moments for various loading conditions are examined as shown in Figures 4 to 6. Loading conditions are classified as numbers in the figure as follows:

Legend	Description
Design	: Full load departure condition at design stage
#102	: Departure condition (October 10, 1993) assuming without overloads such as pickled anchovies, excessive passengers, cargo on upper deck and gravel
#103	: In addition to #102 load condition, 7.3 tons of gravel were loaded at steering gear room
#104	: In addition to #103 load condition, cargo (pickled anchovies) were loaded on the upper deck
#105	: In addition to #104 load condition, 143 excessive passengers were onboard
#106	: In addition to #105 load condition, 4 ton passengers' cargo was loaded (departure condition on Oct. 10, 1993)
#107	: 7.3 tons of gravel were removed from #106 load condition

Figure 4 shows the calculated righting moment in still water condition. Figure 5 presents the minimum righting moment in stern quartering wave condition (wave incidence angle = 45degrees) when wave height is 2m and wave length is 30m. Figure 6 shows the maximum righting moment for the same wave condition as in Figure 5. It can be seen that righting moments decrease as loads on the deck increase. It is believed that increase of draft due to cargo results in decrease of freeboard, which consequently causes wave runs over the upper deck earlier. Though gravel is loaded below the center of gravity location, it has bad influence on the righting moment except for small heeling angle range. Comparing load conditions #104 and #105, it can be seen that excessive passenger is one of decisive factors for loss of stability. When the crest of wave arrives at the middle of ship, righting moments dramatically decrease due to excessive passengers and gravel. Among considered load conditions, the worst case is obtained for loading condition #106 which is assumed for departure condition at the accident. From these investigations, excessive passengers and gravel are main factors for the loss of stability.

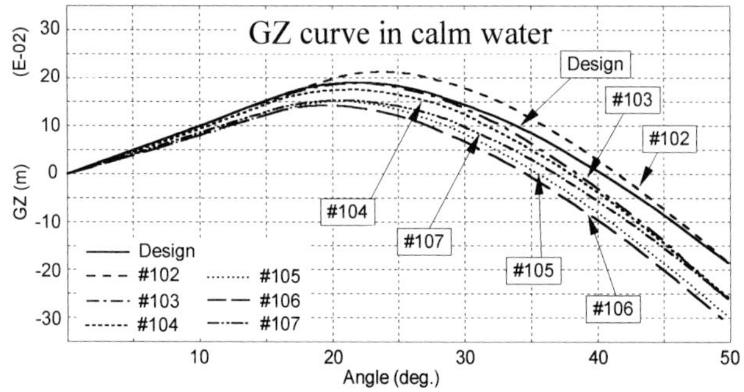

Figure 4: Righting arm curves of a passenger ship for various loading conditions in calm water

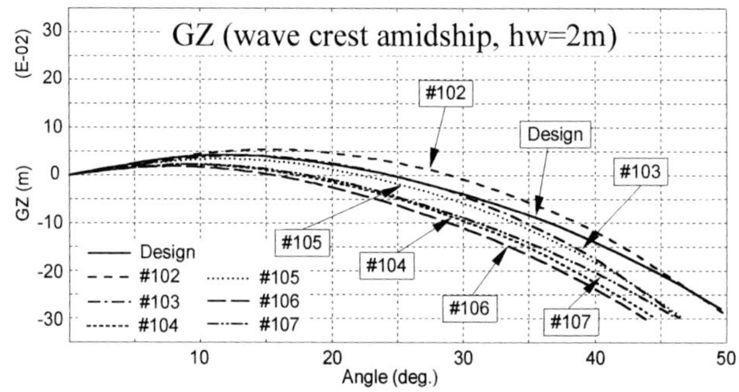

Figure 5: Minimum righting arm curves of a passenger ship for various loading conditions in waves(stern quartering wave)

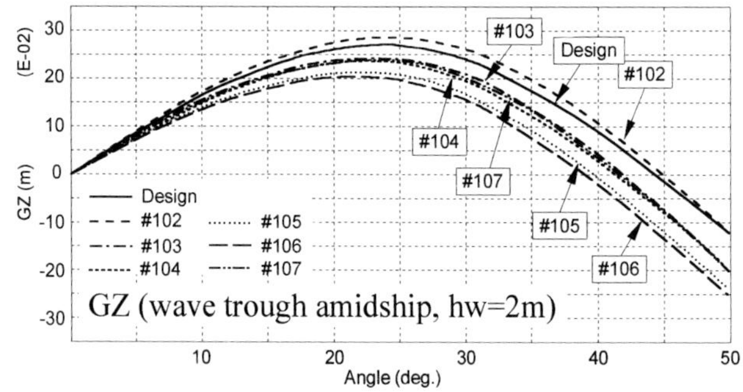

Figure 6: Maximum righting arm curves of a passenger ship for various loading conditions in waves(stern quartering wave)

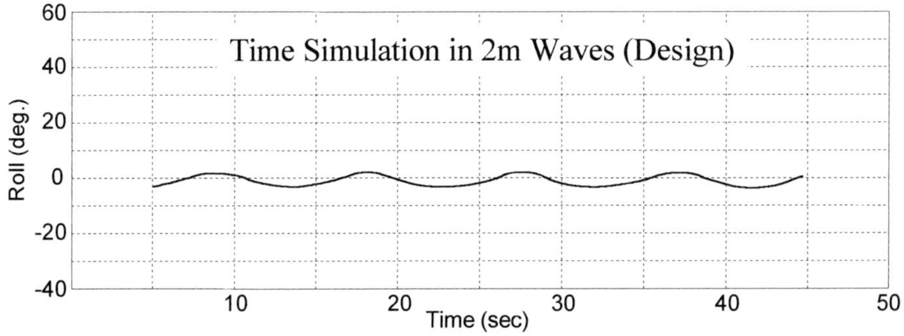

Figure 7: Time simulation of roll motion in 2m waves (Wave incidence angle = 45°)

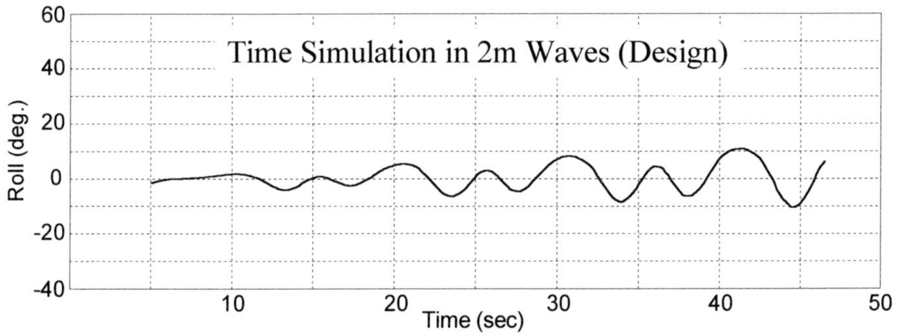

Figure 8: Time simulation of roll motion in 2m waves (Wave incidence angle = 30°)

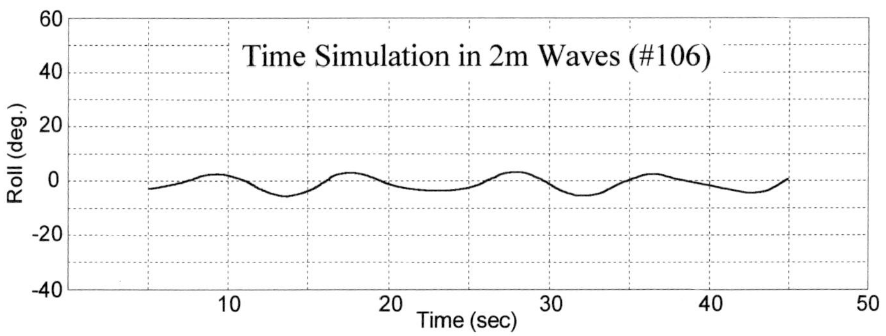

Figure 9: Time simulation of roll motion in 2m waves (Wave incidence angle = 45°)

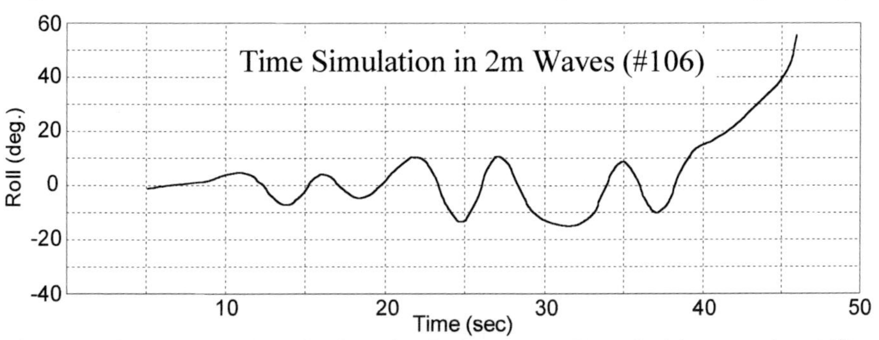

Figure 10: Time simulation of roll motion in 2m waves (Wave incidence angle = 30°)

Capsizing Simulation

In order to investigate the possibility of capsizing under design and departure loading conditions and considered wave condition, the simulation were carried out for various wave incidences from zero to 90 degrees, but results are given for the following conditions which were considered as most severe cases:

Ship speed : 10 knots
Average wind speed : 5.5 m/sec.
Max. wind speed : 10.5 m/sc.
Wave height : 2.0 m
Wave length : 30 m
Wave incidence angle : 30 and 45 degrees (head sea = 180 degrees)
Loading condition : Design load and departure condition(#106)

Figures 7 and 8 show the simulation results for design load condition when wave incidence angle is 45 degrees and 30 degrees, respectively. These results show that capsizing does not occur for design load condition for all wave incidence angles. Figures 9 and 10 show the simulation results for departure load conditions. These results show the significance of wave incidence angle on the capsizing of a ship. Within a minute after simulation, capsizing occurs when the wave incidence angle is 30 degrees. This result demonstrates that the deterioration of designed righting moment causes the ship to capsize in wave conditions such as the case when the high wave and stern wave direction occur simultaneously.

Model test

Since the test conditions were different from those of calculations due to the limitation of tank capability, direct comparison was not made. Experimental results, however, gave good information on understanding of the accident situation of the Seohae Ferry. One important fact found in the experiment was that the accident ship had insufficient freeboard at stern and wave flooding was frequently observed, seemed to have bad influence on the stability of the ship.

Figure 11 compares the measured and calculated heeling angle vs. righting arm. Free model method is used in this measurement. Very good correlation is obtained between measured heeling angles for given heeling moments and the calculated righting arm for given heeling angles. Figure 12 presents the effects of ship speed on the change of heeling angle and sinkage for given heeling moments. It can be found that the sinkage slightly increases and the heeling angle slightly decreases as the ship advances in still water. This implies that a steady Kelvin wave pattern plays a role to increase righting moment and it can be explained that steady wave pattern has its crest at the bow and stern occasionally for moderate to high Froude number (Figure 14). Steady wave of which crests at bow and stern clearly gives good effects on ship stability in waves. This effect could also be found in righting moments measured in stern wave condition for the captive model as shown in Figure 13. Loss of stability occurred at a heeling angle of about 26 degrees under the condition of wave height of 2 meters, wave length of 35 meters and ship speed of 10 knots in following sea condition. Considering the results of

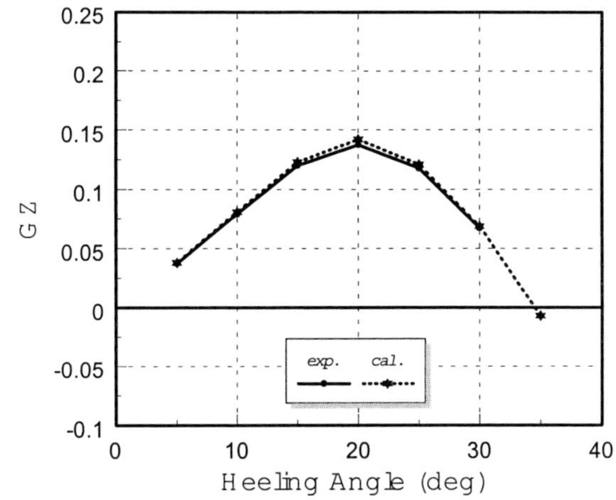

Figure 11: Comparison of Righting moment curve in calm water

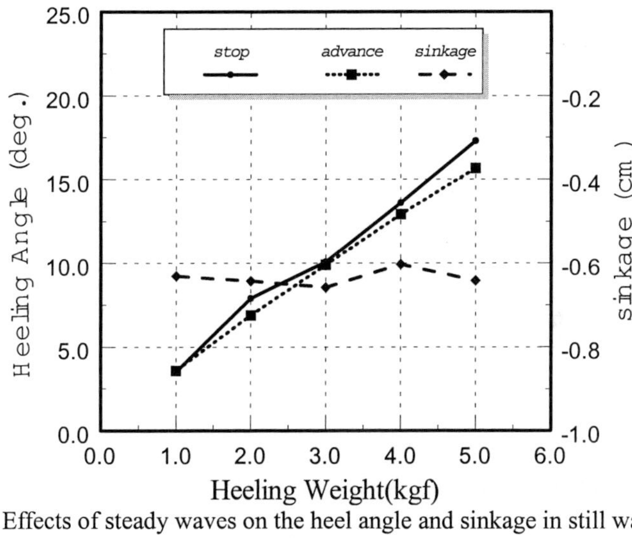

Figure 12: Effects of steady waves on the heel angle and sinkage in still water(Vs= 10 knots)

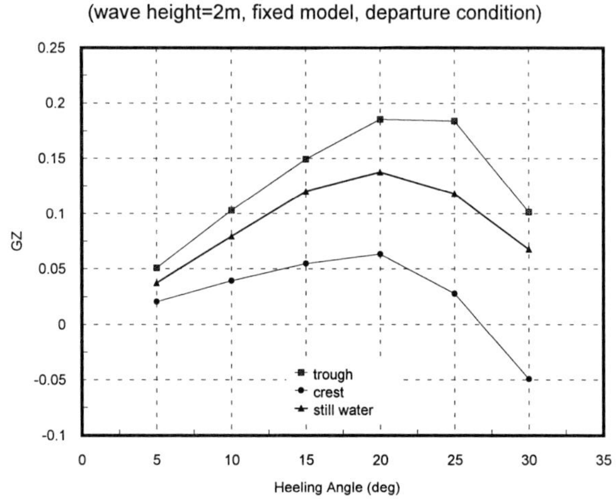

Figure 13: Righting moment curve in following regular wave
(Hs=2m, λ =35m, Vs= 10knots)

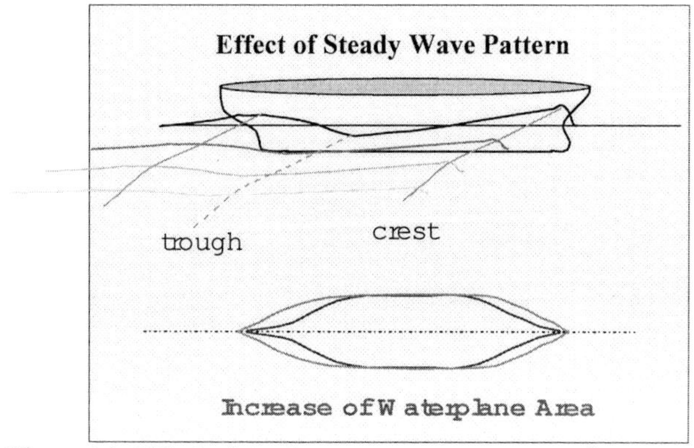

Figure 14: Effect of steady wave pattern on waterplane area

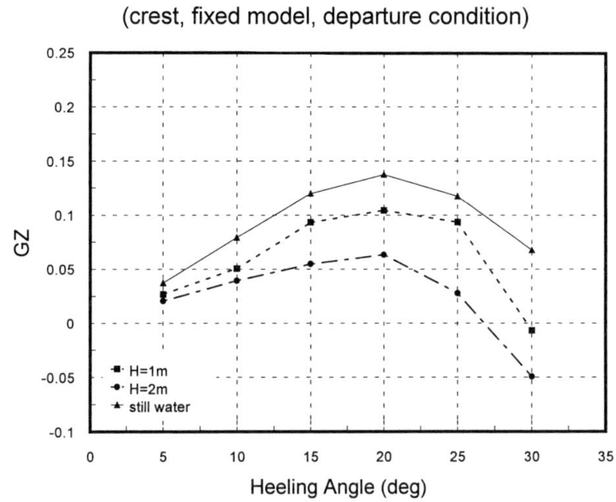

Figure 15a Effects of wave height on the minimum restoring moment in following regular wave (λ =35m, Vs = 10 knots)

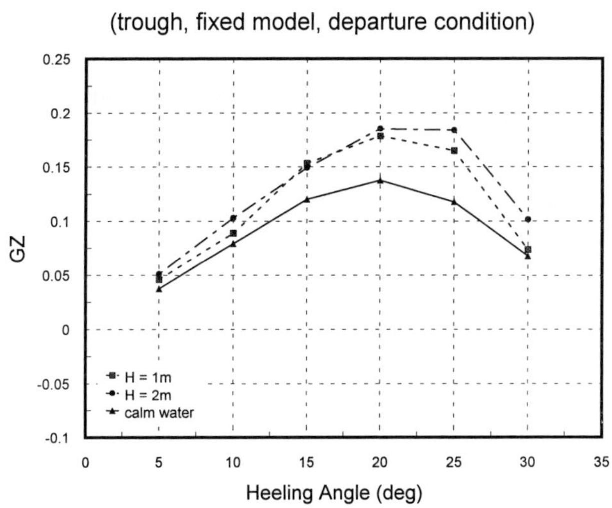

Figure 15b Effects of wave height on the maximum restoring moment in following regular wave (λ =35m, Vs = 10 knots)

theoretical calculation shown in Figures 4 and 5, even though wave incidence and length is different from that of experiment but qualitative trend might be similar, it seems that loss of stability occurred at relatively high roll angle in the experiment. Figure 15 shows the effects of wave heights on righting moment. Increase of maximum righting moment is not so significant as wave height increases, while a decrease of minimum righting is noticeable as wave height increases.
The following effects were considered through experiments which calculation could not fully consider:

- the effect of steady wave pattern which contributes to increase of righting moment
- the nonlinear effects of green water and its flood like flows on the deck
- the effects of freeing port

CONCLUSIONS

From the investigation on the stability of the accident passenger ship Seohae Ferry in waves through numerical and experimental study, the following conclusions can be drawn.

1. It was found that the decrease of designed GM caused the passenger ship Seohae ferry to capsize in assumed wave condition; stern quartering sea, wave height is 2 meters. Therefore deterioration of initial GM could lead to a dramatic decrease of restoring moment in waves, which may cause a ship to capsize in following or stern quartering sea condition even if the wave height is not so extremely high.
2. It was found in model experiment of the accident ship that insufficient freeboard at stern had a bad influence on the capsizing of ship in waves, and that shape of the stern is important in wave flooding in stern waves.
3. Measurements of restoring moment in following waves show that numerical estimation of restoring moment in waves might excessively predict loss of stability. Steady wave flow seems to be an important factor to resist the decrease of righting moment in waves.

References

Lee, J.T. et al. (1994). *Stability and Safety Analysis of a Coastal Passenger Ship*, KIMM Report UCK020-1812.D, Taejon, Korea (in Korean).

Hamamoto, M. and Tsukasa, Y.(1992). An Analysis of Side Force and Yaw Moment on a Ship in Quartering Waves, *J. of Soc. of Naval Arch. of Japan(SNAJ)* **171**, 99-108 .

Kan, M. and Taguchi, H.(1992). Capsizing of a Ship in Quartering Seas(Part 4, Chaos and Fractals in Forced Mathieu Type Capsize Equation), *J. of SNAJ*, **171**, 83-98.

Hamamoto, M., Kim, Y.S., Matsuda, A. and Kotani, H.(1992). An Analysis of a Ship Capsizing in Quartering Sea, *J. of SNAJ* **172**, 135-146.

Hamamoto, M., Matsuda, A. and Ise, Y.(1994). Ship Motion and the Dangerous Zone of a Ship in Severe Following Seas, *J. of SNAJ* **175**, 69-78.

Salvesen, N., Tuck, E.O. and Faltinsen, O.(1970). Ship Motions and Sea Loads, *SNAME Trans.* **78**, 250-287.

Contemporary Ideas on Ship Stability
D. Vassalos, M. Hamamoto, A. Papanikolaou and D. Molyneux (Editors)

PHYSICAL AND NUMERICAL SIMULATION ON CAPSIZING OF A FISHING VESSEL IN HEAD SEA CONDITION

Tsugukiyo Hirayama[1] and Koji Nishimura[2]

[1]Department of Naval Architecture and Ocean Engineering, Yokohama National University
Tokiwdai 79-5, Hodogaya-ku, Yokohama, 240-8501, Japan
[2]Technical Research and Development Institute, Japan Defense Agency, Japan

ABSTRACT

A fishing boat seemed to have capsized in head seas off Japan main island. The displacement of that ship was about 180 tons. Comparing to the case of capsizing in following or quartering seas, capsizing in head sea condition was not investigated enough in the past. So this paper deals with the estimation of the possible capsizing process in head waves both by experiments and by numerical simulations including the effect of manoeuvring.

KEYWORDS

Capsizing, Head Sea, Net Shift, Fishing Boat, Water on Deck, Manoeuvring, Directional Seas, Simulation

INTRODUCTION

There was a sea disaster that a fishing vessel (displacement was about 180 tons) seems capsized in head seas. Comparing to the capsizing in following seas, capsizing in head seas were not investigated enough in the past. This will be because that except the case of direct capsizing in pitch direction (pitch pole) of a very small boat like sailing yacht in breaking waves, capsizing in head seas seems impossible. This paper investigated the possible capsizing process of a not so small fishing vessel both by physical simulation(model experiment) and by numerical simulation, considering composite effects of wave, ship motion and net shift, including the effect of ship maneuvering on rolling.

THE CAPSIZED FISHING VESSEL

In this report, we investigated into the capsized ship in head seas near Japan Main Island. Her length (Lpp) was 23m, and displacement was 179 tons. This fishing vessel was mainly used for carrying fishing nets as roundhaul netter (makiami gyosen in Japanese) and had cabins under the upper deck for crew. Principal dimensions and general arrangements are shown in Table 1 and Fig.1. According to the three times experiments carried out in different occasions, three kinds of experimental values are written in Table 1. Body plan used for numerical simulation is shown in Fig.2. For carrying large amount of nets (about 14.5% of displacement), after deck is relatively flat and most distinguished point is that the deck house is shifted to the left (port side)for giving working space. This is shown by thick line in Fig.1. The thickness of the bulwark of used ship model is relatively thick comparing to the scale ratio from actual ship, so the influence of that thickness on stability curve was also evaluated, but in numerical simulation, this thickness was neglected in case of considering actual ship.

Table 1 Principal Dimensions

ITEMS		SHIP	1/23 MODEL			
		Specified Value		Experiment Value 1	Experiment Value 2	Experiment Value 3
Loa	(m)	29.220	1.270			
Lpp	(m)	23.000	1.000			
B	(m)	5.900	0.257			
D	(m)	2.150	0.0934			
d (fore)	(m)	1.500	0.06522			
d (aft)	(m)	2.586	0.1124			
d (mid)	(m)	1.922	0.08357			
Δ		178.82 (t)	14.339 (kg)	14.336 (kg)	14.519 (kg)	14.265 (kg)
Cb		0.629	0.629			
KG	(m)	1.920	0.0836	0.0740	0.0897	0.0831
GM	(m)	1.450	0.0630	0.0726	0.0558 (G1)	0.0639
KB	(m)	1.31	0.0570	0.0570	0.0570	0.0570
BM	(m)	2.06	0.0896	0.0896	0.0885	0.0900
G-Midship	(m)	-0.960	-0.0420	-0.042		
Kxx/B		0.497	0.497	0.307	0.319	0.334
Kyy/Lpp		0.250	0.250		0.275	0.281
Kzz/Lpp		0.250	0.250	0.285	0.280	0.283
Natural Period (sec)	Roll			0.80	0.81	0.81
	Pitch			0.79	0.80	
	Heave				0.79	

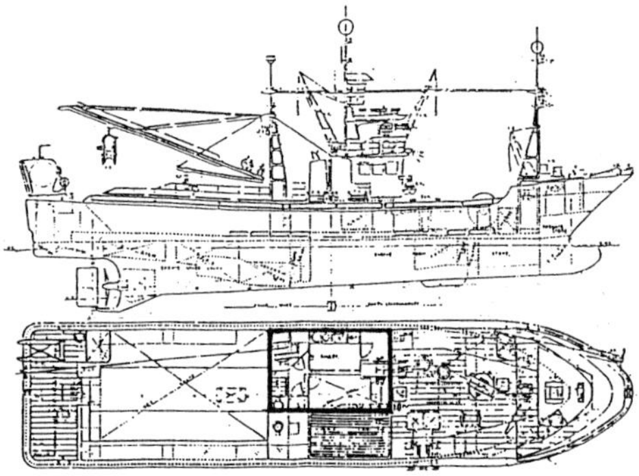

Fig.1 General Arrangement

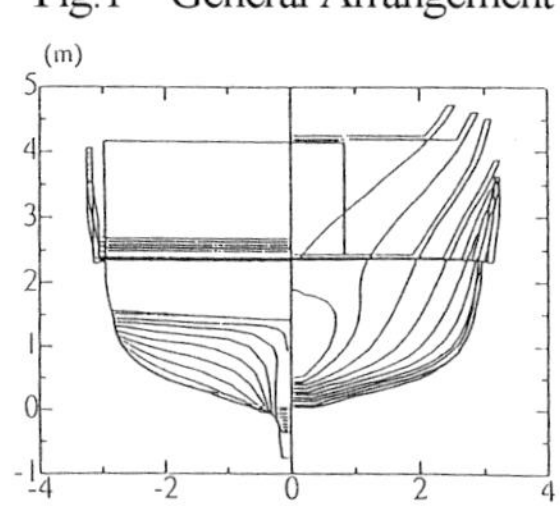

Fig.2 Body Plan for numerical simulation

This ship had capsized on the return course to her mother port (to North East), so the all fishing nets were

on her deck, but were not lashed following to their usual treatment. Adding this, the displacement of this ship was 13% larger than that of designed value, and GM was 20% smaller. About the sea state, wind speed was estimated about 10m/sec from North East direction. Mean wave period was about 7seconds, and significant wave height was about 2 meters. Her heading was estimated toward her mother port, so it was estimated that she was sailing in head waves about by 4 knots(Fn=0.137) speed.
Those estimations were made based on the measured results by weather station and also based on the information from survived crews (two out of twenty) who was in their cabins in the lower deck.

PHYSICAL EXPERIMENTS

Physical experiments were conducted using 1/23 scale model of the capsized vessel in long crested and short crested irregular waves of the towing tank (100m*8m*3.5m)of Yokohama National University. Model scale was decided considering the possible wave length generated by our multi-directional wave generator, and in the result adjusting room for gyradius and the height of the center of gravity became small, so those values became a little different from that of actual ship estimated. Usual measurement system used in our towing tank could not be mounted on this small ship model, so except sensors for roll and pitch angle ,target lights were mounted on the deck for detecting ship attitude by CCD camera. Model was self propelled by an electric motor ,and its rotation (rpm) and rudder angle were remotely controlled through FM radio controller by manual control.

For simulation of net shift by the effect of static heel or dynamic roll, we equipped with movable weight on the deck . This weight could be freely rolled by unlocking a trigger, by FM radio controller considering the unlocking timing to waves. Model propeller used is similar to that of actual ship. Diameter is 76 mm and pitch ratio is 0.71. Measured data in the model was sent out by thin electrical wire not to affect her maneuvering motion. But the roll or heel by large turning motion was finally evaluated only by numerical simulation.

Static Stability

Static stability is the basic characteristics for considering ship capsizing, so firstly we measured that curve by instrument (Hirayama et. al.(1994,1985,1983)). This results is also used for validating the numerical calculation. Stability curve is strongly affected by the existence of bulwark with freeing port closed, water tight deck house or other buoyant compartments. Fig.3 is an example of the case without deck house. Left figure shows both results from experiment and calculation in model scale. Freeing ports are closed, so its effect on stability curve clearly appears. Furthermore, calculation is valid except small errors. Right figure show the influence of the thickness of bulwark in full scale. This case, freeing ports are open and deck house is excluded. In the numerical simulation corresponds to model ship , the effect of the bulwark thickness on stability curve was included even though its effect seems small.

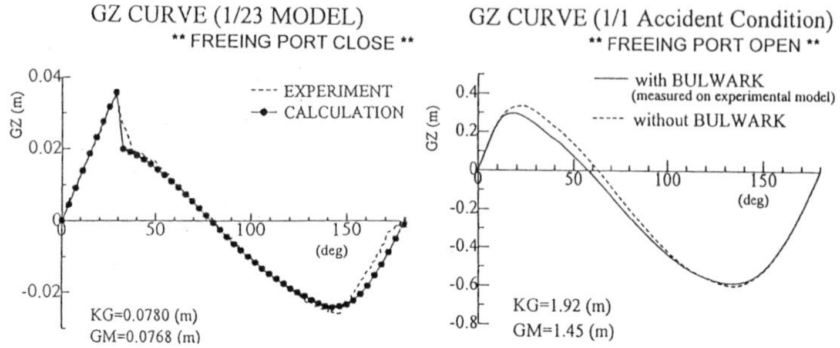

Fig.3 GZ curve without deck house . Freeing port are closed (left) and opened (right).

Stability range is about 60 degrees (Fig.3, without deck house) in case of KG is corresponds to that of capsizing and the so called "C" coefficient of this ship, one kind of safety factor for capsizing, was estimated as 1.058 in home ward condition at the design stage of this ship. This means that this ship is not so poor in stability from the view point of conventional criteria.

Experiments in long crested irregular waves

In Fig.4, we show the frequency transfer functions of this ship in Transient Water Waves (Takezawa et.al.(1971)) by the mark of white circle. Wave steepness was around 1/60. Numerically simulated results in regular waves are also shown by the mark of black circle. Solid lines show theoretical results by New Strip Method (NSM). Ship speed is zero. Result of roll is in beam sea and pitch, heave are in head sea. Calculation of roll by NSM show some different tendency from experiments and numerical simulation. Damping from free rolling is used in both NSM and numerical simulation. From this results ,numerical simulation adopted in this study will be reliable.

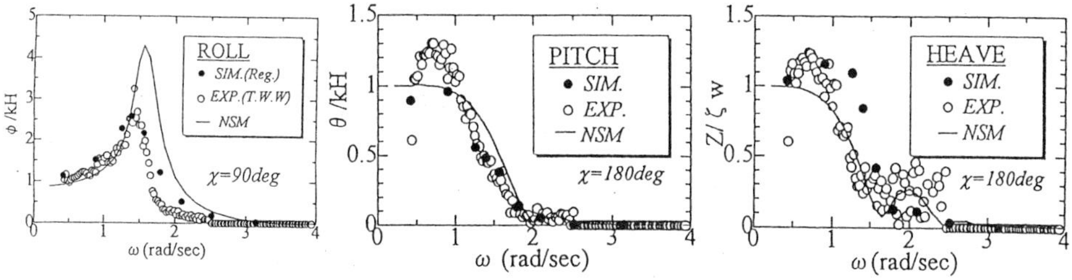

Fig. 4 Transfer Function of roll, pitch and heave(V=0)

In the next, self propelling experiments were carried out in long crested irregular head waves. In the wave condition that the actual ship seemed capsized (T02=1.5sec(7sec in full scale),H1/3=8.7cm(2m)) the model ship did not show large transverse motion, so the experiments near resonance case (natural period of roll and pitch is similar each other for this ship and around 0.8sec(3.8sec))was also conducted .

In this experiment shipping water occurred and in case of freeing port closed, that water is accumulated in starboard side because deck house is on the shifted position to port side. So, starboard side-down heel gradually increased and also rolling motion is gradually increased, but not tends to capsizing. For estimating the occurrence of shipping water phenomena, we calculated significant value of relative wave height along ship length (Fig.5). This time, calculation was made in full scale and in the wave of long crested irregular waves with ITTC spectrum. Considered mean wave period are 3.84sec and 5.07sec. Significant wave height was chosen as 2.0 m. The mark of white circle show the height of bulwark, and the mark of black circle shows the significant relative wave height (double amplitude). It is estimated that the shipping water will become largest at S.S.8 in head waves.

In Fig.6,significanr wave heights in which the probability of occurrence of shipping water is 1/100 are shown on the base of mean wave period. In this calculation based on NSM, we considered that shipping water occurs when relative wave single amplitude excess the freeboard. From this figure, it can be seen that when the wave mean period is 3.84 seconds ,shipping water occurs at around 2 meters significant wave height.

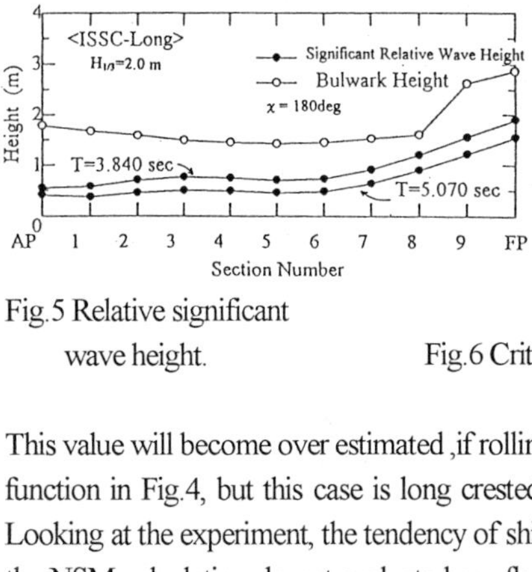

Fig.5 Relative significant wave height.

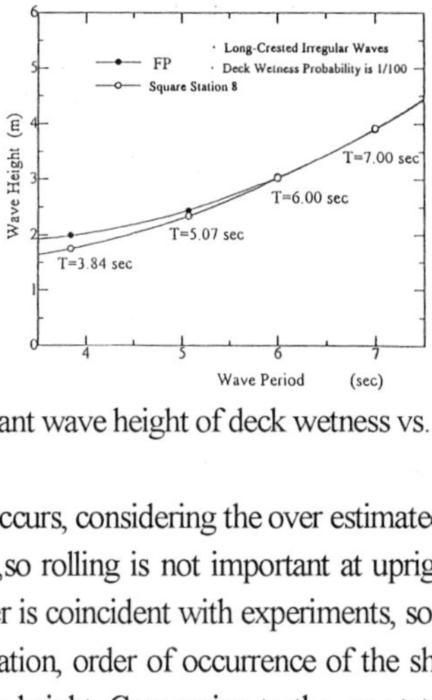

Fig.6 Critical significant wave height of deck wetness vs. wave period

This value will become over estimated ,if rolling motion occurs, considering the over estimated roll transfer function in Fig.4, but this case is long crested head sea ,so rolling is not important at upright condition. Looking at the experiment, the tendency of shipping water is coincident with experiments, so even though the NSM calculation do not evaluate bow flare configuration, order of occurrence of the shipping water seems to be estimated by NSM considering relative wave height. Concerning to the sea state with longer wave period without changing wave height, probability of shipping water is small, but of course it does not mean that shipping water never occurs.

Experiments in multidirectional irregular waves

Considering the situation that no capsizing occurred in the case of long crested irregular waves even shipping water was observed, we introduced another effect and confirmed by experiments. First of all, we introduced short crested irregular waves which generate rolling even in head seas, and also introduced the mechanism that can reappear the net shift phenomena as already described . Shifted weight is about 20 tons (12% of displacement) in full scale and 1.71kg in model scale. A triangular trigger is inserted for automatic movement of weight when the heel angle over the specified value . Another trigger that can be controlled arbitrary by manual remote control using radio was also useed. After weight is shifted, 13 degrees heel occurs when balanced.

For estimating the capsizing condition of actual ship including net shifting phenomena, model ship condition was changed systematically, as follows.

(a) The KG value was changed by five steps beyond the estimated value of capsize occurred.

(b) The shifting weight condition considered were three.

(1) Fixed on the center line of ship.

(2) Movement was started by static heel over 20 degrees.

(3)Movement was started at arbitrary timing by manual moving of trigger through radio control.

(c) Freeing port was both open or closed.

(d) Significant wave height were 1.54m and 2.16m in full scale, considering the estimated wave height at capsizing, mean wave period was also selected as 5 seconds in full scale. Measured directional wave spectra are shown in Fig.7 by model scale.

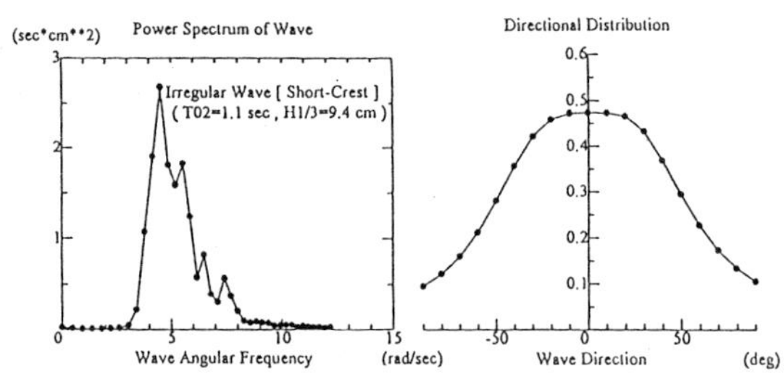

Fig.7 Measured directional wave spectrum used for experiments

During measurements, changing the combination of those four conditions, we could catch the capsizing. Time histories of motion ,rudder, heading angle and ship trajectory are shown in Fig.8. After the net is shifted ,shipping water occurred and finally capsizing occurred.

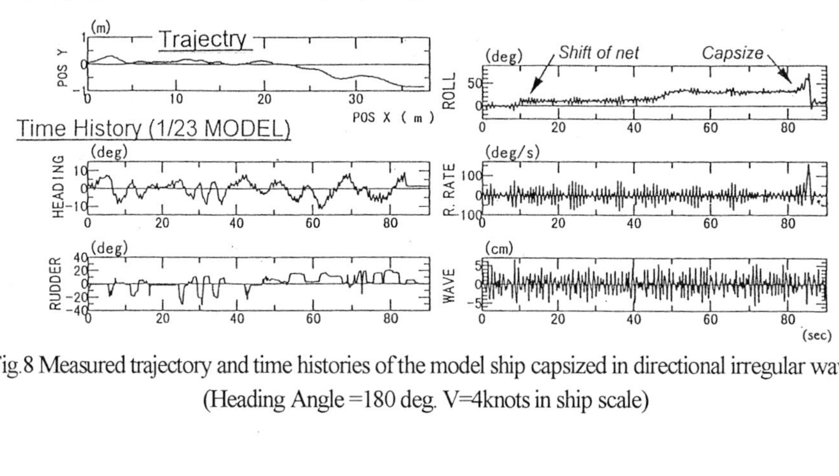

Fig.8 Measured trajectory and time histories of the model ship capsized in directional irregular waves (Heading Angle =180 deg. V=4knots in ship scale)

As already described, prepared model could not perfectly reappear the specified full scale condition as capsizing occurred, but by experiments in around the estimated sea condition, capsizing occurred only at the condition that KG is 15% or more larger than specified one , and at the condition that the weight shifting ,namely net shifting, occurred by the heel from accumulated water. In that condition, shipping water often occurred, so the capsizing process of this ship vessel in head sea condition was estimated as follows (see Fig.9)

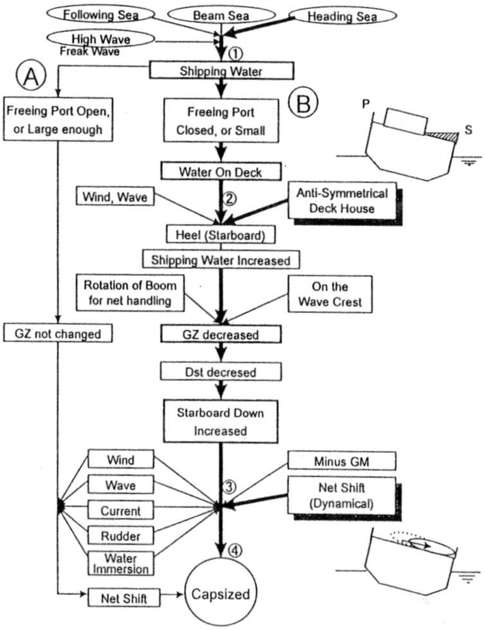

Fig.9 Possible process to capsizing in head waves. Thick lines show the estimated process. Thick lines show the estimated process.

In head seas, large relative wave height cause shipping water. Even if the capacity of freeing port is enough at design stage, there was some possibility of reducing that capacity by the barrier on the deck, and by this reason not-discharged water was gathered to starboard side, because of shifted deckhouse effect. As the results ,heel of starboard down occurs. By the heel, asymmetric roll motion excited, and fishing nets start to shift to starboard. By the movement of fishing nets, additional shipping water occurs, and finally make large heel and resulted in capsizing. At this stage, if there are openings on the deck or deck house, capsizing will be accelerated.

In the Fig.9, possible process to capsizing was also shown, but the estimated results by experiments was shown by thick allows. Process to capsizing include many elements but freeing port capacity and the possibility of nets shift ,in another words non-lashing of fishing nets, are seemed most important.

NUMERICAL SIMULATION

For evaluating the effect of maneuvering on roll or capsizing of this ship, physical simulation was not enough because of the model used was so small and some possibility of disturbances by electrical cable. So, we also carried out numerical simulation in regular waves including rudder effects. This simulation was also planned for detecting capsizing condition from series calculation.

In order to treat large roll motion ,we evaluated fluid pressure at instantaneous wetted surface by wave and ship motions. As the pressure by water, only Froude-Krylov force was taken into account, because the specified wave length of this case was relatively long comparing to ship length.

This assumption seems realistic looking at the results of numerical simulation compared with experiments. Added mass and damping coefficient are obtained from NSM and Motora's Charts(1959). Derivatives for maneuvering equation of motion was quoted from that of similar ship by Karasuno (1990), but some tuning was done to meet with her basic characteristics in still water experiments.

In case of the bulwark was immersed, buoyancy of bulwark or the thickness of the bulwark was neglected considering full scale ship, and the effect of accumulated water on deck was taken into account by equivalent weight shifting.

Equation of motion including maneuver

For expressing ship motion including maneuvering motion, we adopted so called "Horizontal Body Axes" system introduced by Hamamoto et al.(1993). The origin is taken at the center of gravity of the ship, and moves according to ship's heaving ,surging, swaying and yawing keeping initial horizontal plane as horizontal. Pitching and rolling are expressed by the rotation relative to this horizontal plane.

This system combine the both utilize conventional co-ordinate system for maneuvering and seakeeping. In the equation of motions ,interacting terms appear ,but those terms can be evaluated by conventional information both from maneuverability and seakeeping field .

Results of numerical simulation

For confirming the calculation, comparison was made with experiments in still water. Results were all shown in full scale. Comparing to the physical simulation, numerical simulation in still water showed a little deviation in heading angle, and smaller rolling angle, but it will be said that even though relatively large and quick movement of rudder angle, excited roll angle is very small. Furthermore it will be said that this numerical simulation is relatively reliable.

Fig.10 show the comparison with experiment in turning in regular waves (Hw=1.85m,T=7sec,V=4knots). Initial heading is head sea, and then moved rudder to 35 degrees like step function. Phase shift of calculated rolling or pitching angles from experiment are seen because of heading angle time history is different, so the value of corresponding heading angles are written in this figure. If we look at the roll angle at the same heading angle, both calculation and experiment show good coincidence. This means that the Froude-Krylov force is appropriate as external force. This case, excited roll angle is large but not enough to capsize.

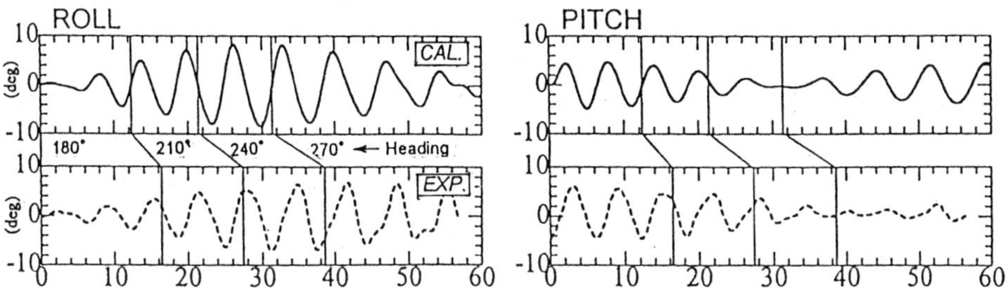

Fig.10 Measured and simulated ship turning motion in regular waves.
(Hw=1.85m,Tw=7sec,Initial heading angle=180deg, Rudder angle=35deg, V=4knots)

Fig.11 shows the change of the attitude of this ship. The condition is the same as Fig.10. The cross section the ship is expressed by wire frame.

Around the specified sea condition, numerical simulation seems reliable ,so we used our numerical simulation code for the evaluation of capsizing process estimated from experiments.
In heading regular waves, parametric roll oscillation will occur by the change of GM in wave trough or crest . For the case of relatively long wave like this time, this change become small and wave period do not fulfill the parametric resonance condition.

Another possible phenomena is excitation of asymmetric roll by the coupling of shifted weight or shifted shipping water and heaving. Initial heel is 9 degrees(starboard down) by shifted weight ,and this corresponds to the effect of shipping water. This time, also the good coincidence between calculation and experiment can be seen, but not enough to capsizing.

Next, according to the estimated process to capsizing, we simulated shipping water effect and net shifting effect by the transverse movement of a weight on the deck. Initial heel by shipping water is set about 9 degrees and the heel angle that the weight start to move was set as the same as that of experiment (20 degrees). Of course, this simplification cannot reappear the dynamic effect of
shipping water precisely.

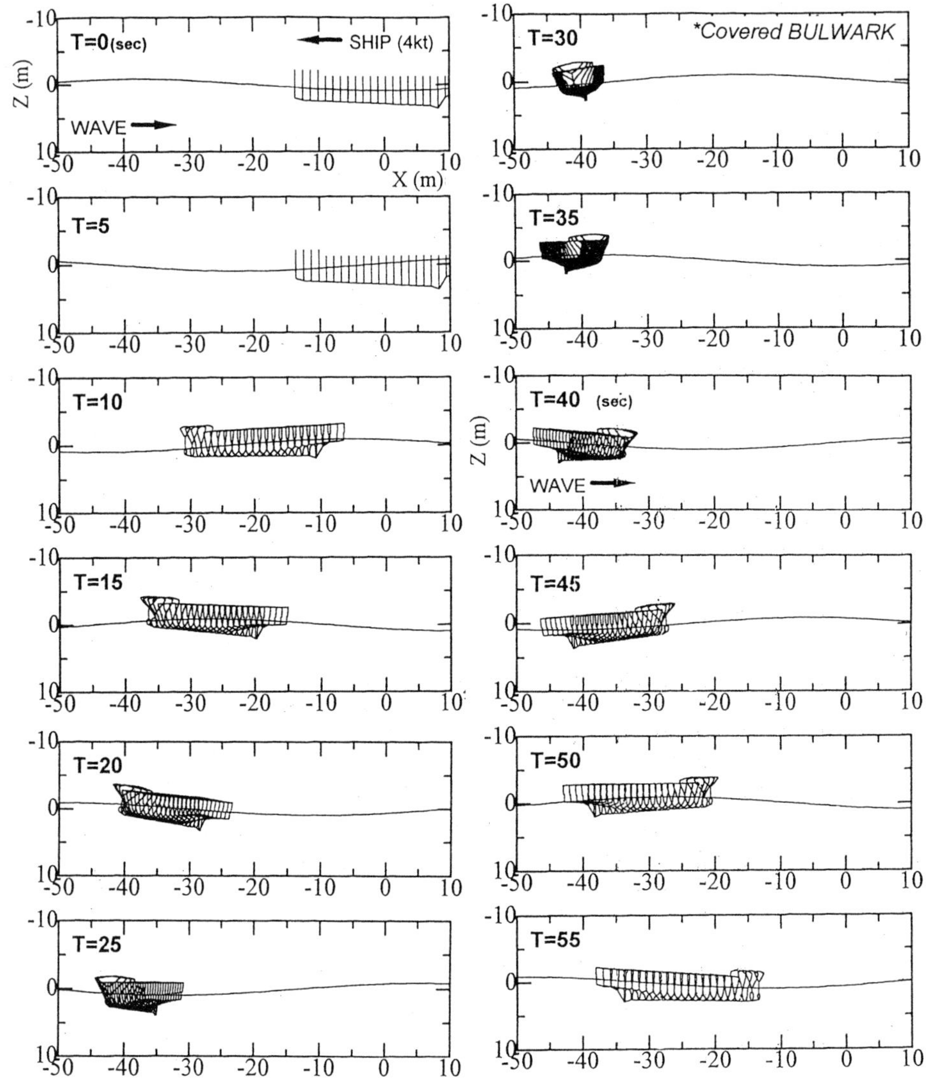

Fig.11 Ship turning motion in regular waves by numerical simulation corresponding to the case of Fig.10 (Rudder fixed (35deg), Initial heading angle=180deg. Initial ship speed =4 knots., Tw=7sec, Hw=1.85m)

In the Fig.12, 7 seconds wave correspond to the specified condition but capsizing did not occur, because further weight shift correspond to net shift did no occur. But in the case of 5 seconds wave, large roll motion excited and weight shift occurred, and finally capsized. Wave condition that capsize occurred is not that of specified, but this wave condition was the same as capsized condition in physical experiment. The main reason will be that the model condition was not the same as that of specified from actual ship.

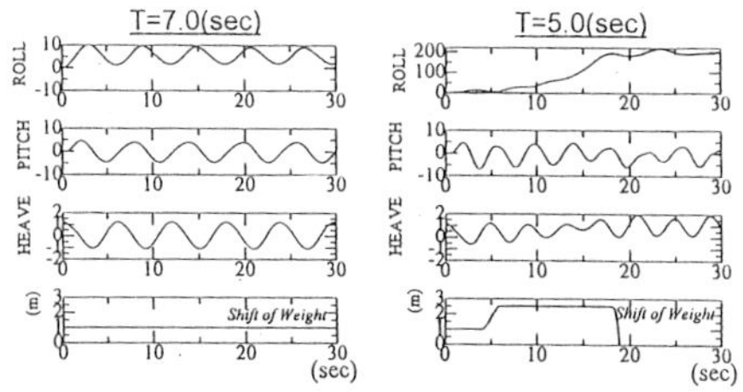

Fig.12 Numerical simulation of rolling in regular head waves with weight shifting.
(Hw=2.1m, Tw=7sec(left), 5sec(right, capsized))

As the same results was obtained comparing to experiment, we carried out systematic calculation and obtained the critical combination of H(m) and GM(m) that tend to capsizing. This result is shown in Fig.13. Experimental condition means that the gyradius was used as that of experiment. Arrow point was confirmed in experiment. From this ,for specified condition, capsizing will occur in smaller wave height than that of experimental condition.

Estimated GM at capsized ship was 1.45 m, and so if the wave period is 6 sec, then the critical wave height become about 2m ,and this coincide with that of specified by the information of weather station with wave sensor. From these physical and numerical simulation ,estimated process to capsizing seems to be confirmed to some extent.

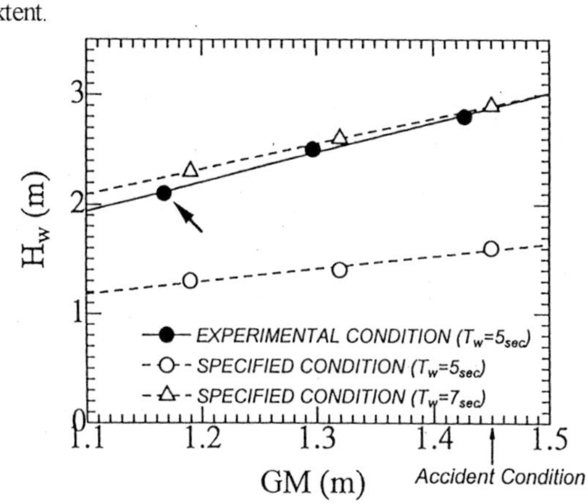

Fig.13 Critical regular wave height for capsizing estimated from numerical simulation and experiments

CONCLUSION

In this paper, we studied about the possible process of capsizing of a fishing vessel in head waves, both by physical and numerical simulation including the effect of maneuver. Summarizing this study, we can introduce the following conclusions.

1) Comparing to large vessel, the influences of shipping water on deck and succeeding cargo shift on ship motions are large to small vessel. So, considering those effect, we could estimate a possible capsizing process of the given ship.
2) Concerning to this ship, shifted deck house arrangement to port side and some prevention to freeing port capacity initialized the heel of starboard down. So, the similar fishing vessel have some dangerous tendency.
3) For numerical simulation, we adopted the so called Horizontal-Body-Axis System for constructing equation of motion both including maneuvering and seakeeping motion. This was introduced by Hamamoto et al (1993). From the numerical simulation, rapid maneuver did not cause large rolling for the given ship.
4) In our numerical simulation, shipping water on deck and net shift was simulated by the weight shift. From this simulation, we could confirm the capsizing process estimated from experiment, and also presented the critical wave height tend to capsizing at the given GM, by numerical simulation.

Acknowledgment: The authors want express their gratitude to Mr. K. Miyakawa and T. Takayama who managed difficult experiments, and to the graduated student Mr. M. Fukushima and those other students who contributed to this experiment. Furthermore to associate prof. N.Ma who gave daily advise and support to the students.

References

Hirayama,T.(1983).Experimental Study on the Probability of Capsizing of a Fishing Vessel in Beam Irregular Waves, *Journal of The Society of Naval Architects of Japan*, **154**, 173- 184

Hirayama,T. et.al.(1985).On the Capsizing Process of a Fishing Vessel in Breaking Waves, *Journal of The Kansai Society of Naval Architects*, Japan, **196**,19-30 (in Japanese)

Hamamoto,M & Kim,Y.S.(1993).A New Coordinate System and the Equations Describing Maneuvering Motion of a Ship in Waves, *Journal of The Society of Naval Architects of Japan,* **173**, 209-220 (in Japanese)

Hirayama,T., et.al.(1994). Capsizing and Restoring Characteristics of a Sailing Yacht in Oblique and Breaking Waves, *Journal of The Kansai Society of Naval Architects, Japan,* **221**,117-122 (in Japanese)

Hirayama,T. and Nishimura,K.(1997). Study on Capsizing Process and Numerical Simulation of a Fishing Boat in Heading Waves, *Journal of The Society of Naval Architects of Japan,* **181**,169-180 (in Japanese)

Karasuno,K., et al.(1990).Physical-Mathematical Models of Hydro-or-Aero-Dynamic Forces Acting on Ship Moving in Oblique Direction, *MARSIM & ICSM'90*, 393-400

Motora,S.(1959).On the Measurement of Added Mass and Added Moment of Inertia for Ship Motions (Part1-3) , *Journal of The Society of Naval Architects of Japan*, **105** (Part 1,.83- 92),**106** (Part 2,59-62, Part 3, 63-68) (in Japanese)

Takezawa, S .and Hirayama,T.(1971).On the generation of Arbitrary Transient Water Waves, *Journal of The Society of Naval Architects of Japan,* **129**, (in Japanese)

D. Vassalos, M. Hamamoto, A. Papanikolaou and D. Molyneux (Editors)

THE INFLUENCE OF LIQUID CARGO DYNAMICS ON SHIP STABILITY

N.N.Rakhmanin and S.G.Zhivitsa

Krylov Shipbuilding Research Institute
St.-Petersburg 196158, Russia

ABSTRACT

This paper presents the analysis of the heeling moment caused by a liquid cargo sloshing in conditions of regular forced angular oscillations of a compartment and has the purpose of demonstrating the liquid free surface dynamics influence on the stability of a ship performing steady motions in a seaway. The already known formerly received results concerning the dynamics of a ship with liquid on board were used .

KEYWORDS

Liquid on board, ship motions, liquid sloshing, heeling moment, ship stability, restoring moment.

INTRODUCTION

Two problems are to be solved when the motions of a ship with liquid cargo in a seaway are examined. One problem is to describe the liquid oscillations in a compartment, when the latter motion is given (sloshing problem). Another problem deals with formulating and solving the equations, that reflect combined ship and liquid oscillations during motions in a seaway.

The first problem is a key one. There are two principal approaches to its solution. One of them is based upon using the method of liquid motion velocity potential for description of liquid volume oscillations, Moiseev(1953). Kochin, Khibel & Rose(1955); The second approach is connected with numerical integrating of the equations of hydrodynamics,Armenio(1997). In both cases the boundary conditions on the walls of a mobile compartment and on the free surface are satisfied, and also the condition of liquid continuity is fulfilled.

Each approach has its own merits and drawbacks. The potential method permits to present the problem solution in the analytical form and makes possible to consider the problem of a ship with liquid on board analytically as a whole. Numerical methods are convenient for constructing the free surface form in any moment of time, for calculation of local pressures and impact loads under the conditions of liquid sloshing within a compartment, the latter is important for the compartment constructive element local strength evaluation. In addition the solution of the problem can be obtained without any limitations for the range of main parameter alteration, namely: the relative depth (f/b_t) and the relative liquid sloshing amplitude. Such limitations exist in the potential method.

In present time the analytical method, in which the liquid velocity potential is constructed using the main coordinates of the compartment liquid sloshing problem, is considered to be the most developed and convenient from the point of view of its employment for practical calculations, Moiseev(1953); Maltsev(1962). As an example of such coordinates a free surface wave slope angle for arbitrary principle mode of liquid sloshing could be taken. The number of modes in this case will be determined by the number of free surface fixed points, Kochin, Khibel & Rose(1955), that are observed during its oscillations. In any choice the main coordinates, being independent, permit the presentation of the velocity potential and the liquid free surface as infinite series of the eigen functions of the problem with variable in time coefficients, which are determined by the solution of a system of differential equations that reflect ship motions and liquid sloshing in a closed form, Rakhmanin & Zhivitsa(1994). In practice the specific features of the problem permit to evaluate the ship dynamics with preserving only the first principle form (mode) of the liquid sloshing. Further on, the liquid dynamics influence on the ship stability is considered with the above-mentioned approximation.

THE HEELING MOMENT INDUCED BY THE LIQUID

If we confine to the angular oscillations of the compartment around a fixed axis, it could be shown that the total heeling moment caused by free surface liquid cargo mobility consists of three components

$$M_{Sl}\left(\phi;\dot{\phi};\ddot{\phi}\right)=M_1(\phi)+M_2(\dot{\phi})+M_3(\ddot{\phi}) \tag{1}$$

The first component $M_1(\phi)$ is related with the compartment angle of inclination; it alters in time in the same phase with this angle and is caused by the liquid cargo centre of gravity shifting in the direction of the board inclination due to liquid cross-flow, which manifests itself outwardly in the form of free surface plane rotation relatively to the compartment walls, which is characteristic for the first oscillation mode. The third component $M_3(\ddot{\phi})$ is caused by liquid mass inertia; it is manifested in its centre of gravity additional shift in the plane of inclination and it may alter in time either in opposite phase with free surface slope angle or in the same phase depending on mutual position of liquid centre of gravity in the state of rest and the axis of compartment rotation. The second component $M_2(\dot{\phi})$ is related with the angular velocity, i.e. alters in phase with the free surface slope rate of rotation and characterises the degree of liquid cargo flow lagging behind the compartment inclination process. This component is explained by the energy loss of liquid volume motion due to the friction forces and due to additional wave mode formation, that shows itself in free surface curvature.

Experimental data, Rakhmanin(1966); Van den Bosch & de Zwaan(1970), and numerical calculations, Armenio(1997), show that in common case the free surface form and the heeling moment coming from liquid cross-flow during the compartment oscillation manifest strong features of non-linearity. Nevertheless it follows from the experiments with models of ships in flooded conditions, Rakhmanin(1962), that the non-linearity of the moment created by a liquid cargo practically does not influence upon the linearity of the whole system. What is more, specific features of liquid sloshing in a closed compartment are of such kind, that it is possible to calculate liquid heeling moment within the limits of linear scheme, taking into consideration only the first principle mode of a liquid volume oscillations, which is characterised by a free surface flat plane rotation. It appears to be possible to connect the damping of such oscillations with liquid mass energy loss for wave mode formation of the second and higher orders, Rakhmanin & Zhivitsa(1994).

All mentioned above is rather convincingly illustrated with Figure 1, which shows the dependence of non-dimensional total heeling moment amplitude M_{S10} due to water in a compartment that performs angular oscillations around the fixed axis of rotation:

$$C_m = \frac{M_{S10}}{\rho g V_t b_t \phi_0} \tag{2}$$

for given frequency $\omega = 0{,}6\sigma_1$ and relative liquid depth $f = 0{,}25\ b_t$, but for different positions of the rotation axis relative to the bottom of a compartment. In formula (2) the used symbols designate: ϕ_0 - compartment angular amplitude; b_t - compartment breadth; f - liquid depth; V_t - liquid volume; σ_1 - first liquid sloshing mode natural frequency and ρ - and g - correspondingly liquid mass density and specific gravity force.

As it is seen from Figure 1 the completely non-linear method, Armenio(1997), based on straight solution of Navier-Stokes equations in Reynolds averaged form, leads to the same results as the linear method based on the above mentioned scheme. The calculation results are presented for two different amplitudes of angular oscillations and demonstrate that the moment coefficient weakly depends upon the amplitude. On the contrary, alteration of the mutual position of liquid mass centre and compartment rotation axis influences rather essentially on the value of the total heeling moment. Besides, liquid mass location below the rotation axis leads to the moment coefficient decrease, and raising above this axis noticeably

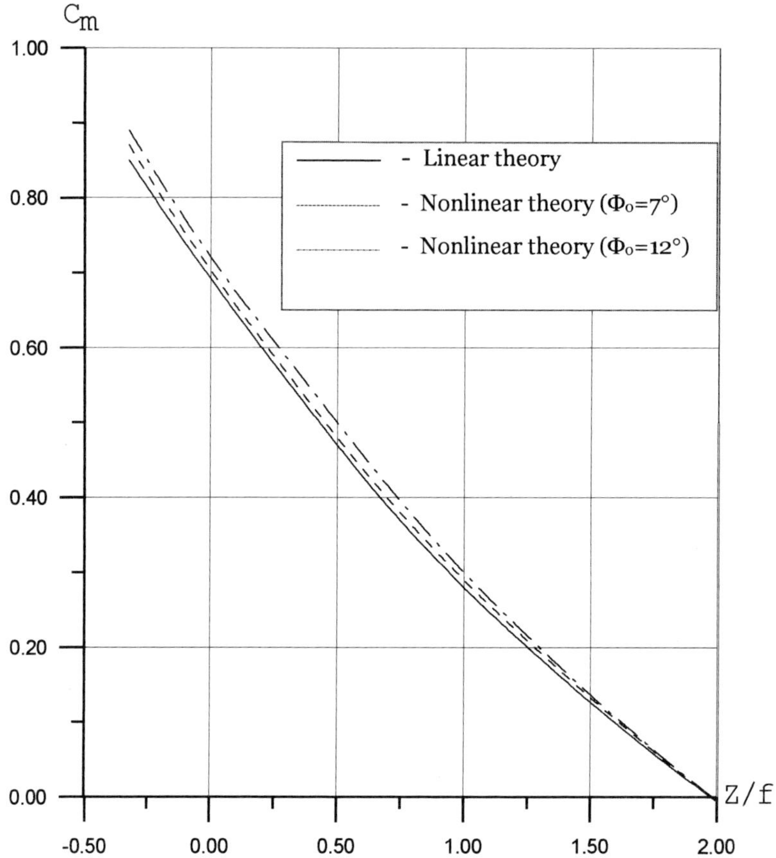

Figure 1. Non-dimensional liquid heeling moment amplitude at angular compartment oscillations versus rotation axis position.

increases the coefficient. For example, when rotation axis coincides with a compartment bottom ($Z/f=0$) the heeling moment is more than two times higher than its value for the case of axis location in the free surface plane ($Z/f=1$).

THE RESTORING MOMENT OF A SHIP WITH LIQUID ON BOARD

Influence of liquid cargo on stability of a ship in conditions of motion may be taken into consideration within the limits of linear approximation by analogy with hydrostatics using an increment for the metacentric height. Transformation of dynamics equations for a ship with liquid on board, Rakhmanin & Zhivitsa(1994), permits to present restoring moment in the form of the following generalised metacentric formula

$$M_{Stab}(\phi) = \gamma V(GM_0 - \Delta GM) \cdot \phi \tag{3}$$

where V is volumetric ship displacement and GM_0 - metacentric height without accounting for liquid free surface influence. Liquid mobility finds its reflection in the expression for the metacentric height increment, which becomes a complex function of motion frequency. For a one-compartment case this generalised stability reduction increment will be determined by the formula

$$\Delta GM(\omega) = \Delta GM_1 \cdot \left(1 + \frac{z_l \omega^2}{g}\right) \cdot K_1(\omega) \cdot \cos \varepsilon_1(\omega). \tag{4}$$

In the latter expression co-ordinate Z_l characterises the vertical distance between the liquid mass centre and the compartment rotation axis, i.e. the ship's centre of gravity; the values $K_1(\omega)$ and $\varepsilon_1(\omega)$ designate correspondingly
- the amplitude - frequency characteristic:

$$K_1(\omega) = \frac{\sigma_1^2}{\sqrt{(\sigma_1^2 - \omega^2) + 4\nu_1^2 \omega^2}} \tag{5}$$

- and the phase - frequency characteristic:

$$\mathrm{tg}\,\varepsilon_1 = \frac{2\nu_1 \omega}{\sigma_1^2 - \omega^2} \tag{6}$$

of the first mode of liquid volume oscillations in case when the axis of compartment rotation iscoincident with the liquid centre of mass at standstill. In formulae (5) and (6) symbols σ_1 and $2\nu_1$ represent the first mass natural frequency and the linear damping dimensional coefficient, which are determined independent of absolute and relative dimensions of the volume occupied with liquid. Besides, coefficient $2\nu_1$ essentially depends upon the amplitude of angular oscillations and upon the compartment permeability coefficient, Rakhmanin(1966); Rakhmanin & Zhivitsa(1994). The value of ΔGM_1 designates hydrostatic metacentric height increment

$$\Delta GM_1 = \frac{i_x}{V},$$

where i_x is the intrinsic moment of liquid free surface area inertia.

In the limiting cases evident and well-known conclusions follow from formulae (4)-(6). For low frequency ship motion $\omega \to 0$, when it may be considered $\frac{z_l \omega^2}{g} << 1$ and $\omega << \sigma_1$, and consequently $K_1(\omega) = \cos\varepsilon_1 \to 1$, the generalised increment for a liquid free surface with consideration of its dynamics comes to a hydrostatic one

$$\Delta GM \cong \Delta GM_1.$$

In case of resonant liquid sloshing when $\omega = \sigma_1$ we get $\varepsilon_1 = \pi/2$, $\cos\varepsilon_1 = 0$ and correspondingly the generalised increment

$$\Delta GM = 0.$$

In the general case the generalised increment (4) has a complex character as a frequency function, that is very sensitive to the alterations of the compartment permeability coefficient and the compartment vertical position on the ship. The above said is illustrated by Figure 2 and Figure 3 where expression (4) is given in non-dimensional diagrams, in which the generalised increment ΔGM is represented in parts of the static one ΔGM_1, and ship motion frequency ω relates to the liquid volume natural frequency in the first mode σ_1.

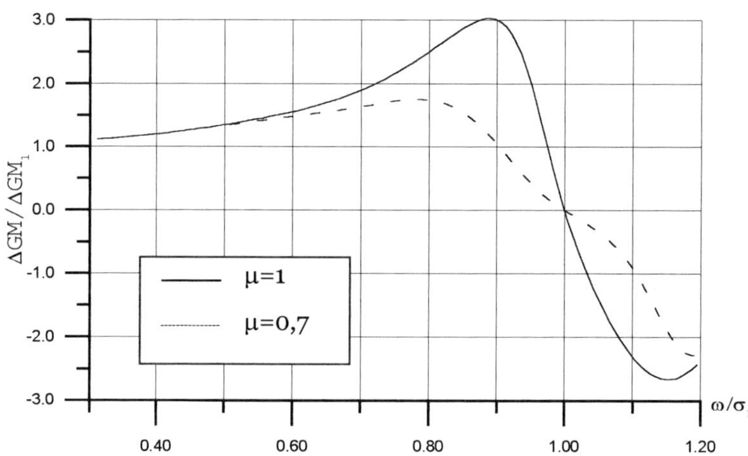

Figure 2. The generalised metacentric height versus the rolling motion relative frequency
- The influence of permeability coefficient (μ)

It is characteristic that in comparatively large relative frequency range the dynamic increment for free surface exceeds the hydrostatic one by more than one and a half times.

The reason for it is that in some range of frequencies close to resonance of liquid volume oscillations the free surface rotates relative to the compartment for an angle that exceeds the angle of compartment inclination. As a consequence the liquid mass centre shifts and the heeling moment due to liquid cross - flow on board becomes noticeably larger than in

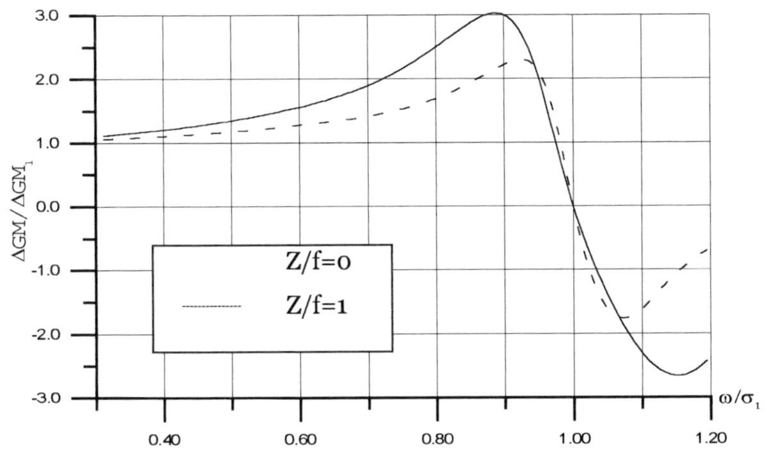

Figure 3. The generalised metacentric height versus the rolling motion relative frequency
- The influence of the vertical compartment position on a ship (Z)

hydrostatics consideration. The above mentioned has clearly manifested itself in the Netherlands experiments, Van den Bosch& de Zwaan(1970).

ON THE PRACTICAL UTILIZATION OF THE RESULTS DISCUSSED

It is evident that ship motions is the natural reason inducing liquid sloshing in a compartment. Consequently, if on the diagrams (Figures 2 and 3) it is shown the rolling motion frequency range, which corresponds to roll amplitudes of a given level (for example not less than an average wave slope angle α_0),then one may judge depending upon the position and length of this range on the relative frequencies ω/σ_1 scale about how much the influence of liquid cargo dynamics in the examined compartment is essential for the stability of a ship sailing in a seaway. The specified rolling motion frequency range depends from one hand upon the spectral frequencies range (periods), that ship may meet at sea, and from the other hand, depends upon natural periods of roll and the first mode of liquid sloshing in the compartment.

The spectral period's range does not correlate with the ship dimensions and is determined from the statistic data about the wave climate in the regions of ship navigation. For example we may refer to the wave statistic data from the reference book of the Shipping Register of the USSR "Winds and Waves at Oceans and Seas"(1974) according to which 80% of time of the ship navigation at open sea is connected with Sea state 3 to 6. At these Sea states the range of prevailing spectral periods spreads from 2,5 to 15 seconds. As for liquid sloshing and rolling motion natural periods, they can be easily calculated according to the Reference book for Ship Theory(1985) or according to other sources, if the dimensions of the free surface, the relative liquid depth and also the dimensions and loading of a ship are known. In this case the band of spectral periods filtered by a ship is established from the condition

$$0.75 < T / T_{\phi} < 1.33,$$

according to which the relative rolling motion amplitude ϕ_0 / α_0 does not decrease below 1. Symbol T_{ϕ} above designates natural roll period of a ship. When these given data are present the calculation of rolling motion frequency range for the relative frequencies ω/σ_1 scale can be easily done, if the period of liquid cargo free oscillations is known.

To orientate the reader the following Table 1 demonstrates the results of an exclusively approximate evaluation of the discussed forced rolling motion frequency range for three dimensional categories of ships and two cases of the liquid mass (cargo or flooded water) vertical position at an up-right ship.

TABLE 1
CHARACTERISTIC PERIODS AND FREQUENCIES FOR A SHIP AND LIQUID IN A FLOODED COMPARTMENT

Ship category	Large Ships	Medium Ships	Small Ships
Ship Length, m	150 ÷ 250	75 ÷ 140	25 ÷ 60
Characteristics of ship rolling motion and liquid sloshing:			
Rolling motion natural period, sec	~ 17	~ 13	~ 7
Forced motion periods range, sec	12,7 ÷ 15	9,7 ÷ 15	5 ÷ 9
Liquid natural sloshing period:			
a) Low position of liquid cargoes, sec	5,0	3,9	2,8
b) High position of liquid cargoes, sec	8,9	6,9	4,9
Forced rolling motion relative frequency band:			
case (a)	0,33 ÷ 0,39	0,26 ÷ 0,40	0,31 ÷ 0,56
case (b)	0,59 ÷ 0,70	0,46 ÷ 0,79	0,54 ÷0,98

The first case (a) is characteristic for a ship in full loading condition, when operational liquids were partly (not more than one half) used and have the largest free surface areas. From the fluid dynamics point of view the low position of liquid cargo, when its main mass is located below the rolling axis of a ship, appears as an important fact for this case. This answers the requirements of designing in traditions of good sea practice, for it leads to essential decrease of the dynamic effect from liquid cross-flow (see Figure 1). Damaged ships with flooded compartments, when the compartment centre locates below the ship centre of gravity, may be classified under this case.

The second case (b) is characteristic for damaged flooding of high-positioned ship spaces as Ro-Ro ferry cargo deck, hangar and other upper decks, and finally open upper decks limited with high bulwark and so on. For such cases practically the whole liquid mass is located higher than the rolling axis of a ship, and its relative depth usually makes not more than 20% of the flooded volume breadth.

Figure 4 shows the correlation of the calculated rolling motion relative frequencies characteristic ranges for ships of the above mentioned categories with generalised stability increment diagram. As it could be seen, for the first case of liquid position on board (Figure 4a), which is characteristic for undamaged ships first of all, the difference between the generalised and the static increments for free surface does not exceed 10% for large and medium vessels and 25% - for small vessels.

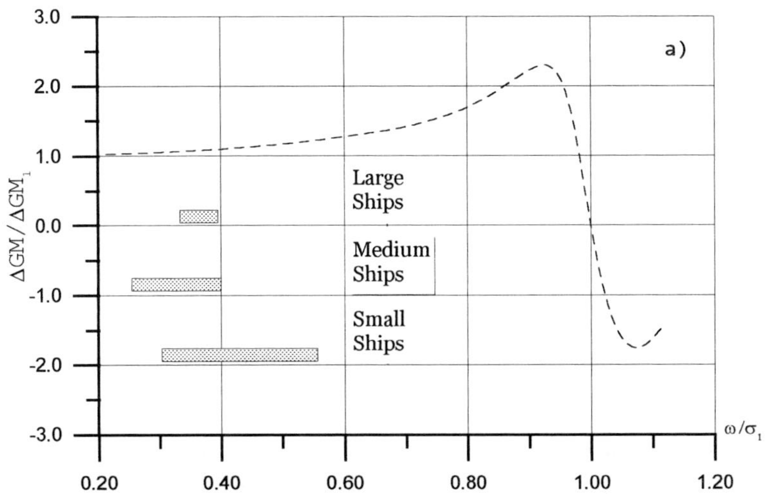

Figure 4. Rolling motion relative frequency band in comparison with generalised metacentric height increment diagram
a) Low position of liquid mass

At high positions of liquid mass (Figure 4b) which is characteristic mainly for severe cases of ship flooding, the situation changes for the worst because the dynamic effect of the liquid sloshing noticeably increases the heeling effect. The generalised increment for large and medium ships can increase for 80-100% in comparison with the static increment, and for smaller ships this value can even be more than that.

The presented results mean that during the motions in a seaway the liquid sloshing in a compartment increasing the intensity of liquid cross-flow from board to board will lead to decrease of ship's resistance to the effect of external heeling moments. This decrease will be especially noticeable for the ships in a damaged condition with high positioned partially flooded decks or platforms. In a low Sea state this danger can at first place show itself in

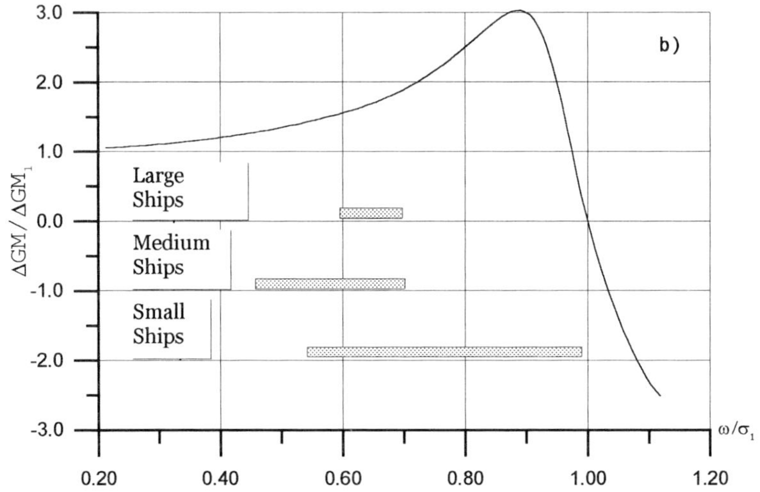

Figure 4. Rolling motion relative frequency band in comparison with generalised metacentric height increment diagram
b) High position of liquid mass

manoeuvring of the ship in damage condition. The ideas discussed in the paper give the possibility to evaluate the degree of such danger in each specific case.

References

Armenio V.(1997). An Improved Mac Method (SIMAC) for Unsteady High Reynolds Free Surface Flows. *International Journal for Numerical Methods in Fluids* **24**, pp185-214.

Van den Bosch J. and de Zwaan(1970). Roll Damping by Free Surface Tanks with Partially Raised Bottom. *Technical Report* **280**, *Delft Technical University,* Netherlands, 134p.

Kochin N.E., Khibel I.A. and Roze N.V.(1955). *Theoretical Hydromechanics, part* **1**, Moscow: Gosizdat, TTL, 550p.

Maltsev N.Y.(1962). On the Question of Dynamics of a Ship with Liquid Cargoes. *Collection of the USSR Academy of Science "Variation Methods in the Problems of Liquid Sloshing and Motions of a Solid Body with Liquid, Moscow*, pp 237-246.

Moiseev N.N.(1953). The Problem of Motions of a Solid Body which Contains Liquid Masses with Free Surface. *Mathematics Collection of the USSR Academy of Science* **32(74)**, pp 61-96.

Rakhmanin N.N.(1962). Rolling Motion of a Ship with Compartments Partially Filled with Liquid. *Proceedings of Krylov Shipbuilding Research Institute* **191**, pp52-72.

Rakhmanin N.N.(1966). The Analysis of Hydrodynamic Pressures on the Walls of a Rectangular Compartment Performing Forced Motions with Liquid Cargo. *Proceedings of Krylov Shipbuilding Research Institute* **232**, pp96-113.

Rakhmanin N.N. and Zhivitsa S.G.(1994). Prediction of Motion of Ships with Flooded Compartments in a Seaway. *Proceedings of the 5th International Conference on Stability of Ships and Ocean Vehicles* **2**,*Florida,* pp103-120.

The Reference Book for Ship Theory(1985), vol.**2:** *Ship Hydrostatics and Motions,* Leningrad: Sudostroenie, 440p.

USSR Shipping Register(1974). *Wind and Waves in Ocean and Seas, the Reference Data,* Leningrad: Transport, 359p.

Contemporary Ideas on Ship Stability
D. Vassalos, M. Hamamoto, A. Papanikolaou and D. Molyneux (Editors)

EXPLORING THE POSSIBILITY OF STABILITY ASSESSMENT WITHOUT REFERENCE TO HYDROSTATIC DATA

Richard Birmingham

Department of Marine Technology
The University of Newcastle upon Tyne, England NE1 7RU

ABSTRACT

The initial stability of a vessel can be evaluated by the established procedure of the inclining experiment, provided that the vessel's lines or hydrostatics are available. For many small working craft however no record exists of these particulars, and therefore the inclining experiment can only be undertaken if it is accepted that the lines will also have to be taken from the vessel in order to generate the necessary data. This paper explores the relationship between the roll period and the metacentric height and demonstrates that in theory it is possible to evaluate the initial stability from the change in roll period as a pair of weights are moved vertically and horizontally above the deck, without requiring reference to any hydrostatic data. A procedure requiring only five observations is described, and the necessary calculations detailed, however the difficulties in implementing such a procedure are also discussed. Further work to identify extensions to this procedure which could be of practical use to a surveyor are outlined.

KEYWORDS

Stability evaluation, inclining experiment, roll period, moment of inertia, initial stability, metacentre, small craft, working craft, fishing vessels.

INTRODUCTION

The conventional procedure for evaluating a vessel's initial stability is the inclining experiment, in which weights are moved across the deck and the angle of heel measured. By reference to the hydrostatic particulars of the vessel in the trial condition the metacentric

TABLE 1
NOMENCLATURE

B	breadth of vessel	k	y offset of ellipse centre
d	vertical distance weights moved	m	combined mass of weights
Δ	displacement of vessel	T	vessel's natural period of roll
f	a correlation factor	u	minor demi-axis of ellipse
g	acceleration due to gravity	v	major demi-axis of ellipse
$\overline{GM}$	metacentric height (initial stability)	x	horizontal co-ordinate
h	x offset of ellipse centre	y	vertical co-ordinate
I'	effective mass moment of inertia		

height and position of the centre of gravity can be found. However for many small craft, such as fishing and other work boats, neither the hull lines nor the hydrostatic data are available. The results of an inclining experiment can only be interpreted, therefore, if the lines are also taken off the vessel and the necessary data derived. Clearly this adds considerably to the cost and practicality of the inclining experiment as a way of assessing the initial stability for such craft.

Methods for obtaining an estimate of initial stability for such vessels, based on the natural period of roll, have long been recognised. These rely on the use of a stopwatch to measure the roll period, and assumptions concerning the transverse mass moment of inertia of the vessel. In the first part of this paper it will demonstrated however, that in theory it is possible to derive the vessel's initial stability (together with the vertical centre of gravity and displacement) by a direct calculation which uses as its only input the periods of roll as a pair of weights are positioned in pre-defined locations on and above the deck. The success of such a procedure depends on the ability to measure the period of roll to a high degree of accuracy, and while modern computer technology provides more than adequate precision, interference from hydrodynamic factors, and the associated difficulty of defining an effective mass moment of inertia, have impeded a practical implementation. These difficulties are described in the latter part of this paper, which culminates with a discussion of how further developments may lead to a practical procedure that could be adopted by surveyors.

INITIAL STABILITY AND THE PERIOD OF ROLL

The link between stability and a vessel's natural period of roll has been acknowledged for several centuries, indeed it was reported that in 1628 a roll experiment aboard the *Vasa* was halted due to the alarming results (Marchaj, 1986). Unfortunately this did not prevent the vessel's loss due to capsize on her maiden voyage.

The period of roll is still used today to assess the stability of fishing vessels. For example the Marine Safety Agency suggests that by obtaining an average time for the period of roll (T) by stopwatch, the metacentric height of a vessel ($\overline{GM}$) can be estimated from the following formula (HMSO, 1993):

$$T = \frac{fB}{\sqrt{\overline{GM}}} \tag{1}$$

This expression uses the breadth of the vessel (*B*), and a correlation factor (*f*). The value of the correlation factor is dependent on the vessel type. In metric units it varies form 0.95 for double boomed beam trawlers, to 0.60 for vessel's with a live fish well. If an appropriate value of the correlation factor can be identified, then the initial stability $\overline{GM}$ can be estimated from the results of a basic roll test. While this may be satisfactory in many cases, it can be envisaged that on occasion there could be considerable error, especially where the vessel does not conform to a recognised class of craft, or where it has been substantially modified.

It is possible however to define the precise relationship between stability and the period of roll which can be defined by considering the righting moment at small angles of heel, and assuming simple harmonic motion. It can then be shown (Muckle and Taylor, 1987) that the period of roll, *T*, is given by:

$$T = \frac{2\pi\sqrt{I'}}{\sqrt{g\Delta\overline{GM}}} \tag{2}$$

In this expression there are three vessel related variables: the displacement (Δ), the effective transverse mass moment of inertia, which includes the effects of added mass (I'), and the metacentric height ($\overline{GM}$). The last of these is considered the measure of initial stability, and it is the relationship of this with the period of roll, *T*, which is of interest. However it is the presence of the other two variables that complicates the direct assessment of stability from the period of roll. (It is worth noting that Equations 1 and 2 are of the same form if it is accepted that the radius of gyration is a function of the vessel's beam).

GEOMETRY AND THE PERIOD OF ROLL

In the inclining experiment weights are moved across a vessel's deck and the induced angle of heel noted. The movement of weights also impacts on the period of roll, but as an angle of heel is not desired for a roll based experiment pairs of weights must be moved symmetrically about the centre plane of the vessel. The period of roll will be affected by moving such pairs of weights horizontally away or toward the centre plane, or by moving them vertically, as both actions affect either the position of the vertical centre of gravity (and hence $\overline{GM}$), or the mass moment of inertia, or both. By considering how changes in the geometric position of a pair of such weights influences the period of roll, it is possible to devise a procedure in which both the metacentric height and the mass moment of inertia can be changed in a systematic way. In theory the observed changes in the period of roll can then be used with Equation 2 to find the initial stability of the vessel, without recourse to correlation factors.

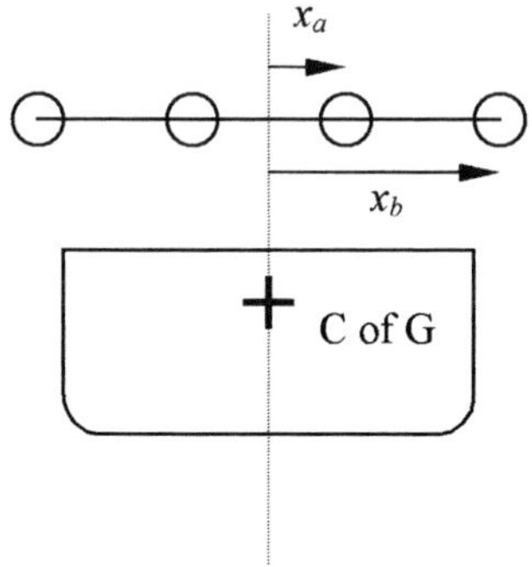

Figure 1: Horizontal definition for two positions of a pair of weights placed symmetrically about the centre plane.

Moving the weights only horizontally has no impact on the vertical centre of gravity (and therefore does not alter $\overline{GM}$), but it does alter the mass moment of inertia. By using Pythagoras's theorem it can be shown that regardless of where the vertical centre of gravity is located this change in the mass moment of inertia (I') is given by Equation 3, where the mass of the combined weights are m, and the dimensions x_a and x_b are as defined in Figure 1. Moving the weights away from the centre plane increases the mass moment of inertia, and vice versa.

$$\delta I' = m\left(x_b^2 - x_a^2\right) \tag{3}$$

Vertical movement of the weights clearly changes the vertical centre of gravity, and hence the metacentric height. This effect is given by Equation 4, where d is the vertical distance that the weights are moved:

$$\delta\overline{GM} = \frac{md}{\Delta} \tag{4}$$

A purely vertical movement of the weights also alters the mass moment of inertia, but it is impossible to establish the magnitude of this change without knowing the position of the vertical centre of gravity. It is evident however that a downward movement of the weights will reduce the mass moment of inertia, provided that the weights remain above the vessel's centre of gravity. As a movement of the weights outboard increases the mass moment of inertia it is possible to both lower the weights and simultaneously move them horizontally such that the net change in the mass moment of inertia is zero. For this to be the case as the weights are lowered they must follow an elliptical path, which starts on the centre plane, and is furthest form the centre plane when the weights are at the same height as the vessel's own centre of gravity, as shown in Figure 2.

The fact that the locus of the weight positions for constant I' follows an elliptical path can be used to derive another equation. If the centre of gravity did not change as the weights were

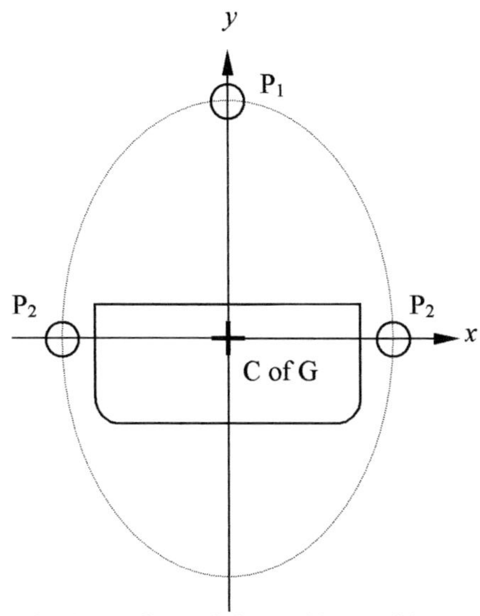

Figure 2: The locus for weight positions with a constant mass moment of inertia.

moved, then the locus for constant I' would describe a circle. The distortion of the circle into an ellipse is due to the downward drift of the centre of gravity as the weights are lowered. The difference between the major and minor demi-axes of the ellipse, v and u, is a measure of the change in the vertical position of the centre of gravity:

$$\delta \overline{GM} = v - u \tag{5}$$

THE BASIS OF DIRECT ASSESSMENT OF STABILITY

It is the existence of this locus for weight positions which will maintain a constant mass moment of inertia which enables a direct evaluation of initial stability. If the period of roll is measured when the weights are on the centre line and high above the deck, and then the position is identified which corresponds to the maximum outboard location of weights with the same mass moment of inertia, and the roll period is again measured, sufficient data will have been gathered to calculate the metacentric height.

The theory can be easily summarised as follows. For the two cases observed (with the weights in positions 1 and 2 as shown in Figure 2) both the vessel's displacement (Δ) and mass moment of inertia (I') are unknown, although they have the same value in each case. The metacentric height ($\overline{GM}$) however, is different in each case. There are therefore four unknowns: Δ, I', $\overline{GM}_1$ and $\overline{GM}_2$. To solve this problem four equations are needed. The basic equation for the period of roll, Equation 2, can be used twice with the appropriate values T_1 and T_2 being used with $\overline{GM}_1$ and $\overline{GM}_2$ respectively. In addition the difference between $\overline{GM}_1$ and $\overline{GM}_2$ is defined in two independent ways in Equations 4 and 5. These can be used to provide the third and fourth equations which can be re-stated as in Equations 6 and 7 below:

$$\overline{GM_2} - \overline{GM_1} = \frac{m(y_1 - y_2)}{\Delta} \tag{6}$$

$$\overline{GM_2} - \overline{GM_1} = y_1 - y_2 - x_2 \tag{7}$$

With four equations for four unknowns the problem is solvable, however this theory presupposes that the ellipse describing weight positions with a constant value of I' can be found. To demonstrate how this can be achieved it is necessary to consider again how the geometry of the weight positions effects the relevant variables. As already discussed the weights can be moved horizontally only, so keeping $\overline{GM}$ constant, or they can be moved along an elliptical path which maintains I' constant. There is however a third path which will alter both $\overline{GM}$ and I' in such a way that the period of roll, T, is kept constant, as shown in Figure 3.

This third locus, where T is constant, has a property which can be used to find the locus for constant I'. This can be shown by first manipulating Equation 2 to obtain a general expression for the partial derivative of I' with respect to $\overline{GM}$:

$$\frac{\partial I'}{\partial \overline{GM}} = \frac{T^2 g \Delta}{4\pi^2} \tag{8}$$

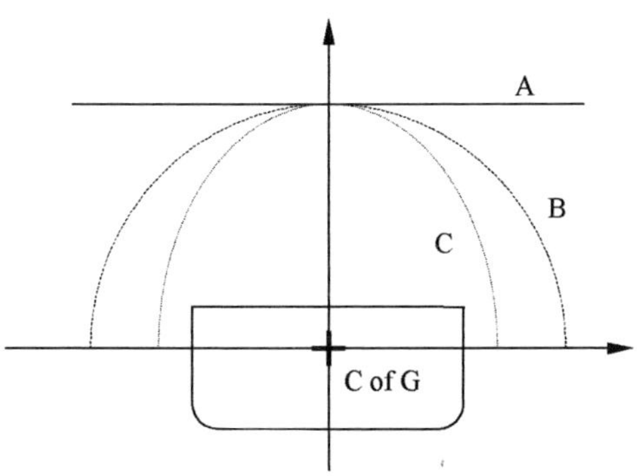

Figure 3: The loci of weight positions for three alternative conditions: A for constant metacentric height; B for constant period of roll; C for constant mass moment of inertia.

Evidently when the weights are moved along the locus that maintains the period of roll, T, constant, then this partial derivative is also a constant. If weights positioned on the centre plane are taken as a first case, and a second position is found below this, where the weights are positioned outboard such that the two periods of roll are the same, then the change in I' from the first case to the second can be found from the product of the partial differential of I' with respect to $\overline{GM}$ and the actual change in $\overline{GM}$:

$$\delta I' = \frac{\partial I'}{\partial \overline{GM}} \delta \overline{GM} \tag{9}$$

which, by substituting from Equations 8 and 4, yields:

$$\delta I' = \frac{T^2 gmd}{4\pi^2} \tag{10}$$

From this expression the change in the mass moment of inertia between the first position (high on the centre plane) and the second position (with weights lower and outboard as shown in Figure 4) can be calculated. As it is possible to calculate the change in the mass moment of inertia when the weights are moved purely horizontally, from Equation 3, so the distance toward the centre plane that the weights should be transferred in order to restore the mass moment of inertia to the same value as for the first position can be calculated.

Identifying points on the ellipse that defines the locus of constant mass moment of inertia can therefore be achieved. Having shown that this is possible it follows that the direct evaluation of initial stability on the basis outlined above is theoretically valid.

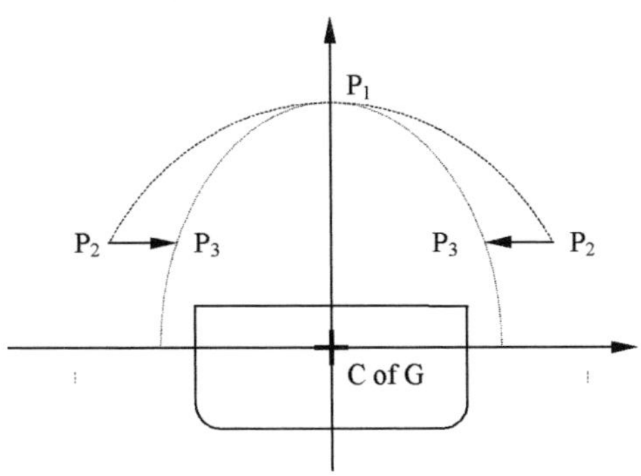

Figure 4: The procedure to find a point on the ellipse with constant mass moment of inertia: by experiment P_2 is found where the period of roll is as for P_1, then P_3 is calculated such that δI from P_2 to P_3 is the same as δI from P_1 to P_2.

A CALCULATION FOR THE MINIMUM OF OBSERVATIONS

The direct evaluation of initial stability is based on the identification of two positions which are on the major and minor axes of an ellipse which is centred on the centre of gravity of the vessel. The period of roll when the weights are located at these positions must be established in order to provide the four equations necessary to proceed with the theoretical calculation. In this section a calculation will be described which uses five observations, the theoretical minimum number, however in the next section the practical difficulties of implementing this simplest of procedures will be discussed.

In this procedure the pair of weights are first positioned together on the centre plane well above the deck. This position is taken to be the extreme vertical point of an ellipse around which the weights can be moved without altering the mass moment of inertia. By positioning the weights in four more locations, and recording the period of roll, it is theoretically possible to find all the desired results, including the initial stability, by using a nine stage calculation summarised below.

The five weight positions necessary are as follows: the first is with both weights together on the centre plane, as high as is practicable above the deck, the second at about half the height of the first (again on the centre plane), and the third on or near the deck (also on the centre plane). The fourth and fifth positions have the weights located at the same heights as the second and third respectively, but at some distance from the centre plane. These positions are shown in Figure 5, as are all the additional positions described below for which data is derived. The calculation uses the first position to define the ellipse of interest, as it is taken to be the extreme point of the ellipse's major axis. This point is also taken as the origin in defining the x and y co-ordinates of the other positions. The calculation to find the extreme point of the minor axis then proceeds in stages 1 to 5 below. The vertical position of this point is the same as the vessel's centre of gravity. Stages 6 to 9 below demonstrate how the mass moment of inertia, the displacement and the metacentric height are also derived.

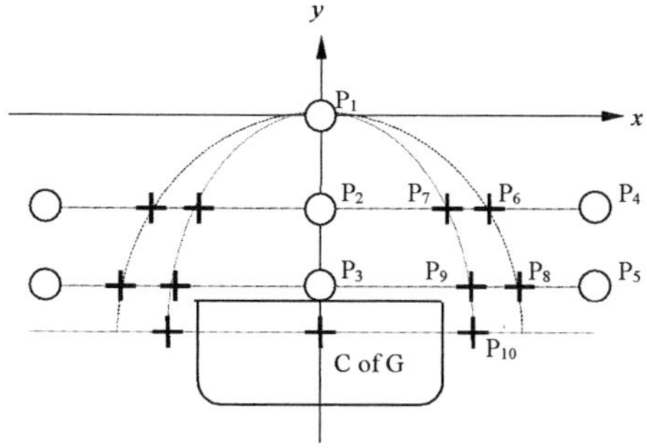

Figure 5: Weight positions for the procedure to calculate initial stability: Positions P_1 to P_5 are observed experimentally, positions P_6 to P_{10} are derived and the period of roll calculated.

1. Positions 2 and 4 are used to find position 6 (which is at the same height as position 2). This will have the same period of roll as position 1. This is achieved by reducing Equation 2 to a simple form which defines the relationship between the period of roll and the distance of the weights from the centre line, given that the weights can only be moved horizontally, and therefore that $\overline{GM}$ will not change. If the constants c_1, c_2 and c_3 represent $\left(2\pi/\sqrt{g\Delta\overline{GM}}\right)$, $\left(I' + my^2\right)$ and (m) respectively, then Equation 2 can be written as:

$$T = c_1\sqrt{c_2 + c_3x^2} \tag{11}$$

   By further combining the constants this relationship reduces to an expression with only two unknown constants, a, and b:

$$T = ax^2 + b \tag{12}$$

   By substituting for T_2, x_2 and T_4, x_4 it is possible to find a and b. Setting $T_6 = T_1$ it is possible to find x_6 from the same equation.

2. Position 7 (also at the same height as 2) which will have the same mass moment of inertia as position 1 is then found from Equations 10 and 3 as discussed above, and the period of roll at this position again interpolated form the observed periods for positions 2 and 4. From Equation 10 the increase in I' can be calculated:

$$\delta I'_{(1\,\text{to}\,6)} = \frac{T_1^2 gm(y_1 - y_6)}{4\pi^2} \tag{13}$$

   And from Equation 3 the decrease in I' can be calculated:

$$\delta I'_{(6\,\text{to}\,7)} = -m\left(x_6^2 - x_7^2\right) \tag{14}$$

   As $\delta I'_{(1\text{ to }6)} = -\delta I'_{(6\text{ to }7)}$ it is possible to find x_7. Interpolate to find T_7.

3. Positions 3 and 5 are used in a similar way to establish the position and period of roll for positions 8 and 9. Position 9 has the same mass moment of inertia as positions 1 and 7.

4. Positions 1, 7 and 9 are three points on the required ellipse and can be used in the standard equation for such a curve to calculate the major and minor demi-axes, v and u. The standard equation for an ellipse, centred at (h,k), and with the minor and major demi-axes defined as u and v, is given by:

$$\frac{(x-h)^2}{u^2} + \frac{(y-k)^2}{v^2} = 1 \tag{15}$$

If position 1 is defined as (0,0) and is on the major axes with the ellipse below it, it can be shown that $h = 0$ and $k = -v$. In this case the equation for the ellipse simplifies to:

$$\frac{x^2}{u^2} + \frac{(y+v)^2}{v^2} = 1 \tag{16}$$

Use (x_7, y_7) and (x_9, y_9) to find u and v.

5. The dimensions of the demi-axes fully define the ellipse. They therefore give the position of the vessel's centre of gravity in relation to position 1, and the extreme horizontal position possible for the weights such that the mass moment of inertia is the same as in position 1, i.e. position 10. The vessel's centre of gravity is given by $(0,-v)$, and position 10 is given by $(u,-v)$.

6. The mass moment of inertia for all points on the ellipse is found from any two points on the ellipse at which the period of roll is known, such as positions 1 and 9. This is achieved by using Equation 2 for both positions, and then subtracting one from the other to obtain an expression for the change in metacentric height. Equation 4 provides an alternative expression for this, and if the two are equated it is found that displacement (the other unknown variable) is cancelled from the equation, and so the mass moment of inertia can be found. From Equation 2:

$$\delta\overline{GM}_{(1\text{ to }9)} = \frac{4\pi^2 I'}{g\Delta T_9^2} - \frac{4\pi^2 I'}{g\Delta T_1^2} \tag{17}$$

From Equation 4:

$$\delta\overline{GM}_{(1\text{ to }9)} = \frac{m(y_9 - y_1)}{\Delta} \tag{18}$$

From which I' can be found:

$$I' = \frac{mg(y_9 - y_1)}{4\pi^2}\left(\frac{T_9^2 T_1^2}{T_1^2 - T_9^2}\right) \tag{19}$$

7. The change in the metacentric height between positions 1 and 10 can be expressed in terms of moments, Equation 4, or in terms of the geometry of the ellipse, Equation 5. As position 10 is now defined, as are the demi-axes of the ellipse, Equations 4 and 5 can be used to find the displacement of the vessel. From Equation 4:

$$\delta\overline{GM}_{(10\text{ to }1)} = \frac{m(y_{10} - y_1)}{\Delta} \tag{20}$$

From Equation 5:

$$\delta\overline{GM}_{(10\text{ to }1)} = v - u \tag{21}$$

From which Δ can be found:

$$\Delta = \frac{m(y_{10} - y_1)}{v - u} \tag{22}$$

8. In stages 6 and 7 above the mass moment of inertia and the displacement have been found. Therefore Equation 2 can be used to find the metacentric height for any point on the ellipse for which the roll period is known, such as position 1. From Equation 2:

$$\overline{GM}_1 = \frac{4\pi^2 I'}{g\Delta T_1^2} \tag{23}$$

9. The metacentric height at position 1 can be corrected to that of position 10, when the weights are at the same vertical height as the vessel's own centre of gravity, by adding the difference between v an u to $\overline{GM}_1$. This then gives the initial stability of the vessel, which is the principal objective of the calculation.

$$\overline{GM} = \overline{GM}_1 + (v - u) \tag{24}$$

THE DIFFICULTIES OF IMPLEMENTATION

The calculation detailed above is valid when tested against the output of a mathematical model of the roll period of a vessel. This is only to be expected as both the model and the calculation are based on the equation describing the period of roll (Equation 2). However attempts to use this procedure to evaluate the initial stability of a model barge in a test tank produced results that were quite unacceptable, with errors on occasion in excess of 100%.

Before seeking the explanation for these unsatisfactory results it is necessary to consider the precision required in this procedure. By deliberately introducing an error into the calculations the impact of inaccuracies in the period of roll can be assessed. Such calculations for a hypothetical vessel with a displacement of 5 tonnes, an initial metacentric height of 1 metre, and an effective mass moment of inertia of 10 tonne metres^2 were carried out, using weights with a combined mass moment in the order of 2.5 tonne metres. The calculations established that the precision required is in the order of one thousandth of a second for acceptable results to be obtained, and if a precision is attainable of only one tenth of a second then the results can be nonsensical. Fortunately such precision is attainable using an electronic inclinometer linked to a portable PC, with roll period accuracy of 5 x 10^{-4} seconds being attainable.

As the necessary precision is attainable in the collection of experimental data it is necessary to look to the assumptions underlying the calculation to explain the unsatisfactory results calculated from experimental data. The theory outlined above assumes that the change in the mass moment of inertia $\delta I'$ of the vessel equal to the change in the mass moment of inertia of the pair of weights. In fact the movement of the pair of weights can cause a change in the mass moment of inertia through a number of other mechanisms. Firstly the change in the centre of gravity of the vessel causes a change in the radius of gyration of the vessel, so altering the mass moment of inertia of the vessel itself. Secondly the change in the period of roll could itself change the dynamic response of the vessel in terms of added mass and fluid damping. Damping is ignored in the derivation of the fundamental equation of roll, Equation 2, while the added mass moment of inertia is assumed to remain constant for any roll period. Finally it is assumed that the vessel is rolling about its own centre of gravity, when in reality the motion is much more complex with a centre of roll motion that is changing for each angle of heel. If an 'effective' centre of roll could be defined, it would not be at the centre of gravity, so the mass moment of inertia in roll must be larger than the mass moment of inertia for the vessel about its own centre of gravity. While the errors attributable to each of these erroneous assumptions may result in extremely small changes to the mass moment of inertia, the precision required for this calculation suggests that the difficulties in implementing this procedure in practice can be attributable to these simplifications.

CONCLUSION

The assessment of initial stability by reference to the roll period is only necessary where the hydrostatic particulars of vessels are not available, as if this information is to hand a conventional inclining experiment could be undertaken. However, the large number of older fishing and other work boats around the world for which no record of lines of hydrostatics exists suggests that such a procedure should be developed, especially as national and international bodies increasingly include smaller vessels in their regulatory requirements. It will be necessary to assess the stability of these work boats in a manner which is both practically and economically viable, otherwise the owners will try to avoid regulation by working without certification and therefore both illegally and unsafely.

In this paper a procedure has been described whereby the metacentric height, vertical centre of gravity, and displacement of a vessel can theoretically be established without reference to the vessel's lines or hydrostatic data. This is achieved by a direct calculation which uses as its only input the periods of roll, measured with computer assisted precision, as a pair of weights are moved to predefined locations on and above the deck. The basis of this procedure is that as the metacentric height and mass moment of inertia are changed in a systematic way, all the unknown variables in the basic equation for the period of roll can be evaluated. However experimental results have shown that the second order effects of moving the pair of weights also influence the vessel's mass moment of inertia to a significant extent. Further work is being undertaken to incorporate these secondary mechanisms into the calculations.

References

HMSO (1993). 'Code of Practice for Vessels up to 24 Meters Load Line Length'. Surveyor General's Organisation. HMSO, London.

Marchaj, C.A. (1986) 'Seaworthiness – The Forgotten Factor'. Adlard Coles, London.

Muckle, W. and Taylor, D.A. (1987). ' Muckle's Naval Architecture', 2nd edition. Butterworths, London.

Contemporary Ideas on Ship Stability
D. Vassalos, M. Hamamoto, A. Papanikolaou and D. Molyneux (Editors)

STABILITY OF HIGH SPEED CRAFT

Y. Ikeda [1] and T. Katayama [1]

[1] Department of Marine System Engineering, Osaka Prefecture University,
1-1, Gakuen-Cho, Sakai, Osaka, 599-8531, Japan

ABSTRACT

The investigations on transverse and longitudinal instability of the motions of a high-speed craft in calm water carried out by the authors are reviewed.

It is experimentally confirmed that the transverse stability significantly decreases at high Froude numbers, and this causes large heel of the craft. The stability loss, which depends on trim angle, also generates roll motion when a craft has pitching motion.

It is found that porpoising of the craft is a self-exciting motion due to the different sign of the coupling restoring coefficients between pitch and heave motions. Finally prediction methods for the roll and heave damping including vertical lift force contribution are proposed.

KEYWORDS

High-Speed, Planing Craft, Stability, Damping, Unstable Phenomena, Porpoising, Pure Stability Loss

INTRODUCTION

A high-speed vessel has different motion characteristics from those of a conventional displacement-type vessel. For example it is well known that in following seas a high-speed vessel can capsize by broaching, parametric rolling or stability reduction by waves. Even in calm water a high-speed craft shows unique motion characteristics, such a chine walking, sudden large heel and porpoising. In the present paper the investigations on the stability and unstable motions of a high-speed craft carried out by the authors for over five years are introduced.

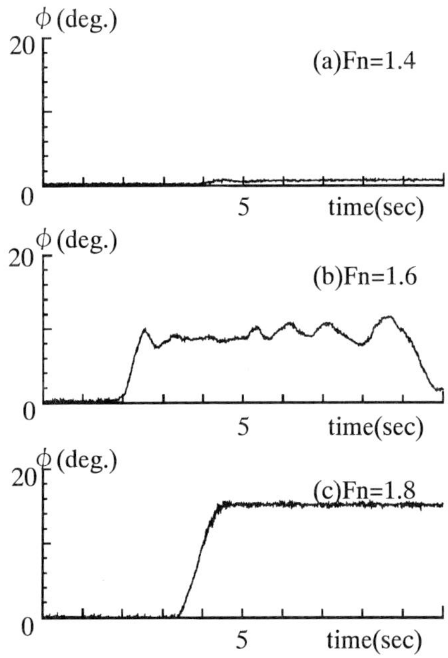

Figure 1: Time histories of roll motion of ShipB-45 at trim angle of 2deg. measured by free rolling test, without initial heel

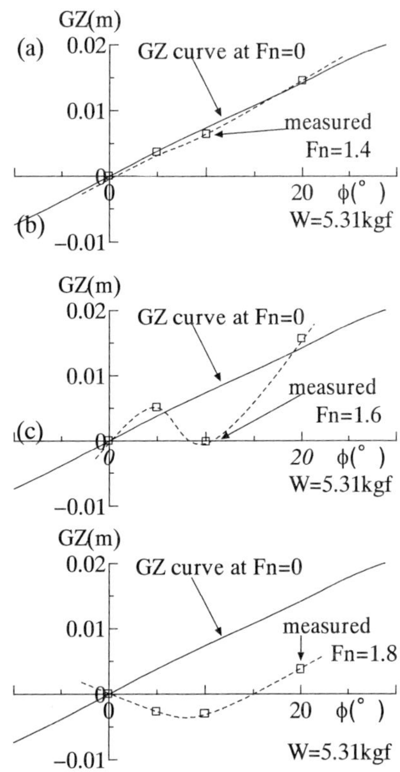

Figure 3: GZ curve of ShipB-45 at trim angle of 2deg., center of gravity 0.074m, ship weight 5.31kgf for several advance speeds :(a)Fn=1.4 (b)Fn=1.6 (c)Fn=1.8

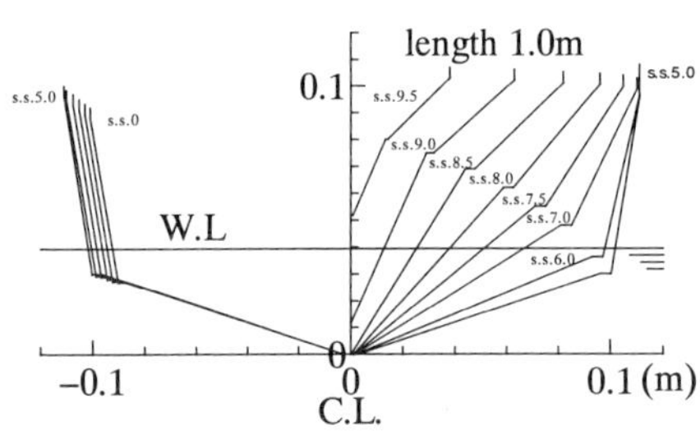

Figure 2: Body Plan of ShipB-45

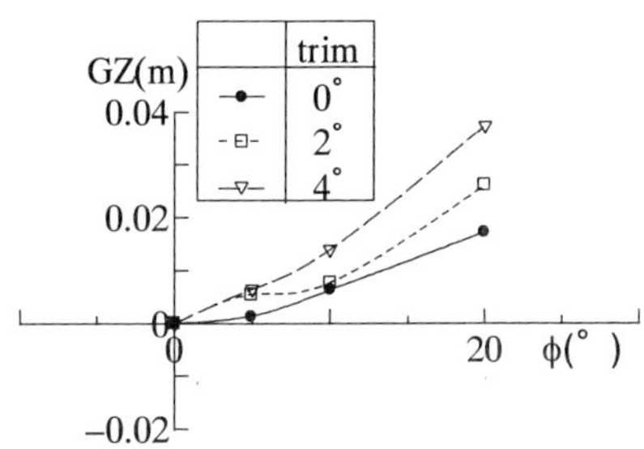

Figure 4: GZ curve of ShipB-45 at center of gravity 0.04m, ship weight 5.31kgf and Fn=1.6

STABILITY LOSS AT HIGH SPEED

It was pointed out that the transverse stability of a displacement-type ship running in calm water at relatively high speeds decreases over Froude number of 0.3 and that the decrease may be caused by the wave pattern on the side of the hull.

The same phenomena are observed for a high-speed ship of semi-displacement type and of planing type. Figure 1 shows the experimental results of roll motion of a planing craft, the body plan of which is shown in Figure 2, in calm water without any disturbance. The results demonstrate that large heel occurs at high Froude numbers. The measured restoring moment of the craft is shown in Figure 3. The results confirm that the large heel shown in Figure 1 is generated by the transverse stability loss due to high advanced speed. The stability loss depends on speed advance, location of centre of gravity and trim angle as well as hull shape. The effect of trim angle on the stability of a craft at high speed is shown in Figure 4.

UNSTABLE ROLLING INDUCED BY PITCHING MOTION

As pointed out in the previous section, the transverse restoring moment significantly varies with trim angle at high speed. This suggests that an unstable rolling motion is induced by pitching motion and that the motion is a parametric oscillation system, because the restoring coefficient depends on trim. In Figure 5 simulation results of a free roll motion test with an initial heel angle of the planing craft in calm water are shown. In each simulation a sinusoidal pitching motion with constant period and amplitude is given. The roll natural period is fixed to be 2.0sec., and the forced pitching periods are changed as 2.0, 1.0 and 0.667sec respectively. When the period of forced pitching motion is half of the roll natural period, roll motion increases with time as shown in Figure 5. These results suggest that pitching motion due to waves or porpoising may cause unstable roll motion when the pitch period is half of the roll natural period. The authors investigated the effects of moment of inertia, roll damping, restoring moment and pitch amplitude on the pitch induced roll motion. In Figure 6, the cases of occurrence of unstable roll motion are plotted by black circles. If the roll damping is larger than these black circles, no rolling occurs. The solid line in the figure shows the predicted roll damping of the craft. In the region where the solid line is larger than the circles, no roll motion occurs and the motion is stable.

UNSTABLE PITCH AND HEAVE COUPLING MOTION

Unstable pitch and heave coupling motion of a high-speed craft in calm water is called 'porpoising'. There are many investigations on the mechanism of porpoising, and several proposals of the prediction method to find the stability criterion. However it is not clear which coefficients in the equation of motion have significant influence on the occurrence of porpoising.

The authors measured restoring forces and moments acting on a craft, and obtained the coefficients in the motion equation for the heave-pitch coupling motion as shown in Figure 7.

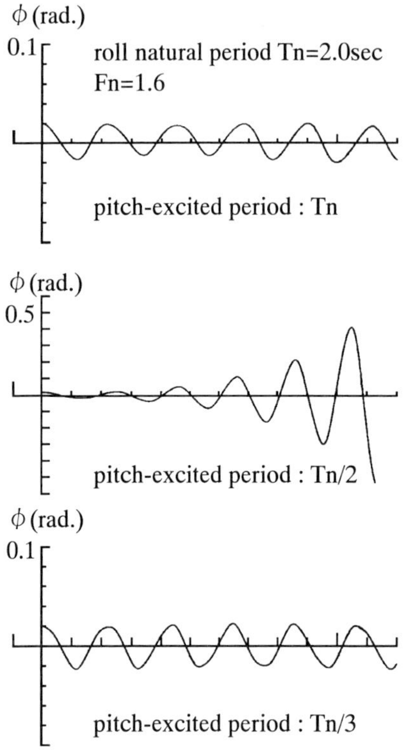

Figure 5: Time histories of roll motion induced by forced pitching motion

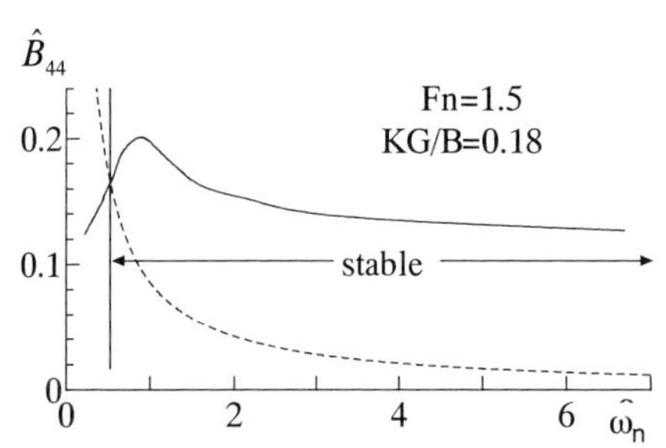

Figure 6: Judgement of whether ShipB-45 is stable or not

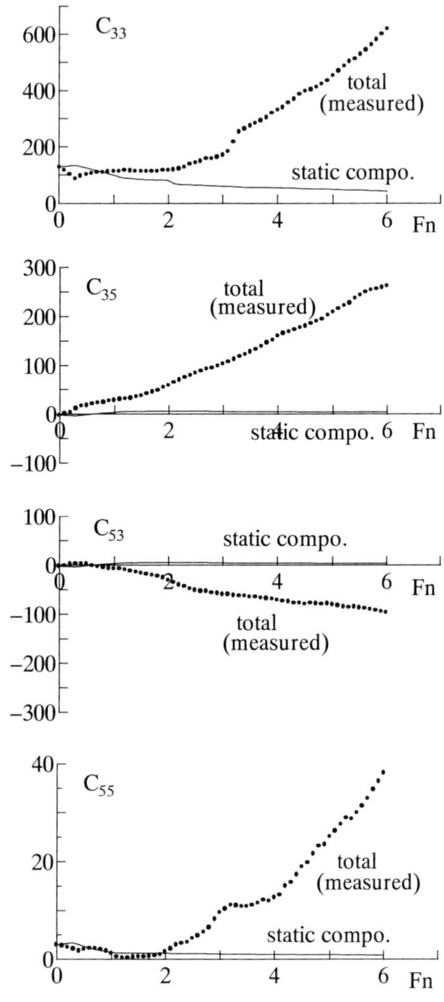

Figure 7: Coefficients of restoring forces and moments of motion equations for heave and pitch; heave restoring force C_{33}, coupled restoring force from pitch to heave C_{35}, coupled restoring moment from heave to pitch C_{53} and pitch restoring moment C_{55}

The model used in the measurement is a hard chine planing craft similar to that shown in Figure 2. Note that the coupling restoring coefficients between heave and pitch motions, C_{35} and C_{53}, have different signs from each other. As is well known, when the coupling restoring terms have different sign in an oscillation system of two degrees of freedom, a self-excited oscillation occurs. Therefore these experimental results suggest that porpoising is a self-exciting motion due to the different sign of the coupling restoring coefficients between pitch and heave motions.

Forced heave and pitch tests of the craft model were also carried out in order to find the effects of hydrodynamic inertia and damping forces on porpoising. The results show that significant nonlinear effects can be seen in these hydrodynamic forces. This fact suggests that a linear prediction theory to find porpoising limit may not be sufficient and that a nonlinear theory should be used to obtain accurate prediction of porpoising.

A simulation using a nonlinear equation with nonlinear hydrodynamic and hydrostatic coefficients is carried out to obtain porpoising of the craft. The Runge-Kutta method of the fourth order is used in the simulation. The results of the simulation are shown in Figure 8 with experimental results. The heaving and pitching amplitudes, Z_{amp} and θ_{amp}, are overestimated near Froude number of 3.0. This is because of lack of data of hydrodynamic forces. Over Froude number of 4.0, the agreement between simulated and measured results is fairly good.

The heave and pitch coupling restoring coefficients, C_{35} and C_{53}, of the craft are shown in Figure 9. We can see that the dynamic components generated by advance speed are significant, and that these coefficients have different sign over Froude number of 0.6. Over this Froude number, porpoising can occur if there is no damping. As damping increases the Froude number where porpoising begins also increases.

Figure 10 shows the resistance increase of the craft running in calm water due to porpoising. The solid line in the figure denotes the simulation result of steady running condition without any motions, and the circles denote the measured resistance.

At low advanced speed where no porpoising occurs, the agreement between them is fairly good, but after occurrence of porpoising, or in the unstable region, the measured resistance with porpoising is larger than the resistance at steady condition without motions. These results may suggest that the energy for porpoising is given by the thrust of the craft and that if porpoising occurs, the resistance increase would reduce the advanced speed of a running craft.

ROLL AND HEAVE DAMPING CREATED BY VERTICAL LIFT FORCE

As described in the previous sections, unstable motions due to the instability of the motion equation can be avoided by increasing motion damping as shown in Figure 6. However the characteristics of the damping of semi-planing and planing craft have not been clarified yet, and there is no practical prediction method for such a craft. In this section, the roll damping and the heave damping will be discussed.

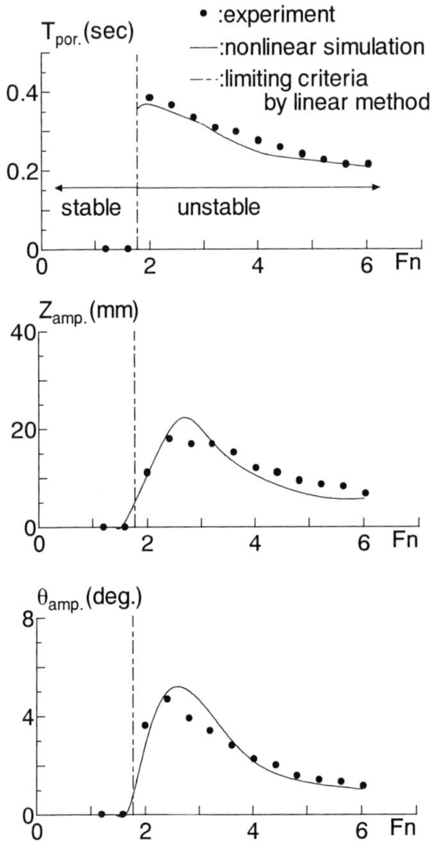

Figure 8: Result of nonlinear motion calculations

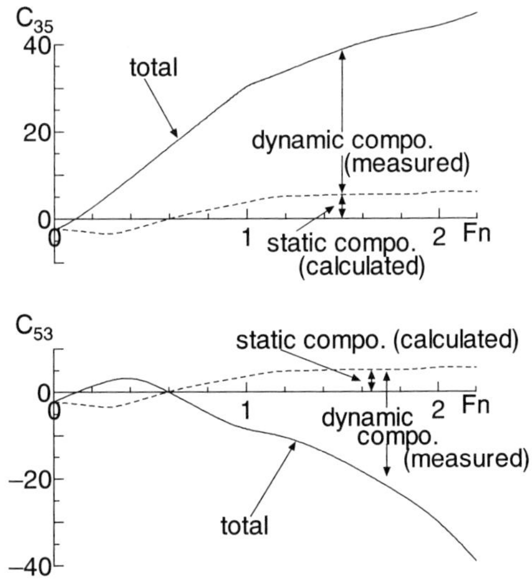

Figure 9: Heave and pitch Coupling restoring coefficients, C_{35} and C_{53}

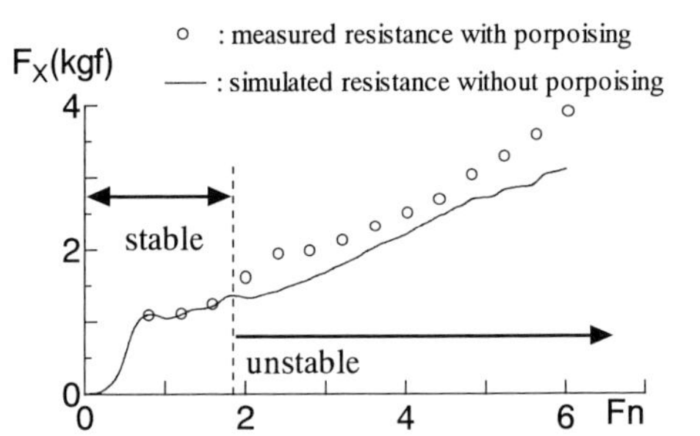

Figure 10: Resistance increases during porpoising

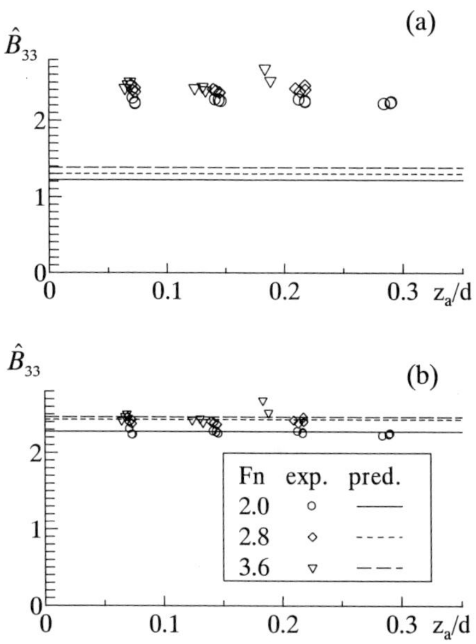

Figure 11: Predicted heave damping coefficient due to heave motion; $\hat{B}_{33}$: (a) potential theory (b) potential theory included lift component due to heave motion

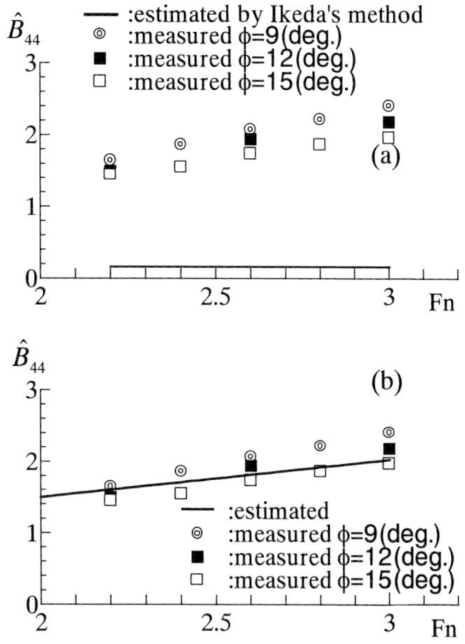

Figure 12: Comparison of roll damping coefficient between measured and estimated by Ikeda's formula; (a) and by present method; (b)

As shown by Ikeda et al., the lift component of the roll damping is dominant at high advance speed. The horizontal lift force created by the horizontal pressure acting on a hull generates the lift component, and prediction methods for a displacement-type ship and a semi-displacement-type ship have been proposed by Ikeda et al..

For a planing craft, however, the projected area of the side hull is usually very small particularly when it runs at very high speed. Therefore instead of the transverse lift force, the vertical lift force acting on the bottom of the craft may play an important role in the damping. On the basis of quasi-steady assumption, the roll and heave damping can be obtained as follows,

$$\text{Roll damping : } B_{44L} = \frac{1}{24}\rho B_{w.l}{}^{4} U k_L(\tau_1) \, ,$$

$$\text{Heave damping : } B_{33L} = \frac{1}{2}\rho B^2 U^2 k_L \alpha = \frac{1}{2}\rho B^2 U k_L \, ,$$

where B denotes the breadth of water plane, k_L the differential coefficient of vertical lift coefficient with respect to trim angle, β deadrise angle in degrees. In Figures 11 and 12 the predicted results of the heave and roll damping coefficients by the previous prediction method and the present one are shown.

Although the previous method including only transverse lift contribution underestimates damping, the present method including the vertical lift contribution gives fairly good prediction results.

CONCLUSIONS

Stability loss due to high advance speed, unstable roll motion induced by pitching motion and longitudinal instability called 'porpoising' of a high speed planing craft are investigated experimentally and theoretically, and the following conclusions are obtained:

1. The transverse stability of a craft decreases at high advance speed.
2. The stability loss causes a large heel at high advance speed.
3. The dependence of the stability loss on trim angle generates unstable rolling motion induced by pitching motion.
4. Porpoising of a craft is a self-exciting motion due to the different sign of the coupling restoring coefficients between pitch and heave motions.
5. The damping plays an important role to avoid the occurrence of unstable motions.
6. A prediction method of the roll and heave damping including the vertical lift contribution is proposed, and it is confirmed that the agreement between predicted and measured is fairly good.

References

Katayama T. and Ikeda Y. (1995). An Experimental Study on Transverse Stability Loss of Planing Craft at High Speed in Calm Water. *Journal of the Kansai Society of Naval Architects, Japan.* **224**, 77-85.

Katayama T. and Ikeda Y. (1996). A Study on Unstable Rolling Induced by Pitching of Planing Crafts at High Advance Speeds. *Journal of the Kansai Society of Naval Architects, Japan.* **225**, 141-148.

Katayama T. and Ikeda Y. (1996). A Study on Transverse Instability of Planing Craft at High Speeds in Calm Water. *Proceedings of Third Korea-Japan Joint Workshop on Ship and Marine Hydrodynamics.* 117-124.

Katayama T. and Ikeda Y. (1996). A Study on Coupled Pitch and Heave Porpoising Instability of Planing Craft (1st Report). *Journal of the Kansai Society of Naval Architects, Japan.* **226**, 127-134.

Katayama T., Ikeda Y. and Otsuka K. (1997). A Study on Coupled Pitch and Heave Porpoising Instability of Planing Craft (2nd Report). *Journal of the Kansai Society of Naval Architects, Japan.* **227**, 71-78.

Katayama T. and Ikeda Y. (1997). A Study on Coupled Pitch and Heave Porpoising Instability of Planing Craft (3rd Report). *Journal of the Kansai Society of Naval Architects, Japan.* **228**, 127-134.

Ikeda Y., Katayama T. and Tajima S. (1998). A Prediction Method of Vertical Lift Component of the Roll and Heave Damping. *Journal of Kansai Society of Naval Architects, Japan.* **229**, 127-134.

Contemporary Ideas on Ship Stability
D. Vassalos, M. Hamamoto, A. Papanikolaou and D. Molyneux (Editors)

NONLINEAR ROLL MOTION AND BIFURCATION OF A RO-RO SHIP WITH FLOODED WATER IN REGULAR BEAM WAVES

Sunao Murashige[1], Motomasa Komuro[2], Kazuyuki Aihara[1] and Taiji Yamada[3]

[1] Department of Complexity Science and Engineering, Graduate School of Frontier Sciences, The University of Tokyo, Japan
[2] Department of Electronics and Information Science, Teikyo University of Science and Technology, Japan
[3] AIHARA Electrical Engineering Co., Ltd., Japan

ABSTRACT

This paper describes nonlinear motion of a Ro-Ro type ship with water-on-deck in regular beam seas. Model experiments demonstrate that a flooded ship can exhibit nonlinear roll motion in waves of relatively moderate height. We investigate this nonlinear response using a mathematical model for coupled motion of roll and flooded water of a box-shaped ship in waves. This model includes not only static but also dynamic effects of flooded water. The Jacobian matrix of the model equations has a discontinuous property. We present the method to determine bifurcation values for this type of system. An example of the bifurcation analysis shows that, when a ship is flooded, both small harmonic and large subharmonic motion can coexist in a wide range of a parameter.

KEYWORDS

roll motion, flooded ship, nonlinear motion, coupled motion, stability, bifurcation.

INTRODUCTION

It is well known that fluid with a free surface inside a ship has some influences on the ship motion (Dillingham (1981) and Caglayan and Storch (1982)). In this paper, we consider nonlinear roll response of a Ro-Ro type ship with water-on-deck in regular beam seas, both experimentally and theoretically. In the experiments, we measured the roll angle ϕ of a ferry model with flooded water in regular beam waves. The results indicate that different types of roll motion can coexist in the same incident waves even in waves of relatively moderate amplitude. In addition, observations of the experiments suggest that this motion was almost two-dimensonal in the vertical section parallel to the direction of progress of incident waves, and that coupled motion of roll and flooded water was dominant. Then we derive a mathematical model for the coupled motion on some assumptions.

In the previous work, we experimentally studied the coupled motion of roll and flooded water in regular waves in more detail (Murashige and Aihara (1998)). We measured the roll angle ϕ of a box-shaped model in regular beam waves, and reconstructed possible attractors directly from the time series data $\phi(t)$ using delay time coordinates $(\phi(t), \phi(t+\tau), \phi(t+2\tau))$ where t and τ denote the time and the delay time, respectively. The experimental results showed that the box-shaped model with flooded water could exhibit some interesting nonlinear roll response. In particular, we found not only periodic but also complicated roll motion even in regular waves. This complicated motion had some typical properties of low-dimensional deterministic chaos. Namely, the power spectrum had the broad band characteristics, the maximum Liapunov exponent was positive, and the stroboscopic plots clearly displayed the stretching, folding, and compressing process. It should be noted that the amplitude of subharmonic and chaotic motion was larger than that of regular harmonic motion with the same period as incident waves, and that the nonlinear roll response was found in a wide range of parameters.

Nonlinear roll motion including ship capsizing has been one of important and challenging subjects in the field of not only naval architects but also nonlinear dynamics (Virgin(1987), Soliman and Thompson(1991), Thompson, Rainey and Soliman(1992), Thompson(1997)). For roll motion of a flooded ship in waves, Kan and Taguchi(1992) showed numerical solutions of chaotic motion, and Falzarano, Shaw, and Troesch(1992) applied global analysis techniques. They used an equation of a single degree of freedom which neglects dynamic effects of flooded water. On the other hand, in this paper, we examine nonlinearly coupled dynamics of roll and flooded water using coupled equations of motion. This mathematical model allows us to examine not only static but also dynamic effects of flooded water.

Bifurcation sets can be the base for understanding of complicated mechanism of the nonlinearly coupled dynamics. This paper describes a discontinuous property of the model equations, presents the method to determine bifurcation values of this type of system, and shows an example of results of the bifurcation analysis.

EXPERIMENTS

Model and experimental method

Experiments were performed using a Ro-Ro model in the wave tank 8m wide, 50m long, and 4.5m deep at Ship Research Institute. The overview and principal particulars of the model are given in fig.1 and table 1. The model was placed at a position 18.75m away from a wave maker of the wave tank so that incident waves came from starboard to port. A closed vessel (the shaded area in fig.1) was fixed in the model and imitates a vehicle deck. We put a prescribed amount of water on the vehicle deck and measured motion of the model in regular beam seas. We also observed behavior of water-on-deck using a video camera attached in the vehicle deck.

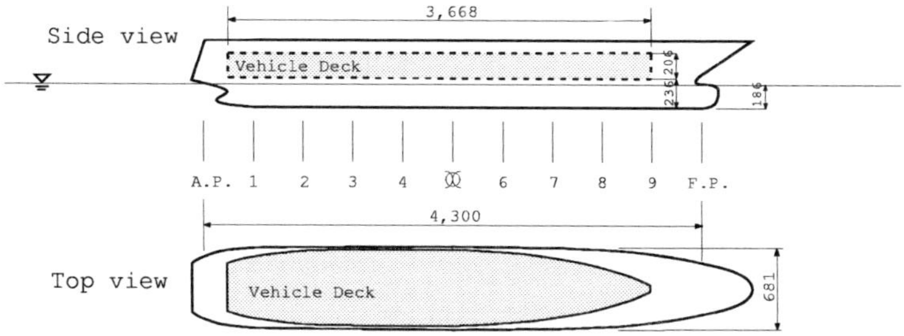

Figure 1: Ro-Ro model
(unit:mm. The shaded area represents the vehicle deck.)

TABLE 1
Principal particulars of the Ro-Ro model
(scale ratio:1/23.5, without water-on-deck)

Length L_{pp}	4.3 m	Vehicle deck height h	0.206 m
Breadth B	0.681 m	Displacement W	272.69 kg
Depth D	0.236 m	Height of center of gravity $\overline{KG}$	0.25 m
Draft d	0.186 m	Metacentric height $\overline{GM}$	0.069 m
Freeboard f	0.050 m	Natural period of roll motion T_r	1.94 sec.

Experimental results

The experiments were started after the model was placed at the statically balanced positions in still water. There are two static equilibrium points of the model with water-on-deck in the lee and the weather sides. Figures 2.(a) and (b) show examples of the time history of the measured roll angle $\phi(t)$ when the amount of water-on-deck w was 20% of the displacement of the model W and the height H_i and the period T_i of incident waves were 13.0cm and 1.44sec., respectively. In this case, the static equilibrium angle is ±19.0deg.. The roll angle $\phi(t)$ is defined to be positive when the model heels to the lee side.

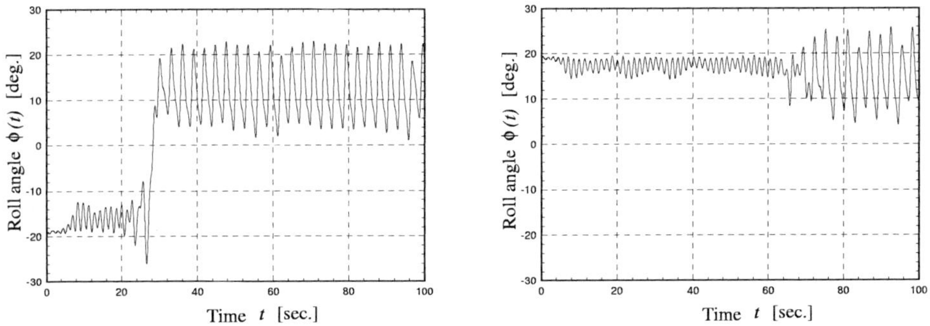

(a) $\phi(t=0) = -19.0$deg.(weather side) (b) $\phi(t=0) = +19.0$deg.(lee side)

Figure 2: Time history of the measured roll angle $\phi(t)$
(The height of incident waves H_i=13.0cm, the period of incident waves T_i=1.44sec., and the ratio of the amount of water-on-deck w to the displacement of the model W is w/W=0.2. In the case (b), the model was lightly impinged by a stick at $t \simeq$65sec..)

From these figures, we can see the following:

- Case (a)
  - At t=0, the model is balanced in the weather side, $\phi(t=0)=-19.0$deg..
  - For t <20sec., the model rolls in the weather side with the average amplitude of about 2 deg. and the same period as that of the incident waves.
  - At $t \simeq$25sec., the roll motion is changed from the weather side to the lee side.
  - For t >30sec., the model rolls in the lee side with the average amplitude of about 9deg. and the average period nearly equal to twice that of the incident waves.
- Case (b)
  - At t=0, the model is balanced in the lee side, $\phi(t=0)=+19.0$deg..
  - For t <65sec, the model rolls with the average amplitude of about 2 deg. and the same period as that of the incident waves.
  - After the model is lightly impinged by a stick at $t \simeq$65sec., the roll motion is changed to the other mode of large amplitude.

These results indicate that some different modes of roll motion coexist for regular incident waves of constant period and amplitude. This is a typical example of nonlinear oscillations.

MODEL EQUATIONS

Modelling the coupled motion of a ship and flooded water

Observations of the experiments suggested that the measured nonlinear motion is dominated by the coupling of a ship and water-on-deck. This section derives model equations for the coupled motion. For simplicity, we consider the two-dimensional motion of a box-shaped floating body as shown in fig.3.

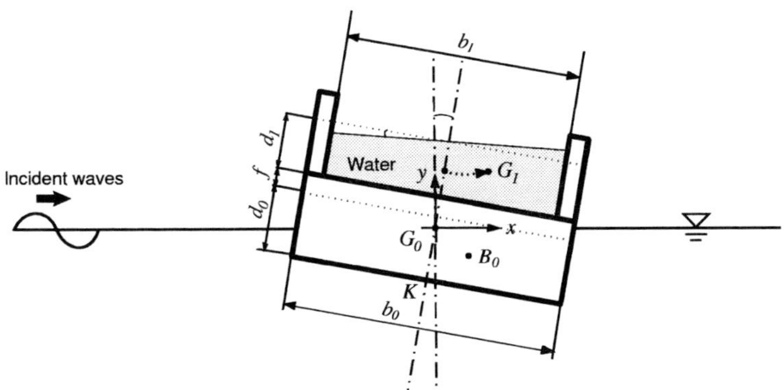

Figure 3: Roll motion of a box-shaped floating body in beam sea

We assume that 1) the coupling of roll motion and water-on-deck is dominant, and sway and heave modes can be neglected, 2) the surface of water-on-deck is flat with the slope χ, 3) the motion of water-on-deck can be approximated by that of a material particle located at the center of gravity G_1, 4) the exciting roll moment varies sinusoidally with the same angular frequency as the incident waves Ω, and 5) the damping forces on a ship and water-on-deck vary linearly with $\dot{\phi}$ and $\dot{\chi}$, respectively. Set the origin at the center of gravity of the floating body G_0 and take the coordinates as shown in fig.3. In order to derive equations of the coupled motion, it may be convenient to use Lagrange's equations of motion with the kinetic energy K, the potential energy P, and the dissipation energy D as follows:

$$\begin{aligned} \frac{d}{dt}\left(\frac{\partial L}{\partial \dot{\phi}}\right) - \frac{\partial L}{\partial \phi} + \frac{\partial D}{\partial \dot{\phi}} &= 0 \, , \\ \frac{d}{dt}\left(\frac{\partial L}{\partial \dot{\chi}}\right) - \frac{\partial L}{\partial \chi} + \frac{\partial D}{\partial \dot{\chi}} &= 0 \, , \end{aligned} \tag{1}$$

where the Lagrangian $L = K - P$. The kinetic energy K and the potential energy P are given by sums of them of each system, namely $K = K_0 + K_1$ and $P = P_0 + P_1 + P_e$ where the subscripts $_0$, $_1$, and $_e$ denote the floating body, the water-on-deck, and the exciting roll moment, respectively. They can be written in the form

$$
\begin{aligned}
&K_0 = \frac{1}{2}(I+\delta I)\dot{\phi}^2 = \frac{1}{2}M\kappa^2\dot{\phi}^2 \ , \ K_1 = \frac{1}{2}m\left(\dot{x_{G_1}}^2 + \dot{y_{G_1}}^2\right) \ , \\
&P_0 = -\rho V g y_{B_0} = -(M+m) g y_{B_0} \ , \ P_1 = m g y_{G_1} \ , \ P_e(\phi,t) = -\phi A_m \sin(\Omega t + \psi) \ , \\
&D = \frac{1}{2}s_0\dot{\phi}^2 + \frac{1}{2}s_1\dot{\chi}^2 \ ,
\end{aligned}
\tag{2}
$$

where I and δI denote the moment and the added moment of inertia about the axis of roll, M and m the masses of the body and of the water-on-deck, κ the radius of gyration, $\boldsymbol{x}_{G_1} = (x_{G_1}, y_{G_1})$ the position of the center of gravity of the water-on-deck G_1, $\boldsymbol{x}_{B_0} = (x_{B_0}, y_{B_0})$ the position of the center of buoyancy of the floating body B_0, A_m the amplitude of the exciting roll moment, ψ the phase difference between the incident waves and the exciting roll moment, and s the damping coefficent, respectively. The positions of G_1 and B_0 can be geometrically obtained as follows:

$$
\begin{aligned}
&\begin{cases}
x_{B_0} = \overline{G_0M_0}\sin\phi + \frac{1}{2}\overline{B_0M_0}\tan^2\phi\sin\phi \\
y_{B_0} = -\frac{1}{2}\overline{B_0M_0}\tan\phi\sin\phi - (\overline{B_0M_0} - \overline{G_0M_0})\cos\phi
\end{cases}
&& \text{for } |\phi| < \phi_1 \\
&\begin{cases}
x_{B_0} = \mathrm{sgn}(\phi)\frac{b_0}{2}\cos\phi - \mathrm{sgn}(\phi)\frac{2}{3}\sqrt{b_0 d_0}\frac{\cos 2\phi}{\sqrt{\sin 2|\phi|}} - \overline{KG_0}\sin\phi \\
y_{B_0} = -\mathrm{sgn}(\phi)\frac{b_0}{2}\sin\phi + \frac{2}{3}\sqrt{b_0 d_0}\sqrt{\sin 2|\phi|} - \overline{KG_0}\cos\phi
\end{cases}
&& \text{for } |\phi| > \phi_1
\end{aligned}
\tag{3}
$$

$$
\begin{aligned}
&\begin{cases}
x_{G_1} = \overline{B_1M_1}\tan\chi\cos\phi + \left(\overline{G_0B_1} + \frac{1}{2}\overline{B_1M_1}\tan^2\chi\right)\sin\phi \\
y_{G_1} = -\overline{B_1M_1}\tan\chi\sin\phi + \left(\overline{G_0B_1} + \frac{1}{2}\overline{B_1M_1}\tan^2\chi\right)\cos\phi
\end{cases}
&& \text{for } |\chi| < \chi_1 \\
&\begin{cases}
x_{G_1} = \mathrm{sgn}(\chi)\frac{b_1}{2}\cos\phi - \mathrm{sgn}(\chi)\frac{2}{3}\sqrt{b_1 d_1}\frac{\cos(\phi+\chi)}{\sqrt{\sin 2|\chi|}} \\
\qquad + (d_0 + f - \overline{KG_0})\sin\phi \\
y_{G_1} = -\mathrm{sgn}(\chi)\frac{b_1}{2}\sin\phi + \mathrm{sgn}(\chi)\frac{2}{3}\sqrt{b_1 d_1}\frac{\sin(\phi+\chi)}{\sqrt{\sin 2|\chi|}} \\
\qquad + (d_0 + f - \overline{KG_0})\cos\phi
\end{cases}
&& \text{for } |\chi| > \chi_1
\end{aligned}
\tag{4}
$$

where d_0 denotes the draft, d_1 the depth of water-on-deck, b_0 the width of the floating body, b_1 the width of the vehicle deck, and f the freeboard of the floating body as shown in fig.3, and $\overline{B_0M_0} = b_0^2/(12d_0)$, $\overline{B_1M_1} = b_1^2/(12d_1)$, $\tan\phi_1 = 2d_0/b_0$, and $\tan\chi_1 = 2d_1/b_1$, respectively. The center of buoyancy of free water B_1 is located at the same position as G_1. Substitution of eqs.(2)~(4) into eq.(1) produces model equations. They can be rewritten in the form

$$\frac{d\boldsymbol{x}}{dt} = \boldsymbol{F}(t, \boldsymbol{x}, \boldsymbol{\lambda}) \,, \tag{5}$$

where $\boldsymbol{x} = [\phi, \dot{\phi}, \chi, \dot{\chi}]^T$, ^{T} denotes the transpose, and $\boldsymbol{\lambda}$ a set of parameters, respectively.

Numerical solutions of the model equations

The model equations (5) were numerically solved using the 4th order Runge-Kutta method. Figures 4.(a),(b), and (c) represent some examples of the computed results under the conditions of T_i=1.44sec., b_0=0.681m, d_0=0.186m, κ=0.253m, $\overline{GM}$=0.069m, b_1/b_0=0.72, m/M=0.19, $A_m/(M\kappa^2)$=0.02, $s_0/(M\kappa^2)$=0.03, $s_1/(M\kappa^2)$=0.08, and ψ=0.0, and with the different initial conditions ($\phi(0)$,$\chi(0)$,$\dot{\phi}(0)$,$\dot{\chi}(0)$). The time history and the phase portrait of the roll angle $\phi(t)$ and the slope of the surface of water-on-deck $\chi(t)$ are displayed. These numerical solutions indicate that different modes of motion can coexist in the system of the model equations (5). This nonlinear phenomenon is similar to the experimental result very well.

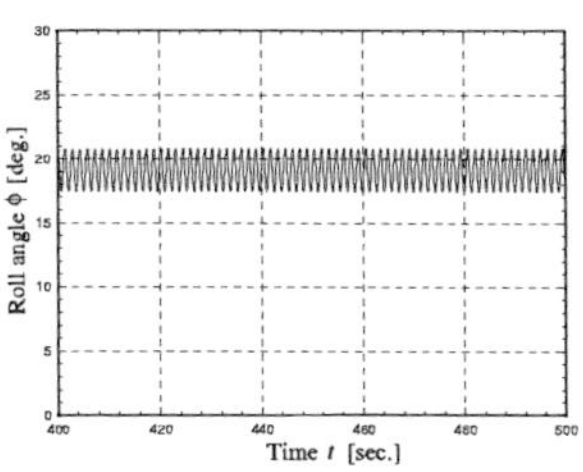

(a.1) Time history of ϕ

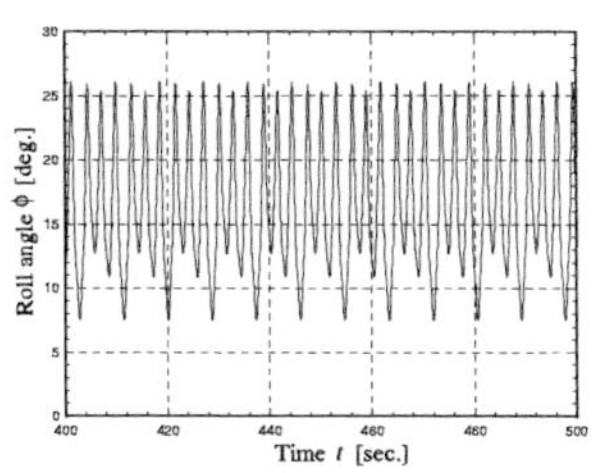

(b.1) Time history of ϕ

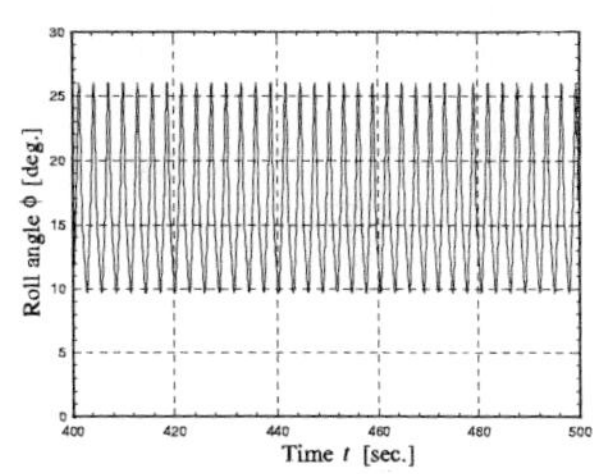

(c.1) Time history of ϕ

Figure 4: Computed results (ϕ:roll angle. χ:slope of the surface of water-on-deck.) ((a)$\phi(0)=\chi(0)$=24.0deg., (b)$\phi(0)=\chi(0)$=28.8deg., (c)$\phi(0)=\chi(0)$=28.9deg.. ($\dot{\phi}(0)$,$\dot{\chi}(0)$)=(0.0,0.0) for all.)

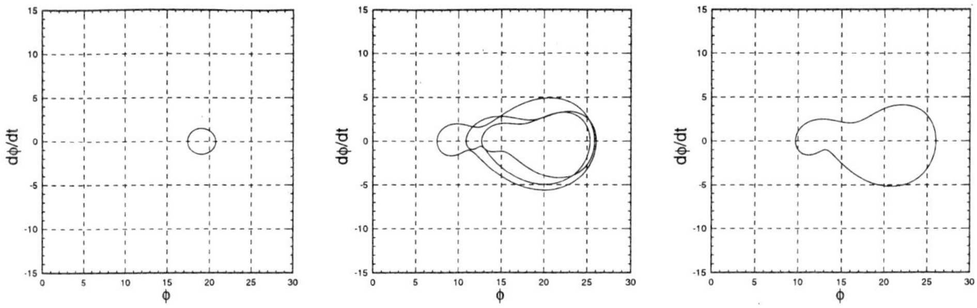

(a.2) Phase portrait $(\phi,\dot{\phi})$ (b.2) Phase portrait $(\phi,\dot{\phi})$ (c.2) Phase portrait $(\phi,\dot{\phi})$

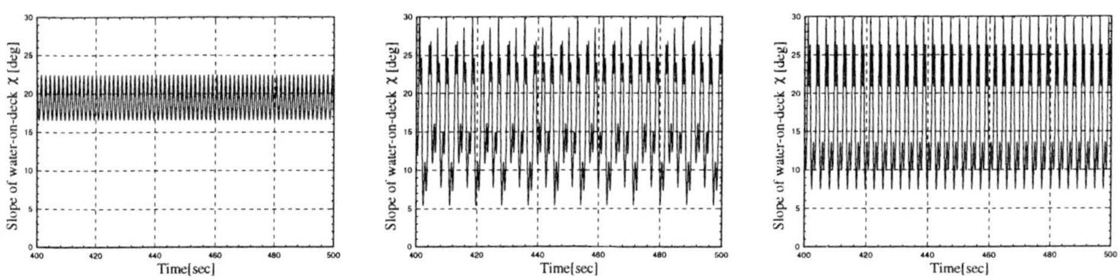

(a.3) Time history of χ (b.3) Time history of χ (c.3) Time history of χ

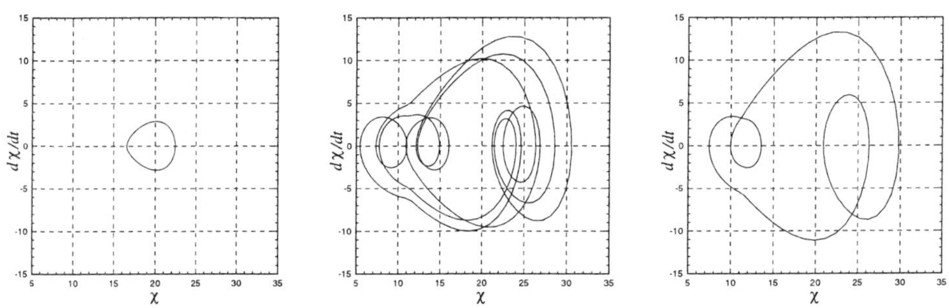

(a.4) Phase portrait $(\chi,\dot{\chi})$ (b.4) Phase portrait $(\chi,\dot{\chi})$ (c.4) Phase portrait $(\chi,\dot{\chi})$

Figure 4: Continued.
((a)$\phi(0)=\chi(0)=24.0$deg., (b)$\phi(0)=\chi(0)=28.8$deg., (c)$\phi(0)=\chi(0)=28.9$deg..
$(\dot{\phi}(0),\dot{\chi}(0))=(0.0,0.0)$ for all.)

BIFURCATION OF THE MATHEMATICAL MODEL

Bifurcation analysis of a system with discontinuous properties

The sectional shape of the ship under the still water surface and that of the flooded water are trapezoidal for $|\phi| < \phi_1$ and $|\chi| < \chi_1$, and triangular for $|\phi| > \phi_1$ and $|\chi| > \chi_1$, respectively. It should be noted that the trajectories of $\boldsymbol{x}_{B_0}$ and $\boldsymbol{x}_{G_1}$ with changing ϕ and χ are continuous at ϕ_1 and χ_1, but that the derivatives with respect to ϕ, χ, $\dot{\phi}$, and $\dot{\chi}$ can be discontinuous there. Thus, the matrix of the partial derivatives of the 4-dimensional vector $\boldsymbol{F}$, namely the Jacobian matrix $\dfrac{\partial \boldsymbol{F}}{\partial \boldsymbol{x}}$, is discontinuous at $\phi = \phi_1$ and $\chi = \chi_1$. This property should be carefully treated in the stability analysis. In order to present the method for finding bifurcation sets of this type of system, consider the following system:

$$\frac{d\boldsymbol{x}}{dt} = \boldsymbol{F} = \begin{cases} \boldsymbol{F}_1(t, \boldsymbol{x}, \boldsymbol{\lambda}) & \text{for } g(\boldsymbol{x}) > 0 \\ \boldsymbol{F}_2(t, \boldsymbol{x}, \boldsymbol{\lambda}) & \text{for } g(\boldsymbol{x}) < 0 \end{cases} \tag{6}$$

where $\dfrac{\partial \boldsymbol{F}}{\partial \boldsymbol{x}}$ is discontinuous at $g(\boldsymbol{x}) = 0$. The solution of this system with the initial values $\boldsymbol{x}(t = t_0) = \boldsymbol{x}_0$ can be expressed as

$$\boldsymbol{x} = \boldsymbol{\varphi}(t, \boldsymbol{x}_0, \boldsymbol{\lambda}) \ . \tag{7}$$

For stability analysis of N-periodic solutions with the period $N\dfrac{2\pi}{\Omega}$, we define the Poincaré map $\boldsymbol{T}$ defined by

$$\boldsymbol{T}: \ \boldsymbol{x}_0 \ \mapsto \ \boldsymbol{T}(\boldsymbol{x}_0) = \boldsymbol{\varphi}(t = t_0 + \frac{2\pi}{\Omega}, \boldsymbol{x}_0, \boldsymbol{\lambda}) \ . \tag{8}$$

The N-periodic solutions satisfy the fixed point condition as follows:

$$\boldsymbol{x}_0 = \boldsymbol{T}^N(\boldsymbol{x}_0, \boldsymbol{\lambda}) \ . \tag{9}$$

We can examine stability of the N-periodic solutions by eigenvalues μ_i (i=1~4) of $\dfrac{\partial \boldsymbol{T}^N}{\partial \boldsymbol{x}_0}$. The eigenvalues μ_i (i=1~4) are solutions of the characteristic equations

$$\left| \mu \boldsymbol{I} - \frac{\partial \boldsymbol{T}^N}{\partial \boldsymbol{x}_0} \right| = 0 \ , \tag{10}$$

where $\boldsymbol{I}$ denotes the 4×4 identity matrix. Assume that the Jacobian matrix $\dfrac{\partial \boldsymbol{F}}{\partial \boldsymbol{x}}$ is discontinuous at $t = t_K$ ($t_0 < t < t_M$ and $K = 1, 2, \cdots, M-1$), namely $g(\boldsymbol{x}) = 0$ at $t = t_K$, and that $t_M = t_0 + N\dfrac{2\pi}{\Omega}$. Then $\dfrac{\partial \boldsymbol{T}^N}{\partial \boldsymbol{x}_0}$ can be written in the form

$$\begin{aligned} \frac{\partial \boldsymbol{T}^N}{\partial \boldsymbol{x}_0} &= \frac{\partial \boldsymbol{x}_M}{\partial \boldsymbol{x}_0} \\ &= \frac{\partial \boldsymbol{x}_M}{\partial \boldsymbol{x}_{M-1}} \cdot \frac{\partial \boldsymbol{x}_{M-1}}{\partial \boldsymbol{x}_{M-2}} \cdots \frac{\partial \boldsymbol{x}_1}{\partial \boldsymbol{x}_0} \\ &= \frac{\partial \boldsymbol{T}_{M-1}}{\partial \boldsymbol{x}_{M-1}} \cdot \frac{\partial \boldsymbol{T}_{M-2}}{\partial \boldsymbol{x}_{M-2}} \cdots \frac{\partial \boldsymbol{T}_0}{\partial \boldsymbol{x}_0} \ , \end{aligned} \tag{11}$$

where

$$\begin{cases} \boldsymbol{T}_0; & \boldsymbol{x}_0 \longmapsto \boldsymbol{x}_1 = \boldsymbol{\varphi}_0(t_1(\boldsymbol{x}_0, \boldsymbol{\lambda}), \boldsymbol{x}_0, \boldsymbol{\lambda}) \\ \boldsymbol{T}_1; & \boldsymbol{x}_1 \longmapsto \boldsymbol{x}_2 = \boldsymbol{\varphi}_1(t_2(\boldsymbol{x}_1, \boldsymbol{\lambda}), \boldsymbol{x}_1, \boldsymbol{\lambda}) \\ & \vdots \\ \boldsymbol{T}_{M-1}; & \boldsymbol{x}_{M-1} \longmapsto \boldsymbol{x}_M = \boldsymbol{\varphi}_{M-1}(t_M = t_0 + N\frac{2\pi}{\Omega}, \boldsymbol{x}_{M-1}, \boldsymbol{\lambda}) \end{cases} \tag{12}$$

Here it should be noted that the time t_K depends on the initial values $\boldsymbol{x}_{K-1}$ and the parameters $\boldsymbol{\lambda}$ for $K = 1, 2, \cdots, M-1$. Thus $\dfrac{\partial \boldsymbol{T}_K}{\partial \boldsymbol{x}_K}$ is given by

$$\begin{aligned} \frac{\partial \boldsymbol{T}_K}{\partial \boldsymbol{x}_K} = \frac{\partial \boldsymbol{x}_{K+1}}{\partial \boldsymbol{x}_K} &= \frac{\partial}{\partial \boldsymbol{x}_K}\boldsymbol{\varphi}_K(t_{K+1}(\boldsymbol{x}_K, \boldsymbol{\lambda}), \boldsymbol{x}_K, \boldsymbol{\lambda}) \\ &= \frac{\partial \boldsymbol{\varphi}_K}{\partial \boldsymbol{x}_K} + \frac{\partial \boldsymbol{\varphi}_K}{\partial t}\frac{\partial t_{K+1}}{\partial \boldsymbol{x}_K} \\ &= \frac{\partial \boldsymbol{\varphi}_K}{\partial \boldsymbol{x}_K} + \boldsymbol{F}\frac{\partial t_{K+1}}{\partial \boldsymbol{x}_K}, \end{aligned} \tag{13}$$

where $\dfrac{\partial \boldsymbol{\varphi}_K}{\partial \boldsymbol{x}_K}$ and $\dfrac{\partial t_{K+1}}{\partial \boldsymbol{x}_K}$ can be obtained using

$$\begin{aligned} \frac{d}{dt}\left(\frac{\partial \varphi_{Ki}}{\partial x_{Kj}}\right) &= \frac{\partial F_i}{\partial x_k}\frac{\partial \varphi_{Kk}}{\partial x_{Kj}} \quad \text{with} \quad \left.\frac{\partial \varphi_{Ki}}{\partial x_{Kj}}\right|_{t=t_K} = \begin{cases} 1 & \text{for } i = j \\ 0 & \text{for } i \neq j \end{cases}, \\ \frac{\partial t_{K+1}}{\partial x_{Ki}} &= -\frac{1}{\dfrac{\partial g}{\partial x_k}F_k}\frac{\partial g}{\partial x_j}\frac{\partial \varphi_{Kj}}{\partial x_{Ki}}. \end{aligned} \tag{14}$$

We can compute the eigenvalues μ_i (i=1~4) for the N-periodic solutions using eqs.(10), (11), (13), and (14).

An example of bifurcation analysis

Numerical solutions in fig.4 showed coexistence of the two periodic solutions, 'period-1' ($\boldsymbol{x}_0 = \boldsymbol{T}(\boldsymbol{x}_0)$) and 'period-2' ($\boldsymbol{x}_0 \neq \boldsymbol{T}(\boldsymbol{x}_0)$ and $\boldsymbol{x}_0 = \boldsymbol{T}^2(\boldsymbol{x}_0)$). Figure 5(a) shows variation of strobed values $\phi(t_0 + mT_i)$ of 'period-1' and 'period-2' in fig.2 with A_1. We can see that these two types of motion coexist for $0.0067 < A_1 < 0.0294$.

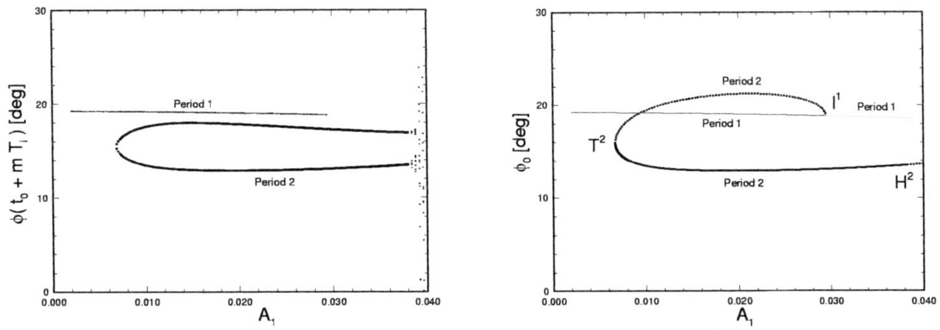

(a) Variation of strobed values of ϕ with A_1 (m=1,2,$\cdots$,30)

(b) Variation of the fixed point $\boldsymbol{x}_0 = \boldsymbol{T}^N(\boldsymbol{x}_0, \boldsymbol{\lambda})$ with A_1 (solid line: stable ($|\mu|_{max} < 1$), dotted line: unstable ($|\mu|_{max} > 1$))

Figure 5: Bifurcation of the N-periodic solutions with A_1 (N= 1 and 2) (I^N: inverse (period doubling), T^N: tangent (saddle-node), H^N: Hopf, Wave period: T_i=6.98.)

We can follow the periodic solutions and determine the bifurcation points by the fixed point condition eq.(9) and the Newton method (Kawakami(1984)). This algorithm enables us to trace not only stable but also unstable solutions. Figure 5(b) shows the computed results, namely variation of the fixed point $\boldsymbol{x}_0 = \boldsymbol{T}^N(\boldsymbol{x}_0)$ with A_1. In fig.5(b), the solid and the dotted line show $|\mu|_{max} < 1$ and $|\mu|_{max} > 1$, namely stable and unstable solutions, respectively. We can see that the inverse or period doubling bifurcation of 'period-1' occurs at I^1 ($A_1 = 0.0294$), the tangent or saddle-node bifurcation of 'period-2' at T^2 ($A_1 = 0.0067$), and the Hopf bifurcation of 'period-2' at H^2 ($A_1 = 0.0383$). At H^2, the 'period 2' solution bifurcates to a quasi-periodic solution with increase of A_1. It should be noted that both 'period-1' and 'period-2' coexist in a wide range of A_1 as shown in figs.5(a) and (b). In many cases, the amplitude of subharmonic motion is larger than that of 'period-1' as shown in fig.4. Thus, for safety of ship motion, a control technique for maintaining 'period-1' may be desired. In addition, the quasi-periodic solutions for $A_1 > 0.0383$ can develop into chaotic ones with further increase of A_1. We will study such a bifurcation elsewhere.

CONCLUSIONS

We have investigated nonlinearly coupled motion of a ship and water-on-deck in regular beam seas. Experiments using a model ship showed coexistence of different modes of roll motion even in regular waves of moderate amplitude. We derived the mathematical model for coupled motion of roll and flooded water in regular waves. This model produced numerical solutions similar to some of experimental results. The Jacobian matrix of this model has a discontinuous property. We presented the method to find bifurcation points of this type of system. An example of the computed results showed that both small harmonic and large subharmonic motion can coexist in a wide range of a parameter. We need to proceed the bifurcation analysis for further understanding of complicated phenomena of this nonlinearly coupled dynamics.

References

Dillingham, J. (1981). Motion Studies of a Vessel with Water on Deck, *Marine Technology*, **18 : 1**, 38-50.

Caglayan, I. and Storch, R.L. (1982). Stability of Fishing Vessels with Water on Deck: A Review, *J. Ship Res.*, **26 : 2**, 106-116.

Murashige, S. and Aihara, K. (1998). Experimental Study on Chaotic Motion of a Flooded Ship in Waves, *Proc. R. Soc. Lond. A*, **454**, 2537–2553.

Virgin, L.N. (1987). The Nonlinear Rolling Response of a Vessel including Chaotic Motions leading to Capsize in Regular Seas, *Applied Ocean Research*, **9 : 2**, 89–95.

Soliman, M.S. and Thompson, J.M.T. (1991). Transient and Steady State Analysis of Capsize Phenomena, *Applied Ocean Research*, **13 : 2**, 82–92.

Thompson, J.M.T., Rainey, R.C.T., and Soliman, M.S. (1992). Mechanics of Ship Capsize under Direct and Parametric Wave Excitation, *Phil. Trans. R. Soc. Lond. A* **338**, 471–490.

Thompson, J.M.T. (1997). Designing against Capsize in Beam Seas: Recent Advances and New Insights, *Appl. Mech. Rev.*, **50**, 307–325.

Kan, M. and Taguchi, H. (1992). Chaos and Fractals in Loll Type Capsize Equations, *Trans. West-Japan Soc. of Naval Arch.*, **83**, 131–149.

Falzarno, J.M., Shaw, S.W., and Troesch, A.W. (1992). Application of Global Methods for Analyzing Dynamical Systems to Ship Rolling Motion and Capsizing, *Intl. J. Bifurcation and Chaos*, **2**, 101–115.

Kawakami, H. (1984). Bifurcation of Periodic Responses in Forced Dynamic Nonlinear Circuits: Computation of Bifurcation Values of the System Parameters, *IEEE Trans. Circuits and Systems*, **CAS-31**, 248–260.

ACKNOWLEDGEMENTS

The authors thank Professor Hiroshi Kawakami, Dr. Tetsuya Yoshinaga, and Dr. Tetsushi Ueta of Tokushima University for their helpful discussions.

Contemporary Ideas on Ship Stability
D. Vassalos, M. Hamamoto, A. Papanikolaou and D. Molyneux (Editors)

EFFECTS OF SOME SEAKEEPING/ MANOEUVRING ASPECTS ON BROACHING IN QUARTERING SEAS

N. Umeda*

National Research Institute of Fisheries Engineering,
Ebidai, Hasaki, Ibaraki, 314-0421, Japan

ABSTRACT

The author has already proposed a method for predicting critical conditions of broaching by applying non-linear dynamical system approach to steady states of a surge-sway-yaw-roll model in quartering seas. (Umeda & Renilson, 1992; Umeda & Vassalos, 1996) Reminding that this method involves several seakeeping and manoeuvring aspects, the author, in this paper, investigates effects of these seakeeping/ manoeuvring aspects on broaching. As a result, several conclusions are presented: 1) for predicting broaching, it is essential to consider the hydrodynamic lift due to wave particle velocity; 2) an empirical prediction for manoeuvring coefficients is not sufficient to predict broaching; 3) the broaching prediction depends on a prediction method for forward speed effect of roll damping.

KEYWORDS

broaching, seakeeping, manoeuvring, nonlinear dynamics, quartering seas, roll damping

NOMENCLATURE

A_R	rudder area
B	ship breadth
C_b	block coefficient
d	ship mean draught

* Address correspondence to: N. Umeda, Department of Naval Architecture and Ocean Engineering, Osaka University, 2-1 Yamadaoka, Suita, Osaka, 565-0871, Japan

d_a	ship aft draught
d_f	ship fore draught
D	ship depth
D_p	propeller diameter
GM_0	metacentric height
H	wave height
K_R	rudder gain
l.c.b.	longitudinal position of centre of buoyancy
L	ship length between perpendiculars
p	roll rate
r	yaw rate
S_F	wetted surface area
T_D	time constant for differential control
T_E	time constant for steering gear
T_ϕ	natural roll period
u	surge velocity
v	sway velocity
δ	rudder angle
κ_{yy}	gyro radius in pitch
κ_{zz}	gyro radius in yaw
λ	wave length
Λ	rudder aspect ratio
ξ_G	longitudinal position of centre of gravity from a wave crest
ϕ	roll angle
χ	heading angle from a wave direction

INTRODUCTION

In naval architecture, like many other engineering, phenomena have been categorised and a suitable methodology, which has been developed for each category, provides useful outputs. However, it is noteworthy that some phenomena existing on a borderline among established categories remain unclarified. One example of these phenomena is "broaching" of a ship. Broaching is a phenomenon that a ship cannot maintain her desired constant course in spite of the maximum steering effort and then suffers a violent yaw motion. Because this violent yaw motion may cause capsizing, broaching is crucial to ship stability. This phenomenon is often experienced when a ship runs in quartering seas at relatively high speed. Since wave action is a main cause of broaching, broaching can be regarded as a subject of seakeeping. On the other hand, broaching can be dealt as a matter of manoeuvring, because steering and yaw motion are essential to broaching. Therefore, a mathematical model for realising broaching involves several elements from both seakeeping and manoeuvring fields.

Another essential character of broaching is "non-linear." While broaching does not occur with small wave steepness, it may occur with larger wave steepness. This non-linearity is different from both that in seakeeping and that in manoeuvring. In seakeeping field, non-linearity often means hydrodynamic one due to a non-linear free surface and hull surface conditions, which may result in wave breaking, impact pressure and so on. In manoeuvring field, non-linearity usually means hydrodynamic one due to a curvature of shed vortex layer from a hull, which results in the rudder dead band found in a directionally unstable ship. Non-linearity causing broaching is a dependence of horizontal displacements on wave forces. Non-dimensional wave exciting forces treated in a seakeeping theory are functions of time only. If we assume them to be functions of time and displacements, broaching can be realised.

As a result of these substantial non-linearity, whether broaching occurs or not depends on an initial condition very much. (Motora et al., 1982) Therefore, it is very difficult to assess a global picture of broaching by simply repeating numerical simulations or model experiments. To overcome this difficulty, Umeda & Renilson (1992; 1993; 1994) carried out non-linear dynamic analysis focusing on an equilibrium point as one of the steady states of ship motion in quartering seas. The equilibrium point corresponds to a surf-riding. Their work showed that the equilibrium point can be easily unstable if steering effort is not enough. It is further pointed out that an invariant manifold from this unstable equilibrium point represents a typical trajectory of broaching. However, the existence of unstable equilibrium point does not directly lead to broaching. If a stable periodic motion also exists, the ship does not meet a danger of broaching. Thus, Umeda & Vassalos (1996) investigated a periodic motion, its stability and outstructure as one of other steady states in the same mathematical model, by making use of an averaging method. This work demonstrated that the periodic motion becomes unstable when the encounter frequency becomes small. Therefore, combining the above two procedures, we can explain broaching as follows. (Umeda, 1996) *When a ship suffering a periodic motion in following and quartering seas increases her propeller revolution, stability of the periodic motion decreases and surf-riding equilibria potentially emerge. Eventually the ship may leave the periodic motion and be attracted by a surf-riding equilibrium. In other words, the ship is accelerated up to almost the wave celerity. Since this equilibrium is a saddle, the ship is firstly attracted by the equilibrium and then repelled. Even if the maximum opposite rudder angle is commanded immediately after approaching to the equilibrium, the ship's yaw angle may increase at a wave downslope near a wave trough.* Furthermore, the numerical results based on the above method were compared with existing results (Hamamoto et al., 1996) with a free running model. (Umeda et al., 1997) This comparison demonstrated that critical condition can be well predicted with a surge-sway-yaw-roll model. It was also suggested that the wave effect on the roll restoring moment contributes to improve the agreement, although an improvement in modelling this effect is necessary.

This combined method enables us to quantitatively predict critical condition for broaching. However, since the method involves several seakeeping and manoeuvring aspects, final prediction results may depend on these aspects. Therefore, this paper investigates effects of several seakeeping and manoeuvring aspects on broaching. They covers wave force prediction, manoeuvring coefficients prediction and roll damping prediction. This investigation aims at contributing toward a practical use of the broaching prediction method mentioned above.

OUTLINE OF PREDICTION METHOD

The prediction method for broaching used in this paper has been already proposed by the author. The details of the method and their background are available in the reference (Umeda & Renilson, 1992; Umeda & Vassalos, 1996; Umeda, 1999B). The symbols are defined in the nomenclature.

The state vector **x** of this system is defined as follows:

$$\mathbf{x} = \left\{\xi_G / \lambda,\ u,\ v,\ \chi,\ r,\ \phi,\ p,\ \delta\right\}^T \quad (1).$$

The dynamical system can be represented by the following state equation:

$$\dot{\mathbf{x}} = \mathbf{F}(\mathbf{x}) = \left\{f_1(\mathbf{x}),\ f_2(\mathbf{x}),\ \cdots,\ f_8(\mathbf{x})\right\}^T \quad (2)$$

where the functions of $f_i(\mathbf{x})$ $(i = 1,\cdots,8)$ are shown in the companion paper (Umeda, 1999B). Since the external forces are functions of the horizontal displacement but not time, this equation is non-linear and autonomous.

At an equilibrium point of this system all elements of the state vector do not change with time. Thus the ship in this equilibrium is required to keep a certain relative position to wave with a certain drift angle, heading angle, heel angle and rudder angle. This is known as so-called "surf-riding." The stability of this equilibrium can be assessed with eigenvalues of locally linearised state equation at this equilibrium point. Further, the outstructure of this equilibrium can be obtained with eigenvectors and invariant manifold analysis.

One of other important steady states of the system described in Eqn. (2) is a periodic motion whose period is equal to the encounter one. This motion, known as a harmonic motion, is rather common. Although a linear strip theory can deal with the harmonic motion only, such linear theories are not useful enough to assess its stability. There are several geometrical or analytical methods to determine periodic motions, their stability and outstructure. For practical use it is desirable that a method does not involve tedious computations of time series. Thus the author applied an averaging method to harmonic motions in quartering seas. The averaging theorem guarantees that a fixed point of the averaged equation and its eigenvalues correspond to a periodic motion of the original equation and its stability. In the method that the author proposed for broaching both equilibrium points and periodic motions are treated in parallel.

If periodic motions become unstable and a stable equilibrium does not exist, the ship may experience some motions other than these. If an unstable equilibrium exists, the ship can be attracted by the equilibrium point and then repelled with a violent yaw motion. This is definitely broaching. Since a helmsman should command the maximum opposite rudder angle for preventing broaching, whether an unstable equilibrium point exists with the maximum opposite rudder angle or not can be a criterion. Here it is noteworthy that broaching is a transient state. The process of broaching sometimes ends in capsizing. If not, the broaching may end in large heading angle. As a result, the relative velocity of a ship to waves becomes too large to keep equilibrium and then the ship returns to a periodic motion. After that the ship can suffer broaching again. This series of events has been often reported and can be regarded as a kind of periodic attractor of the system, which may be called as "repeating broaching".

NUMERICAL RESULTS AND DISCUSSION

The above method was applied to a 135 GT purse seiner, the principal particulars of which are shown in Table 1. The wave-induced sway force, yaw moment and roll moment are theoretically estimated as sums of the Froude-Krylov components and hydrodynamic lift due to wave particle velocity. (Umeda et al., 1995) The propulsive and manoeuvring coefficients are assumed to be independent of waves, and obtained with captive model experiments, such as a computer-controlled circular motion test, in a seakeeping and manoeuvring basin in National Research Institute of Fisheries Engineering (NRIFE). The added masses and moments of inertia were estimated theoretically or empirically. The roll damping moment was estimated with a linear component of the roll damping moment measured without a forward velocity and corrected for forward velocity with Takahashi's empirical method. (Lewis, 1989) The roll restoring moment in still water is fitted with a fifth order polynomial. To simulate a PD auto pilot used in the model experiments, auto pilot parameters are provided: K_R=1.0, T_D=1.24 sec and δ_{max}=15 degrees. Throughout this paper, the wave condition recorded in the experiment is used: H/λ=1/15 and λ/L=1.5.

The numerical results are shown in Figure 1 with results of free running model experiments by Hamamoto et al. (1996) The periodic motion is explored in sequence from lower speed to higher speed. The step of increasing the Froude number is 0.001. The Froude number, where a real part of eigenvalues reaches zero or the amplitude of rudder angle reaches its limit, is regarded as the upper limit of stable periodic motions and shown with a solid line. The equilibrium points are explored in sequence from χ=0, the pure following seas. The dashed line indicates a limit of stable equilibrium and the dotted line does unstable equilibrium with the maximum rudder angle. The estimated zone of stable periodic motions includes the observed periodic motions and excludes other motions observed in the experiments. It involves a jump when the heading angle is about 10 degrees. In the zone where both periodic and equilibrium points are unstable, unstable equilibrium with the maximum opposite rudder angle exists. This theoretically indicates that broaching can occur near the unstable equilibrium point in the zone. As shown in this figure, the experiments support this theoretical prediction. Similar results were presented in a separate paper (Umeda et al., 1997) but equilibria were obtained with a surge-sway-yaw model in that paper. In this paper, both periodic motions and equilibrium points are obtained with the surge-sway-yaw-roll model. In the separate paper, the wave effect on roll restoring moment was also discussed but it is excluded throughout this paper for a simplicity sake.

TABLE 1 PRINCIPAL PARTICULARS OF THE PURSE SEINER.

L	34.5 [m]	κ_{yy}/L	0.316
B	7.6 [m]	κ_{zz}/L	0.316
D	3.07 [m]	GM_0	0.75 [m]
d_f	2.50 [m]	T_ϕ	8.9 [sec]
d_a	2.80 [m]	A_R	3.49[m²]
C_b	0.597	Λ	1.84
l.c.b. (aft)	1.31 [m]	T_E	0.47[sec]
S_F	324. [m²]	K_R	1.0
D_p	2.60[m]	T_D	1.24[sec]

Effect of hydrodynamic lift due to wave particle velocity

Although the comparison in forces and moments shows that hydrodynamic lift is essential for wave forces, it has not yet been fully established whether hydrodynamic lift due to wave particle velocity is essential for prediction of broaching or not. In this line of thinking, Spyrou & Umeda (1996) showed that equilibrium points and their stability depend on the hydrodynamic lift, in other words, the diffraction component, very much and this paper discussed both equilibria and periodic motions and their stability.

Figure 2 shows numerical results without the hydrodynamic lift due to wave particle velocity. A comparison of Figures 1-2 indicates that no significant difference exists in the upper limit of stable periodic motion. On the other hand, the results without the hydrodynamic lift enlarges a zone for a stable equilibrium point of surf-riding, which is overlapped with a zone for a stable periodic motion. In this overlapped region, a ship motion depends on initial condition. (Umeda, 1990) By ignoring the hydrodynamic lift due to wave particle velocity, an unstable equilibrium point with the maximum opposite rudder angle disappears. Thus, Figure 2 does not indicate a possibility of broaching at all. These are because a wave-induced sway force and yaw moment are significantly reduced. Therefore, the hydrodynamic lift due to wave particle velocity is essential for broaching prediction.

Effect of manoeuvring coefficients prediction

Even nowadays it is still difficult to theoretically predict manoeuvring coefficients. The author has mainly relied on captive model experiments to obtain them for broaching prediction. However, it is desirable to predict them without making a model, especially in an initial design stage. For this purpose several empirical methods have been proposed. Thus this paper examines whether one of the most reliable empirical methods can be used for broaching prediction or not. The methods used here are Inoue's method for hull forces and Kijima's method for rudder forces, which have been developed with results of captive model tests with many cargo ship models and are often utilised for a practical ship design. (Kijima et al., 1990) In addition, for coefficients related to roll, the existing experimental data for container ships (Hirano & Takashina, 1979) are used in place of those for this purse seiner. These empirical prediction were compared with the captive model tests, as shown in Table 2. Except for N_{Hv}', x_H/L and z_H/d, fairly good comparisons are obtained. Reminding that experimental data used for these empirical methods do not cover fishing vessel at all, these results should be regarded as exceptionally good.

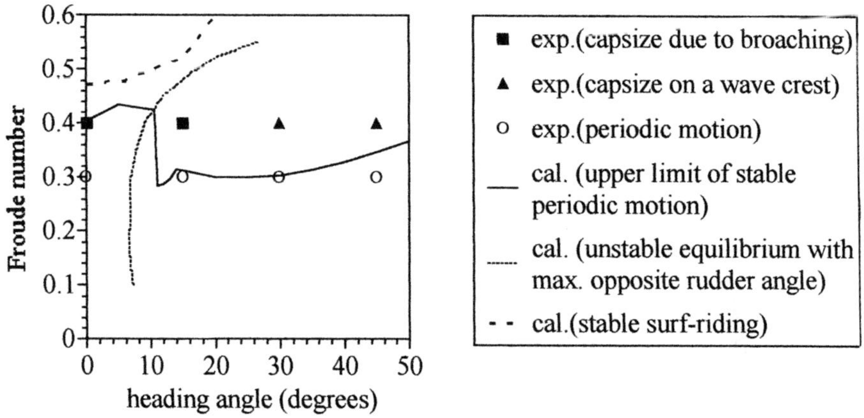

Figure. 1 Critical conditions for broaching

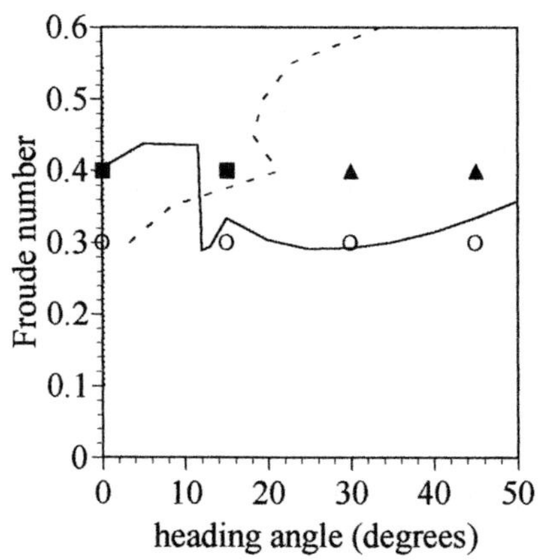

Figure 2 Critical conditions without hydrodynamic lift due to wave particle velocity

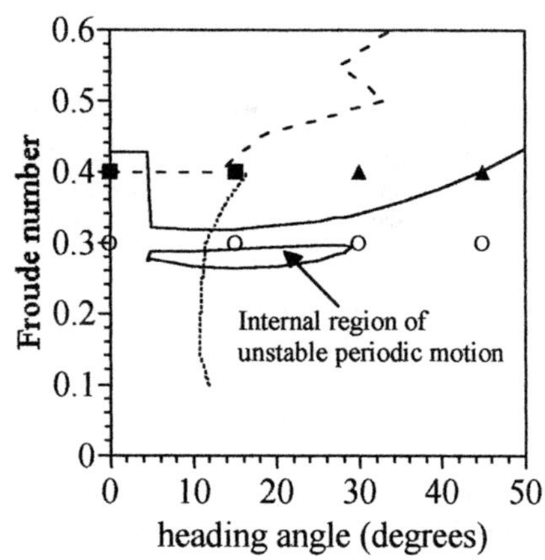

Figure 3 Critical conditions with manoeuvring coefficients estimated empirically

Figure 3 shows, however, that numerical results with empirically predicted manoeuvring coefficients are not only quantitatively but also qualitatively different from those with directly measured manoeuvring coefficients. The zone for a stable periodic motion shrinks at smaller heading angle but enlarges at larger heading angle. In addition, an internal region of an unstable periodic motion emerges. The zone for a stable equilibrium significantly enlarges and the unstable equilibrium with the maximum opposite rudder angle does not exist when the Froude number is greater than 0.4. This means that the broaching dealt here cannot occur when the Froude number is greater than 0.4. Broaching is rather sensitive to even small change in manoeuvring coefficients. Therefore, we should deliberately use empirical method for manoeuvring coefficients to predict critical condition of broaching. While manoeuvring in still water is crucial for a blunt ship, such as a supertanker, broaching occurs mainly for a ship running at a relatively high-speed, such as a fishing vessel. Thus, it is necessary to develop an empirical method for manoeuvring coefficients with experimental data of such small and high-speed ships.

TABLE 2 MANOEUVRING COEFFICIENTS

	items	captive experiment	empirical method
Y_{Hv}'	non-dimensional partial derivatives of hull sway force with respect to sway velocity	-0.588	-0.586
Y_{Hr}'	non-dimensional partial derivatives of hull sway force with respect to yaw rate	0.179	0.175
N_{Hv}'	non-dimensional partial derivatives of hull yaw moment with respect to sway velocity	-0.0431	-0.111
N_{Hr}'	non-dimensional partial derivatives of hull yaw moment with respect to yaw rate	-0.0909	-0.0760
a_H	interaction factor between hull and rudder	0.312	0.313
x_H/L	longitudinal position of centre of interaction force between hull and rudder	-0.321	-1.44
κ_P	interaction factor between propeller and rudder	0.595	0.5
ε_R	wake ratio between propeller and rudder	0.825	1.01
γ_R	flow-straightening effect coefficient	0.296	0.296
l_R/L	correction factor for flow-straightening effect due to yaw rate	-0.55	-0.90
z_H/d	height of centre of hull sway force	0.428	0.750

Effect of roll damping prediction

While most of damping forces in seakeeping almost consist of a wave-making component only, roll damping consists of a wave-making, eddy-making, lift and frictional components. In particular, the eddy-making component mainly induces non-linearity in roll damping. This non-linearlity makes capsizing predictions difficult. However, in the case of broaching, where forward velocity is very high and encounter frequency is small, this explanation is not always applicable. As an experimental work of Ikeda et al. (1988), showed, measured roll damping can be regarded as linear when the Froude number is greater than 0.2. Since eddies flow away at high speed, the eddy-making component, that is non-linear, disappears. In addition, a wave-making component is not significant because of small encounter frequency. Therefore, roll damping relating to broaching consists of mainly a lift component, which is linear and depends on forward velocity.

Based on the above thinking, the author estimates roll damping with the measured linear component and corrected with Takahashi's method for forward velocity. Here a question, whether final results for broaching prediction depend on the empirical method for forward velocity, is raised. Thus the author carried out a numerical study for broaching with Ikeda's method for the lift component as well as that without forward velocity effect. In Ikeda's method the lift component is estimated by using a reverse flow theorem with measured value of Y_{Hv}'. (Ikeda et al., 1988)

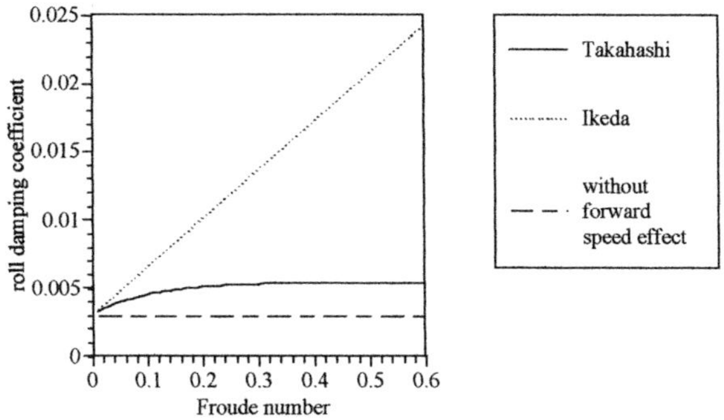

Figure. 4 Non-dimensional roll damping moment estimated empirically.

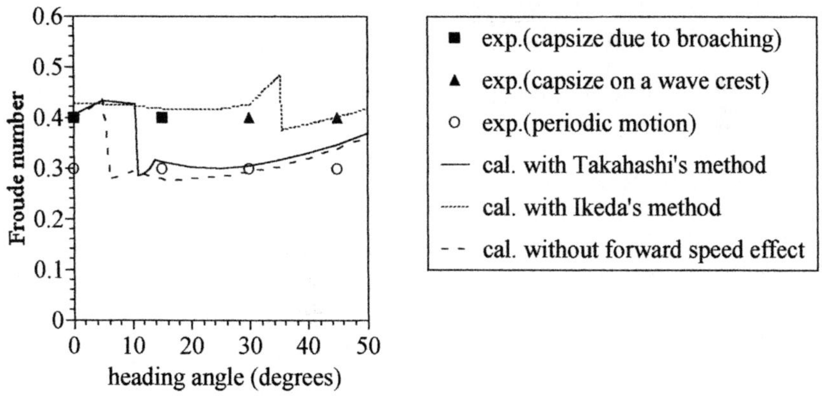

Figure 5 Effect of roll damping on stability of periodic motions

Figure 4 shows a comparison in roll damping itself among three methods. Here the roll damping is non-dimesionalised as follows:

$$\hat{B}_{44} = B_{44}\sqrt{B/(2g)}\Big/\left(mB^2\right) \tag{3}$$

where B_{44} is the roll damping coefficient and m is the ship mass; B and g are the ship breadth and the gravitational acceleration, respectively. Figures 5-6 show the upper limit of stable periodic motions and equilibria, respectively. While no significant difference in equilibria among three methods exists, some differences in periodic motions among three exist. When the heading angle is smaller than 10 degrees, the method without forward speed effect shows significant difference from other two methods. When the heading angle is larger than 10 degrees, Ikeda's method shows significant difference from other two methods. Thus we can conclude that the broaching prediction depends on the roll damping prediction. Since the roll decay test with a forward velocity are not available for this ship, further experimental investigation are required to prove a more reliable guidance for broaching prediction.

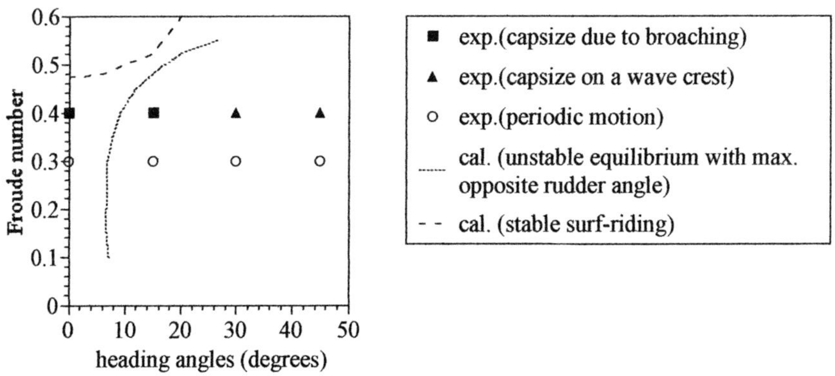

Figure 6 Effect of roll damping on surf-riding equilibria

ACKNOWLEDGEMENTS

This research was partly supported by the Grant-in-Aid for Scientific Research of the Ministry of Education, Science and Culture, Japan, and the author would like to acknowledge the support of the project leader, Professor M. Hamamoto. The author are also grateful for appropriate discussion from Professor Y. Ikeda.

REFERENCES

Hamamoto M., Enomoto T., Sera W., Panjaitan J.P., Ito H., Takaishi Y., Kan M., Haraguchi T. and Fujiwara T. (1996). Model Experiments of Ship Capsize in Astern Seas –Second Report-. *Journal of the Society of Naval Architects of Japan*. **179**, 77-87.

Hirano M. and Takashina J (1979). A Calculation of Ship Turning Motion Taking Coupling Effect due to Heel into Consideration. *Transaction of the West-Japan Society of Naval Architects*. **59**, 71-81.

Ikeda Y., Umeda N. and Tanaka N. (1988). Effect of Forward Speed on Roll Damping of a High-Speed Craft (in Japanese). *Journal of the Kansai Society of Naval Architects, Japan*. **208**, 27-34.

Kijima K., Katsuno T., Nakiri Y. and Furukawa Y. (1990). On the Manoeuvring Performance of a Ship with the Parameter of Loading Condition. *Journal of Society of Naval Architects of Japan*. **168**, 141-148.
Lewis E.V. (1989). *Principal Naval Architecture*, Soc of Nav Archit and Mar Eng, Jersey City, USA, **3**, 82.
Motora S., Fujino M. and Fuwa T. (1982). On the Mechanism of Broaching-To Phenomena. In: *Proceedings of the 2nd International Conference on Stability of Ships and Ocean Vehicles*, Soc of Nav Archit of Japan, Tokyo, 535-550.
Spyrou K.J. and Umeda N. (1995). From Surf-Riding to Loss of Control and Capsize: A Model of Dynamic Behaviour of Ships in Following / Quartering Seas. In: *Proceedings of the 6th International Symposium on Practical Design of Ships and Mobile Units*, Soc Nav Archit of Korea, Seoul, **1**, 494-505.
Umeda N. (1990). Probabilistic Study on Surf-Riding of a Ship in Irregular Following Seas. In: *Proceedings of the 4th International Conference on Stability of Ships and Ocean Vehicles*. University Federico II of Naples, Naples, Italy, 336-343.
Umeda N. (1996). Some Remarks on Broaching Phenomenon. In: *Proceedings of the 2nd International Workshop on Stability and Operational Safety o f Ships*, Osaka Uni, Osaka, Japan, 10-23.
Umeda N. (1999A). Nonlinear Dynamics of Ship Capsizing due to Broaching in Following and Quartering Seas. *Journal of Marine Science and Technology*. **4:1**, (in press).
Umeda N. (1999B). Application of Nonlinear Dynamical System Aproach to Ship Capsize due to Broaching in Following and Quartering Seas. In: Vassalos D. (editor) *Contemporary Ideas on Ship Stability*, Elsevier Science, Amsterdam, the Netherland, (to be appeared).
Umeda N, Matsuda A., Hamamoto M. and Suzuki S. (1999). Stability Assessment for Intact Ships in the Light of Model Experiments *Journal of Marine Science and Technology*. **4:2**. (in press).
Umeda N. and Renilson M.R. (1992). Broaching - A Dynamic Behaviour of a Vessel in Following Seas -. In: Wilson P.A. (editor) *Manoeuvring and Control of Marine Craft*. Computational Mechanics Publications, Southampton, UK, 533-543.
Umeda N. and Renilson M.R. (1993). Broaching in Following Seas –A Comparison of Australian and Japanese Trawlers. *Bulletin of National Research Institute of Fisheries Engineering*. **14**, 175-186.
Umeda N. and Renilson M.R. (1994). Broaching of a Fishing Vessel in Following and Quartering Seas. In: *Proceedings of 5th International Conference on Stability of Ships and Ocean Vehicles*. Florida Tech, Melbourne, USA, **3**, 115-132.
Umeda N. and Vassalos D. (1996). Non-Linear Periodic Motions of a Ship Running in Following and Quartering Seas. *Journal of the Society of Naval Architects of Japan*. **179**, 89-101.
Umeda N., Vassalos D. and Hamamoto M. (1997). Prediction of Ship Capsize due to Broaching in Following / Quartering Seas. In: *Proceedings of the 6th International Conference on Stability of Ships and Ocean Vehicles*, Varna, **1**, 45-54.

Contemporary Ideas on Ship Stability
D. Vassalos, M. Hamamoto, A. Papanikolaou and D. Molyneux (Editors)

SHIP MANOEUVRING PERFORMANCE IN WAVES

K. Kijima and Y. Furukawa

Department of Marine Systems Engineering, Faculty of Engineering, Kyushu University, 6-10-1, Hakozaki, Higashi-ku, Fukuoka, 812-8581, Japan

ABSTRACT

External forces acting on a ship induced by wave will be one of the important factor to predict ship manoeuvring performance. This paper deals with influence due to wave on ship manoeuvrability. Furthermore the parameters of metacentric height and ship speed also have much influence on ship manoeuvrability. Generally a container ship or RORO ship may often have small $\overline{GM}$ depending on loading condition and roll motion would be induced by steerage. From the numerical calculation in this paper, it is found that ship manoeuvring performance would be different depending on $\overline{GM}$ and ship speed and performance indices such as advance, tactical diameter and overshoot angles vary due to influence of wave.

KEYWORDS

Manoeuvrability, Waves, Numerical Simulation, Roll Motion, $\overline{GM}$, Criteria

INTRODUCTION

Interim standards of ship manoeuvrability adopted by International Maritime Organization (IMO) provides ship manoeuvring performance. Sea trial is one of the useful methods to evaluate ship manoeuvring performance and the sea trial should be carried out in full load and even keel condition at deep and unrestricted water as provided in the imterim standards. The explanatory note in the interim standards of ship manoeuvrability shows sea state 4 as allowable condition. However it is not still clarified that how much are the performance indices, such as advance, tactical diameter or overshoot angles, influenced by wave. Then the influence due to wave should be investigated. Furthermore conditions of $\overline{GM}$ or ship speed are not provided though these parameters also will have much influence on ship manoeuvrability. Generally a container ship or

RORO ship may often have small $\overline{GM}$ and large $\overline{KG}$ depending on loading conditions and roll motion would be induced by steerage. Then we examined ship manoeuvring motion in waves by numerical simulations including effect of roll motion.

EQUATIONS OF SHIP MANOEUVRING MOTION AND MATHEMATICAL MODEL

Equations of Ship Manoeuvring Motion

Figure 1 shows the coordinate systems for formulation of ship manoeuvring motion. o - x_0y_0 shows coordinate system fixed on earth, and G - xy shows coordinate system fixed on the center of gravity of ship. χ_0 indicates initial incident angle of wave. The equations of ship motion including roll motion can be written in following forms,

$$\left.\begin{aligned}
(m'+m'_x)\left(\frac{L}{U}\right)\left(\frac{\dot{U}}{U}\cos\beta-\dot{\beta}\sin\beta\right)+(m'+m'_y)r'\sin\beta &= X',\\
-(m'+m'_y)\left(\frac{L}{U}\right)\left(\frac{\dot{U}}{U}\sin\beta+\dot{\beta}\cos\beta\right)+(m'+m'_x)r'\cos\beta &= Y',\\
(I'_{zz}+i'_{zz})\left(\frac{L}{U}\right)^2\left(\frac{\dot{U}}{L}r'+\frac{U}{L}\dot{r}'\right) &= N',\\
(I'_{xx}+i'_{xx})\left(\frac{L}{U}\right)^2\left(\frac{\dot{U}}{L}p'+\frac{U}{L}\dot{p}'\right) &= K'.
\end{aligned}\right\} \quad (1)$$

The superscript " $'$ " refers to the nondimensional quantities as follows,

$$\left.\begin{aligned}
&m',m'_x,m'_y=\frac{m,m_x,m_y}{\frac{1}{2}\rho L^2 d}, \quad I'_{zz},i'_{zz},I'_{xx},i'_{xx}=\frac{I_{zz},i_{zz},I_{xx},i_{xx}}{\frac{1}{2}\rho L^4 d},\\
&X',Y'=\frac{X,Y}{\frac{1}{2}\rho L d U^2}, \quad N',K'=\frac{N,K}{\frac{1}{2}\rho L^2 d U^2}, \quad r',p'=\frac{r,p}{U}L,
\end{aligned}\right\} \quad (2)$$

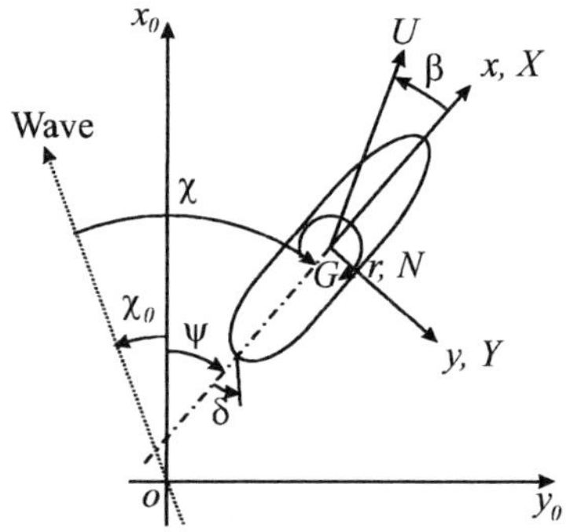

Figure 1: Coordinate systems

where,

L, d	:	ship length and draft,
m	:	ship mass,
m_x, m_y	:	x and y-axis components of added mass of ship,
I_{zz}, I_{xx}	:	moment of inertia according to z and x-axes,
i_{zz}, i_{xx}	:	added moment of inertia according to z and x-axes,
U, β	:	ship speed and drift angle,
r, p	:	yaw rate and roll rate of ship ($r = \dot{\psi}, p = \dot{\phi}$),
ψ, ϕ	:	heading angle and heel angle,
X, Y	:	x and y-axis components of external force acting on a ship,
N, K	:	yaw and roll moments acting on a ship,
ρ	:	density of fluid.

As for the expression of external forces and moments shown in the right-hand side of the Eqn. (1), we assumed that X', Y', N' and K' consist of components of hull, propeller, rudder and wave,

$$\left.\begin{aligned} X' &= X'_H + X'_P + X'_R + X'_W, \\ Y' &= Y'_H + Y'_R + Y'_W, \\ N' &= N'_H + N'_R + N'_W, \\ K' &= K'_H + K'_R + K'_W. \end{aligned}\right\} \tag{3}$$

In the Eqn. (3), the subscripts "$_H$", "$_P$" and "$_R$" symbolize ship hull, propeller and rudder respectively. The subscript "$_W$" denotes hydrodynamic forces induced by wave.

Mathematical Model for Hydrodynamic Forces Acting on Ship Hull

For the longitudinal component of the force acting on a ship hull, the following expression was assumed,

$$X'_H = X'_{\beta r} r' \sin\beta + X'_{uu} \cos^2\beta. \tag{4}$$

$X'_{\beta r}$ and X'_{uu} are hydrodynamic derivatives.

We assumed that the lateral force and yaw moment acting on a ship hull consist of two terms. The first terms, Y'_{H0} and N'_{H0}, in the right-hand side of the Eqn. (5) express the force and moment acting on a ship without roll motion and the second terms, Y'_{H1} and N'_{H1}, indicate effect of roll motion,

$$\left.\begin{aligned} Y'_H &= Y'_{H0} + Y'_{H1}, \\ N'_H &= N'_{H0} + N'_{H1}. \end{aligned}\right\} \tag{5}$$

We used following expression for Y'_{H0} and N'_{H0} presented by the authors (Kijima et al. (1990)),

$$\left.\begin{aligned} Y'_{H0} &= Y'_\beta \beta + Y'_r r' + Y'_{\beta\beta} \beta|\beta| + Y'_{rr} r'|r'| + (Y'_{\beta\beta r} \beta + Y'_{\beta rr} r')\beta r', \\ N'_{H0} &= N'_\beta \beta + N'_r r' + N'_{\beta\beta} \beta|\beta| + N'_{rr} r'|r'| + (N'_{\beta\beta r} \beta + N'_{\beta rr} r')\beta r'. \end{aligned}\right\} \tag{6}$$

$Y'_\beta, Y'_r, \cdots$ are hydrodynamic derivatives. As for Y'_{H1} and N'_{H1}, we constructed a mathematical model to express the change of hydrodynamic forces due to roll motion as follows (Kijima et al. (1997)),

$$\left.\begin{aligned} Y'_{H1} &= Y'_\phi\phi + Y'_{\beta\phi}\beta\phi + Y'_{r\phi}r'\phi + Y'_{\beta\beta\phi}\beta|\beta||\phi| + Y'_{rr\phi}r'|r'|\phi + Y'_{\beta\beta r\phi}\beta^2 r'|\phi| + Y'_{\beta rr\phi}\beta r'^2\phi, \\ N'_{H1} &= N'_\phi\phi + N'_{\beta\phi}\beta\phi + N'_{r\phi}r'|\phi| + N'_{\beta\beta\phi}\beta|\beta|\phi + N'_{rr\phi}r'|r'|\phi + N'_{\beta\beta r\phi}\beta^2 r'|\phi| + N'_{\beta rr\phi}\beta r'^2\phi. \end{aligned}\right\} \quad (7)$$

The roll moment acting on a ship hull can be written as follows,

$$K'_H = \left\{ -\frac{2\pi(I_{xx}+i_{xx})}{T_R}\kappa\dot{\phi} - W\cdot\overline{GZ} - Y_H\cdot z_H \right\} / \frac{1}{2}\rho L^2 dU^2, \quad (8)$$

where,

κ	:	extinction coefficient,
T_R	:	period of roll motion,
W	:	displacement of ship,
$\overline{GZ}$	:	righting lever,
z_H	:	vertical distance between the center of gravity of ship and the center of lateral force Y_H acting on ship hull.

The first and second terms in the right-hand side of the Eqn. (8) represent damping moment and restoring moment respectively. The last term is moment induced by the lateral force Y_H acting on a ship hull.

Hydrodynamic Forces and Moments Induced by Wave

The hydrodynamic forces and moments induced by wave can be represented by dividing into two components as follows,

$$\left.\begin{aligned} X'_W &= X'_{WD}, \\ Y'_W &= Y'_{WE} + Y'_{WD}, \\ N'_W &= N'_{WE} + N'_{WD}, \\ K'_W &= K'_{WE} + K'_{WD}. \end{aligned}\right\} \quad (9)$$

The subscripts "$_{WE}$" and "$_{WD}$" denote wave exciting force and wave drifting force respectively. We assumed that the wave exciting forces consist of only Froude-Krilov force and other components are negligible small. We also assumed that wave is regular wave. Under these assumptions, the wave exciting forces can be calculated by following equations (Tasai (1966)),

$$\left.\begin{aligned} Y'_{WE} &= \int_{-L/2}^{L/2} S(x)g\Theta_W S_W(x)\sin\chi\sin(kx\cos\chi - \omega_e t)dx / \frac{1}{2}LdU^2, \\ N'_{WE} &= \int_{-L/2}^{L/2} S(x)g\Theta_W S_W(x)\sin\chi\sin(kx\cos\chi - \omega_e t)xdx / \frac{1}{2}L^2dU^2, \\ K'_{WE} &= Y'_{WE}\left\{\overline{OG} - l(x)\right\}/L, \end{aligned}\right\} \quad (10)$$

$$\left.\begin{aligned} l(x) &= \frac{\{R(x)-P(x)\}\,d(x)}{S(x)}, \\ S(x) &= \frac{2}{S_W(x)k\sin\chi}\int_0^{d(x)} e^{-kz}\sin(ky\sin\chi)dz, \\ P(x)-R(x) &= \frac{2}{d(x)S_W(x)k\sin\chi}\left\{\int_0^{B(x)/2} e^{-kz}\sin(ky\sin\chi)ydy\right. \\ &\qquad \left. -\int_0^{d(x)} e^{-kz}\sin(ky\sin\chi)zdz\right\}, \end{aligned}\right\} \quad (11)$$

where,

B	:	breadth of ship,
Θ_W	:	maximum wave slope,
S_W	:	sectional area under water plane,
χ	:	angle of encounter,
k	:	wave number,
ω_e	:	frequency of encounter,
g	:	gravitational acceleration,
t	:	time,
$\overline{OG}$	:	vertical distance between the center of gravity and the water plane.

As for wave drifting forces, we adopted Newman's prediction formulae for a slender ship (Newman (1967)). We can calculate wave drifting forces using these formulae as function of angle of encounter χ and wave length/ship length ratio λ/L. In the Eqn. (12), asterisks denote the complex conjugate,

$$\left.\begin{aligned} X'_{WD} &= \frac{gk^3}{2LdU^2}\int_{-L/2}^{L/2} B(x)\left(A_W e^{ikx\cos\chi}+i\zeta_3-ix\zeta_5\right) \\ &\quad\times\int_{-L/2}^{L/2} B(\xi)\left(A_W e^{-ik\xi\cos\chi}-i\zeta_3^*-i\xi\zeta_5^*\right) \\ &\quad\times\{J_0(kx-k\xi)\cos\chi+iJ_1(kx-k\xi)\}\,d\xi dx, \\ Y'_{WD} &= \frac{gk^3}{2LdU^2}\int_{-L/2}^{L/2} B(x)\left(A_W e^{ikx\cos\chi}+i\zeta_3-ix\zeta_5\right) \\ &\quad\times\int_{-L/2}^{L/2} B(\xi)\left(A_W e^{-ik\xi\cos\chi}-i\zeta_3^*-i\xi\zeta_5^*\right) \\ &\quad\times J_0(kx-k\xi)\sin\chi d\xi dx, \\ N'_{WD} &= -\frac{gkA_W^2\sin\chi}{L^2dU^2(I_0I_2-I_1^2)}\int_{-L/2}^{L/2} xB(x) \\ &\quad\times\int_{-L/2}^{L/2} B(\xi)(I_2-I_1x+I_0x\xi)\sin\{k(x-\xi)\cos\chi\}\,d\xi dx, \\ K'_{WD} &= Y'_{WD}\cdot z_W/L, \end{aligned}\right\} \quad (12)$$

$$\left.\begin{aligned}\zeta_3 &= \frac{iA_W}{I_0I_2-I_1^2}\int_{-L/2}^{L/2} B(x)(I_2-I_1x)e^{ikx\cos\chi}dx,\\ \zeta_5 &= \frac{iA_W}{I_0I_2-I_1^2}\int_{-L/2}^{L/2} B(x)(I_1-I_0x)e^{ikx\cos\chi}dx,\\ I_n &= \int_{-L/2}^{L/2} B(x)x^n dx \qquad \text{for} \quad i=0,1,2,\end{aligned}\right\} \tag{13}$$

where,

A_W	:	wave amplitude,
z_W	:	vertical distance between the center of gravity and the center of lateral drifting force,
J_0, J_1	:	the Bessel functions of the first kind,
ζ_3, ζ_5	:	heave and pitch amplitude,
I_n	:	waterline moments.

NUMERICAL SIMULATIONS OF SHIP MANOEUVRING MOTION IN WAVES

Verification of Mathematical Model

We carried out numerical simulations of ship manoeuvring motion such as turning motion and zigzag manoeuvre including influence of wave and roll motion. Table 1 shows principal particulars of the model ship used in the numerical simulations. In the table, principal particulars for full scale ship are also shown for reference.

TABLE 1
PRINCIPAL PARTICULARS OF SHIPS

	Model ship	Full scale ship	Model ship (Hirano et al.)
L(m)	3.0	175.0	5.0
B(m)	0.435	25.4	0.761
d(m)	0.1629	9.5	0.212
C_b	0.5717	0.5717	0.660

First of all, we made a comparison of results of the numerical simulations and experimental results performed by Hirano et al. (1980) to verify the mathematical model for numerical simulation. Hirano et al. carried out model experiments of turning motion in regular wave using a 5.0m long model ship shown in Table 1. The model ship is different from the model ship used in our numerical simulations. Then we can't compare these results directly because both ships originally have inherent manoeuvring performance respectively. However there is few experimental results of ship manoeuvring motion in wave. We considered that characteristic of the

TABLE 2

CONDITIONS OF SHIP SPEED, RUDDER ANGLE AND WAVE

	Case 1		Case 2	
	Author's cal. Figure 2(a)	Hirano's exp. Figure 2(b)	Author's cal. Figure 3(a)	Hirano's exp. Figure 3(b)
Froude number	0.26			
Rudder angle δ	$-15.0°$			
Initial incident wave angle χ_0	180° (head sea)			
Wave height h_W	0.06m	0.10 m	0.06m	0.10m
h_W/L	1/50			
λ/L	1.0		0.75	

effect of wave on ship manoeuvrability would be comparable.

Figure 2 and Figure 3 show comparison of turning trajectories of the model ships for two conditions shown in Table 2. The calculated turning trajectories in regular wave and calm water are shown with a solid line and a break line respectively. On the other hand, Hirano's experimental results of the turning trajectories in regular wave and calm water are shown with black and white circles respectively.

Figure 2 shows turning trajectories in Case 1. The ratio of wave length λ to ship length L is equal to 1.0. It is shown in Figure 2(a) that the calculated turning trajectory deviate slightly from the turning trajectory in calm water. It seems that there exists the same tendency in devi-

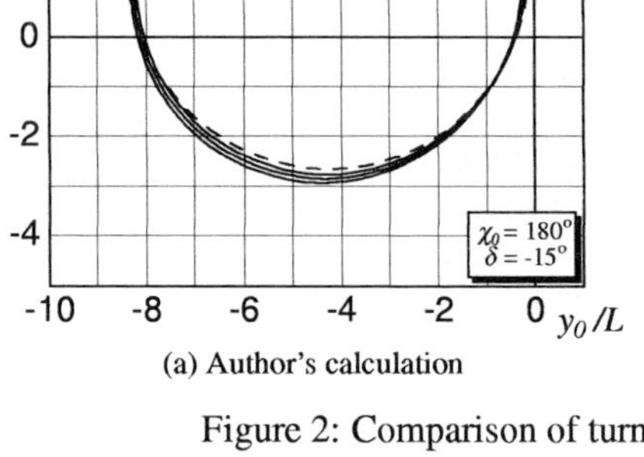

(a) Author's calculation

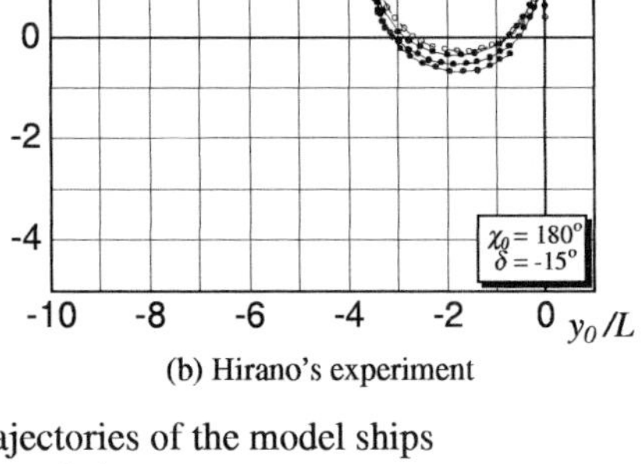

(b) Hirano's experiment

Figure 2: Comparison of turning trajectories of the model ships ($\chi_0 = 180°$, $h_W/\lambda = 0.02$, $\lambda/L = 1.0$)

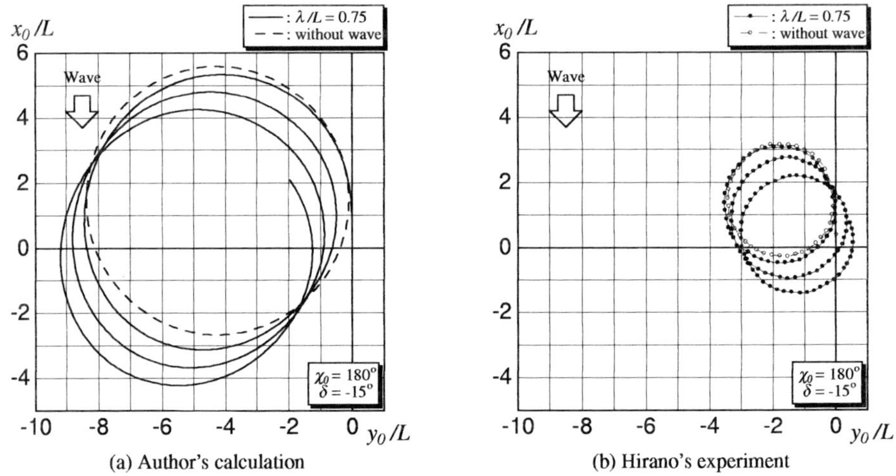

Figure 3: Comparison of turning trajectories of the model ships
($\chi_0 = 180°$, $h_W/\lambda = 0.027$, $\lambda/L = 0.75$)

ation of turning trajectory in Figure 2(b) which shows Hirano's experimental results. Figure 3 shows turning trajectories in Case 2. The ratio of wave length λ to ship length L is equal to 0.75. As the value of wave length/ship length ratio λ/L becomes smaller, it can be seen that deviation of the turning trajectory becomes larger than that in Figure 2. There is discrepancy in direction of deviation of ship trajectory between numerical simulation and experimental results. As we mentioned before, we used different ship model and direction of deviation would be much influenced by inherent manoeuvring performance of ship. However both drifting distances from turning trajectory in calm water shown in Figure 3(a) and Figure 3(b) are not particularly different. Then we considered that our numerical simulation gives appropriate prediction of ship motion in regular wave.

Influence of $\overline{GM}$ and Ship Speed on Ship Manoeuvring Performance

In order to investigate the effect of $\overline{GM}$ and ship speed on ship manoeuvring performance, numerical simulations with various values of $\overline{GM}$ and ship speed were carried out using the mathematical model. We changed values of $\overline{GM}$ from 0.0086m to 0.0343m every 0.0086m and values of ship speed U from 1.078m/sec to 1.617m/sec every 0.135m/sec. These values correspond to following values for full scale ship, such as $\overline{GM}_S$ = 0.5m to 2.0m every 0.5m and U_S = 16kt to 24kt every 2kt. The subscript "S" denotes that the parameters with this subscript express converted values for full scale ship. Hereafter conditions for numerical simulations using these converted values for full scale ship will be presented.

Figure 4(a) shows turning trajectories in calm water with various values of $\overline{GM}_S$. Rudder angle δ and ship speed U_S used in the numerical simulations are 35° and 20kt respectively. In the figure, turning trajectory calculated without effect of roll motion ($\overline{GM}_S$ = 5.0m) is also shown with dotted line for reference. Figure 4(b) also shows turning trajectories for three conditions

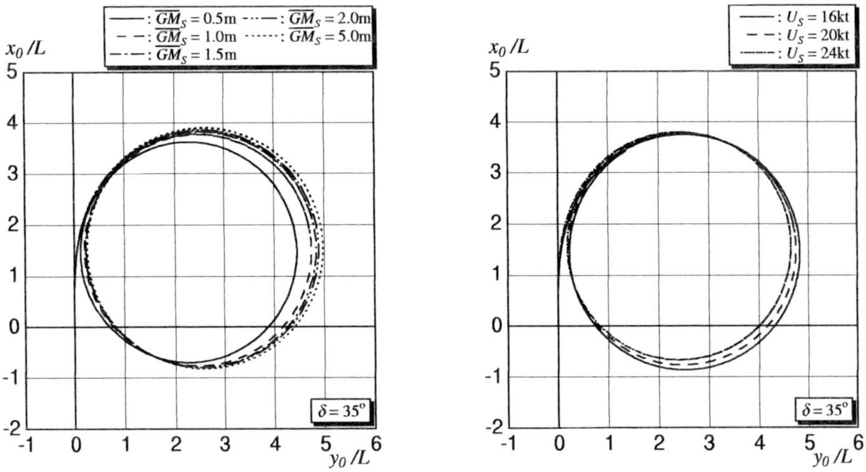

(a) Variation of $\overline{GM}_S$ (U_S = 20kt) (b) Variation of U_S ($\overline{GM}_S$ = 1.0m)

Figure 4: Turning trajectories with variation of either $\overline{GM}_S$ (U_S = 20kt) or ship speed U_S ($\overline{GM}_S$ = 1.0m)

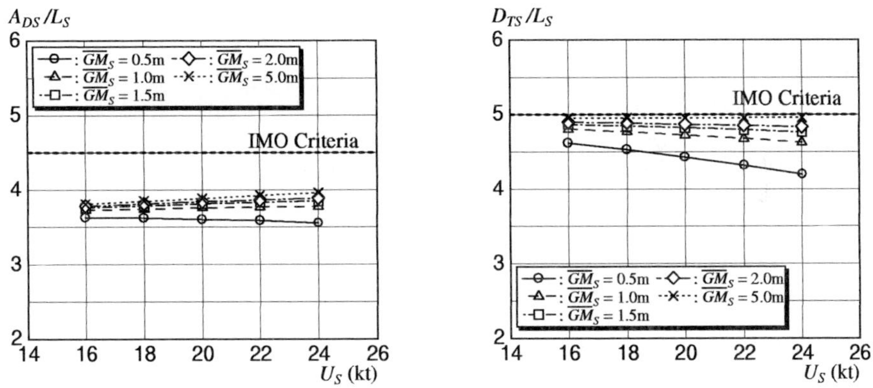

(a) Advance (b) Tactical diameter

Figure 5: Values of advance A_{DS} and tactical diameter D_{TS} as function of U_S and $\overline{GM}$

of ship speed namely U_S is equal to 16, 20 and 24kt respectively. It is understood from these figures that turning radius becomes smaller as value of $\overline{GM}_S$ decreases and ship speed becomes faster.

Nondimensional values of advance A_{DS}/L_S and tactical diameter D_{TS}/L_S in turning motion as function of ship speed U_S for five conditions of $\overline{GM}_S$ are shown in Figure 5 based on above numerical simulations. IMO criteria is also shown in the figures by thick broken line. These figures show that both values of advance and tactical diameter become smaller as values of $\overline{GM}_S$ becomes smaller. Effect of change of ship speed U_S appears remarkably on tactical diameter

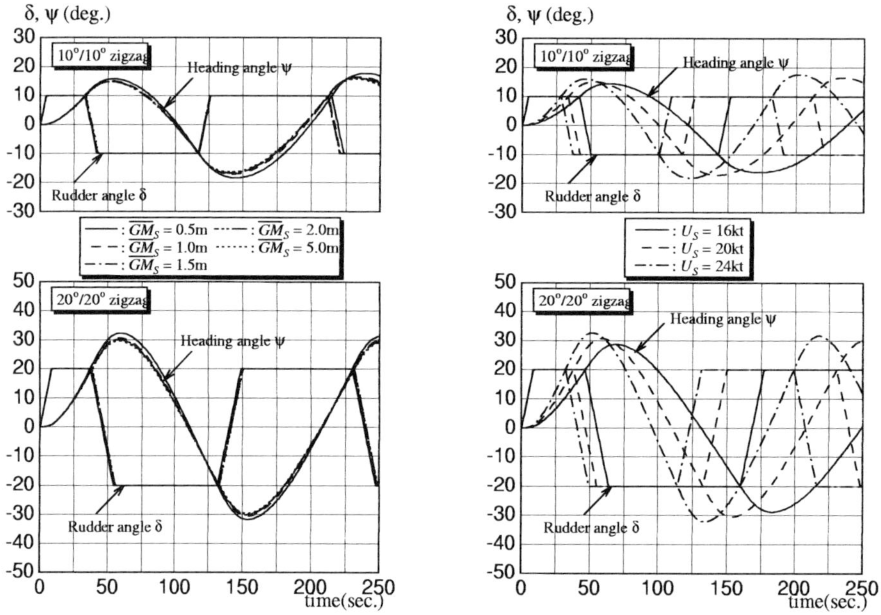

(a) Variation of $\overline{GM}_S$ (U_S = 20kt) (b) Variation of U_S ($\overline{GM}_S$ = 1.0m)

Figure 6: Time histories of heading angle ψ and rudder angle δ with variation of either $\overline{GM}_S$ (U_S = 20kt) or ship speed U_S ($\overline{GM}_S$ = 1.0m)

except for the condition $\overline{GM}_S = 5.0$m.

We also performed numerical simulations of $10°/10°$ and $20°/20°$ zigzag manoeuvre similarly. Time histories of heading angle ψ and rudder angle δ are shown in Figure 6(a) and Figure 6(b) for various values of $\overline{GM}_S$ and U_S respectively. It is found that both 1st and 2nd overshoot angles become larger as value of $\overline{GM}_S$ decreases or ship speed U_S becomes faster. Furthermore phase differences in rudder and heading angles appear remarkably depending on ship speed U_S.

In order to examine the influence of these changes in overshoot angle as performance index in interim manoeuvring standards, Figure 7 shows values of 1st overshoot angle ψ_1 and 2nd overshoot angle ψ_2 for $10°/10°$ and $20°/20°$ zigzag manoeuvre as function of L_S/U_S. It is found that values of overshoot angle become larger as value of $\overline{GM}_S$ becomes smaller and these tendency appears remarkably as ship speed becomes faster, namely value of L_S/U_S becomes smaller.

It is understood from these numerical simulations that the manoeuvring performance changes remarkably depending on $\overline{GM}$ or ship speed. Then $\overline{GM}$ and ship speed are important factors at sea trial to evaluate ship manoeuvring performance precisely. Furthermore, it is also important to take influence of roll motion into account if one evaluates ship manoeuvring performance by

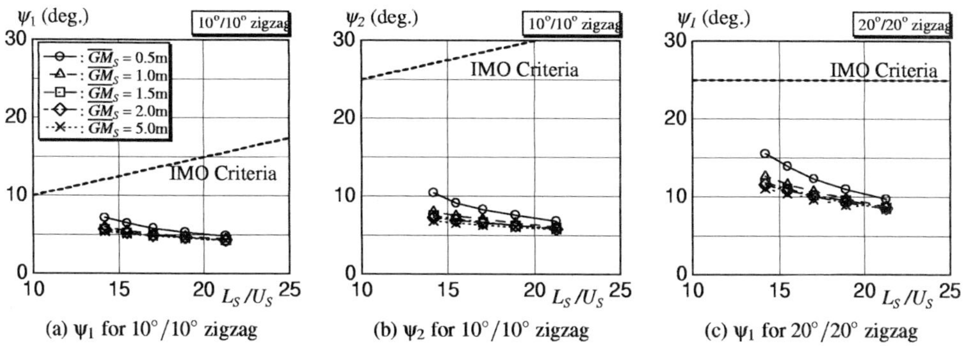

Figure 7: Values of 1st overshoot angle ψ_1 and 2nd overshoot angle ψ_2 for 10°/10° and 20°/20° zigzag manoeuvre as function of L_S/U_S

numerical simulation at design stage.

Influence of Wave on Ship Manoeuvring Performance

As we mentioned in the section "*Verification of Mathematical Model*", turning trajectory in wave will deviate from that in calm water due to influence of wave. It will be considered that performance indices, such as advance, tactical diameter or overshoot angles, should change depending on wave condition. Then numerical simulations in regular wave were done for many wave conditions shown in Table 3.

TABLE 3
CONDITIONS OF SHIP SPEED, RUDDER ANGLE AND WAVE

	Case A	Case B	Case C	Case D
Ship speed U_S	16kt, 18kt, 20kt, 22kt, 24kt			
$\overline{GM}_S$	0.5m, 1.0m, 1.5m, 2.0m, 5.0m			
Initial incident angle of wave χ_0	0° to 360° every 30°			
Wave height h_{WS}	2.0m			
Wave length λ_S	80m	120m	160m	200m
h_{WS}/λ_S	1/40	1/60	1/80	1/100
λ_S/L_S	0.457	0.686	0.914	1.143

Figure 8 shows nondimensional values of advance A_{DS}/L_S and tactical diameter D_{TS}/L_S in turning motion as function of initial incident angle of wave χ_0 for the cases shown in Table 3. Rudder angle δ and ship speed U_S used in the numerical simulations are 35° and 20kt respectively. In the figures, IMO criteria and calculated results without influence of wave and roll motion are also shown by thick broken line. It can be seen that values of advance and tactical diameter fluctuate as function of initial incident angle of wave χ_0 and wave with small value

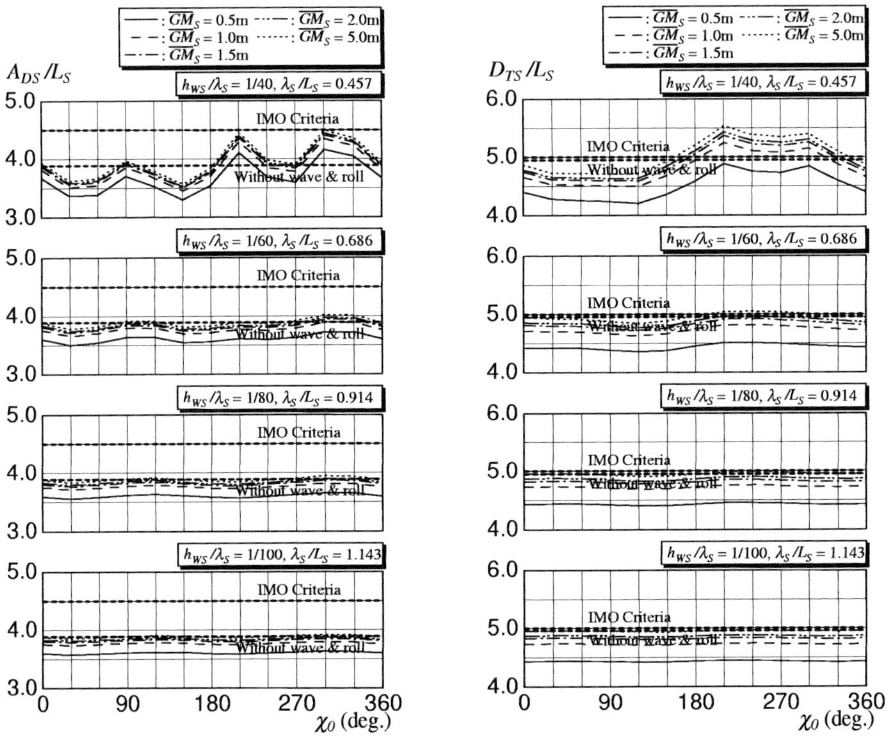

Figure 8: Values of advance A_{DS} and tactical diameter D_{TS} as function of initial incident angle of wave χ_0 ($U_S = 20$kt)

of wave length/ship length ratio λ_S/L_S gives much difference between the values of A_{DS} and D_{TS} in wave and that in calm water. However the values of advance and tactical diameter for Case A may be too large even if the ship manoeuvring motion is much influenced by wave. Possible reason for this may rest with the accuracy of prediction of wave drifting forces acting on ship. The prediction formulae shown in the Eqn. (12) tend to give overestimated lateral drifting forces for small value of λ_S/L_S because of assumption of slender body. Consequently ship manoeuvring motion in wave with small value of λ_S/L_S should be too much affected by wave. On the contrary, we consider that numerical simulations for large value of λ_S/L_S gives reasonable prediction of ship maneuvering motion as shown in Figure 2 and Figure 3. As for advance in wave, the model ship has sufficient margin to IMO criteria. On the other hand, some values of tactical diameter exceed IMO criteria according to χ_0.

Figure 9 indicates values of 1st overshoot angle ψ_1 and 2nd overshoot angel ψ_2 as function of χ_0. There is the same tendency for overshoot angle to vary with χ_0 and the quantity of fluctuation increases with decreasing value of λ_S/L_S. Though wave with large value of λ_S/L_S affects on zigzag manoeuvre, there is a little influence on evaluation of manoeuvring performance be-

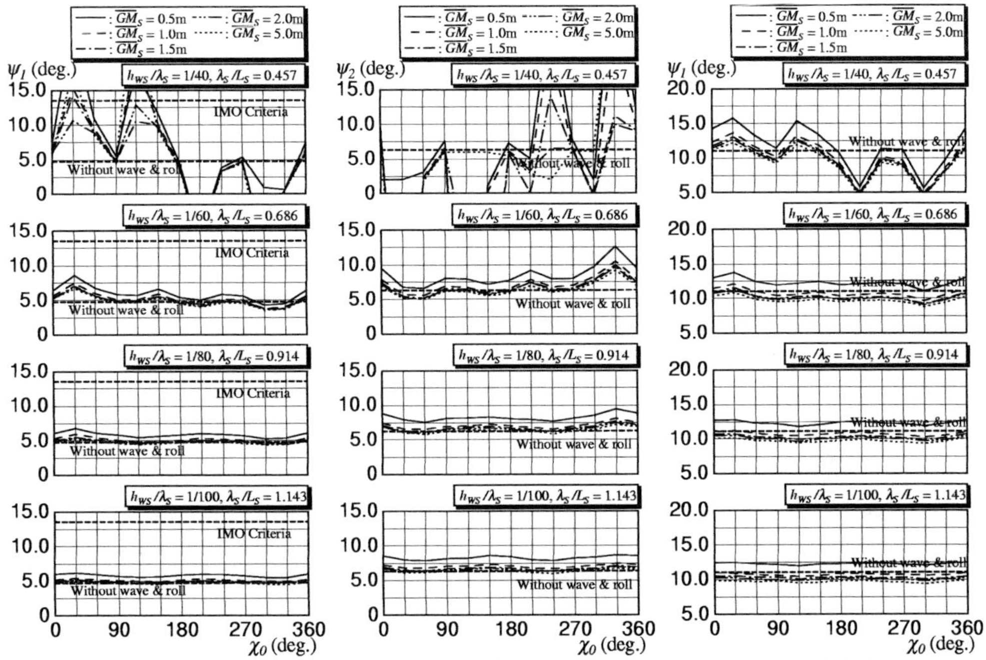

(a) ψ_1 for 10°/10° zigzag (b) ψ_2 for 10°/10° zigzag (c) ψ_1 for 20°/20° zigzag

Figure 9: Values of 1st overshoot angle ψ_1 and 2nd overshoot angle ψ_2 for 10°/10° and 20°/20° zigzag manoeuvre as function of initial incident angle of wave χ_0 (U_S = 20kt)

cause the model ship has sufficient margin to IMO criteria.

It is found from the figures that performance indices such as advance, tactical diameter or overshoot angles change depending on the influence of wave and the indices may exceed IMO criteria according to circumstances. The values of advance in starboard side turning has a tendency that it becomes larger than that in calm water when ship is in quartering sea from port side at initial position. On the other hand, the values of tactical diameter in starboard side turning tends to become larger when the ship get wave from port side at initial position. As for overshoot angles, there is a tendency that value of the 1st overshoot angle becomes larger than that in calm water when ship get wave obliquely. Then conditions of wave should be important factor to evaluate manoeuvring performance precisely by sea trial in waves.

CONCLUSIONS

We carried out numerical simulations of ship manoeuvring motion in regular wave including effect of roll motion. From the numerical simulations, it is found that ship manoeuvring performance would be different depending on $\overline{GM}$ and ship speed U and the performance indices such as advance, tactical diameter and overshoot angles vary due to influence of wave. Especially tactical diameter may exceed IMO criteria according to wave which have small value of wave length/ship length ratio λ/L. Then it will be important to take the influence of roll motion and wave on ship manoeuvrability into account to evaluate the ship manoeuvring performance precisely. However further investigation on prediction of wave drifting force will be necessary to estimate ship manoeuvring motion in waves by numerical simulation from practical point of view.

REFERENCES

Hirano M., Takashina J., Takaishi Y. and Saruta T. (1980). Ship Turning Trajectory in Regular Waves, *Trans. of the West-Japan Society of Naval Architects* **60**, 17-31.

Kijima K., Katsuno T., Nakiri Y. and Furukawa Y. (1990). On the Manoeuvring Performance of a Ship with the Parameter of Loading Condition, *Jour. of the Society of Naval Architects of Japan* **168**, 141-148.

Kijima K., Furukawa Y., Eshima M. and Yamamoto K. (1997). Influence of Roll motion on Ship Manoeuvrability, *Trans. of the West-Japan Society of Naval Architects* **93**, 35-46.

Tasai F. (1966). On the Swaying, Yawing and Rolling Motions of Ships in Oblique Waves, *Trans. of the West-Japan Society of Naval Architects* **32**, 25-40.

Newman J.N. (1967). The Drift Force and Moment on Ships in Waves, *Jour. of Ship Research* **11:1**, 51-60.

Contemporary Ideas on Ship Stability
D. Vassalos, M. Hamamoto, A. Papanikolaou and D. Molyneux (Editors)

STABILITY OF A PLANING CRAFT IN TURNING MOTION

Y. Ikeda [1], H. Okumura [1] and T. Katayama [1]

[1] Department of Marine System Engineering, Osaka Prefecture University, 1-1, Gakuen-Cho, Sakai, Osaka, 599-8531, Japan

ABSTRACT

The stability of a hard-chine planing hull in turning at high speed is examined on the basis of experiments. Six-components of hydrodynamic forces acting on a fully captured model are measured by an oblique towing test at Froude number of 2.0 to 4.4 in which attitudes; trim, rise and heel angle, are changed systematically.

The experimental results show that all hydrodynamic forces are proportional to the square of advanced speed at planing speed. The results also demonstrate that the effects of attitude on the hydrodynamic forces are significant, and that the restoring roll moment decreases rapidly with increasing yaw angle. The reduction of the restoring moment may cause large heel in turning motion.

KEYWORDS

High-Speed, Planing Craft, Maneuverability, Hydrodynamic Forces, Stability, Oblique

INTRODUCTION

For a planing craft in turning at high speed, the stability quality is a very important factor for its safety. However, the characteristics of hydrodynamic forces acting on a planing craft in turning condition have not been clarified yet.

In the present study, six-component hydrodynamic forces acting on a model obliquely towed at constant high speed are measured for various attitudes, rise, heel and trim, in the towing tank of Osaka Prefecture University. The characteristics of the hydrodynamic forces acting on a planing hull are investigated.

EXPERIMENTAL SETUP

The model used in the experiment is a 1/4-scale model of a personal watercraft with water-jet propulsion. The principal particulars of the model are shown in TABLE 1. The hull is hard-chine type, and has a duct without any impeller.

TABLE 1

PRINCIPAL PARTICULARS OF MODEL

length(m)	L	0.630
breadth(m)	B	0.223
depth(m)	D	0.100
draft(m)	d	0.055
ship weight(kgf)	W	5.796
KG(m)		0.107
LCG from transom(m)		0.255
Deadrise angle(degree)		22

The experimental setup is shown in Figure 1. The model is captured by 6-component load cell, and towed by an unmanned carriage the maximum speed of which is 15m/s. The rise H (mm), heel angle ϕ (deg.), trim angle τ (deg.) and yaw angle β (deg.) are systematically changed as shown in TABLE 2. The rise here is defined by vertical displacement of the center point of rotation to change trim angle, which is located at 0.178m from the keel line at midship. The zero levels of all measured forces are set at rest just before starting of the carriage. The measured roll, pitch and yaw moments around the load cell are converted into the values about the standard location of the center of gravity of the craft.

The measured forces are non-dimensionalized as follows,

$$C_{Fx} = \frac{Fx}{0.5\rho SyU^2} \tag{1}$$

$$C_{Fy} = \frac{Fy}{0.5\rho SyU^2} \tag{2}$$

$$C_{Mx} = \frac{Mx}{W \cdot B} \tag{3}$$

$$C_{My} = \frac{My}{0.5\rho SxLU^2} \tag{4}$$

$$C_{Mz} = \frac{Mz}{0.5\rho SyLU^2} \tag{5}$$

$$C_{Fz} = \frac{Fz}{0.5\rho SzU^2} \tag{6}$$

where, *Fx*: resistance, *Fy*: transverse force, *Fx*: vertical lift, *Mx*: roll moment, *My*: trim moment, *Mz*: yaw moment, *Sy*: projected area of wetted body from side, *Sz*: water plane area, *L*: overall length of a craft, B: breadth, W: displacement and ρ: density of water. In them, *Sy* and *Sz* are calculated for each attitude without any disturbance on free surface.

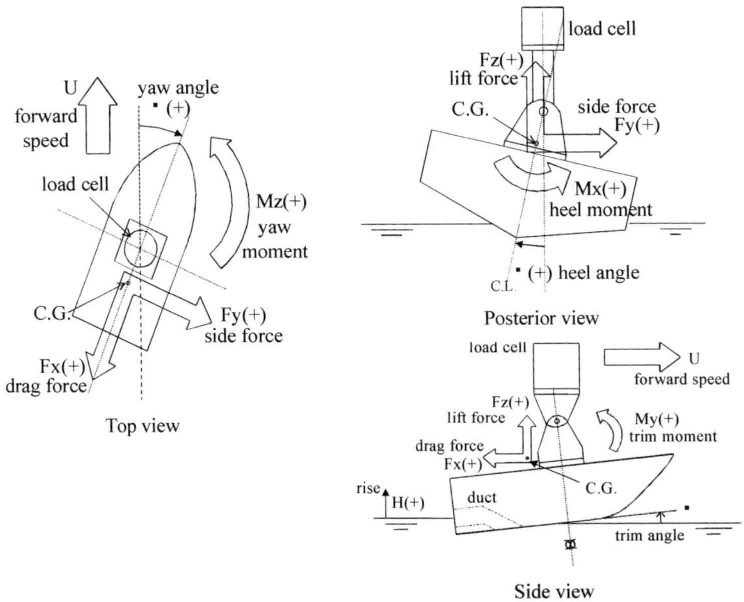

Figure 1: Schematic views of experimental setup and coordinate system

TABLE 2
EXPERIMENTAL CONDITIONS

Condition	A	B	C
Rise H(mm)	20	30	40
Heel ϕ (degree)	20	30	10
Trim τ (degree)	2	4	4
Yaw β (degree)	10	10	10

EXPERIMENTAL RESULTS

Effect of advance speed

Measurements of hydrodynamic forces acting on the hull are carried out for various advanced speeds in the range of Froude number between 2.0 and 4.4. Measured transverse force and roll moments are shown in Figure 2. The results of transverse force are proportional to the square of advance speed as shown in this figure. The results of resistance, vertical lift, trim moment and yaw moment show the same tendency too. Measured roll moment, however, is not proportional to square of speed at high advance speed.

Experimental Results

Some typical results of measured hydrodynamic forces are shown in Figures 3-8.

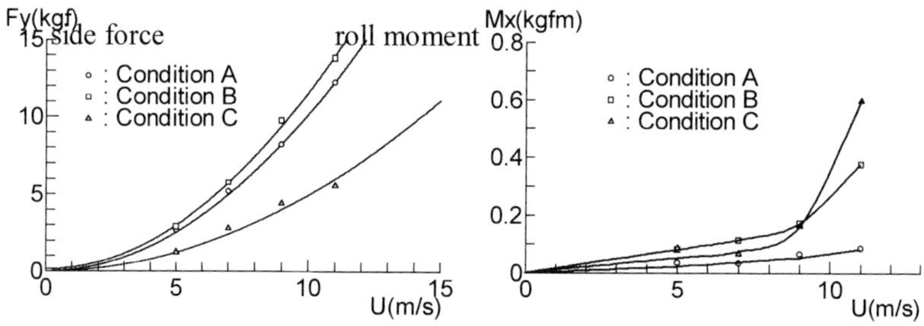

Figure 2: Effects of advance speeds on forces acting on fully captive model

DISCUSSION

Yaw Moment

Yaw moment significantly affects the maneuverability of a craft. The measured results are shown in Fig.3, in which the value is positive when the restoring moment is acting. The results shown in Fig.3 show that the yaw moment is usually positive, and the positive value increases as the craft rises. This suggests that a planing craft has a good course keeping ability in planing condition. This fact is in good agreement with the conclusion by Kobayashi et.al.(1995). The experimental results show that for attitude with zero trim and small rise the yaw moment becomes negative. This suggests that the turning ability of such craft becomes good for such attitude.

The experimental results at zero yaw angles demonstrate that yaw moment is generated by heel angle.

Transverse Force

The measured transverse force is shown in Figure 4. The force is proportional to yaw angle, or angle of attack. The effect of rise on transverse force is not significant except when trim angle is zero. The effect of trim angle on it is significant as shown in this figure.

Resistance

Resistance may affect speed reduction in maneuvering motion. The measured force is shown in Figure 5. The results show that resistance in oblique towing condition gradually decreases with increasing yaw angle.

Vertical Lift Force

Vertical lift may affect planing condition in turning motion. Measured results are shown in Figure 6. The results show that vertical lift force mainly depends on trim angle, and the effect of rise on it is small. When yaw angle is small (β<10deg.), the lift is independent of heel angle. When yaw angle is large, however, the lift force increases with increasing heel angle. At large yaw angle, the lift force decreases with yaw angle when heel angle is small, and increases up to several times of the value at β=0. This may be because the increase of heel angle works as increasing angle of attack.

Trim Moment

Trim moment is one of most important factor for planing because vertical lift force depends on it. The measured trim moment is shown in Figure 7. The trim moment is almost constant without any effect of yaw angle when heel angle is moderate (ϕ<20deg.). The moment significantly depends on rise. At small rise, bow up moment is acting on the hull. As rise increases, the bow up moment decreases, and when rise is large bow down moment start acting on it. At large heel condition, the trim moment is significantly affected by trim and yaw angles.

Heel Moment

The measured heel, or roll moment is shown in Figure 8. The values in this figure show the dynamic component of roll restoring moment, which does not include the static component at Fn=0. The results demonstrate that the roll moment decreases with increasing yaw angle, and becomes negative at large yaw angles.

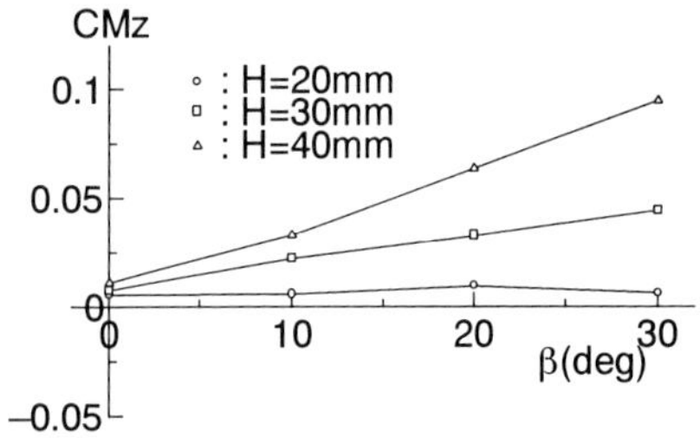

rise at heel Φ =20deg., trim τ =4deg.

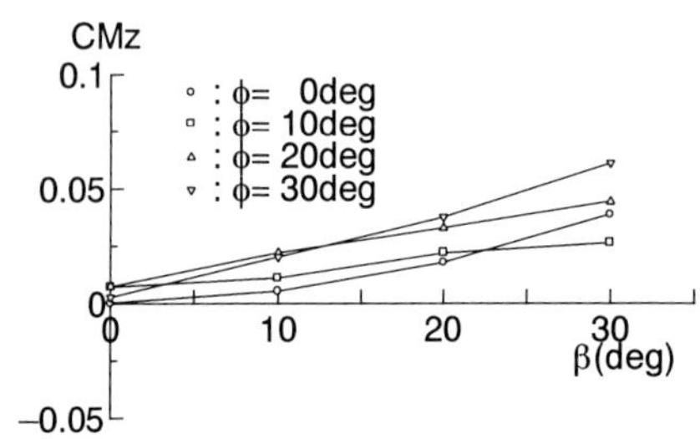

heel at rise H=30mm, trim τ =4deg.

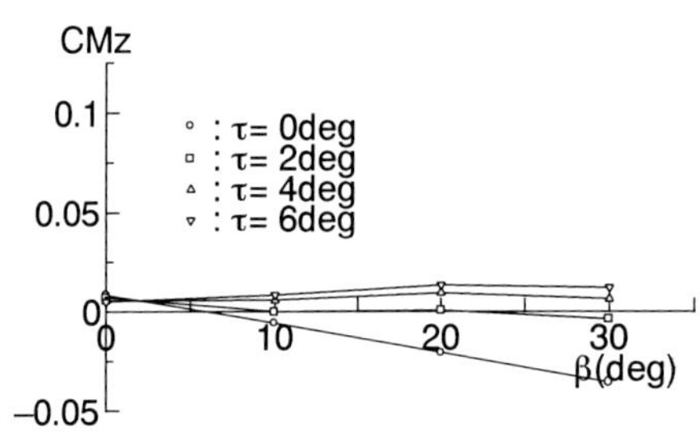

trim at rise H=20mm, heel Φ =20deg.

Figure 3: Effect of running attitude on yaw moment coefficient

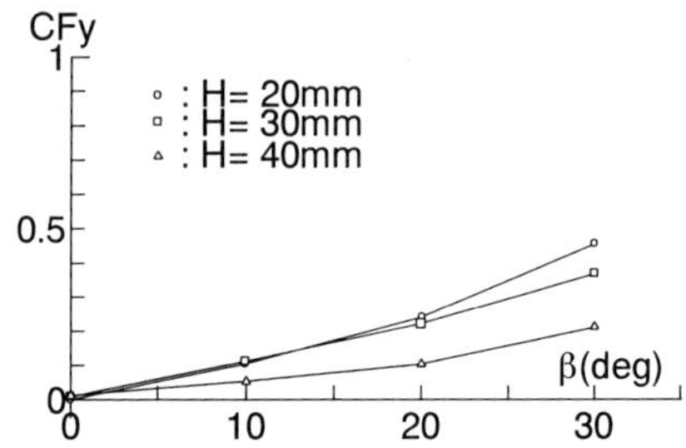

rise at heel Φ =10deg., trim τ =0deg.

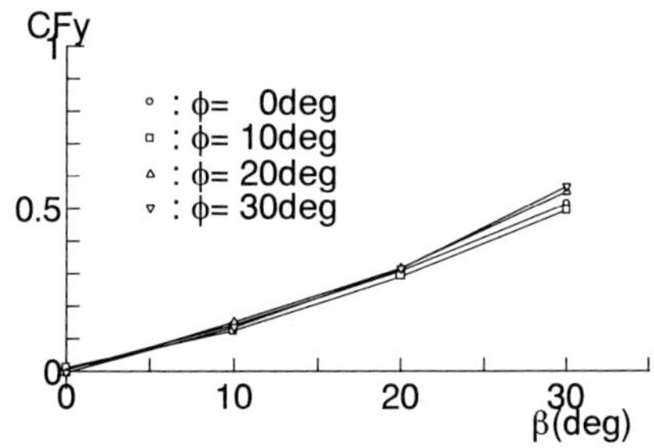

heel at rise H=20mm, trim τ =2deg.

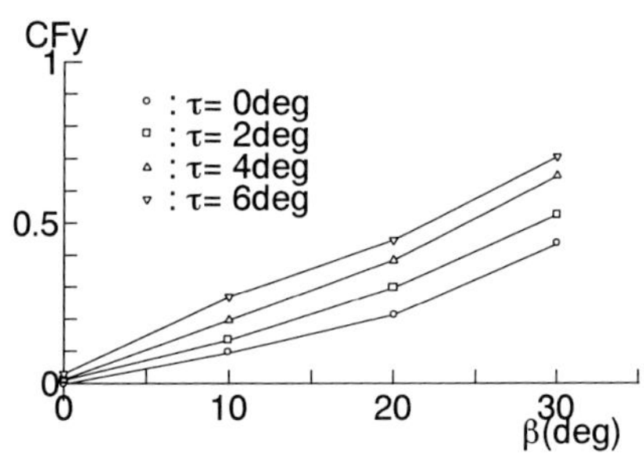

trim at rise H=30mm, heel Φ =20 deg.

Figure 4: Effect of running attitude on side force coefficient

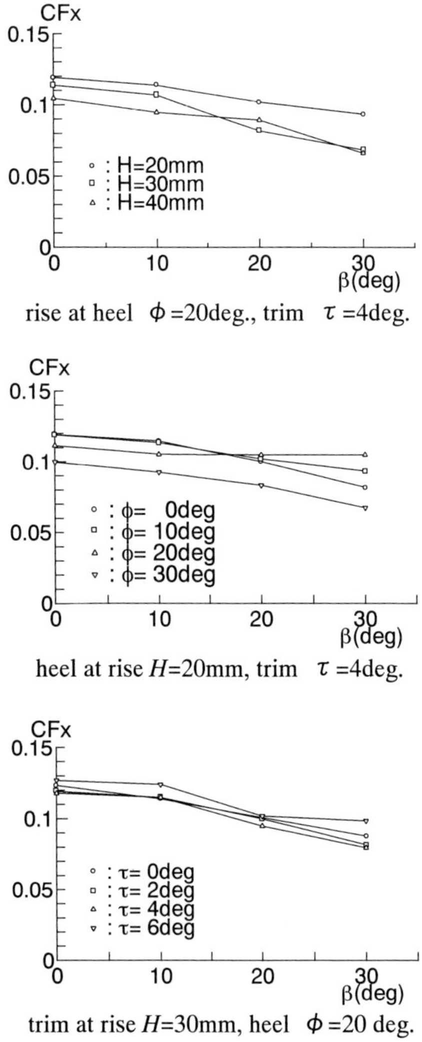

Figure 5: Effect of running attitude on drag coefficient

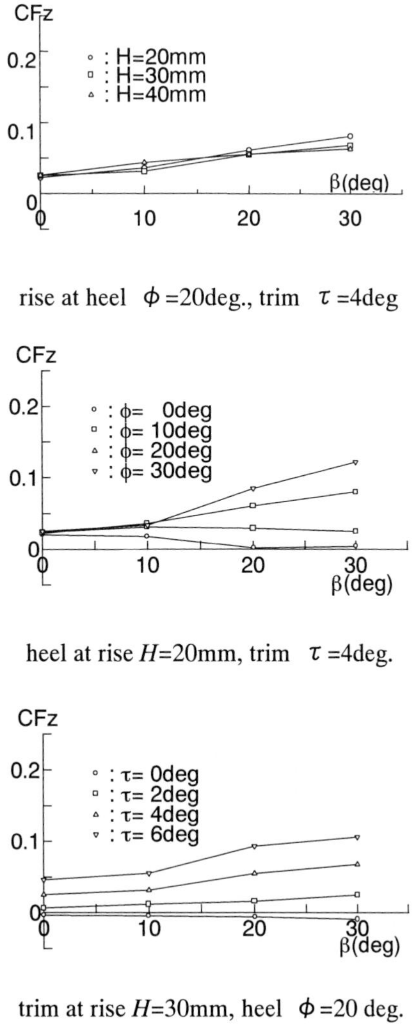

Figure 6: Effect of running attitude on vertical lift coefficient

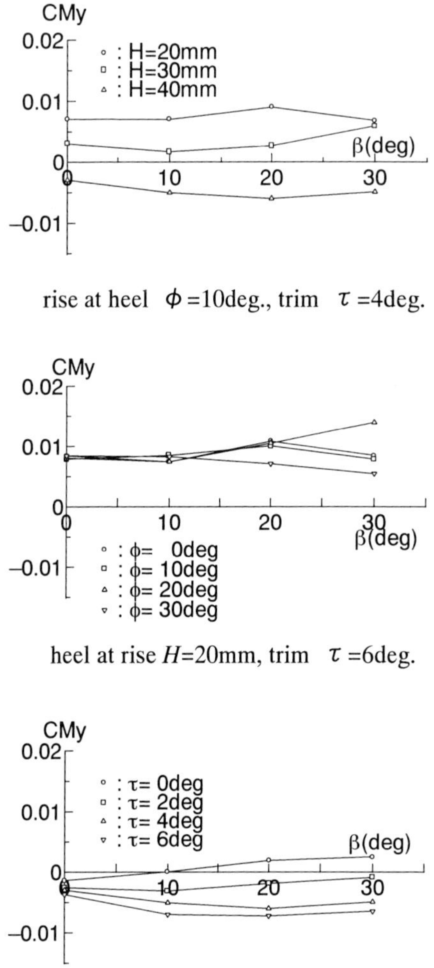

rise at heel ϕ =10deg., trim τ =4deg.

heel at rise H=20mm, trim τ =6deg.

trim at rise H=40mm, heel ϕ =10 deg.

Figure 7: Effect of running attitude on trim moment coefficient

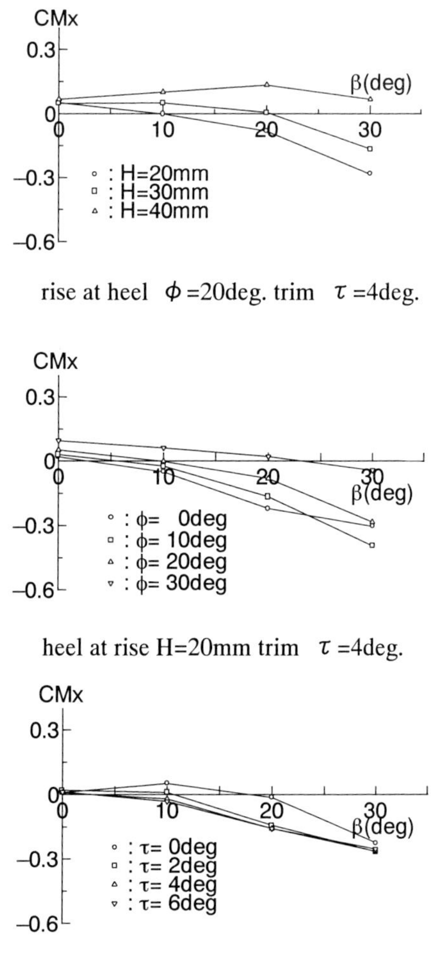

rise at heel ϕ =20deg. trim τ =4deg.

heel at rise H=20mm trim τ =4deg.

trim at rise H=30mm heel ϕ =0 deg.

Figure 8: Effect of running attitude on heel moment coefficient

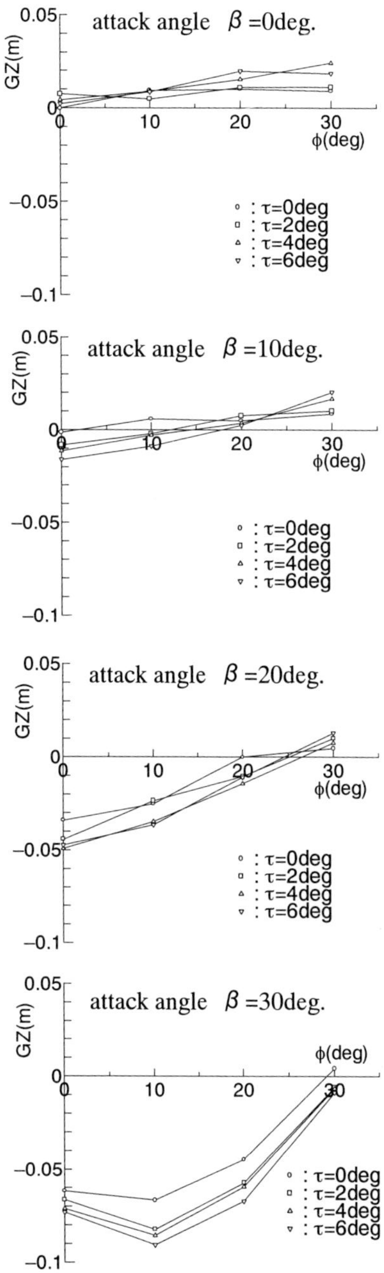

Figure 9: GZ curve at H=20mm

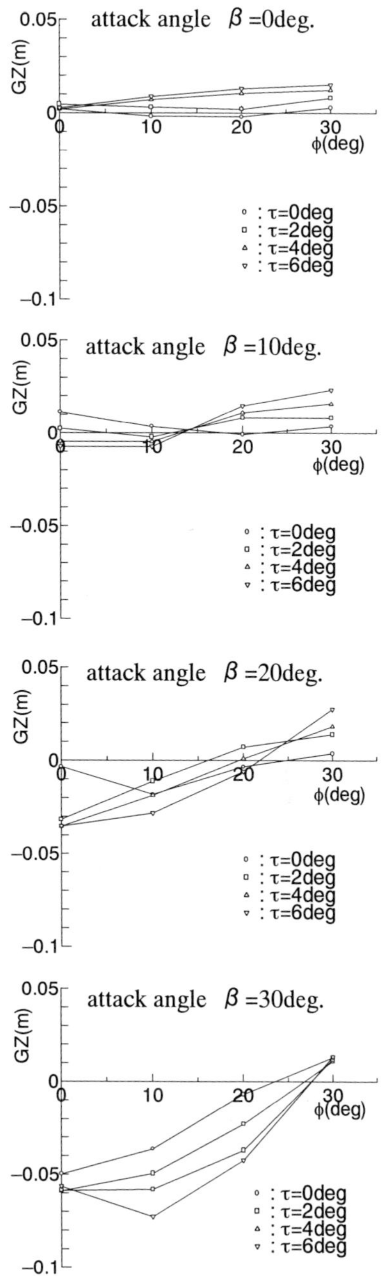

Figure 10: GZ curve at H=30mm

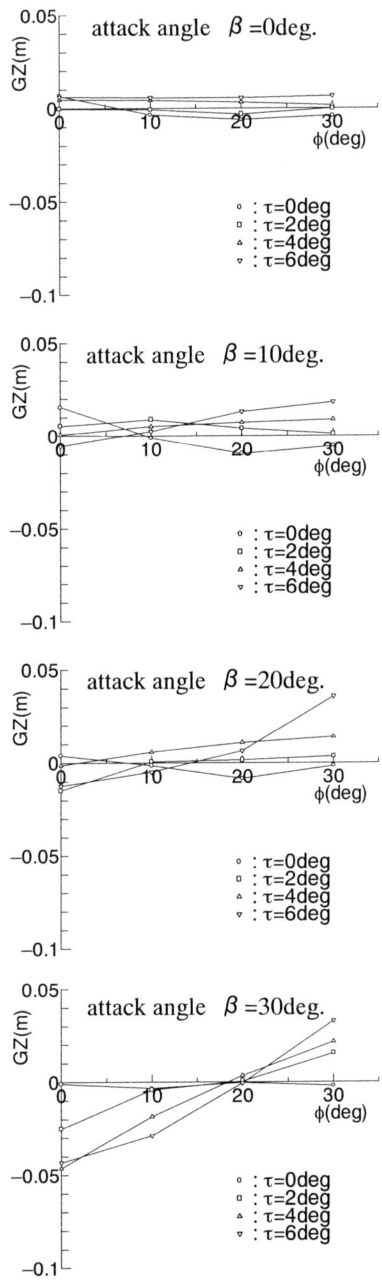

Figure 11: GZ curve at H=40mm

STABILITY AT HIGH SPEED TURNING

GZ curves are calculated from the measured heel moment shown in the previous chapter and calculated static restoring moment as shown in Figures 9-11. The results demonstrate that large negative moment is acting on the hull at large yaw angles. The forces acting on the hull in turning motion are shown in Figure 12. Large negative moment can cause large heel in high speed turning.

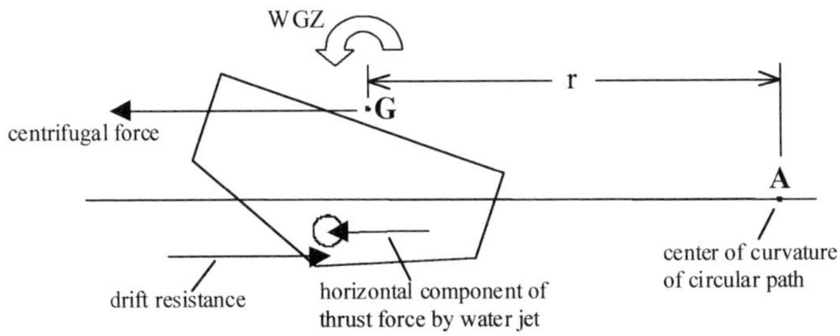

Figure 12: Forces acting on a planing hull during steady turning motion

CONCLUSIONS

Measurements of six-component hydrodynamic forces acting on a planing hull obliquely towed in constant speed are carried out for various attitudes in a towing tank and the following conclusions are obtained.

1. Hydrodynamic forces acting on a planing hull obliquely towed significantly depend on running attitudes. Therefore, in maneuvering of such planing craft it should be needed to take into account the effects of attitude on hydrodynamic derivatives.
2. At planing condition, strong restoring moment in yaw is acting on the hull. Therefore, the craft have good course keeping performance at high speed.
3. Dynamic force decreases roll restoring moment at large yaw angle. Therefore large heel can occur in high speed turning for a planing craft.

References

Kobayashi H., Arai Y., Ishibashi A., Okuda S., Okamoto Y. and Takeuchi A. (1995). A Study on the Maneuverability of High-Speed Boat. *Journal of the Kansai Society of Naval Architects, Japan.* **223**, 81-90.

D. Vassalos, M. Hamamoto, A. Papanikolaou and D. Molyneux (Editors)

AN EXPERIMENTAL STUDY ON THE IMPROVEMENT OF TRANSVERSE STABILITY AT RUNNING FOR HIGH-SPEED CRAFT

Y. Washio[1], K. Kijima[2] and T. Nagamatsu[3]

[1] Shimonoseki Shipyard and Machinery Works, Mitsubisihi Heavy Industries, Ltd., 16-1, Hikoshima, Enoura-cho, 6-chome, Shimonoseki, 750, Japan
[2] Department of Naval Architecture, Faculty of Engineering, Kyushu University, 6-10-1, Higashi-ku, Hakozaki, Fukuoka, 812-8581, Japan
[3] Faculty of Fisheries, Kagoshima Universisy, 1-21-40, Korimoto, Kagoshima, 890-0065, Japan

ABSTRACT

It is well known for high-speed craft that transverse instability might occur when running at higher than a certain speed. In this study model experiments in the towing tank were carried out to examine the transverse instability at various speeds and $\overline{GM}$ values with a hard chine hull form and to develop a device for improvement of transverse stability. Captive model tests were carried out to search for the range of instability, and were done at various speeds in three different conditions of $\overline{GM}$ values to examine hydrodynamic forces such as heel moments and sway forces acting on the hull in specified instability conditions. As a result, it is confirmed that this ship model loses transverse stability at higher speed and smaller $\overline{GM}$ as expected. For the improvement of transverse stability, three types of effective spray strip including an ordinary one were added to the hull as appendages respectively and their effect was examined. As a result, the so-called "Reaction Flap" on the fore part of the hull which is one of three types, showed remarkable improvement. The existence of "Reaction Flap" results in almost negligible resistance. This "Reaction Flap" makes the conventional mono-hull form possible to be more stable at relatively higher speed.

KEYWORDS

Stability, High-Speed Craft, Experiment, Improvement

INTRODUCTION

It is known for high-speed craft that transverse instability might occur when running at higher than a certain speed even though they posses adequate static stability satisfying relevant criteria, Marwood et al. (1968), Millward (1979), Schwanecke et al. (1992), Suhrbier (1978).

Generally a hard chine hull is relatively more stable than a round bilge hull at the same running conditions. However even for a hard chine hull, an effective device for improvement of transverse stability is required because there is a possibility of instability at higher running speeds.

In this study model experiments in the towing tank were carried out to examine the transverse instability at various speeds and $\overline{GM}$ values for a high-speed craft with conventional hard-chine. Then, for the improvement of transverse stability, three types of spray strip including an ordinary one were added to the hull as appendages respectively and their effectiveness examined for transverse stability.

The resistance test with the most effective appendages chosen among them was also carried out in order to determine the influence upon the propulsive performance.

EXPERIMENTS

Ship Model

A typical example of a hard chine hull form is shown in Photo. 1 of a large high-speed passenger craft, taken by one of the authors.

In this study model experiments in the towing tank were carried out to examine the transverse instability at various speeds and $\overline{GM}$ values. A ship model was used of the typical type of a hard chine hull form shown in Fig.1 in order to search the range and limit of instability for higher speed or smaller $\overline{GM}$ in comparison with the original planned values.

Photo.1: Conventional Hard-Chine type High-Speed Craft at running

The principal dimensions of the ship model are 3.8m in length, 0.63m in breadth, 0.14m in draft, and the scale ratio is 1/12.3 to the full scale ship.

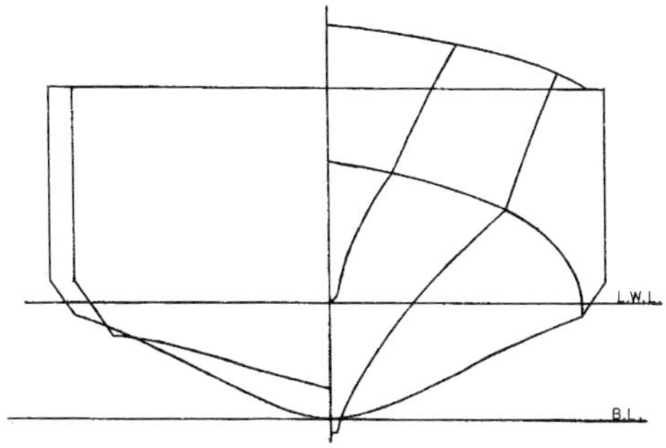

Figure 1: Hard-chine type hull form used in the present experiment

In order to improve the transverse stability, it seems easy to increase the breadth of the ship because it is known by the results of recent research that wider breadth is so effective as larger $\overline{GM}$ value, or lower center of gravity.

However, too much breadth might give bad effects upon other characteristics such as propulsive performance and seakeeping quality etc. So generally it is easier way for designers to change the type of ships from mono-hull to twin hull, so-called catamaran to improve transverse stability while keeping the other performance as it is.

It must be important for naval architects to know how to improve the hull form and optimize the total performance of ships at a practical design stage.

It will be preferable to improve the transverse stability without changing to other types from mono-hull because mono-hull has many advantages in comparison to other types of hull forms. From this point of view, some ideas for a mono-hull are already proposed such as special types of stabilized fins and outriggers added to the outside of hull, but most of them do not always give better results to keep the same performance of the original hull.

In this study methods of improvement which keep the advantages of a mono-hull are proposed and experiments by using a towing tank were carried out to examine the effects as expected.

As it is known by the recent research, Suhrbier (1978), spray strips are effective for the improvement of transverse stability, three types of spray strip including an ordinary one were added to the hull as appendages individually as shown in Fig.2 and Fig.3.

These appendages are intended to make use of the restoring force against transverse instability which is produced by bow waves creeping up above the still waterline at high-speed and colliding against these appendages as shown in Photo. 1.

A-type and B-Type are given the name of "Reaction Flap" which form an upset-U-shaped cross section by themselves and the side hull platings so as to produce a restoring force against bow waves.

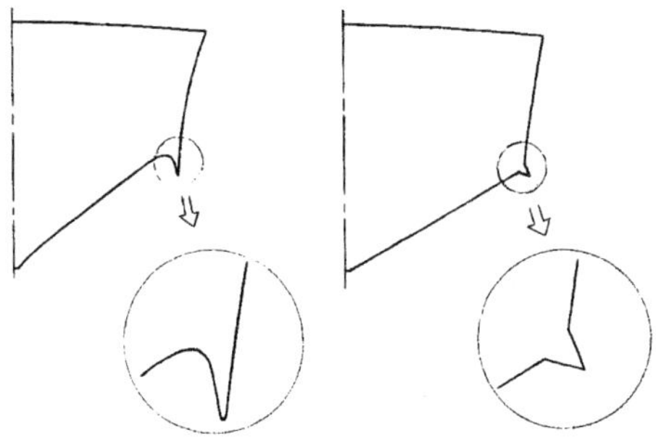

Figure 2: Section of "Reaction Flap" and spray strip

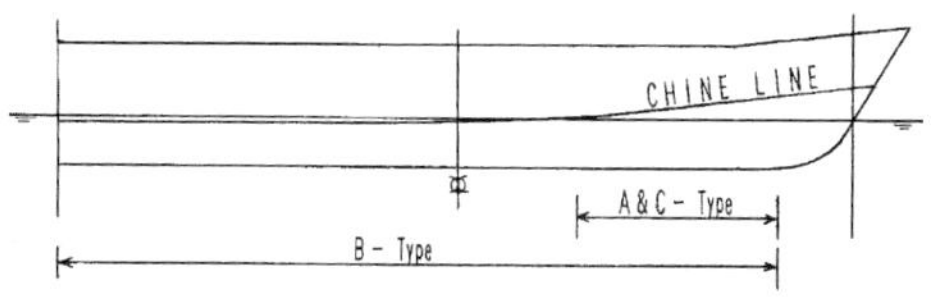

Figure 3: Applied zone of "Reaction Flap" and spray strip

A-Type is provided at the fore parts of the hull alongside the chines above the still waterline, and B-Type is the extension of A-Type to the aft end of the hull alongside the chines which submerge below the still waterline at the after part of the hull. Photo 2 shows the model with "Reaction Flap" of A-type.

C-Type is an ordinary spray strip which has wedge shaped cross section only at the fore parts of the hull alongside the chines above the still waterline as A-Type.

Photo. 2: Ship model with "Reaction Flap"

Method of Measurement

For high-speed craft transverse instability might occur when running at higher than a certain speed.

This phenomenon is observed as the ship begins to heel firstly by instability at running, and then to turn not to keep the course and finally capsizes in the worst case. Therefore it is very difficult to simulate this phenomenon perfectly by using a towing tank for the restriction of measurements, so many experiments were carried out by using a ship model restricted motion, Baba et al. (1982), Millward (1979).

In this study firstly the experiments were carried out under the fixed condition of sway and yaw motion, and finally done for various speeds in three different $\overline{GM}\,s$ chosen under the fixed condition of all motion.

In the experiments under the fixed condition of sway and yaw motion, the degree of change of running trim and heel, and the range of instability were measured as a change of $\overline{GM}$ value and speed.

The experiments under the fixed condition of all motion were carried out to examine hydrodynamic forces such as heel moments and sway forces acting on the hull for specified instability conditions. The heel moments and the side forces were measured at various heeling angles under the same conditions such as running trim and dipping gained from the results of experiments under fixed condition of sway and yaw motion.

For the experiments of the ship model with all motions fixed, the schematic diagram of measurement is shown in Fig.4, and the coordinate system and symbols in Fig.5.

The model has twin hanging rudders, but not propellers. All experiments were carried out at the Towing Tank of Nagasaki Research and Development Center, Mitsubishi Heavy Industries, Ltd.

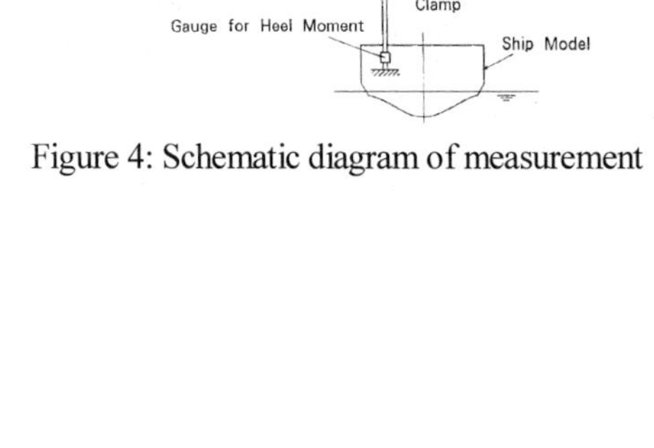

Figure 4: Schematic diagram of measurement

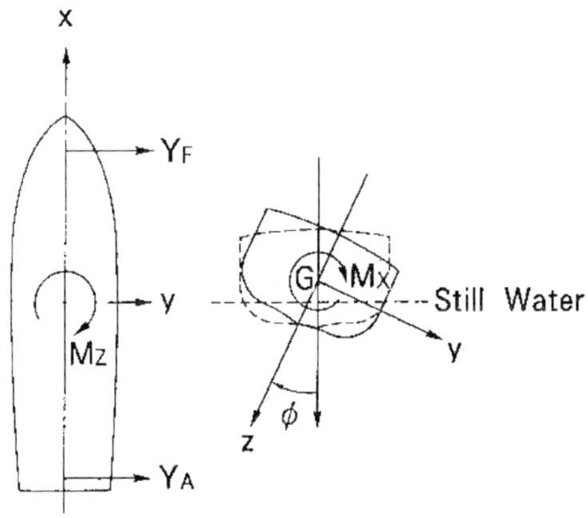

Figure 5: Coordinate system and symbols

RESULTS OF EXPERIMENTS

Results of Experiments for Transverse Stability

At first, experimental results for the ship model without appendages for improvement of instability are described. This model is called "the original hull" hereinafter. By the experiments of using the original hull fixed sway and yaw motion, it was observed to be stable by maintaining certain heel angle at relatively lower speed, but not to be stable at higher speed and to result in capsizing finally at further higher speed. The tests were started from searching for such limits of $\overline{GM}$ value and speed which might cause a transverse instability resulting in capsizing finally, and then tests were performed at various speeds for three different conditions of $\overline{GM}$ values chosen. The results of experiments, measured heel angles ϕ which were saturated values for each individual speed are shown in Fig.6(a). (The values of figures hereinafter correspond to the model scale.)

As shown in Fig.6(a), the heel angle ϕ increases with increasing speed and finally the model capsizes for all $\overline{GM}$ *s* chosen.

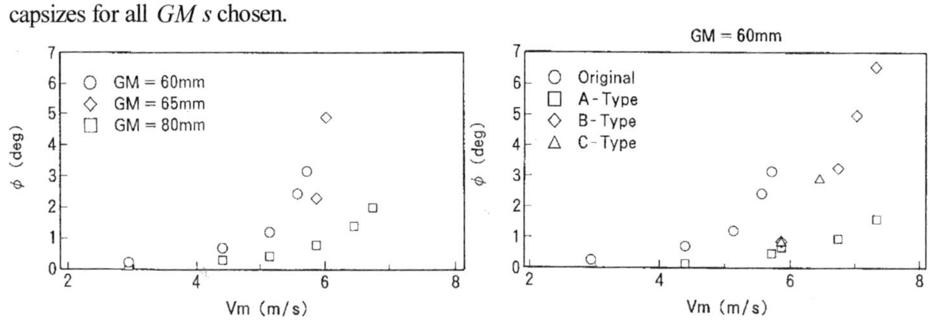

Figure 6 (a) and (b): Measured heel angle

Also the time histories of heel, heave and pitch measured at running in the case of $\overline{GM}$ =60mm are shown as Fig.7(a) and Fig.7(b) for reference.

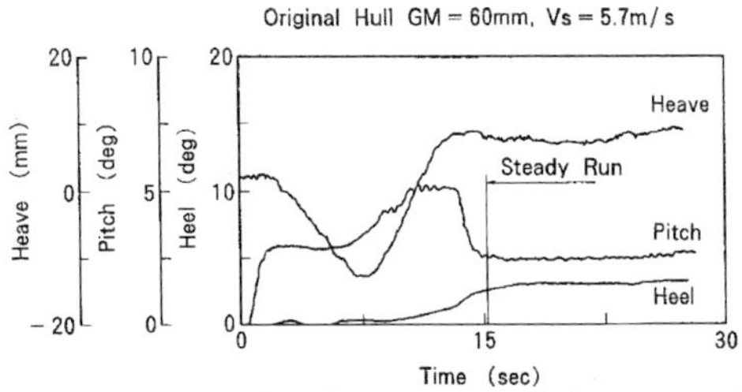

Figure 7(a): Time history of heel, heave and pitch measured at running

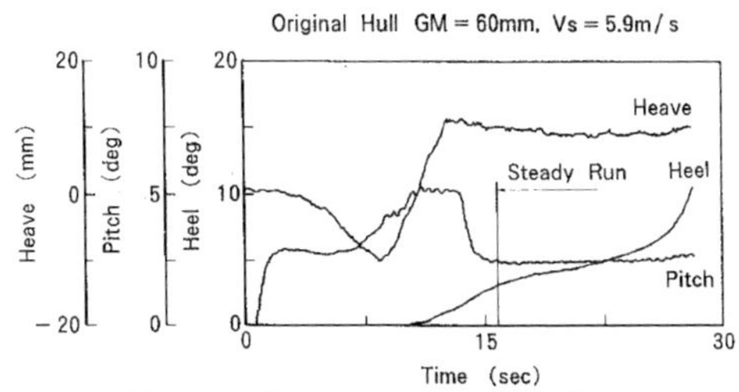

Figure 7(b): Time history of heel, heave and pitch measured at running

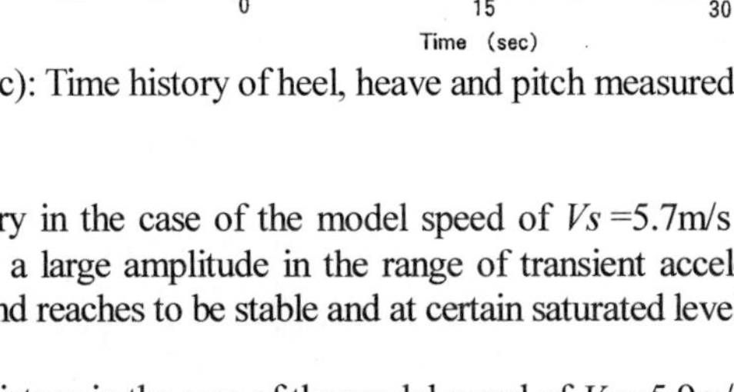

Figure 7(c): Time history of heel, heave and pitch measured at running

Fig.7(a) is a time history in the case of the model speed of *Vs* =5.7m/s which shows heave and pitch motion vary with a large amplitude in the range of transient acceleration before the model reaches to run steady, and reaches to be stable and at certain saturated levels.

Fig.7(b) is also a time history in the case of the model speed of *Vs* =5.9m/s which shows heave and pitch motion to have reached steady state and a saturated certain value when the model reaches to run steady as in the case of *Vs* =5.7m/s, but the heel angle does not reach to be stable and grows rapidly until resulting in capsizing.

The speed of the model at capsizing was Vs >5.9m/s in the case of $\overline{GM}$ =60mm, Vs >6.1m/s in the case of $\overline{GM}$ =65mm and Vs >7.0m/s in the case of $\overline{GM}$ =80mm.

Thus it is found by the experiments that the hull form used in this study becomes unstable as the speed becomes higher and $\overline{GM}$ value smaller.

Fig.6(b) shows the measured results under the conditions of models with appendages such as "Reaction Flap" and spray strips with $\overline{GM}$ =60mm.

Comparing with the results of the original hull at the same speed, heel angles are reduced by adding all the above appendages of A-Type, B-Type and C-Type. In particular, A-Type, the so-called "Reaction Flap", extending along both side platings nearly from the bow in a direction towards the stern shows remarkable improvement in comparison with other appendages, in which the heeling angle ϕ is very small even at the speed of Vs =7.3m/s.

Fig.7(c) shows a time history of heel, heave and pitch measured at running of Vs =5.9m/s in the case of "Reaction Flap", A-Type. The heel angle ϕ of the model with A-Type keeps a small constant value at steady running in comparison with the original hull as shown in Fig.7(b) where heel angle is still increasing after steady running.

Side views of the original hull and that with "Reaction Flap", A-Type at running are as shown in Photo. 3.

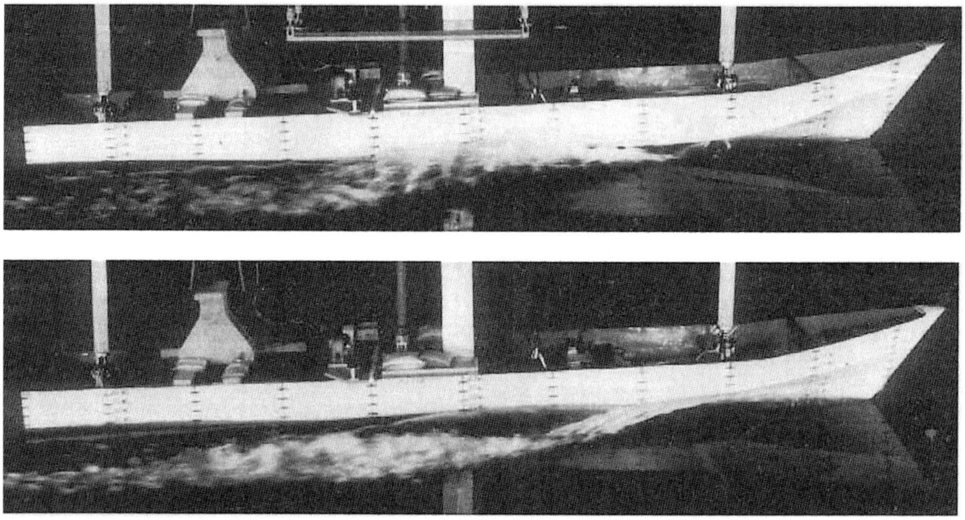

Photo. 3: Side view of original hull and that with "Reaction Flap" at running

By considering the above results, the experiments under the fixed condition of all motions were carried out only for the original hull with and without "Reaction Flap", A-Type. Also tests were performed for a constant value of $\overline{GM}$ =60mm at two cases of speed, one at Vm =4.4m/s and the others at Vm =5.9m/s, which are stable and very unstable respectively for the original hull. Measured heel moments Mx for the above are shown in Fig.8(a) and Fig.8(b).

For the original hull heel moment Mx against heel angle ϕ is relatively very small at the speed of Vm =4.4m/s which tends to become unstable when the model is forced to heel by some disturbance. At Vm =5.9m/s, the heel angle ϕ increases more and more by some disturbance and finally capsizes because the heel moment Mx has a positive slope against heel angle ϕ as shown in Fig8(a).

On the contrary in the case of the hull with A-Type, the model is stable at both speeds of Vm =4.4m/s and 5.9m/s even when heeling because the heel moment Mx has a negative slope against heel angle ϕ as shown in Fig8(b).

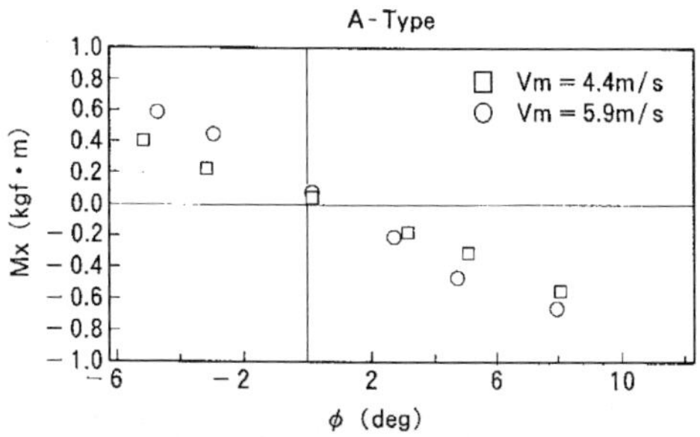

Figure 8(a): Measured heel moment

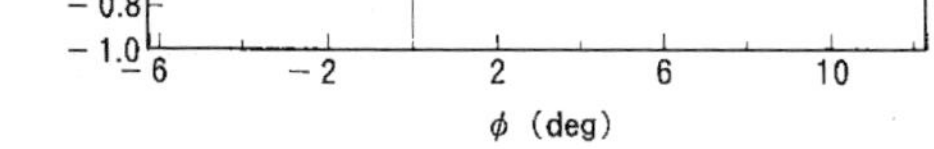

Figure 8(b): Measured heel moment

Then Fig.9(a) and Fig.9(b). show measured sway forces at the positions of fore and aft perpendiculars of the model at the speed of Vm =5.9m/s.

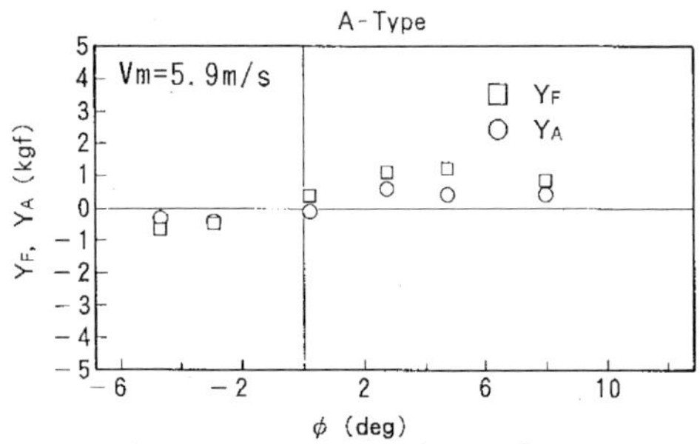

Figure 9(a): Measured sway forces

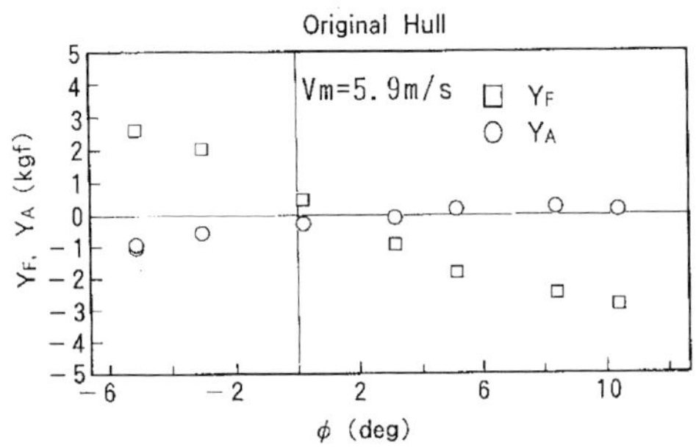

Figure 9(b): Measured sway forces

For the original hull when the model heels to starboard side (ϕ >0), the value of sway force at the fore perpendicular Y_F is negative (direction of the force is from starboard side to port side) and the value of sway force at the aft perpendicular Y_A is zero. Then the hydrodynamic force acts to turn the model to the port side. When ϕ <0, vice versa.

Yaw moment increases as of sway force $|Y_F|$ increases with increasing heel angle ϕ, therefore a turning circle decreases. And then heel angle ϕ increases more and more because the centrifugal force to the hull becomes larger.

However for the hull with A-Type, both absolute values of Y_F and Y_A are small and have a positive slope against heel angle ϕ. This means the model results in being stable at running because yaw moment is small and turning direction is opposite to the original hull which makes heel angle ϕ decrease even if the model might slightly sway to the heeling side.

Concluding the results of the experiment, it is found that "Reaction Flap", A-Type shows remarkable improvement of transverse stability at high running speed.

Result of Resistance Test

A resistance test with "Reaction Flap" of A-Type was carried out in order to examine the influence upon the propulsive performance. The test was performed under the fixed condition of sway, yaw, and roll motion.

The result of the resistance test is shown as compared to that of the original hull in Fig.10. The existence of "Reaction Flap" A-Type on the fore part of the hull, results in having almost negligible effect on resistance.

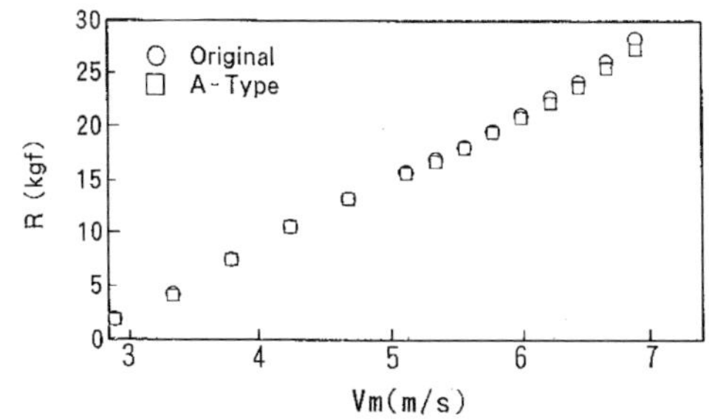

Figure 10: Resistance test result

CONCLUDING REMARKS

It is well known for high-speed craft that transverse instability might occur when running at higher than a certain speed.

Taking the above characteristics into consideration, in this study research on a method to improve transverse stability during running through experiments was undertaken without loosing the advantageous characteristics of mono-hull forms such as propulsive performance, seakeeping quality etc.

At first the range or limits of instability for various speeds and different $\overline{GM}$ values were experimentally examined for the ship model of a hard chine high-speed craft.

Then three types of effective spray strip including an ordinary one were added to the original hull as appendages respectively and were also examined from their transverse stability point of view.

As a result, the so-called "Reaction Flap", one of three types, extending along both side hull platings nearly from the bow in a direction towards the stern shows remarkable improvement in comparison with other appendages although other types of strip are found to be also effective in improving transverse stability.

Furthermore the existence of "Reaction Flap" on the fore part of the hull is also found to have negligible effect on resistance.

This "Reaction Flap" shows one of the possibilities for conventional mono-hull form to become more stable at relatively higher speeds without changing hull form or principal dimensions. Also it is useful to keep transverse stability by adding the "Reaction Flap" in the case of demands for raising the center of gravity or increasing speed when planning conversion of the ship.

References

Baba, E., Asai, S. and Toki, N. (1982). A Simulation Study in Sway-Roll-Yaw Coupled Instability of Semi-Displacement Type High Speed Craft. Proceedings of the 2nd International Conference on Stability of Ships and Ocean Vehicles, Part IV, The Society of Naval Architects of Japan.

Ibaragi, H., Kijima, K. and Washio, Y. (1996). Study on the Transverse Instability of a High-Speed Craft. Transactions of The West Japan Society of Naval Architects No.91.

Kijima, K., Ibaragi, H. and Washio, Y. (1996). Study On the Transverse Instability of a High-Speed Craft. 2nd Workshop on Stability and Operational Safety of Ships.

Marwood, W. J. and Bailey D. (1968). Transverse Stability of Round-Bottomed High Speed Craft Underway. NPL report 98.

Millward, A. (1979). Preliminary Measurements of Pressure Distribution to Determine the Transverse Stability of a Fast Round Bilge Hull. ISP, Vol.26, No.297.

Schwanecke, H. and Muller-Graf, B. (1992). Die Dynamische Querstabilitat Schneller Rundspantund Knickspantboote. Bericht Nr.1201/92, Versuchsanstalt fur Wasserbau und Schiffbau, 02.

Suhrbier, K.R. (1978). An Experimental Investigation on the Roll Stability of a Semi-Displacement Craft at Forward Speed. Symposium on Small Fast Warship and Security Vessels, RINA.

Washio, Y. and Doi, A. (1991). A study on the Dynamical Stability of High-Speed Craft. Transactions of The West Japan Society of Naval Architects, No.82.

Washio, Y, Ibaragi, H. and Kijima, K. (1997). Study on the Transverse Instability of a High-Speed Craft (Continued). Transactions of The West Japan Society of Naval Architects, No.94.

D. Vassalos, M. Hamamoto, A. Papanikolaou and D. Molyneux (Editors)

WATER DISCHARGE FROM AN OPENING IN SHIPS

S. M. Calisal[1], M. J. Rudman[2] , A. Akinturk[1], A. Wong[1] and B. Tasevski[1]

[1]University of British Columbia, Dept. of Mechanical Engineering, 2324 Main Mall, Vancouver, B. C., V6T 1Z4 Canada
[2]CSIRO Building, Construction and Engineering, Highett, Victoria 3190 Australia

ABSTRACT

Water trapped on ship decks can play an important role in the safety of ships. The goal of this study is to understand the governing factors and to model the discharge mechanism of water from an opening, e.g. freeing-ports, onboard of ships. In the formulation of the discharge flow from a flooded deck two major problems exist. One is the form and location of the free surface, and the other discharge rate and how it is connected to the form of the free surface. In the study of water-on-deck flows without freeing ports, only the form of the water-on-deck is of interest. The problem of water-on-deck with discharge can be reduced to a water-on-deck without freeing ports if a relationship between the discharge rate and the free surface form can be established. This is the starting point of this research. The proposed method and approach consists of both the numerical and the experimental study of water discharge from the open deck of a ferry or a fishing vessel through permanently open freeing ports or freeing ports with a flap cover. A two-dimensional model of the discharge was built during the summer of 1997 to visualize the free surface and to measure the flow discharge from a flat bottom. The discharge flow pattern and free surface form were recorded with a digital camera. By image analysis, the location and the form of the free surface were established. From the knowledge of the free surface, the change in volume, discharge rate velocity at the freeing port and various parameters of the discharge kinematics were calculated. The experimental results will be studied to generate numerical algorithms that can be used to calculate the discharge rates from freeing ports. Numerical modeling and preliminary results are also given in this paper. Initial results suggest a very good agreement between the experimental and numerical results.

KEYWORDS

Ship stability, damage stability, ship safety, water discharge, volume tracking method, Navier-Stokes, variable density fluid

NOMENCLATURE

A	Cross sectional area of the exit opening
Fn	Depth Froude number
g	Gravitational acceleration
H	Initial water height inside the tank
Hg	Height of the opening (for water discharge)
Hr	Water height at the rear end of the tank
L	Flooded length of the tank
Q	Volume of the water being discharged per unit time
S	Distance between camera and the tank

INTRODUCTION

Some Canadian ferries operating in British Columbia in relatively sheltered areas have open car decks. Water could possibly collect on deck of these ferries by the scooping action in waves, or as a result of an accident thus causing loss of stability by a process which is generally known as the "free surface" effect. Various researchers in Canada have studied the possibility of water on deck and its drainage from freeing ports. The authors participated in the experimental study of water accumulation on a ferry model in head seas. In this study the model was in moderate waves and a procedure was developed to estimate the minimum freeboard required to avoid water accumulation on deck for a given sea state, Calisal et al. (1997). This procedure consists of experimental and numerical results. During the summers of 1997 and 1998, discharge from a two-dimensional model of a section of a ferry was experimentally studied. The model, initially full of water was drained by instantaneously opening a freeing port. The measurement of the form of the water surface on board and the discharge rates were the primary objectives of this preliminary study. This preliminary experimental study allowed us to measure and document the variation of the free surface and discharge rates of water from the freeing ports during the draining phase. An optical measurement system was developed for this purpose, which uses a video camera, a frame grabber and specialized software developed in-house to calculate the location of the free surface during draining. The data collected are currently under study. This paper presents some of the preliminary results and comparison of the experimental results with the numerical predictions. The visualization showed that in addition to the expected gradual drop in the free surface, some traveling waves are also present on deck. The objective of this study is to develop a numerical procedure for calculating the water collection and discharge rates from the open deck of a ferry or a fishing vessel. The numerical procedures will be validated in experimental work and suitable design procedures and algorithms will be developed for the time-domain calculations of ship stability. The requirement of an estimate of the discharge rate is essential for the numerical calculations, and the use of such an algorithm permits the definition of the boundary conditions necessary for the application of numerical methods such as Boundary Element Method. Defining a boundary condition supported by experimental work is expected to increase the numerical accuracy of the computed flow field, therefore of the critical discharge time. This in turn is expected to permit relatively fast evaluation of damaged stability of open deck vessels in time domain.

Another important objective is the establishment of a "closure" relationship for the completion of the potential flow formulation. That is, knowledge of the discharge rates for unsteady flow is necessary in order to assign the normal velocity boundary condition for the potential flow formulation (BEM). Work done on steady waterfalls suggests that the depth-Froude number of the flow should be equal to one (Fn = 1). However, our recent experimental work with a constant model length but with various initial water heights suggests that the Froude number of the flow discharging from the decks is time dependent and starts at a value of zero. The Froude number then increases to a value of approx. 0.7 and then starts to decrease continuously back to zero. We will study various model lengths and model roll frequencies to establish if numerical algorithms can be developed to predict the discharge velocity from the knowledge of the instantaneous water height. Of course, the form of the water surface will remain an unknown and will be calculated by a numerical procedure. The overall objectives can be listed as:

- Establishment of a relationship between water height and discharge speed.
- We expect to find a time domain expression for the Froude number for unsteady free-surface flows. This algorithm is expected to improve the performance of existing codes on water deck flows, as it will permit a relatively easy and accurate calculation of the discharge rates.
- This result will also be important for the understanding of free-surface flows such as waterfalls, where for steady conditions, the depth-Froude number is usually assumed to be equal to one.

The development of a numerical code to establish water accumulation and discharge volume rates for different ship conditions such as rolling, and stationary, and with and without list, will permit the study of the dynamic stability of damaged ferries.

The study will give design criteria for the minimum free board necessary for open-deck ferries, as this height determines the amount of water which will accumulated on deck for a given wave height condition. As this rate must be smaller than the discharge rate for the number of available freeing ports, a design procedure based on a design wave height, freeboard height and number, and size of freeing ports will result from this study.

EXPERIMENTS

The experimental apparatus consisted mainly of the following items: water discharge tank, data acquisition devices including a high speed camera, VCR, and data analysis devices including video frame grabber, imaging software, and a surface scanner program.

A discharge tank was constructed with 1/2' clear Plexiglas (see Figure 1). The inner dimensions of the tank are 6 feet long by 1.5 feet tall by 1 feet wide or 72 x 18 x 12 inches respectively. The tank is closed on one end and water is only allowed to drain out of the other end. A height adjustable gate was installed on the open end such that the opening area of discharge port can be changed between 0 and 12 inches. An elastic cord mousetrap like device was used to open the piano hinged door to start the discharge very quickly.

A dexion table with four height adjustable feet was also constructed to support the tank. The adjustable feet allowed us to properly adjust the height such that the tank could be perfectly level.

For experiments where shorter tank length was required, a piece of 1/2-inch removable Plexiglas divider was placed at the desired location (see Figure 2). The divider was supported by an angle plate and a 2 x 6 lumber to prevent it from sliding back when the other side was filled with water. A non-permanent rubber gasket tape was also added at the edges to prevent leakage.

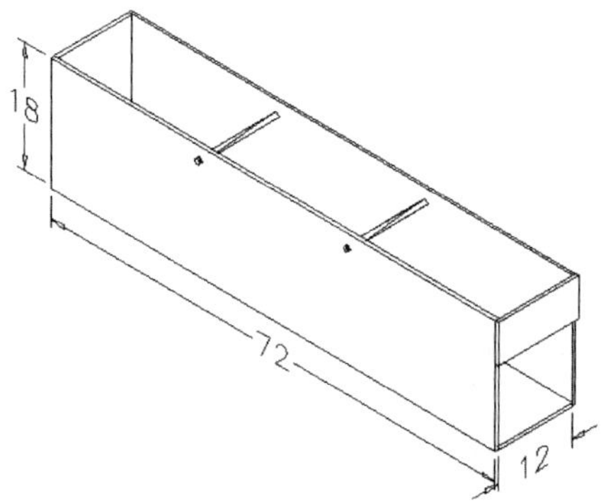

Figure 1: Plexiglas discharge vessel dimensions

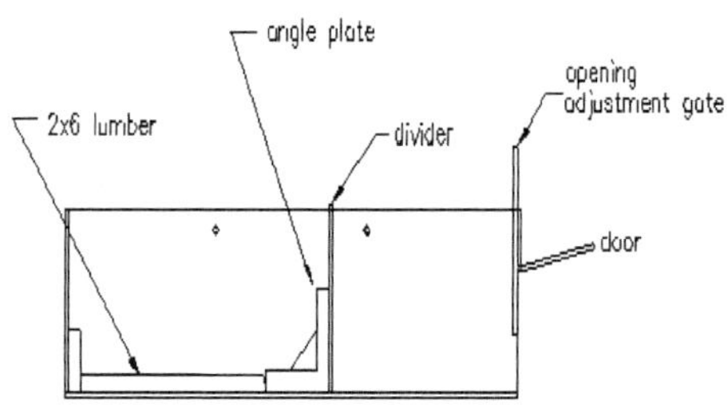

Figure 2: Side-view of discharge vessel

Data Acquisition Devices

The Optikon MotionScope High Speed Video System was used to capture the discharging water profile near the opening of the gate (see Figure 3). The system is a simple to "point" and "shoot" device, with a built-in 5" monochrome CRT video display and a separate video camera. The system has the capability to capture 60, 120, 180, 250, 300, 400, and 500 frames per second. The electronic shutter is also user adjustable and can be set from 1X to 20X of the set recording rate. The images captured were stored temporary in the system's buffer, with a capacity to store up to 2,048 full frames.

The lens, which came with the MotionScope, was removed and a Cosmicar CCTV 1/2-inch manual-iris c-mount lens was used as a replacement (see Figure 4). This lens was used because the software that was used to analyze the captured data had previously been calibrated for distortion with this particular lens. With the irregularity of the MotionScope's camera however, some problems were encountered with the lens geometrical specifications. For one, the Cosmicar lens was located too closely to the CCD of the camera, which caused the lens to provide a very large field of view (FOV) which cannot be focused properly. A 5mm CS to C mount adapter ring, which was supplied with the MotionScope, must be installed onto the lens mount of the sensor head assembly before the c-mount lens could be mounted. However, with the use of the adapter ring, the lens was moved too far away from the CCD, which gave us a very small FOV (5.64 degrees). Nonetheless, this set up was used because no other options were available if the MotionScope was to be used. The small FOV was compensated for by moving the camera further away from our water tank.

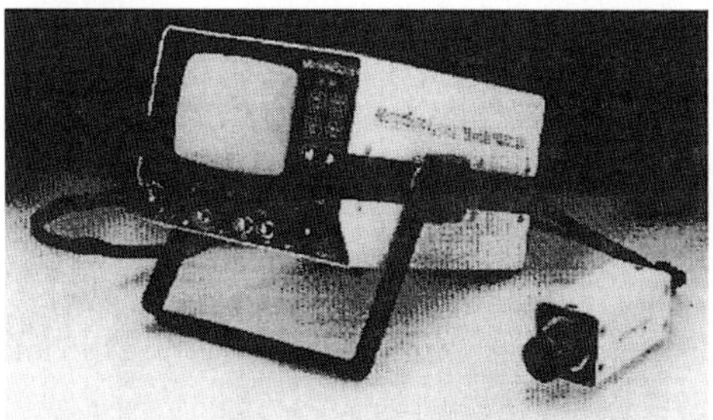

Figure 3: Optikon MotionScope High Speed Video System

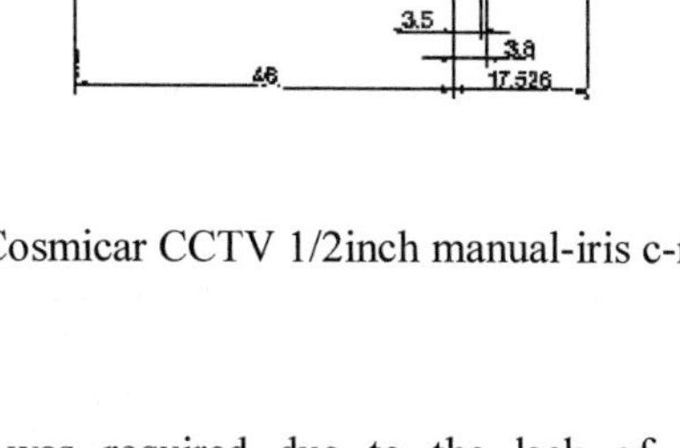

Figure 4: Cosmicar CCTV 1/2inch manual-iris c-mount lens

TV and VCR

A TV and VCR combo was required due to the lack of permanent storage on the MotionScope. The data from the MotionScope was played back through its RS-170 (NTSC compatible) video output and recorded onto a videocassette in the VCR. If a PC computer with a frame-grabbing card was available, then the TV and VCR would not be required as we could directly save the MotionScope playback images as a bitmap on the computer.

A schematic of the experimental setup is shown in Figure 5. The camera was placed in such a way that the line passing through the center of the lens and its focal point is perpendicular to the side of the discharge tank. With the current distance (S), it was able to capture the flooded length of the discharge tank.

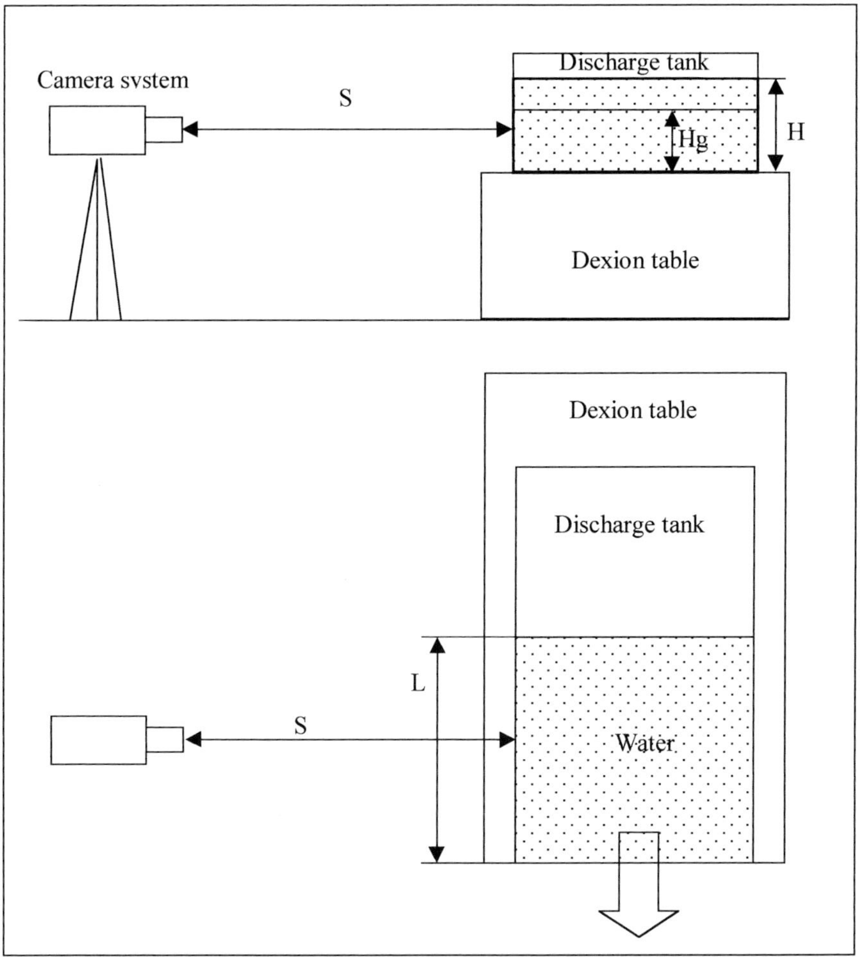

Figure 5: Experimental setup

In this initial study, there were three parameters that were changed: Hg- the height of the opening, H- the initial water height in the tank and L- the flooded length of the tank. For the results presented in this paper, only initial water height was varied as 6, 10 and 14 inches. The flooded length was 2 feet and height of the opening for water discharge was 4 inches.

NUMERICAL METHOD

The numerical method used to simulate the flow of water through the dam sluice gate is the volume tracking method of Rudman (1998). The method is based on the Volume-of-Fluid (VOF) method introduced by Hirt and Nichols (1981) and improved by Youngs (1982). A brief overview of the method is given here, but details of the implementation are beyond the scope of this paper and may be found in Rudman (1998).

The gas-liquid system is treated numerically as a single incompressible fluid whose density and viscosity vary rapidly in the vicinity of physical interfaces. The incompressible Navier-Stokes equations for a variable density fluid are written:

$$\frac{\partial(\rho \boldsymbol{U})}{\partial t}+\nabla\cdot(\rho \boldsymbol{U}\boldsymbol{U})=-\nabla P+\rho \boldsymbol{g}+\boldsymbol{F}_S+\nabla\cdot \mathrm{T}' \tag{1}$$

$$\nabla\cdot \boldsymbol{U}=0 \tag{2}$$

$$\frac{\partial C}{\partial t}+\nabla\cdot(\boldsymbol{U}C)=0 \tag{3}$$

Where ρ is the density, $\boldsymbol{U}$ is the velocity vector, P is the pressure, $\boldsymbol{g}$ is the gravity vector, $\boldsymbol{F}_S$ is the surface tension force and $\mathbf{T}$ is the stress tensor defined as:

$$\mathbf{T}_{ij}=\mu\left(\frac{\partial \boldsymbol{U}_i}{\partial x_j}+\frac{\partial \boldsymbol{U}_j}{\partial x_i}\right) \tag{4}$$

The fractional volume function C is a function that takes a value of one inside the liquid and zero inside the gaseous phase. In computational cells through which the interface passes, the value of C varies between 0 and 1. Local densities are calculated from C using:

$$\rho=C\rho_G+(1-C)\rho_L \tag{5}$$

And local values of the dynamic viscosity μ are determined in a similar manner. The equations are discretised on a rectangular Cartesian mesh.

The numerical method is second order in time and space. It uses the Flux-Corrected Transport (FCT) ideas of Zalesak (1979) to calculate the advective terms in the momentum equations and a multi-grid pressure solver based on the Galerkin coarse grid approach of Wesseling (1992) to solve for pressure and enforce incompressibility. Accurate determination of surface tension forces is often an important part of the solution of free surface flow problems and is achieved here using a kernel-based variant of the Continuum Surface Force (CSF) method of Brackbill et al. (1992). In this method, a continuously varying body force approximates the exact discontinuous surface force over a thin transition region near the interface. Volume tracking (Eqn 3) is undertaken using a Volume-of-Fluid method based on that of Youngs (1982). VOF methods are designed to maintain very thin numerical interfaces, with the transition from gas to liquid occurring across just one mesh cell in most instances. The advantage of VOF methods over more common approaches for interface problems (such as Boundary Integral Methods) is the ability to accurately simulate arbitrarily complicated problems of fluid coalescence and fragmentation without the need of purpose-built algorithms. In the code used here, the only difference to the method discussed in Rudman (1998) is the inclusion of obstacle cells that allow arbitrarily complex internal boundaries to be included in a computation. These obstacle cells are included in the same way as in the original Marker and Cell (MAC) method of Welch et al. (1965)

The basic first-order in time algorithm on which the second-order method is based is as follows:

1. Estimate new values of C:

$$C^{n+1} = C^{n} - \delta t(\nabla \cdot \boldsymbol{U}^{n} C^{n})$$

2. Estimate new densities (ρ^{n+1}) and viscosities (μ^{n+1}) using Eqn 5.
3. Estimate new velocities using old timestep velocities and pressures:

$$\rho^{n+1}\boldsymbol{U}^{*} = \rho^{n}\boldsymbol{U}^{n} + \delta t(-\nabla \cdot \rho^{n}\boldsymbol{U}^{n}\boldsymbol{U}^{n} - \nabla P^{n} + \rho^{n}\boldsymbol{g} + \boldsymbol{F}_{S}^{n} + \nabla \cdot \mathrm{T}^{n})$$

Calculate the pressure correction ρP required to enforce incompressibility:

$$\nabla \cdot \left(\frac{1}{\rho}\nabla \delta P\right) = \frac{1}{\delta t}\nabla \cdot \boldsymbol{U}^{*}$$

4. Adjust velocities and pressure:

$$\boldsymbol{U}^{n+1} = \boldsymbol{U}^{*} - \frac{1}{\rho^{n+1}}\nabla \delta P \qquad\qquad P^{n+1} = P^{n} + \delta P$$

The second-order in time algorithm used in this study performs two passes of steps 1-4. On the first pass, steps 1-3 are performed using a half time step. In the second pass, steps 1-4 are performed again with a full time step, the only other difference being that the pressures and velocities on the right-hand side of step 4 are replaced by the half time estimates calculated in the first pass.

The computational domain was discretised on a uniform mesh of 256 × 192 grid cells with physical dimensions 1000mm × 750mm. The holding tank (dimensions 601mm × 425mm) was then numerically `constructed' by placing a horizontal row of obstacles cells at a height of 200mm above the domain bottom (forming the tank base) and a vertical row 601mm from the left wall of the domain (forming the tank wall). The additional part of the domain outside the tank was required in order to allow the fluid to drain from the tank in a natural way without enforcing arbitrary (and possibly incorrect) boundary conditions on the draining process. The initial condition had the tank filled to a depth of 356mm. The initial velocities were zero and the pressure was set to be equal to the hydrostatic pressure equilibrium that would exist if the tank gate were closed. At zero time, the gate is instantaneously removed and the water flows out of the opening under gravity.

RESULTS

As mentioned earlier, after the discharge gate opens, a travelling wave is observed as the free surface level drops gradually. Figure 6 shows the drop in the free surface level and the formation of the travelling wave as the time progresses. The discharge end of tank corresponds to the location around 0.6 meters in the figure. Numerical results are shown as lines. "x" marks the experimental results in the figure. Generally, there is a very good agreement between the two results. In both of the results travelling wave phenomenon was apparent.

In Figure 7 discharged volumes for both numerical and experimental study are compared. Initially, the agreement is very good between the numerical and experimental volume data. However, as the free surface level decreases considerably, there appears to be some differences between the two cases (see Figure 7).

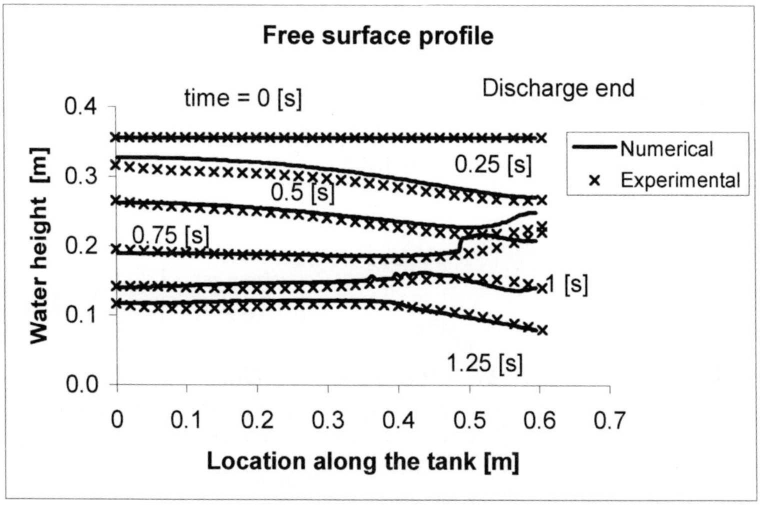

Figure 6: Comparison of the free surface profiles inside the discharge tank at different times

The effects of initial water height on the Froude number (Fn) is shown in Figure 8. The definition of Fn is given as follows:

$$Fn = {(Q/A)} \big/ {\sqrt{g * Hr}} \tag{6}$$

Where Q is the discharged volume per unit time, A is the exit cross sectional area, g is the gravitational acceleration and Hr is the water height at the rear end of the flooded section. As shown in the figure, Froude number initially starts from 0, increases to a certain value (less than 1) and drops to zero as the amount of water reduced in the discharge tank. From the figure, it seems that as the initial water level in the tank increases so as the maximum Froude number for each experiment. Maximum Froude numbers for the experiments with 14, 10 and 6 inches of initial water heights are approximately 0.7, 0.6 and 0.3 respectively.

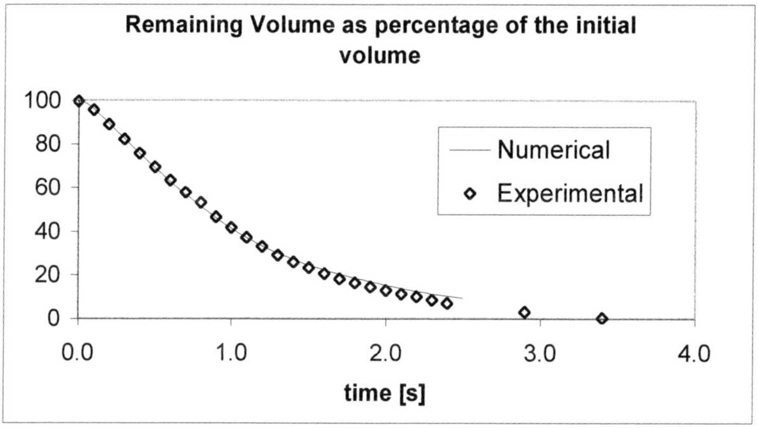

Figure 7: Remaining volume in the discharge tank as the percentage of the initial water volume

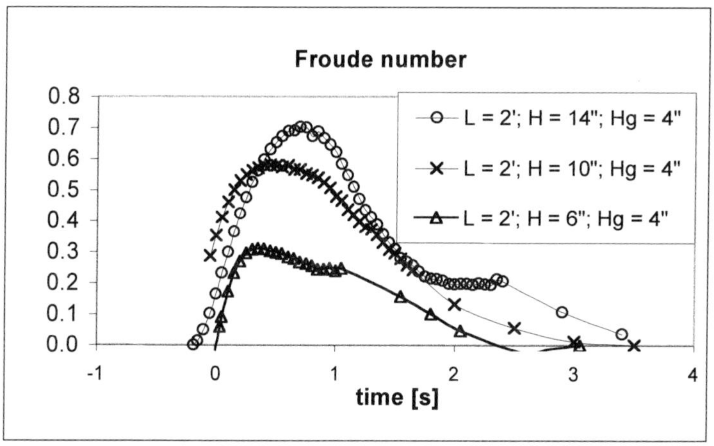

Figure 8: The effects of initial water height on the Froude number

Additional tests conducted by varying the flooded length (L) suggest that the IMO criteria for the damaged hull conditions may not be adequate. In certain cases, i.e. larger beam, there may not be sufficient time for the trapped water to discharge.

SUMMARY

Up to now two model lengths (length of the flooded section in the discharge tank) have been used and discharge data for them at various initial water heights stored. However, the results presented in this paper correspond to 60.96 cm model length only. The discharge tank was filled with water at a prescribed level and the discharge gate was opened to simulate the freeing ports with a flap cover. The discharge flow pattern and free surface form were recorded with a digital camera. The frames were captured on a computer and the location and the form of the free surface established. From knowledge of the free surface the change in volume, discharge rate velocity at the freeing port and various parameters of the discharge kinematics were calculated. In addition to using horizontal bottom conditions we intend to study discharge from listing and periodically rolling decks both with permanently open freeing ports and with freeing ports that have flapped covers. The experimental results will be studied to generate numerical algorithms that can be used to calculate the discharge rates from freeing ports. Time domain results will also be used to validate the numerical studies.

Initial numerical calculations done by Rudman showed that a very good representation of the flow can be predicted by his formulation including wave formation by the opening of the gate. This type of wave formation was observed during the experiments and was successfully predicted using this code.

After completion of the two- dimensional studies we intend to model symmetric, three-dimensional flows and study them experimentally and numerically.

References

Brackbill, J. U., Kothe, D. B. and Zemach, C., (1992). A continuum method for modelling surface tension. *J. Comput Phys.* **100,** 335–354

Calisal, S. M., Akinturk, A., Roddan, G. and Stensgaard, G. N., (1997). Occurrence of water-on-deck for large, open shelter-deck ferries. *STAB'97, Sixth International Conference on Stability of Ships and Ocean Vehicles, Varna, Bulgaria.*

Hirt C. W. and Nichols, B. D., (1981). Volume of Fluid (VOF) Methods for the dynamics of free boundaries. *J. Comput. Phys.* **39,** 201–225

Rudman, M., (1998). A Volume-tracking method for simulating multi-fluid flows with large density variations. *Int. J. Numer. Methods Fluids* **28**, 357-378.

Welch, J. E., Harlow, F. H., Shannon, J. P. and Daly, B. J., (1965). The MAC method: A computing technique for solving viscous, incompressible, transient fluid-flow problems with free surfaces. *Los Alamos Scientific Laboratory Report LA-3425.*

Wesseling, P., (1992). *An introduction to multigrid methods.* John Wiley and Sons. Chichester U.K.

Youngs, D. L., (1982). Time-dependent multi-material flow with large fluid distortion. *Numerical Methods for Fluid Dynamics.* Morton and Baines (eds), Academic Press, New York: 273-285

Zalesak, S. T. (1979). Fully Multi-dimensional Flux Corrected Transport Algorithms for Fluid Flow. *J. Comput. Phys.* **31,** 335–362

4. Impact of Stability on Design and Operation

Contemporary Ideas on Ship Stability
D. Vassalos, M. Hamamoto, A. Papanikolaou and D. Molyneux (Editors)

PASSENGER SURVIVAL-BASED CRITERIA FOR RO-RO VESSELS

D. Vassalos, A. Jasionowski and K. Dodworth

The Ship Stability Research Centre (SSRC), Department of Ship and Marine Technology
University of Strathclyde, Glasgow, UK

ABSTRACT

This paper outlines the research work undertaken at the University of Strathclyde aiming to address the question of "passenger survival" following large scale flooding on the vehicle deck of passenger/Ro-Ro vessels. An extensive investigation was carried out, with the intention of characterising the survival of a vessel in marginal conditions, culminating in a proposal for design criteria which ensure an assumed sufficient survival time for the complete evacuation of passengers and crew following a breach in the hull integrity. The results of the parametric investigation are presented and discussed and the rationale behind the proposed criteria explained.

KEYWORDS

Damaged Ro-Ro vessels, water-on-deck, numerical simulation, capsize boundaries, passenger safety, time-based survival criteria.

INTRODUCTION

The field of naval architecture in the area of ship survivability has, to date, been somewhat lacking in that the inadequate understanding of the complexities of damage stability has lead to over-simplified codes and regulations. Such shortcomings have resulted in several tragic accidents, most recently the Herald of Free Enterprise and the Estonia, as well as a general uneasiness in the industry and the passengers it provides for. One of the starkest realities of both accidents was the catastrophic nature of the accident rather than a slow, controlled loss of stability. The aim of this paper is to ensure that such rapid capsize is both better understood and dealt with in a manner that ensures passenger survival in the event of a breach in the hull integrity. Such an attempt might be proved futile if there is no appreciation of the time it takes to fully evacuate a loaded passenger Ro-Ro vessel. A resolution adopted at the 1995 SOLAS Diplomatic Conference was that the time for the evacuation of passengers and crew should be no greater than 60 minutes [RINA (1996), paper 1]. This might be taken as a likely starting point when determining the minimum time that a vessel should maintain its function as a safe passenger-carrying medium. Recent insights into evacuation have proved that survival times of the aforementioned one hour or even more might be required. A trial evacuation of the Stena Evicta [RINA (1996), paper 2] passenger Ro-Ro vessel carried out by the MSA predicted an evacuation time of about 65 minutes for the mustering and disembarkation of the full passenger and crew complement into a sunlit harbour without the panic environment of a real evacuation. The under-prediction of such trials was demonstrated by the emergency evacuation of the St Malo in the English channel of one-hour and 17 minutes duration for a vessel which was evacuated in eight minutes during its trial. Clearly the field of passenger evacuation has a great distance to cover. However, as steps forward are being made it is imperative that vessel survivability complements these advances.

DEFINITIONS

Before proceeding with a proposal for passenger survival-based criteria, it is considered necessary to clearly state what the terms used in this paper refer to. The definitions provided in the following are not intended to be absolute but indicative of the purpose the relevant terms are used for.

Capsize

The term "capsize" is used to define the status of a vessel with an excessively large angle of roll, defined in this study as the value at which the low frequency roll response (indicating the heel of the vessel) has exceeded twenty degrees. At inclinations larger than this angle it is assumed that evacuation could not take place.

Capsize Band and Boundaries

The capsize boundary defined in the work completed in Vassalos *et al.* (1996) was expressed in terms of limiting sea states above which the test vessel will always capsize and below which the vessel will always survive. This is a useful simplification of the phenomenon of capsize. However, as indicated in Vassalos *et al.* (1996) and shown in this study this limit is not a definitive boundary, rather a region of uncertainty ('capsize band') in which the fate of the vessel cannot be evaluated deterministically, Figure 1. The hypothesis will be put forward that in the capsize band, vessels will always capsize if the simulation or experiment time tends to infinity. The reasoning for this conjecture is that within the band there is enough energy within the environment to capsize the vessel and, within infinitely large time, the vessel will encounter a sufficiently large wave group which will cause capsize. As simulation time and number of runs are both finite, parameters characterising the band, as shown above, can be expected to be random variables. In Figure 1 'Survivability Related Parameter' refers to any design parameter affecting the ability of the vessel to withstand large scale flooding. H_{sig} refers to the significant wave height, chosen to be indicative of environment. The term 'boundary' is used throughout the paper to represent a contour within the capsize band with equal probability of capsize before a fixed time.

Vessel Configurations and Damage

The same hull form and car deck compartmentations as in Vassalos *et al.* (1996) were used in this study, involving open deck, side and central casing and transverse bulkheads. The vessel damage used is of a trapezoidal shape, situated amidships with a damaged waterline width as defined by SOLAS (1989).

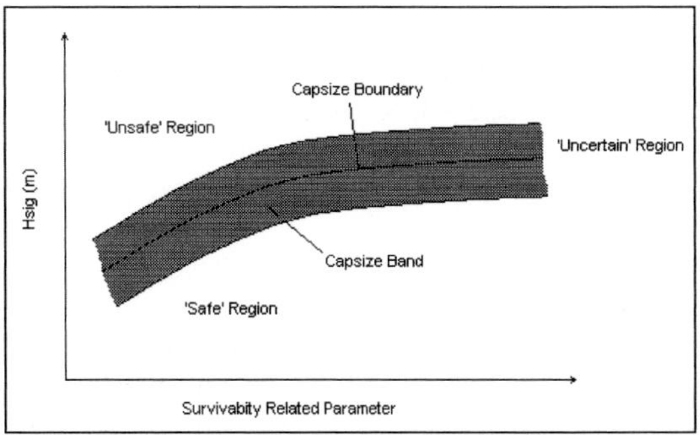

Figure 1: Definition of Capsize Band and Survival Boundaries

MATHEMATICAL MODEL

The mathematical model of a damaged ship motion in a random sea state that was used extensively throughout this study was developed and validated in the early nineties during the UK Ro-Ro Research Programme and the Joint North West European Project. A more detailed description of this can be found in DMI (1990), Dand (1991), Turan (1993), Velschou & Schindler (1994), Vassalos *et al.* (1996), Werenskiold (1996), RINA (1996) and Vassalos *et al.* (1997). The stage has now been reached where it can be confidently used to assess the ability of any vessel type and compartmentation to survive damage in a given environment.

Ship Motion Model

The model relates to a three-degree-of-freedom non-linear system of equations describing ship response to environmental excitation. Random sea state is represented by the JONSWAP wave energy spectrum with peakness parameter γ = 3.3 and mean zero crossing period $T_z = 4.0 \cdot H_{sig}/1.2905$. The model accounts for progressive flooding through any ship compartmentation. The system of equations can be summarised as follows;

$$\{[M(t)]+[A]\}\{\ddot{Q}\} + [B]\{\dot{Q}\} + [C]\{Q\} = \{F\}_{WAVE} + \{F\}_{WOD}$$

with, $[M(t)]$: Instantaneously varying mass and mass moment of inertia matrix.

[A], [B]	:	Generalised added mass and damping matrices, calculated once at the beginning of the simulation at the frequency corresponding to the peak frequency of the wave spectrum chosen to represent the random sea state.
[C]	:	Instantaneous heave and roll restoring, taking into account ship motions, trim, sinkage and heel.
$\{F\}_{WAVE}$	:	Regular or random wave excitation vector, using 2D or 3D potential flow theory.
$\{F\}_{WOD}$	:	Instantaneous heave force and trim/roll moments due to flood water.

The $\{F\}_{WOD}$ is assumed to move in phase with the ship roll motion with an instantaneous free-surface parallel to the mean waterplane. This assumption has been considered acceptable with large ferries since, owing to their low natural frequencies in roll, it is unlikely that floodwater will be excited at a resonant frequency. The resonance is further spoiled as a result of progressive flooding hence when the water volume is sufficiently large to alter the vessel behaviour, small phase differences are expected between the flood water and ship roll motions. During simulation, the centre of gravity of the ship is assumed to be fixed and all undamaged subdivisions watertight.

Water Ingress/Egress Model

The aforementioned water ingress model is described by a simple hydraulic model, where the mode of flow is governed by the sign of $(h_{out} - h_{in})$ and the volume flow rate as a function of hydrostatic head pressure and area of damage hole as shown in Figure 2.

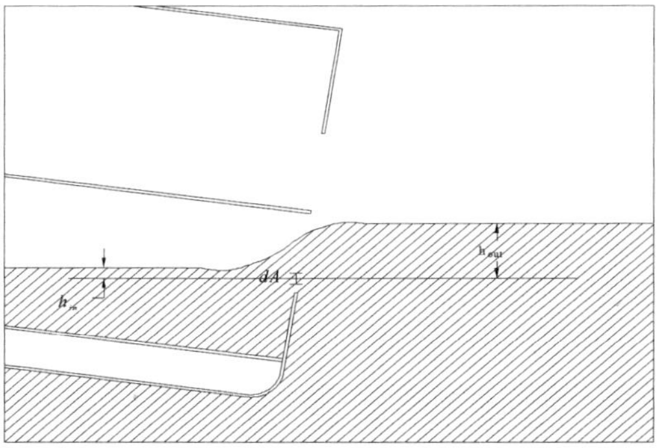

Figure 2: Water Ingress Main Parameters

Any additional phenomena affecting the rate of flow, like edge effect, shape of opening, ship motion and so on are accounted for through the flooding coefficient (K) approximated by studying and analysing model experiments. The expression representing water ingress/egress can be shown to be

$$dQ = K\sqrt{2g(h_{out} - h_{in})}dA$$

where, $dA \equiv$ incremental area; $dQ \equiv$ incremental flow rate through area dA; $h_{out/in} \equiv$ water level outside/inside the damage.

APPROACH ADOPTED

Using the time-domain simulation program, a number of numerical experiments were performed to characterise the random variables involved and to undertake a sensitivity study of key parameters. Based on observed trends and the insight gained as a result of this investigation a proposal for passenger safety-based survival criteria was put forward.

Measurement Procedures

As mentioned above, the processes within the capsize band are random in nature i.e. they cannot be quantified deterministically and some degree of uncertainty can be expected in quantifying them. Care is therefore required in devising measurement procedures. The capsize band is quantified using primarily two parameters, the relative frequency (an approximation of probability) of capsize occurrence and the mean of capsize time (description of the expected value) both of which are defined below

$$\text{Capsize Time Expected Value} = \frac{\dfrac{\sum \text{Capsizal Time}}{\text{Number of Capsizals}}}{\text{Number of Capsizals Within Simulation Duration}}$$

$$\text{Relative Frequency} = \frac{\text{Number of Capsizals Within Simulation Duration}}{\text{Total Number of Runs}}$$

Both these quantities are random variables but the variance can be reduced by multiple trials. The following strategy was adopted to lessen the degree of spread in the results to a level perceived as reasonable:

- Ten computer simulations for initial screening of parameters.
- Twenty computer simulations for the majority of cases.
- One hundred computer simulations to elucidate erroneous results.

A run time length of thirty minutes was used for the initial screening of parameters in keeping with previous research. After these preliminary runs it was concluded that a longer run time would be required to characterise fully the sample for cases close to the lower ('safe') limit of the capsize band. As mentioned in the foregoing, survival times of over one hour are necessary for the safe evacuation of a vessel. For this reason 70 minutes was chosen as a run time for the majority of numerical experiments. The relative frequency in these cases, according to the hypothesis presented above, is an estimation of the probability of the vessel capsizing before the assumed evacuation time.

Summary of Sensitivity Study

In order to develop a sufficiently large database from which to draw conclusions, a comprehensive parametric investigation covering approximately 14,000 numerical experiments was completed. A complete understanding of the rationale behind the criteria proposed in the following does not depend on an appreciation of all the available results, hence only the conclusions of interest will be presented here. The results obtained show a number of prevalent trends which summarise the behaviour and fate of a vessel in a given condition and environment. Starting with an explanation of the capsizal time expected value the following characteristics are noteworthy:

- This statistic increases rapidly with decreasing wave height. This would suggest an asymptotic behaviour of the full population's mean.
- In all cases the survival time decreases with increasing H_{sig} until an approximately constant level is reached.

These attributes can be detected in the sample graphs presented in Figure 3. This regular shaped graph, described here as a 'time template', moves in accordance with the position of the capsize band. The existence of this asymptote at the safe limit of the capsize band is merely implied by the data but assuming a smooth transition in the time population mean between the safe region (where the population mean can be assumed to approach infinity) and the capsize band, it would appear as though this is a logical assumption. This limit is difficult to define with the survival time information alone, hence an examination of the curves of relative frequency is necessary as displayed in Figure 4.

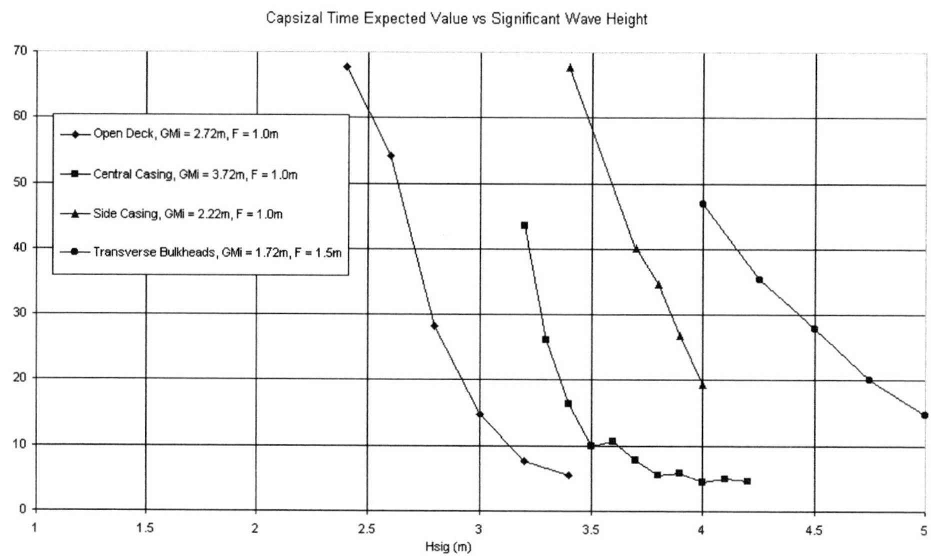

Figure 3: Survival Time vs. Significant Wave Height (H_{sig}) for a Range of Metacentric heights (GM)

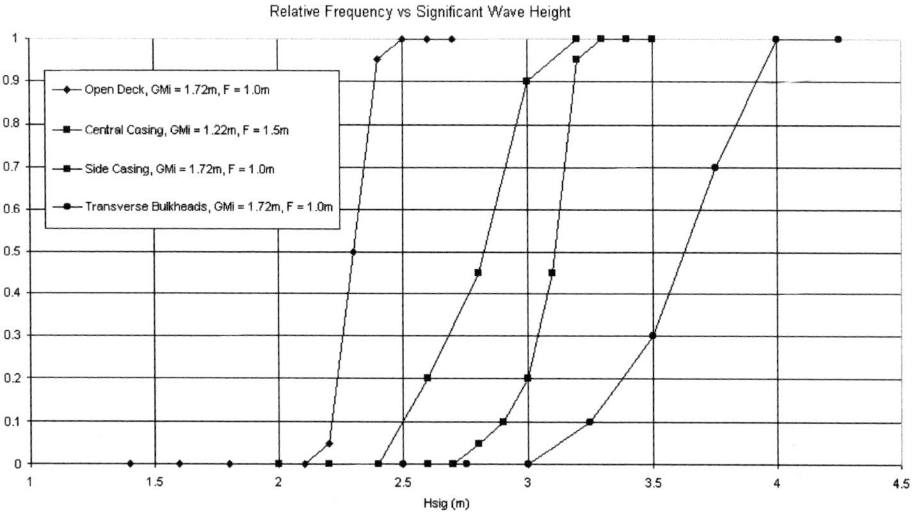

Figure 4: Capsizal Relative Frequency (P(capsize)) vs. Significant Wave Height (H_{sig}) for a Range of Metacentric Heights (GM).

The curves of relative frequency vs. H_{sig} also have a characteristic shape. When the H_{sig} is remote from the capsize band it approaches one of two values; unity when below the band and zero when above the band. It cannot be stated that either boundary curve is **at** these values as there is still a remote probability above the band that the ship survives a set time and similarly a possibility that a vessel below the band does not survive. Instead, it can be said that the capsizal probability approaches these values. Within the band there is a characteristic elongated 'S' shape. Both the time and relative frequency templates are illustrated in Figure 5.

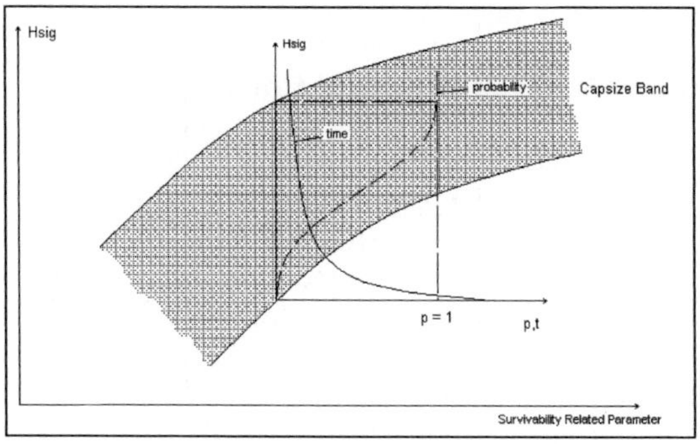

Figure 5: Time and Relative Frequency Templates

The negative correlation between the relative frequency and the Capsize Time Expected Value can be explained by the use of the capsizal time probability density function (p.d.f). As mentioned earlier, all vessels will capsize in infinitely large time in which case it is possible to present the capsizal time random variable by a continuous p.d.f. The parameters measured in this paper are given as follows:

- Relative Frequency ≈ Area of $A_{s/l}$
- Capsize Time Expected Value ≈ Position of centroid of area $A_{s/l}$

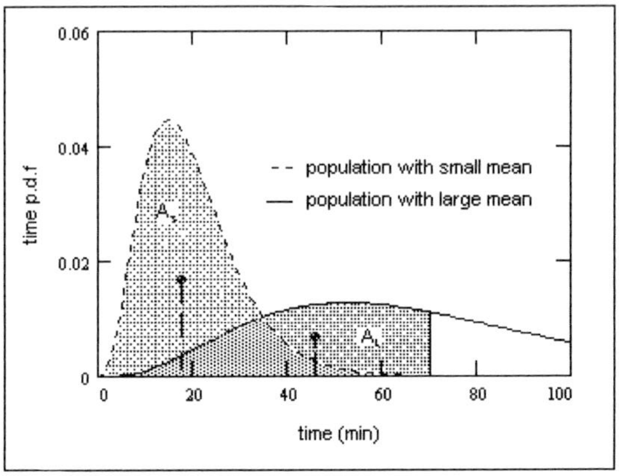

Figure 6: Comparison of p.d.f.'s for two Capsizal Time Populations
[$A_{s/l}$ Area of interest for small/large means. In both cases the area of interest, from basic distribution theory, is the area to the left of the seventy-minute simulation duration]

As the vessel design and environmental parameters come closer to ensuring the survival of the vessel past the survival time, the p.d.f. spreads out as the population mean increases. As demonstrated in Figure 7, the centroid of A_L is positioned at larger time than that of A_s suggesting a larger capsize time expected value. Also, the area of A_L is smaller than A_s which infers a smaller relative frequency value.

Survival Time Boundary Curves and Proposed Criteria

Having collated the extensive data from the parametric study it can now be examined how it relates to the individual time curves and propose criteria based on ensuring 'adequate' survival time to evacuate passengers and crew. Studying these curves it can be seen that the characteristics of the time curves (described in the foregoing) are reflected in Figure 7. The vertical asymptote for all curves is reflected in the convergence of boundary curves at higher times. These higher time boundary curves are what will be important in trying to develop time-based survival criteria, as the time to evacuate a vessel will of the same order as the time of seventy minutes used in this investigation. As the higher time boundaries converge on the lower part of the capsizal band it is more logical to base criteria on this region rather than any one of the individual time boundary curves within this area.

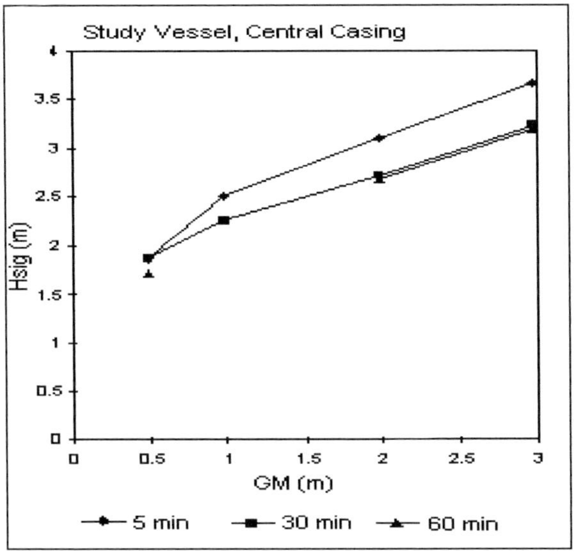

Figure 7: Example Time Boundary Curves (probability of capsize within time ≈ 0.80)

Position of Capsize Band

Having developed a more precise definition of the capsize band and the behaviour of the various parameters within this region what remains is to locate the position of this band's lower, safe boundary where adequate survival time is most likely to be achieved. Recalling the work of Vassalos *et al.* (1996), the probability of survival of a Ro-Ro vessel with water on deck is simply given by the probability of not exceeding the critical wave height at which the vessel will capsize. Following from this, the critical task has been to formulate a connection between the critical sea state and vessel related parameters, which can be readily calculated without resorting to costly, time consuming numerical simulations or experiments. A key observation deriving from this work is that vessel capsizal occurs close to the angle where the righting moment curve has its maximum. The volume of water on deck causing the ship to assume an angle of equilibrium that equals the angle ϕ_{max} was therefore compared with the critical volume of water at the instant of capsize and a good correlation was found. It was subsequently shown that the governing parameter was the height difference between the inner waterline and the sea level at the angle of GZ_{max}, the h parameter. This connection was shown to be universal for all arrangements studied and formed the basis for the calculation of the critical wave height. As a result, the survivability of the vessel was expressed as a function of the critical significant wave height as denoted below:

$$h = f(H_{sig})$$

The h parameter is defined as the critical depth of water on the vehicle deck at the point of capsize which is equivalent to the depth of water on deck required to incline the vessel to the angle at which maximum righting occurs with the additional assumption of the vessel being flooded below the vehicle deck in selected compartments. A comprehensive reasoning for this relationship can be found in the aforementioned work. This finding represents a major advance in that the survivability of a vessel could be summarised in a single parameter. However, the critical wave height used in this equation corresponds closely with the Pr(capsize within evacuation time) ≈50 % contour as shown in Figure 8 which is not an acceptable risk by any definition of the term.

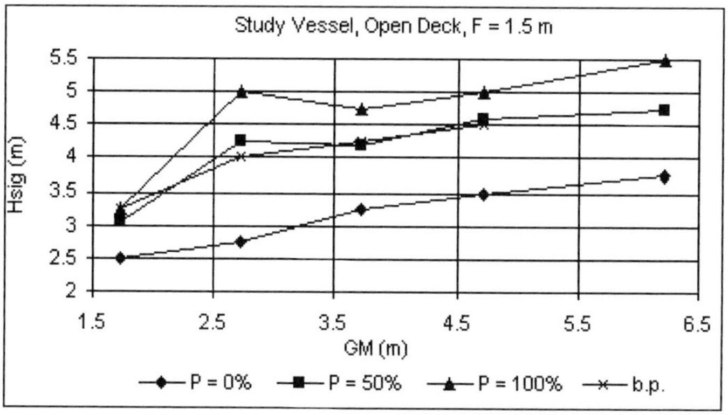

Figure 8: Three probability of capsize within assumed evacuation time contours (P) compared with findings of Vassalos *et al.* (1996)

This formulation forms the basis of the criteria proposed in this paper but the critical significant wave height must be newly defined as the wave height corresponding to the lower limit of the capsize band where the probability of capsize within the assumed evacuation time is sufficiently low. A fresh examination of the data presented in Vassalos *et al.* (1996) shows that the effect of freeboard is not adequately accounted for by the h parameter. In fact, a remarkable improvement in the goodness-of-fit for the trend proposed in this work can be gained from the inclusion of residual freeboard (F) in the relationship between the h parameter and the critical significant wave height. Hence the proposed equation will be of the following form:

$$h = f(H_{sig}, F)$$

A regression analysis was carried out by a least mean squares approach with the contribution from each data point weighted with respect to the estimated variance of each point on the data presented in Figure 9. It was found that the following equation fits the data reasonably well:

$$h = 0.088 \cdot H_{sig}^{\,0.97+0.46 \cdot F} \text{, where}$$

h = Critical Amount of Water on Deck

H_{sig} = Critical Sea State

F = Residual Freeboard

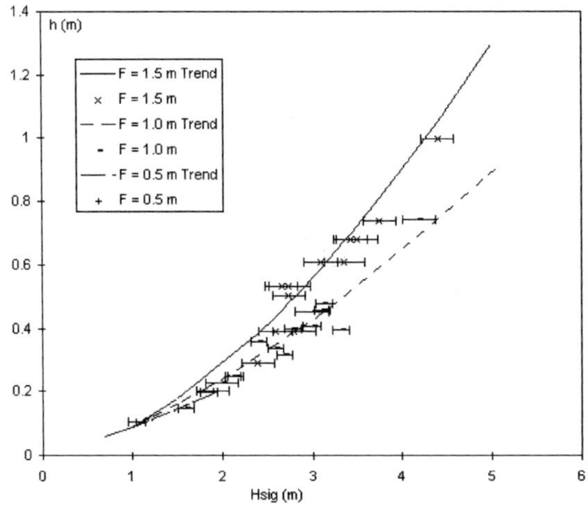

Figure 9: Critical state vs. h parameter for a range of freeboards

According to these results a vessel designed to comply with criteria of this form will have a survival time over one hour.

CONCLUDING REMARKS

One of the most conspicuous conclusions from the results was that existing vessel designs have little prospect of surviving for more than thirty minutes in many of the sea states they are designed to operate in. Evidently, changes to the current state of Ro-Ro vessel design must be effected if passenger safety is to be raised to a satisfactory level. It is intended that the given results and discussion have gone some way to further understanding the phenomenon of capsize in passenger Ro-Ro vessels. The range of vessel configurations studied provides some clues to how the risk of catastrophic capsize might be reduced even if the vessel becomes flooded following a damage. The criteria provided is a suggestion of how passenger safety might be provided for with the smallest impact on the effective business of ferry operators.

References

Dand I.W. (1990). Experiments with a Floodable Model of a Ro-Ro Passenger Ferry, BMT Project Report, for the Department of Transport, Marine Directorate, BMT Fluid Mechanics Ltd., February.

Dand I. W. (1991). Experiments with a Flooded Model of a Ro-Ro Passenger Ferry, 2nd Kummerman Int. Conf. on Ro-Ro Safety and Vulnerability – The Way Ahead, RINA, London, April, paper No. 11.

Danish Maritime Institute (1990). Ro-Ro Passenger Ferry Safety Studies, Model Tests for F10, Final Report of Phase I for the Department of Transport, DMI 88116, February.

International Convention for the Safety of Life at Sea (SOLAS) (1989). Texts of Amendments Relating to Passenger Ro-Ro Ferries Adopted on 21 April and 28 October 1988, IMO, London.

RINA (1996). Escape, Evacuation & Rescue Design for the Future, International Conference in association with The Nautical Institute, 19&20 November, London.

Turan O. (1993). Dynamic Stability Assessment of Damaged Passenger Ships Using a Time Simulation Approach, *Ph.D. Thesis*, Department of Ship and Marine Technology, University of Strathclyde.

Vassalos D., Jasionowski A. and Dodworth K. (1997). Assessment of Survival Time of Damaged RO-RO Vessels, Internal Report, the Ship Stability Research Centre, Uni. of Strathclyde, March.

Vassalos D., Pawlowski M. and Turan O. (1996). A Theoretical Investigation on the Capsizal Resistance of Passenger/Ro-Ro Vessels and Proposal of Survival Criteria, Final Report, Task 5, The Joint R&D Project, March.

Velschou S. and Schindler. M. (1994). Ro-Ro Passenger Ferry Damage Stability Studies - A Continuation of Model Tests for a Typical Ferry", RINA Symp. on Ro-Ro Ship's Survivability - Phase 2, RINA, London, November, paper No. 5.

Werenskiold P. (1996). High speed marine craft, 5th Int. Conf. On High Speed Craft, Bergen 10-13, September.

D. Vassalos, M. Hamamoto, A. Papanikolaou and D. Molyneux (Editors)

NONLINEAR DYNAMICS OF SHIP ROLLING IN BEAM SEAS AND SHIP DESIGN

K. J. Spyrou, B. Cotton and J.M.T. Thompson

Centre for Nonlinear Dynamics and its Applications
University College London, Gower Street, London WC1E 6BT, UK

ABSTRACT

The possibility to use in ship design certain recent results of the nonlinear analysis of beam-sea rolling in order to maximize resistance to capsize is discussed. The loci of transient and steady-state capsize are approximately located on the plane of forcing versus frequency through Melnikov analysis, harmonic balance and use of the variational equation. These loci can be parametrized with respect to the restoring and damping coefficients. The minimization of the capsize domain leads naturally to the formulation of an interesting hull optimization problem.

KEYWORDS

Ship, design , capsize, roll, stability , nonlinear, dynamics, Melnikov, harmonic balance.

INTRODUCTION

Recent efforts to understand the mechanism of ship capsize in regular beam seas have revealed enormous complexity in large amplitude rolling response patterns, even though these investigations have relied on simple nonlinear, single-degree models, Thompson (1997). Whilst the existence of bistability, jumps and subharmonic oscillations near resonance were known from earlier studies based on perturbation-like techniques [see for example Stoker (1950) and Nayfeh & Mook (1979) on the forced oscillator; and Cardo et al. (1981) for a more ship-specific viewpoint] a whole range of new phenomena including global bifurcations of invariant manifolds, indeterminate jumps and chaos have been shown recently to underlie roll models with cubic or quartic potential wells. There are good reasons to believe that such phenomena are generic and their presence should be expected for a wide range of ship righting-arm and damping characteristics.

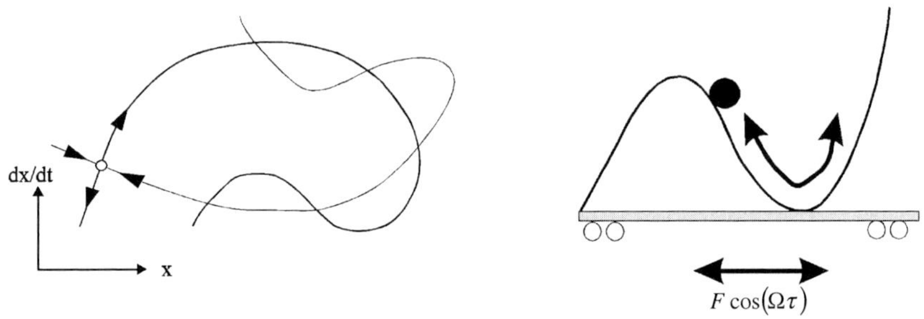

Fig. 1: Intersection of stable and unstable manifolds (left) for a simple system (right) characterised by an "escape" mechanism.

For the practising engineer this new information will be of particular value if it can be utilized effectively towards designing a safer ship. So far, rather than trying to discriminate between good and less good designs in terms of resistance to capsize in beam seas, the current analyses set their focus mainly on developing an understanding of the nature of the nonlinear responses in their various manifestations. However it seems that the time is now ripe for addressing also the design problem. Attempts to develop an interface between nonlinear analysis and ship design are by no means a novelty since they date back, at least, to the discussions about Lyapunov functions in the seventies and early eighties Odabasi (1978), Caldeira-Sarava (1986). Nonetheless, a meaningful and practical connection between nonlinear analysis and ship design is still wanting.

In our current research, the main ideas and some preliminary results of which are presented here, we are exploring the potential of two different assessment methods, based on well known approximate escape criteria of forced oscillators. The first method capitalizes upon the so-called Melnikov criterion which provides a fair estimate of the first heteroclinic tangency (homoclinic for an asymmetric system) that initiates erosion of the safe basin, Fig. 1 [Thomspon et al. (1990), Kan (1992), Falzarano et al. (1992)]. In the second method the key concept is the wedge-like boundary of steady-state escape on the *forcing* - versus - *frequency* plane, Szemlinska-Stupnicka (1988) & (1992), Virgin (1989). The left branch of this boundary is the locus where jumps to capsize from the lower fold take place, Fig 2. As for the right branch, it is generally practical to assume as such the symmetry-breaking locus near resonance (or, the first flip for an asymmetric system).

These two criteria of transient and steady-state escape should be applied in conjunction with general-enough families of restoring and damping curves. A seventh-order polynomial is often seen as a suitable representation of restoring. For damping, however, at this stage we shall confine ourselves to the equivalent linear one. Once the roll equation obtains a specific parametric form, expressions can be developed linking the coefficients of the restoring polynomial with damping, forcing and encounter frequency to the capsize loci. The obvious usefulness of these expressions is that they allow us to assess how hull modifications can affect the thresholds of transient or steady-state capsize. This leads to consider setting up an optimization process with governing objective the definition of a hull characterized by maximum resistance to capsize. The procedure offers also the interesting opportunity to evaluate the steady-state and transient criteria against each other, with the view to establishing whether they lead to similar optimum hull configurations.

KEY FEATURES OF THE SINGLE-WELL OSCILLATOR

Consider the following single-degree model for ship rolling, Thompson (1997):

$$\ddot{x} + D(\dot{x}) + R(x) = B + F\cos(\Omega\tau) \tag{1}$$

Equation (1) is written in terms of the scaled roll angle x. If φ is the actual roll angle and φ_v is the angle of vanishing stability then $x = \varphi/\varphi_v$. The function $D(\dot{x})$ represents the damping. Ω is the ratio of the frequency of encounter ω between the ship and the wave (as we assume a beam-sea this is also the wave frequency) and ω_0 is the natural frequency which can be found from the well known expression $\omega_0 = \sqrt{W(GM)/(I+\Delta I)}$. In conventional notation W is the weight of the ship, (GM) is the metacentric height and I is the second moment of inertia in roll and ΔI is the added component. Also in (1), F is the amplitude of the scaled external periodic forcing expressed as $F = \dfrac{I}{I+\Delta I}\dfrac{Ak\,\Omega^2}{\varphi_v}$ where Ak is the wave slope. Finally, B is a scaled constant excitation, for example due to steady wind; $R(x)$ is a scaled polynomial that approximates the restoring curve and is characterised by unit slope at the origin [$dR(x)/dx = 1$ at $x = 0$]; and τ represents the nondimensional time, $\tau = \omega_0 t$, where t is the real time.

Let us consider for a while an asymmetric escape equation with periodic forcing, linear damping, $D(\dot{x}) = 2\zeta\,\dot{x}$, and a single quadratic, "softening" type, nonlinearity in restoring, $R(x) = x - x^2$. Such an equation, which can be regarded as the simplest possible nonlinear equation akin to the capsize problem, has been studied to considerable depth. Figs. 2 and 3 (Thompson 1996) summarise the most characteristic aspects of the steady and transient behaviour of a system governed by such an equation. Near resonance the response curve exhibits the well known bending-to-the-the left property that creates the lower fold A and the upper fold B. Point A is a saddle-node and a jump towards either some kind of resonant response or towards capsize will take place if the corresponding frequency threshold is exceeded. On the resonant branch

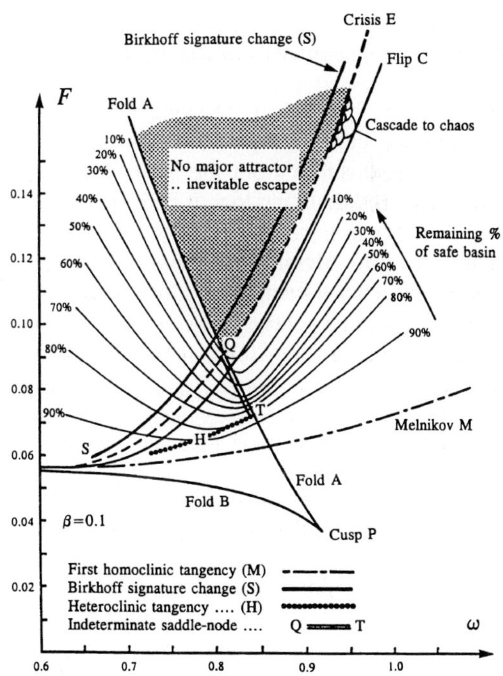

Fig. 2: Bifurcation diagram of the escape equation (Thompson 1996)

different types of instability can arise. If the wave slope Ak is slowly increased, period-doublings (flips) are noticed that usually lead to chaos (a "symmetric" system with cubic instead of quadratic nonlinearity must first go through "symmetry-breaking" at a supercritical pitchfork bifurcation). Further increase in forcing leads ultimately to the so-called *final crisis*, where the chaotic attractor vanishes as it collides with a saddle forming a *heteroclinic chain.* At relatively high levels of excitation there is no alternative "safe" steady-state and subsequently escape is the only option. Long before such high levels of forcing have been attained, however, the "safe" basin has started diminishing after an homoclinic tangency (heteroclinic in the case of a symmetric system). The heteroclinic (homoclinic) tangency is usually considered as the threshold of transient escape. Melnikov analysis allows approximate analytical prediction of the relation between the oscillator's parameters on this threshold.

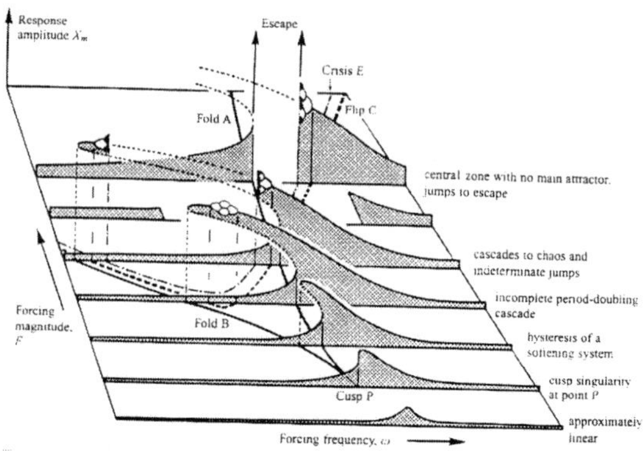

Fig. 3: Resonance response curve (Thompson 1996)

In a diagram of Ak versus Ω (for constant damping), the earlier discussed thresholds appear as boundary curves, Fig. 2. The locus of the first homoclinic tangency can lie at a considerable distance from the "wedge"-like boundary formed by the fold and symmetry breaking/period doubling loci. It is of course desirable that the Melnikov curve lies as high in terms of Ak as possible. It follows that a desirable hull configuration should present the minimum of its Melnikov curve at Ak as high as it can be. Alternatively, it is possible to take into account a range rather than a single frequency, thus seeking to maximize the area below the Melnikov curve between some suitable low and high frequencies, respectively Ω_1 and Ω_2. In the ideal case where the Melnikov curve can be expressed explicitly as $Ak(\Omega)$, one will be seeking to identify the combination of restoring and damping coefficients, representing the connection with the hull, that maximizes the integral $\int_{\Omega_1}^{\Omega_2} Ak(\Omega)\, d\Omega$. More sophisticated criteria based on wave energy spectra, and thus incorporating probabilistic considerations, could also be considered. These are left however for later studies.

A similar type of thinking can be applied for steady-state capsize. Here one could require the lowest point of the wedge to be as high as possible in terms of forcing; or again, the area under the wedge between suitable Ω_1 and Ω_2 to be maximized. One possible way of defining Ω_1 and Ω_2 rationally could be attained by drawing the breaking-wave line on the (Ak,Ω) plane and taking its intersections with the fold and flip curves. Unfortunately for the considered range of frequencies this line may not intersect the flip curve. The rational definition of Ω_1 and Ω_2 needs further consideration.

Assume finally the following "symmetric" representation of restoring:

$$R(x) = x + a_1 x^3 - a_2 x^5 + (-1 - a_1 + a_2) x^7 \tag{2}$$

The main advantage in using the seventh-order polynomial is that it provides two points of inflection, see Appendix. Here a_1, a_2 are the two free parameters of the restoring curve. The coefficient of the seventh-order term is selected so that the saddle points are always at $x = 1$ and -1. Thus we shall be dealing from now on with the following roll equation, Fig. 4:

$$\ddot{x} + 2\zeta\,\dot{x} + x + a_1 x^3 - a_2 x^5 + (-1 - a_1 + a_2) x^7 = F\cos(\Omega\tau) \tag{3}$$

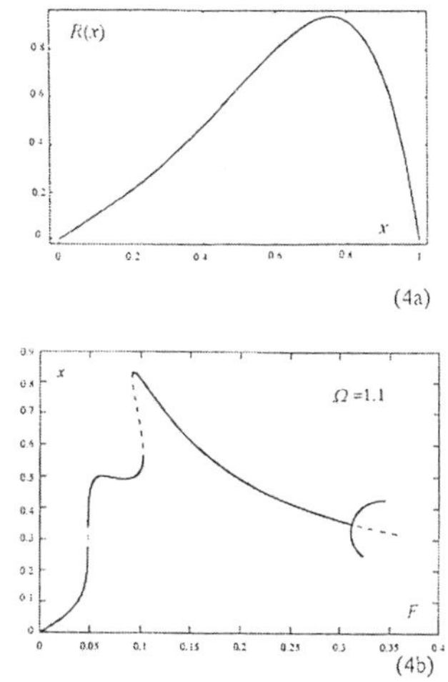

Fig. 4: Restoring curve (upper) and steady response curve (lower), for $a_1 = 1.5$, $a_2 = 1$

MELNIKOV-BASED CRITERIA

Details about Melnikov analysis can be found in a number of texts and no attempt will be made to repeat these here, e.g. Guckenheimer & Holmes (1983), Bikdash et al. (1994), Nayfeh & Balachandran (1995). The method is based on the calculation of the signed distance between the stable and unstable manifolds of one or more saddle equilibrium points when this distance is small. Melnikov analysis can also be regarded as an energy balance method where the total energy dissipated through damping should equal the energy supplied through the external forcing, Thompson (1996). A more sophisticated version of the method can be applied also for highly dissipative systems, Salam (1987).

Melnikov analysis includes basically the following stages. Firstly we calculate the Hamiltonian H of the unperturbed ($\zeta = F = 0$) system and from this the heteroclinic (homoclinic) orbit as $dx/d\tau = p(x)$. Then, we attempt to derive, if possible analytically, the time variation along this orbit: namely to derive expressions for x and $dx/d\tau$ that are functions of time, $x = h_1(\tau)$ and $dx/d\tau = h_2(\tau)$. This often represents the first major difficulty in applying the method. The next step is to calculate the Melnikov function given below:

$$M(\tau_0) = \int_{-\infty}^{+\infty} \mathbf{f}[\mathbf{x}(\tau)] \wedge \mathbf{g}[\mathbf{x}(\tau), \tau + \tau_0]\, d\tau \tag{4}$$

where $\mathbf{x} = [x,\ dx/d\tau]^{\mathrm{T}}$ and $d\mathbf{x}/d\tau = \mathbf{f}[\mathbf{x}(\tau)]$ is the equation of the unperturbed (Hamiltonian) system. The function $\mathbf{g}[\mathbf{x}, \tau]$ is periodic and represents the damping and forcing terms considered as constituting a perturbation. Also, τ_0 is phase lying in the range $0 < \tau_0 < 2\pi/\Omega$.

The symbol $\wedge$ means to take the cross product of vectors. The main objective in this method is to identify those marginal combinations of parameters where the Melnikov function admits real zeros.

Let us apply the above method in respect to equation (3). The equation of the unperturbed system is:

$$\ddot{x} + x + a_1 x^3 - a_2 x^5 + (-1 - a_1 + a_2) x^7 = 0 \tag{5}$$

which can be written further in the form:

$$\begin{aligned} x_1 &= x \\ \dot{x}_1 &= x_2 = \vartheta H / \vartheta x_2 \\ \dot{x}_2 &= -x_1 - a_1 x_1^3 + a_2 x_1^5 - (-1 - a_1 + a_2) x_1^7 = -\vartheta H / \vartheta x_1 \end{aligned} \tag{6}$$

Hamiltonian:

$$H = \frac{1}{2}\left(x_2^2 + x_1^2 + \frac{a_1}{2} x_1^4 - \frac{a_2}{3} x_1^6 + \frac{-1 - a_1 + a_2}{4} x_1^8 \right) \tag{7}$$

Heteroclinic orbit :

$$\dot{x} = \pm \sqrt{\left(\frac{3}{4} + \frac{a_1}{4} - \frac{a_2}{12} \right) - x^2 - \frac{a_1}{2} x^4 + \frac{a_2}{3} x^6 - \frac{-1 - a_1 + a_2}{4} x^8} \tag{8}$$

Let the time variation along the heteroclinic orbit to be: $x = h_1(\tau)$ and $dx/d\tau = h_2(\tau)$. These can be found with appropriate variable transformations, or they can be approximated.

Melnikov function :

$$\begin{aligned} M(\tau_0) &= \int_{-\infty}^{+\infty} x_2 \{ F \cos[\Omega(\tau - \tau_0)] - 2\zeta\, x_2 \} d\tau = \\ &= F \cos(\Omega \tau_0) \int_{-\infty}^{+\infty} h_2(\tau) \cos(\Omega \tau) d\tau - F \sin(\Omega \tau_0) \int_{-\infty}^{+\infty} h_2(\tau) \sin(\Omega \tau) d\tau - 2\zeta \int_{-\infty}^{+\infty} h_2^2(\tau) d\tau \end{aligned} \tag{9}$$

The second integral is expected to be zero because $h_2(\tau)\sin(\Omega\tau)$ is an odd function [$h_2(\tau)$ is expected to be even, $\sin(\Omega\tau)$ is of course odd]. However if the homoclinic orbit is considered it is the first integral that can be zero.

The condition to have simple zeros for the Melnikov function written in terms of Ak is thus:

$$\frac{Ak\Omega^2}{\varphi_v} > 2\zeta \frac{\int_{-\infty}^{+\infty} h_2^2(\tau) d\tau}{\int_{-\infty}^{+\infty} h_2(\tau) \cos(\Omega\tau) d\tau} \tag{10}$$

The threshold Ak that gives rise to equality in (10), $Ak_{\min}$, will mean tangency of manifolds and will thus determine the Melnikov curve $Ak = M(\Omega)$.

Criterion 1:

$Ak_{\min}(\Omega)$ to become maximum in terms of the parameters a_1, a_2, φ_v, ζ. It is understood of course that as $2\zeta = D/\sqrt{W(GM)(I+\Delta I)}$ where D is the true dimensional equivalent linear damping, (GM) and $I+\Delta I$ participate also in the optimization.

Criterion 2 :

The following objective function S should be maximized:

$$S = \int_{\Omega_1}^{\Omega_2} Ak\, d\Omega = 2\int_{\Omega_1}^{\Omega_2} \zeta \frac{\varphi_v \int_{-\infty}^{+\infty} h_2^2(\tau)d\tau}{\Omega^2 \int_{-\infty}^{+\infty} h_2(\tau)\cos(\Omega\tau)d\tau} \tag{11}$$

To ensure that the method produces meaningful alternative design solutions, additional conditions must be supplied. Current IMO or Naval (GZ)-curve shape criteria use as benchmarks the highest point of the curve as well as certain areas under the curve (up to 30 and 40 deg as well as between the two) see for example MOD (1989). The search for maximum of the objective function should thus be constrained by suitable extra conditions that will guarantee that stability criteria in common use are being satisfied (see Appendix).

STEADY-STATE CRITERIA

These criteria require to locate the fold and symmetry breaking boundaries. Firstly, a low-order analytical solution of (3) is found with use of the method of harmonic balance. This solution is subsequently 'coupled' with suitable stability conditions. To identify the fold it is rather straightforward to request $\partial\Omega/\partial x_0 = 0$, where x_0 is the amplitude of roll motion, making sure of course that the lower fold A is the one considered. To approximate the locus of symmetry breaking we derive the variational equation and we find the relation that allows the existence of an asymmetric solution (or of a subharmonic solution in the case of an asymmetric system).

Solution with harmonic balance

We rewrite (3) as follows :

$$\ddot{x} + 2\zeta\,\dot{x} + x + a_1 x^3 - a_2 x^5 + (-1 - a_1 + a_2)x^7 = F\cos(\Omega\tau - \Theta) \tag{12}$$

where Θ is the phase difference between excitation and response that must be identified. We seek a steady-state solution $x = x_0\cos(\Omega\tau)$. We substitute this into (12), expand the trigonometric terms, retain only the terms of harmonic frequency and equate the coefficients of $\cos(\Omega\tau)$ and $\sin(\Omega\tau)$ on both sides of the equation, obtaining finally:

$$x_0 = \frac{F}{\sqrt{N^2 + 4\zeta^2\Omega^2}} \tag{13}$$

$$\Theta = \arctan\frac{-2\zeta\Omega}{N} \tag{14}$$

where

$$N = -\Omega^2 + M \tag{15}$$

$$M = 1 + \frac{3a_1x_0^2}{4} - \frac{5a_2x_0^4}{8} + \frac{9(-1-a_1+a_2)x_0^6}{16} \tag{16}$$

An alternative useful form of the above is obtained by solving for Ω:

$$\Omega = \frac{\sqrt{2}}{2}\sqrt{2M - 4\zeta^2 \mp \sqrt{(2M-4\zeta^2)^2 - 4\left(M^2 - \frac{F^2}{x_0^2}\right)}} \tag{17}$$

With plus we obtain the high-frequency branch and with the minus the low one.

Approximation of the fold

With differentiation of (13) in terms of x_0, imposition of the condition $\vartheta\Omega/\vartheta x_0 = 0$ and some rearrangement, the following relation is derived:

$$\Omega^4 - (2M + x_0M' - 4\zeta^2)\Omega^2 + (M^2 + M\,M'x_0) = 0 \tag{18}$$

An alternative expression based on F can also be derived :

$$F^4 - x_0^3\,M'(x_0 - 4\zeta^2)F^2 + 4M\zeta^2 = 0 \tag{19}$$

where

$$M' = \frac{\partial M}{\partial x_0} = \frac{a_1}{2}x_0 - \frac{5a_2}{2}x_0^3 + \frac{27(-1-a_1+a_2)}{8}x_0^5 \tag{20}$$

Finally x_0 must be eliminated between (17) and (18) and also F must be written in terms of Ak to obtain an expression, say $G(Ak,\Omega)=0$ that defines the fold locus on the $(Ak,\,\Omega)$ plane.

Approximation of the symmetry breaking locus

Consider again (3) and let x be increased by a very small amplitude ξ, such that ξ^2, ξ^3 etc. can be neglected. Then by substituting x with $x+\xi$ in (3) we obtain:

$$\left[\ddot{x}+2\zeta\,\dot{x}+x+a_1x^3-a_2x^5+(-1-a_1+a_2)x^7-F\cos(\Omega\tau)\right]+\ddot{\xi}+2\zeta\,\dot{\xi}+\left[\frac{\partial q(x)}{\partial x}\right]\xi=0 \tag{21}$$

where

$$q(x)=R(x)-F\cos(\Omega\tau) \tag{22}$$

In (21) the quantity inside the first brackets is zero by definition and therefore we are left only with the so-called *variational equation*, Hayashi (1964), McLachlan (1956):

$$\ddot{\xi}+2\zeta\,\dot{\xi}+\left[\frac{\partial q(x)}{\partial x}\right]\xi=0 \tag{23}$$

$$\ddot{\xi}+2\zeta\,\dot{\xi}+\left[1+3a_1x^2-5a_2x^4+7(-1-a_1+a_2)x^6\right]\xi=0 \tag{24}$$

where $x=x_0\cos(\Omega\tau)$. We want to find the threshold where an asymmetric solution first appears, so we consider a perturbation ξ that includes constant term and second harmonic :

$$\xi=b_0+b_{2c}\cos(2\Omega\tau)+b_{2s}\sin(2\Omega\tau) \tag{25}$$

Parenthetically is mentioned that if the asymmetric equation was used we should consider a subharmonic perturbation :

$$\xi=b_{1c}\cos\left(\frac{\Omega\tau}{2}\right)+b_{1s}\sin\left(\frac{\Omega\tau}{2}\right)+b_{3c}\cos\left(\frac{3\Omega\tau}{2}\right)+b_{3s}\sin\left(\frac{3\Omega\tau}{2}\right) \tag{26}$$

With substitution of x and ξ [from (25)] in (21) and application of harmonic balance, where we retain only terms up to second harmonic, we obtain a linear system of algebraic equations in terms of b_0, b_{2c} and b_{2s} :

Coefficient of the constant term:

$$\left[1+\frac{3}{2}a_1x_0^2-\frac{15}{8}a_2x_0^4-\frac{35}{16}(1+a_1-a_2)x_0^6\right]b_0+\left[\frac{3}{4}a_1x_0^2-\frac{5}{4}a_2x_0^4-\frac{105}{64}(1+a_1-a_2)x_0^6\right]b_{2c}=0 \tag{27}$$

Coefficient of $\cos(2\Omega\tau)$:

$$\begin{aligned}&\left[\frac{3}{2}a_1x_0^2-\frac{5}{2}a_2x_0^4-\frac{105}{32}(1+a_1-a_2)x_0^6\right]b_0+\\&\quad+\left[1+\frac{3}{2}a_1x_0^2-\frac{35}{16}a_2x_0^4-\frac{91}{32}(1+a_1-a_2)x_0^6-4\Omega^2\right]b_{2c}+(4\zeta\,\Omega)b_{2s}=0\end{aligned} \tag{28}$$

Coefficient of $\sin(2\Omega\tau)$:

$$(4\zeta\,\Omega)b_0 - (4\zeta\,\Omega)b_{2c} + \left[1 - \frac{3}{2}a_1 x_0^2 - \frac{25}{16}a_2 x_0^4 - \frac{49}{32}(1 + a_1 - a_2)x_0^6 - 4\Omega^2\right] b_{2s} = 0 \qquad (29)$$

The condition $\Delta = 0$ where Δ is the determinant of (27), (28) and (29) provides the sought equation for the symmetry-breaking locus. It is interesting that the expression is analytically solvable for Ω. Again however the elimination of x_0, through combining with (17), is problematic.

Derivation of steady-state criteria

The lowest point of the wedge corresponds obviously to the intersection of the curves $G(Ak,\Omega)$ and $\Delta(Ak,\Omega) = 0$. Let us define this point as (Ak_0, Ω_0). We want to maximize Ak_0 in terms of the coefficients a_1, a_2, φ_v and also ζ [which, it should not be forgotten, includes (GM)]. Also in respect to the area criterion, if $Ak_G(\Omega)$, $Ak_\Delta(\Omega)$ are explicit representations of wave slope in terms of Ω at the fold and flip loci respectively, we want:

$$\int_{\Omega_1}^{\Omega_0} Ak_G(\Omega)d\Omega + \int_{\Omega_0}^{\Omega_2} Ak_\Delta(\Omega)d\Omega$$

to be maximum.

STEADY VERSUS TRANSIENT CAPSIZE CRITERIA

Although the transient and steady-state capsize criteria are dynamically different and the basin erosion begins much earlier than the first period doubling, it is not known how they reflect on the actual optimization parameters. Do they result in similar optima or do they produce considerably different ones? With the earlier developed tools it should be possible to infer to what extent the steady-state and capsize criteria coincide in their predictions of the optimum hull configuration. It is hoped that it will be possible to provide specific answers in a future publication.

References

Birkdash, M.U., Balachandran, B., Nayfeh, A.H.(1994): Melnikov analysis for a ship with a general damping model, *Nonlinear Dynamics*, **6**, 101-124.

Caldeira-Saraiva, F. (1986): A stability criterion for ships using Lyapounov's method, *Proceedings*, The Safeship Project, , Royal Institution of Naval Architects, London.

Cardo, A. Francescutto, A. & Nabergoj, R. (1981): Ultraharmonics and subharmonics in the rolling motion of a ship: Steady-state solution. *International Shipbuilding Progress*, **28:326**, 234-251.

Falzarano, J.M., Shaw, S.W., Troesch, A (1992): Application of global methods for analysing dynamical systems to ship rolling motion and capsizing, *International Journal of Bifurcation and Chaos*, **2**:**1**, 101-115.

Guckenheimer, J. and Holmes, P.J. (1983): *Nonlinear Oscillations, Dynamical Systems and Bifurcations of Vector Fields*, Springer-Verlag, Applied Mathematical Sciences, **42**, New York.

Hayashi, C. (1964): *Nonlinear Oscillations in Physical Systems*, McGraw Hill, New York.

Kan M. (1992): Chaotic capsizing, *Proceedings*, ITTC SKC-KFR Meeting on Seakeeping Performance, Osaka, 155-180.

McLachlan, N.W. (1956): *Ordinary Nonlinear Differential Equations in Engineering and Physical Sciences,* Oxford at the Clarendon Press.

MOD (1989) Stability standards for surface ships, Naval Engineering Standard 109, Sea Systems Controllerate, Issue 3, Ministry of Defense, Bath, UK.

Nayfeh A.H & Mook, D.T. (1979): *Nonlinear Oscillations*, Wiley, New York.

Nayfeh, A.H. and Balachandran, B. (1995): *Applied Nonlinear Dynamics: Analytical, Computational and Experimental Methods*, Wiley Series in Nonlinear Science, New York.

Odabasi, A.Y. (1978): Conceptual understanding of the stability theory of ships, Schiffstechnik, **25**, 1-18.

Salam, F. (1987): The Melnikov technique for highly dissipative systems. *SIAM Journal of Applied Mathematics*, **47**, 232-243.

Szemplinska-Stupnicka, W. (1988): The refined approximate criterion for chaos in a two-state mechanical oscillator, *Ingenieur-Archiv*, **58**, Springer-Verlag, 354-366.

Szemplinska-Stupnicka, W. (1992): Cross-well chaos and escape phenomena in driven oscillators. *Nonlinear Dynamics*, **3**, 225-243.

Stoker, J.J. (1950): *Nonlinear Vibrations in Mechanical and Electrical Systems*, Wiley, New York.

Thompson, J.M.T. (1996): Global dynamics of driven oscillators: Fractal basins and indeterminate bifurcations. Chapter 1 of *Nonlinear Mathematics and its Applications*, P.J. Aston(ed.), Cambridge University Press, 1-47.

Thompson, J.M.T. (1997) Designing against capsize in beam seas: Recent advances and new insights, *Applied Mechanics Reviews*, **50**:**5**, 307-325.

Thompson, J.M.T. , Rainey R.C.T & Soliman, M.S.(1990): Ship stability criteria based on chaotic transients from incursive fractals. *Philosophical Transactions of the Royal Society of London*, A **332**, 149-167.

Virgin, L.N. (1989): Approximative criteria for capsize based on deterministic dynamics, *Dynamics and Stability of Systems*, **4**:**1**, 55-70.

APPENDIX

Consider the following polynomial for restoring:

$$R(x) = x + a_1 x^3 - a_2 x^5 + (-1 - a_1 + a_2) x^7$$

Area under the curve: $$\int_0^1 R(x)dx = \frac{9 + 3a_1 - a_2}{24}$$

The 'true' area under the $GZ(\varphi)$ curve is: $$W(GM)\varphi_v^2 \frac{9 + 3a_1 - a_2}{24}$$

The area up to an angle φ is: $$W(GM)\left[\frac{\varphi^2}{2} + \frac{a_1\varphi^4}{4\varphi_v^2} - \frac{a_2\varphi^6}{6\varphi_v^4} + \frac{(-1 - a_1 + a_2)\varphi^8}{8\varphi_v^6}\right]$$

The maximum of the curve is found by solving for x the equation $\frac{dR(x)}{dx} = 0$:

$$(x^2)^3 - \frac{5a_2}{7(-1 - a_1 + a_2)}(x^2)^2 + \frac{3a_1}{7(-1 - a_1 + a_2)}(x^2) + \frac{1}{7(-1 - a_1 + a_2)} = 0$$

There is one real and positive root which can be found analytically with, for example, *Mathematica*. For the equation $(x^2)^3 - a(x^2)^2 + b(x^2) + cx = 0$ the real and positive root is:

$$x_{\max} = \sqrt{\frac{a}{3} - \frac{\sqrt[3]{2}(-a^2 + 3b)}{3D} + \frac{D}{3\sqrt[3]{2}}}$$

where: $$D = \sqrt[3]{2a^3 - 9ab + \sqrt{4(-a^2 + 3b)^3 + (2a^3 - 9ab - 27c)^2} - 27c}$$

Points of inflection at $\frac{d^2R(x)}{dx^2} = 0$:

$$x_{\text{infl}} = \sqrt{\frac{10a_2 \pm \sqrt{100a_2^2 - 252a_1(-1 - a_1 + a_2)}}{42(-1 - a_1 + a_2)}}$$

Contemporary Ideas on Ship Stability
D. Vassalos, M. Hamamoto, A. Papanikolaou and D. Molyneux (Editors)

SHIP CRANKINESS AND STABILITY REGULATION

N.N.Rakhmanin and G.V.Vilensky

Krylov Shipbuilding Research Institute
St. Petersburg 196158, Russia

ABSTRACT

Ship crankiness usually manifests itself in large heeling angles significantly exceeding roll amplitudes corresponding to the exciting moment at given sea state and sailing conditions. The phenomenon is considered on the basis of modern theory of ship motions and the corresponding numerical measure for this ship quality is found. Namely it is assumed to use the amplitude of steady parametric rolling motion as such a measure. Finally it is suggested the new idea to check ship stability in following seaway condition by means of the criterion which supposes to restrict above mentioned parametric roll amplitude.

KEYWORDS

Crankiness, ship safety, following seaway, parametric roll, capsizing, stability regulation.

NOMENCLATURE

L - ship length,
B - ship breadth,
GM_0 - metacentric height,
ΔGM - metacentric height increment in waves,
GZ_{max} - maximal arm of stability curve,
n_ϕ - natural roll frequency,
n - integer,
$\nu_{\phi 0}$ - non-dimensional roll damping coefficient,
$H_{1/3}$ - significant wave height,
ω_e - encounter frequency,
χ - wave heading angle,

ϕ_o^{par} – parametric roll amplitude,

$\left(\phi_o^{par}\right)_{max}$ – maximal parametric roll amplitude,

ϕ_o^{norm} – agreed margin for parametric roll amplitude.

1. INTRODUCTION

The problem of providing for safe ship navigation while sailing in following seaway is still subject of actual interest today, although it started to draw the attention of the specialists as far back as in the mid-fifties. A great deal of knowledge has been accumulated in the field of stability and ship behaviour dynamics under the conditions of following seaway. The unfavourable and dangerous situations that the navigator may meet at sea have been systematised and certain recommendations, which help the captain to escape some dangers of navigation under such conditions, have been found. However, the variety of mentioned dangerous situations and the difficulty of their mathematical description create not a few obstacles on the way of searching for practically acceptable standards of ship safety in following seaway. These standards must reflect the most essential connections between safety criteria and those ship constructional characteristics the change of which on a design stage permits to eliminate the capsizing.

Even not full enumeration of the names of scientists who dealt with the problem shows its complexity and variety. In Russia - S.N. Blagoveschensky, I.K. Boroday, V.V.Lugovsky, N.V. Sevastyanov, Y.I. Netchaev, D.M. Ananiev, N.Y. Maltsev, V.N. Saltovskaya, Y.L. Makov, in other countries - B. Arndt, K. Vendel, O. Grim, J. Paulling, S. Kastner, S. Motora and in recent years S. Renilson, N. Umeda, G. Thomas, M. Kan have made a great contribution to the investigation of the problem. Reviews of works and publications on the discussed theme are available in books of Lugovsky(1966), Boroday & Netsvetaev(1982) and in proceedings of the International Conference on Stability of Ships and Ocean Vehicles(1990), (1994), e.g. Umeda(1994).

According to Boroday & Netsvetaev(1982) «except the stability decrease and course instability with broaching accompany ship sailing in following seaway, in some cases parametrically excited roll may present a certain danger for the ship safety».

However it was generally accepted Netchaev(1978) in the end of seventies, that with parametric excitation the rolling motion amplitudes do not increase infinitely according to the well-known solution of Mathieu equation for unstable region but stay limited because of nonlinearity of the restoring and the damping moments. In other words the parametric rolling in following seaway was considered only as a circumstance decreasing the ship's resistance to external heeling moments but not as a direct danger for its safety, Netchaev(1978); Boroday & Netsvetaev(1982). So the situation in following waves considered to be important for small fishing vessels or ships of other types with $L \leq 60$ m, the USSR Shipping Register Rules(1977).

The first evidence of that following seaway can create troubles for large modern ships of merchant fleet with the length $L \geq 100$ m have appeared in IMO on the border of seventies and eighties. At that time the Organization started its work on review of the stability requirements for transport ships on the basis of introduction of a weather criterion to the international practice. The German delegation has repeatedly drawn the attention of the IMO Subcommittee on Stability and Load Lines to the dangerous crankiness of containerships in following seaway and to the necessity of this problem to be researched for large ships. The Head of the Ship Safety Division, Germanischer Lloyd Mr. W. Hausler, the Member of German delegation for the Subcommittee has repeatedly spoken about the reports the Division received about container losses in the open sea because of unexpected heeling angles about 30°-45°. These heels without a visible reason (such as a wind squall or riding on wave that is unlikely for large ships) could throw the crew down into panic. And it must be noted that the reason is actually existing.

In the middle of the eighties the sufficiently reliable experimental and gained with numerical simulation theoretical data appeared confirming the possibility of the ship's capsizing in the strictly following seaway only as a result of roll in the regime of parametric resonance, which arises because of periodical stability alterations. In particular Prof. Paulling(1982) demonstrated the danger of the main parametric resonance by means of seaway dynamics analysis of the «Mariner» type ship with the stability curve that meets all the IMO requirements for intact condition. Figure1 gives an idea of the capsizing dynamics, the calculation results are shown with dots, and the model test data are shown with continuous lines. As it could be seen the fatal inclination can occur after 3 - 4 roll double amplitudes and

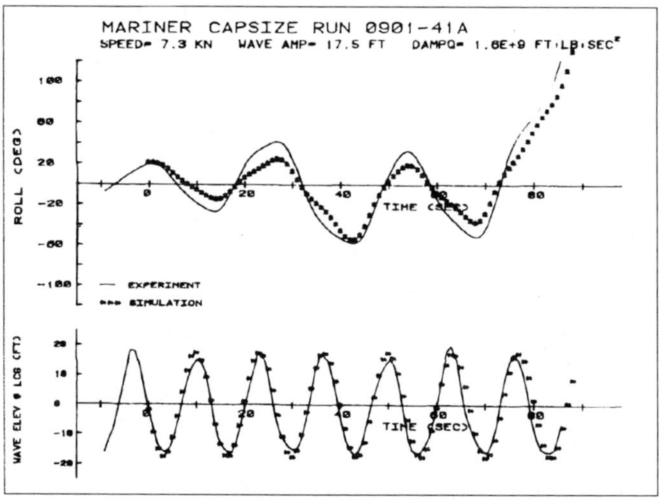

Figure 1. The results of experimental and numerical determinations of rolling motion amplitudes in the following seaway in the regime of main parametric resonance, Paulling(1982): —— - the experiment; ■ - simulation.

it is practically impossible to escape it by changing the ship course. Analogous results were received during the tests with radio-operated self-propelled models in the Sevastopol Bay, Medved(1980) and in seakeeping basin, Allievi, Calisal & Rohling(1986).

In spite of all mentioned above the design situation assuming the ship sailing on the following wave crest with the wave length approximately equal to the ship length, nowadays finds its direct reflection only in the National Standards for the German Navy ships, Grim(1952), Arndt(1965); Arndt, Brandl & Vogt(1982). This situation is not formulated in the Rules of Russian Shipping Register, but it follows from the Annex explaining the principles of composing these Rules(1977), where the lower limit for maximum stability curve arm is recommended. One can find analogous requirements in the Japanese Rules of Stability for passenger ships.

Besides it is necessary to mention a number of proposals, which didn't find their reflection in the practical Rules, but which are instructive from the methodical side, the USSR Shipping Register(1977); Blume & Hattendorf(1982); Helas(1982); Martin,Kuo & Welaya(1982); the Poland Shipping Register(1984); Bogdanov(1993). Most of them are connected with the efforts to create the criteria of sailing safety in following seaway on the basis of using the main weather criterion idea Blagoveschensky(1965). The extreme strictness of criteria that have been found in this way prevented from their practical use.

The second group of works, Blume & Hattendorf(1982); Bogdanov(1993), paid attention to the alterations of ship's hydrostatic characteristics under the conditions of sailing in following seaway without additional external actions. As a whole the criteria established in this case being useful as a generalization of certain experience bear rather relative character and not always help to correctly foresee the danger connected with following seaway. Such approach leads either to unjustified severity of stability requirements, Blume & Hattendorf(1982), or, on the contrary, permits the reduction of stability to rather low limits excusing this possibility with its short duration if the course and speed are properly chosen, Nogid(1967); Bogdanov(1993).

This approach may appear to be inadmissible for a cranky ship. Resonance roll excited by means of short-term but deep alterations of the restoring moment under the conditions of following seaway together with an insufficient level of stability in such situation may itself as it has been mentioned lead the ship to capsizing in the course of several cycles of oscillations. In this case the crew will have no time to alter the course or the speed for safe ones.

2. SHIP CRANKINESS

Crankiness as a characteristic of a ship to show big inclinations to the side without visible external reasons, can be explained fully enough from positions of the modern theory of ship motions, namely, by the ship heeling dynamic instability which originates as a result of periodical alterations of her stability while sailing in seaway. It especially reveals itself when the ship moves in following or quartering waves.

The problem of roll caused by the variability of the restoring moment comes to well-known Mathieu equation, which has been repeatedly discussed in shipbuilding literature, Kerwin(1955); Basin(1969); Boroday & Netsvetaev(1982); Paulling(1982). From the theory of these equations it is known, that under certain combination of its parameters characterising the roll damping $\nu_{\phi o}$, the natural roll frequency n_ϕ and the depth of stability modulation $\Delta GM/GM_o$ the unstable, prone to increase roll oscillations may appear. The regions of unstable equation solutions pointing to the ship crankiness are located in vicinities of the following relative frequencies:

$$\frac{\omega_e}{2n_\phi} = \frac{1}{n} \tag{1}$$

where $n = 1, 2, 3 \ldots$

The case, when $n = \infty$, i.e. the apparent frequency of encounter $\omega_e \rightarrow 0$, corresponds to the static equilibrium condition of a ship with reduced or lossed stability. An evident relation for static instability reflects this

$$\frac{\Delta\,GM}{GM_o} \geq 1. \tag{2}$$

For small values of initial stability $GM_o / B < 0.02\text{-}0.03$, which is characteristic for cranky ships, this relation is realized with a high degree of probability.

When the value of apparent frequency ω_e differs from zero the possibility of realisation of different unstable solutions of the Mathieu equation is not the same. For small roll damping values $\nu_{\phi o}$ and small disturbance levels $\Delta GM/GM_o$ the width of unstable regions is proportional correspondingly to $(\Delta GM/GM_o)^n$, and the depth of stability modulation necessary for unstable roll development (the threshold of parametric roll excitation) appears to be proportional to the 1-st or 1/2 degree of roll damping coefficient. In particular, the excitation threshold for the main parametric resonance $(n=1)$ is determined by condition, Basin(1969)

$$\frac{\Delta GM}{GM_o} \geq 4\nu_{\phi o}. \tag{3}$$

For monohull ships without bilge keels the non-dimensional linear roll damping coefficient $2\nu_{\phi o}$ is within the limits of 0.05-0.10, therefore condition (3) seems to be easier realised than the static instability condition (2) and manifests itself in a rather wide range of the parameter $\Delta GM/GM_o$ values.

Not only the above mentioned results which determine the crankiness presence or absence and the frequency regions where parametric roll may occur are known nowadays, but the calculation techniques to determine the amplitudes of such rolling motion are developed, which give an idea of crankiness degree and its danger Kerwin(1955); Paulling(1982); the Poland Shipping Register(1984); Vilensky(1994).

J. Kerwin calculated the rolling motion amplitudes in the main parametric resonance regime on the basis of Mathieu equation and has taken into consideration the nonlinear character of roll damping by means of binomial formula use with linear and quadratic terms for resistance law. Specialists from Poland (1984) considered the nonlinear character of the restoring moment at the linear law of roll damping. J. Paulling researched the nonlinear in damping and restoring moment roll equation numerically having taken the stability alteration in seaway into consideration, and got satisfactory agreement with the test (see Figure 1). G. Vilensky [5] established general analytical solution of nonlinear roll equation for the case of ship sailing in regular following and quartering waves. In this approach the stability curve and its modulation were expanded successfully into thrigonometrical series, and the disturbing wave moment and static wind moment were taken into consideration.

Calculated research and model tests in seakeeping basin, Vilensky(1994), demonstrated that under the conditions of purely following seaway the parametric roll with frequency ω_e is significantly lower than the roll which occurs with frequency $\omega_e/2$. However, the parametric excitation with frequency ω_e in the stern quartering waves can be summed up with the resonance effect of the exciting moment. This case of combination resonance (see Figure 2) doesn't coincide with known solutions of Mathieu equation and may lead to dangerous heeling angles (~ 60°). The essential part of zero harmonics (a constant component) is

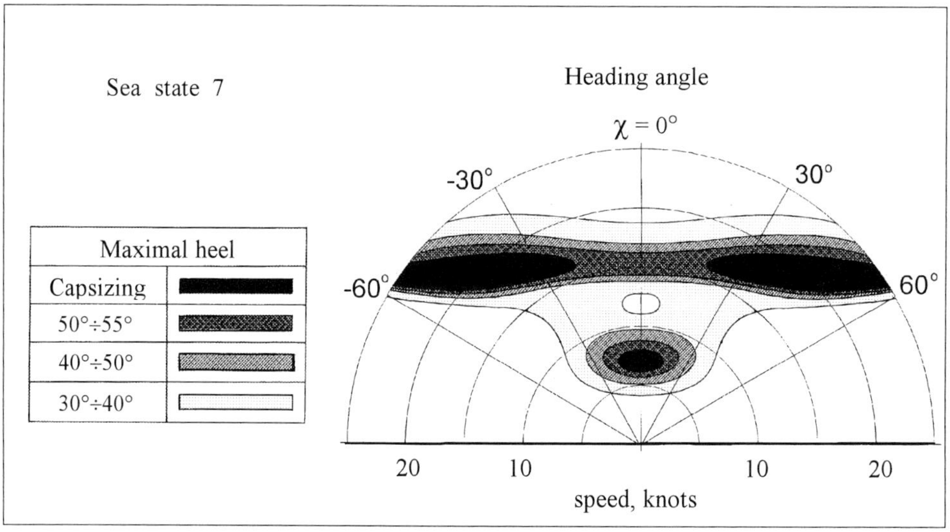

Figure 2. Relation between maximal heeling angles during the parametric roll $(\phi_o^{par})_{max}$ and ship's speed and course angle χ to the wave: GM=0,3m - metacentric height; $H_{1/3}$=6,5m - significant wave height.

characteristic for this mode of resonance. The considerable constant component increasing the ship crankiness appears even without the wind .

Recent experimental and calculation research by means of analytical method of Vilensky(1994) executed in the Krylov Research Institute confirmed the known facts, that rolling motion parametrically excited in following seaway can be developed right up to the capsizing. It was found that with the ratio $\Delta GM/GM_o$ increase and the coefficient $\nu_{\phi o}$ decrease maximal inclinations or crankiness of a ship increases, the range of apparent frequencies of encounter at which the mentioned roll regimes exist widens, and the rate of their amplitudes growth increases.

Known opinion has been confirmed that the parametric resonance in the regime of $\omega_e/2$ is not dangerous in head seas. In this case it arises with a sufficiently high stability and consequently relatively small $\Delta GM/GM_o$, high natural frequency n_ϕ and occurs with small amplitudes or doesn't occur at all. On the contrary, rolling motion that arises in the main parametric resonance regime in following seaway is as a rule several times higher in amplitudes than the usual one caused by the exciting moment and serves as an indication of dangerous ship crankiness.

As an illustration for above said Figure 3 demonstrates the results of a three-meter multipurpose bulkcarrier model test under unfavourable loading case connected with container transportation on the upper deck (GZ_{max} = 0,35 m, vanishing angle of stability curve - 65^0 and GM_o = 0,67 m). The tests were carried out to evaluate ship crankiness with various modifications of the constructional elements and model loading, and also in order to work out the recommendations for limitations of crankiness during sailing in purely following waves of Sea state 7 ($H_{1/3}$ = 6,5 m). The experimental data correlate quite well with the maximal roll amplitude values in the main parametric resonance regime calculated with consideration of Kerwin's(1955) recommendations . Figure 4 demonstrates the variation of parametric rolling motion amplitudes versus the ship's speed, and the calculated values for the amplitudes of 3% exceedance probability of usual forced ship roll in irregular quatering seaway while sailing at resonance course angles.

The analysis of data shown in Figure 3 and Figure 4 demonstrates that:
Firstly, the amplitudes of parametric roll are appreciably higher than the amplitudes of forced rolling motion. At resonance conditions this difference can achieve 10 times, if the rolling motion is not completed with capsizing;
Secondly, it is seen that the amplitudes of parametric roll are rather sensitive to the alteration of parameters characterising the rolling motion excitation threshold (3) , and for this reason it is very convenient to use them as a measure for ship crankiness. The latter can be controlled at the design stage by the rational selection of a hull form and ship load (in order to reduce the relative depth of stability curve modulation), by means of the bilge keel area increase, and during the operation - by means of stability increase and as well by means of rational alterations in due time in speed and course angle;
Thirdly, the possibility and the expediency of maximal parametric roll amplitude values limitation becomes clear.

It is evident, that ship's crankiness may be considered as a safe one, if its measure $\left(\phi_o^{par}\right)_{max}$ does not exceed reasonable limitations (Figure 3).

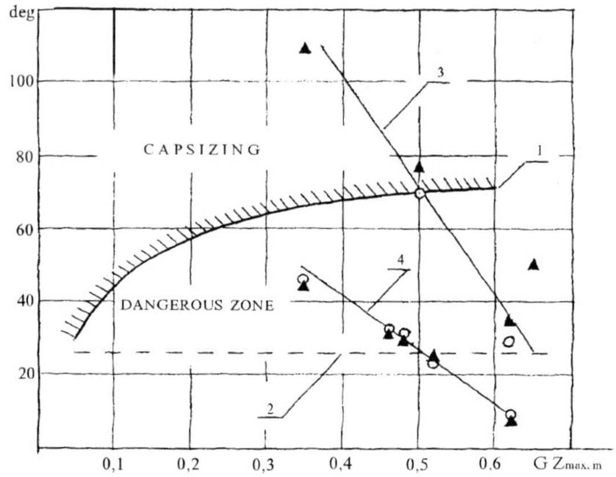

Figure 3. Relation between maximal parametric roll amplitudes in following seaway and the stability curve maximal arm value in calm water according to test — (o) and to calculation, Kerwin(1955), — (▲):
1 - vanishing angle line;
2 - line of agreed heeling angles limitation;
3 - data for models without bilge keels;
4 - data for models with keels.

The authors consider that in this way the real enough and physically well-founded criterion for the following seaway may be achieved as

$$\left(\phi_o^{par}\right)_{max} \leq \phi_o^{norm} . \tag{4}$$

At this the following value can be taken as a norm of crankiness:

$$\phi_o^{norm} = k \cdot \phi_{lim}$$

where ϕ_{lim} - is maximal allowed heeling angle, for instance equal to the stability curve vanishing angle, or to shift cargo angle, or to flooding angle and so on depending on which is less; k - is coefficient which takes into account the inaccuracy of roll amplitudes calculation scheme, in particular the error in roll damping coefficient determination, stability alterations in seaway or test errors in the case of experimental amplitude determination $\left(\phi_o^{par}\right)_{max}$.

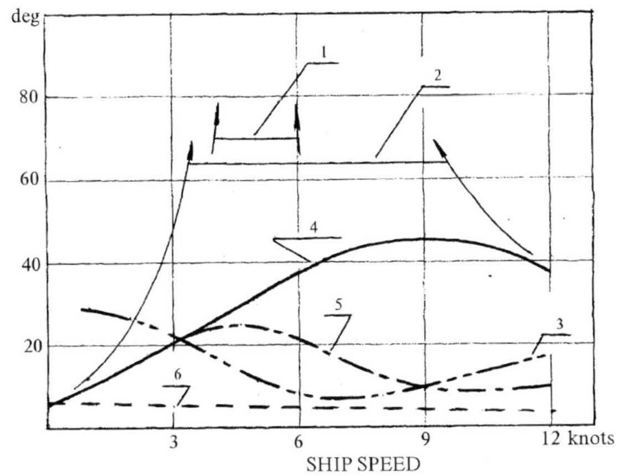

Figure 4. Relation between parametric roll amplitudes in following seaways and ship's speed according to model test data:
1 - the range of capsizing with GZ_{max} = 0,5 m and without keels;
2 - the same with GZ_{max} = 0,35 m;
3 - the curve for a ship without keels with GZ_{max} = 0,62 m;
4 - the ship with keels and GZ_{max} = 0.35 m;
5 - the ship with keels and GZ_{max} = 0.62 m;
6 - maximal forced roll amplitudes at resonance course angles.

The criterion proposed takes into consideration rather important specific character of rolling motion in the following and quatering course angles and head wave course angles either. The application of this criterion does not exclude the possibility of additional ship stability check-up under the conditions of durable decrease of the restoring moment at sailing in following seaway using other rational criteria or the official requirements of Stability Rules. At the same time it may be expected that with reasonably selected crankiness the level of the ship safety will increase to some degree also in other hard situations related to following seas, for example in the situations of broaching or riding on wave.

REFERENCES

Allievi A.G., Calisal S.M., and Rohling G.F.(1986). Motions and Stability of a Fishing Vessel in Transverse and Longitudinal Seaways. *Proceedings of 11th Ship Technology and Research (STAR) Symposium.* **1**.

Arndt B. (1965). Ausarbeitung einer Stabilitatsvorschift fur die Handelsmarine. *Jahrbuch der STG,* **59** Band.

Arndt B., Brandl H. and Vogt K. (1982). 20 Years of Experience Stability Regulations of the West-German Navy. *Proceedings of the 2nd International Conference on Stability of Ships and Ocean Vehicles, Tokyo.* pp 765-775.

Basin A.M.(1969). *Ship Motions* .Moscow: Transport, 272 p.

Blagoveschensky S.N.(1965). National Requirements to Intact Stability. *Proceedings of USSR Shipping Register «Theoretical and Practical Questions of Ocean Going Ship Stability and Subdivision»* Moscow-Leningrad: Transport, pp3-52.

Blume P. and Hattendorf H.(1982). An Investigation on Intact Stability of Fast Cargo Liners. *Proceedings of the 2nd International Conference on Stability of Ships and Ocean Vehicles, Tokyo. pp171-183.*

Bogdanov A.I.(1993). Regression Formula for Calculation of Stability Alteration Coefficient for Following Seaways. *Proceedings of Scientific and Technical Conference «Krylov's Readings-93», St. Petersburg.* pp71-74.

Boroday I.K. and Netsvetaev Y.A.(1982). *Ship Seakeeping.* Leningrad: Sudostroenie, 288p

Grim O.(1952). Rollschwingungen,Stabilitat und Sicherheit im Seegang.*Schiffstechnik,* **1.** pp10-15.

Helas G.(1982). Intact Stability of Ships in Following Waves. *Proceedings of the 2nd International Conference on Stability of Ships and Ocean Vehicles, Tokyo.* pp669-700.

Kerwin J.E.(1955). Notes on Rolling in Longitudinal Waves. *International Shipbuilding Progress* **2:16,** pp3-27.

Lugovsky V.V.(1966). *Nonlinear Problems of Ship Seakeeping,* Leningrad:Sudostroenie, 235p.

Martin J., Kuo Ch. and Welaya Y.(1982). Ship Stability Criteria Based on Time-Varying Roll Restoring Moments. *Proceedings of the 2nd International Conference on Stability of Ships and Ocean Vehicles,Tokyo.* pp227-242.

Medved A.F.(1980). The Research of General Behaviour and Capsizing of Radio-Controlled Models in Natural Seas. *Cybernetics in the Ocean Transport,* Kiev **:9**, pp55-60.

Netchaev Y.I.(1978). *Ship Stability in Following Seaway*, Leningrad: Sudostroenie, 272p.

Nogid L.M.(1967). *Ship Stability and Ship Behaviour in Seaway.* Leningrad: Sudostroenie, 241p.

Paulling J.R.(1982). A Comparison of Stability Characteristics of Ships and Offshore Structures. *Proceedings of the 2nd International Conference on Stability of Ships and Ocean Vehicles, Tokyo.* pp581-588.

Poland Shipping Register(1984). *Research project «The Development of Calculation and Evaluation Methods for Intact Ship Stability in Following Seaway», Topic 5.4:1,Plan 5,* Gdansk.

Sevastyanov N.B.(1970). *Fishing Vessel Stability*, Leningrad: Shipbuilding, 255p.

Umeda N.(1994). Operational Stability in Following and Quatering Seas: A Proposed Guidance and Its Validation. *Proceedings of the 5th International Conference on Stability of Ships and Ocean Vehicles* **2**, *Florida.* pp 71-85.

USSR Shipping Register,(1977). *Rules for Classification and Building of Ocean Going Ships, Part IV «The Stability»,* Leningrad: Transport.

USSS Shipping Register,(1977). *The Method of Stability Evaluation in Following Seaway,* Leningrad: Transport.

Vilensky G.V.(1994). Reasons of Dangerous Roll in Following Seaway. *Proceedings of International Shipbuilding Conference (ISC) on Centenary of Krylov Research Shipbuilding Institute, Section B,*. St. Petersburg: KSRI, pp 265-271.

Contemporary Ideas on Ship Stability
D. Vassalos, M. Hamamoto, A. Papanikolaou and D. Molyneux (Editors)

THE IMPACT OF RECENT STABILITY REGULATIONS ON EXISTING AND NEW SHIPS.

IMPACT ON THE DESIGN OF RO-RO PASSENGER SHIPS

Markku Kanerva

Deltamarin Ltd,
Purokatu 1, FIN-21200 Raisio, Finland

ABSTRACT

The establishment of SOLAS 90 damage stability regulations and the Stockholm Agreement concerning water on deck has led to the development of new ro-ro passenger ferry concepts and configurations.

Utilisation of the space below the main ro-ro deck, freeboard deck, has been studied and a lower hold configuration for trailers has been developed with drive-through ramp arrangement. Machinery arrangements with conventional direct-geared propulsion have been compared.

Side and centre casing arrangements below and above the main ro-ro deck have been developed and comparisons have been made concerning cargo capacity, structural principles and weight, as well as damage stability.

Cargo handling efficiency is discussed for the lower hold arrangement.

An example of recently built ro-ro passenger ferries with large lower hold, side casings and diesel electric machinery is described.

Damage stability and floodable length definitions in connection with the large lower hold are discussed and a methodology is proposed for calculating both including lower hold with all necessary two compartment damages.

Upgrading of existing ferries in accordance with A/Amax, SOLAS 90 and Stockholm agreement requirements is described by explaining two actual cases. Internal and external options for required modifications are described.

KEYWORDS

Ro-ro passenger ferries, damage stability, large lower holds, SOLAS 90 and water on deck requirements, newbuildings and retrofits, diesel-electric machinery, side casings.

INTRODUCTION

There has been tremendous development in passenger ships and ferries since the early eighties. The size of these ships has progressively increased; passenger cruise ships first up to Panamax measures and now even above.

The development for car-passenger ferries has been partly very similar. The large so-called jumbo ferries, also called cruise ferries, were introduced in the early eighties, and new record in size and capacity was introduced by nearly every newbuilding through out the eighties. Development during the nineties has turned more into trailer ferries with some passenger capacity.

Tendency has been to higher capacity that has meant bulkier ships, under water and above water. Typical block coefficient (C_B) of the beginning of eighties was 0,57 today between 0,67-0,71. The target has been to win more space. The additional space in the under water hull is required to be able to carry and to guarantee the additional stability for the more attractive space above. Additional space, however, requires also additional propulsion power in order to reach the same speed. But the tendency lately has been even towards higher speeds.

The challenge for a ro-ro passenger ship designer today is obvious. Safety standards are being continuously upgraded, not only damage stability, fire safety, passenger evacuation and similar, but also requirements for operational safety.

At the same time the cargo transport and passenger markets are again steadily growing but so is also competition. This means: lower investment and operational costs are key features to survive. At the first glance conflicting criteria should be met to succeed: better safety standards, better earning capacity and, of course, at lower costs.

The design approach in the ro-ro passenger ship market is rather contradictory. The use of lower cargo hold under freeboard deck has several different interpretations, not only for damage stability but for other safety features as well. Side casings versus centre casing is another subject very much debated and affecting the overall design of the vessel.

The recent design development of ro-ro passenger ferries is discussed in this paper.
Application of SOLAS 90 and Stockholm agreement is discussed for both newbuildings and for existing ferries by describing typical examples.

The philosophy for attaining improved safety is based on three basic assumptions. First of all the ship should be easy to operate, and loading should be done rationally and with the best stability in mind. Secondly, the configuration itself should provide a better damage stability. Finally, the design and shape of the hull should eliminate the possibility of unsafe operation in harsh weather, thus increasing safety onboard.

SPACE BELOW FREEBOARD DECK

Shortly after the disaster of the 'Herald of Free Enterprise' it became obvious for the ferry industry not to locate passenger cabins below freeboard deck anymore. The volume below the main ro-ro deck is large and leaving it as void would be waste of valuable space. It could be used for cargo and/or for storage. The storage spaces, however, would only accommodate maximum one third of the volume below freeboard deck and outside engine rooms. The only efficient way to utilise this space is to use it for cargo, to apply the so-called lower hold configuration. The hold is typically limited by longitudinal bulkheads inside B/5 line and with deck above B/10 line. The damage stability for the first modern lower hold ro-pax vessels was simply calculated disregarding the lower hold from any damage cases and the floodable lengths were calculated applying the principle of equivalent bulkheads with the hold area applying the 'B/5 rule'. Another option was to use the A265 method specifically intended for vessels with longitudinal subdivision.

Later investigations proved that the B/5 limit was actually not an adequate limit for collision damages. The use of large bulbous bows in modern commercial vessels had changed the collision pattern. The B/5 limit still exists in the SOLAS but most of the authorities have difficulties in accepting it as physical limit for damage cases.

The A265 method was tested in a few ro-pax ferries but it was soon understood that it was clearly limiting the maximum number of passengers and transverse bulkheads may be required in the lower hold destroying the cargo flow.

It was necessary to have a wider approach to the lower hold configuration. SOLAS 90 gave a starting ground. The lower cargo hold in ro-ro passenger ferry 'Normandie' was still limited with bulkheads inside B/5 and deck above B/10, but both damage cases and floodable lengths were considered with damages extended into the lower hold. Two compartment damage cases (without lower hold) were calculated in accordance with SOLAS 90 criteria. And two compartment damage cases together with the lower hold were calculated in accordance with the SOLAS 90 intermediate stage criteria (the intermediate stage criteria were applied as criteria for the final flooding stage also). The floodable lengths were compensated with the direct damage stability calculations including lower hold. This approach has been used since 'Normandie' for several newbuildings.

The obvious further step was to check if it would be possible to fulfil even full SOLAS 90 criteria with all two compartment damages together with the lower hold. This principle is now being applied in the recent ro-ro passenger ferry newbuildings for Stena.

It is possible to construct a modern, efficient ro-ro passenger ship without using doubtful limitations for the damage definitions. This is unfortunately not the industry practice yet and there seem to be at least six different interpretations on the market for lower hold vessels. This is certainly not an item to be proud of.

CENTRE OR SIDE CASINGS

Typical arrangement for ro-ro passenger ship has been a centre casing accommodating funnels, lifts, stair cases, garbage room, fire stations and similar. Some stores and cargo offices have been located in the corners of the main cargo deck (freeboard deck). Side casing arrangement was used only in exceptional cases. One central casing was typically considered to be cheaper and less complex to build.

SOLAS 90 damage stability requirements, however, changed the situation. Stability characteristics required after flooding and during intermediate flooding stages urged for additional buoyancy volume compared to previous requirements. This could be arranged by increasing the height of the freeboard deck and increasing the beam of the vessel. Raising freeboard deck is not a very effective measure; vertical centre of gravity (KG) is raised at the same time. Increasing beam increases also damage volume. SOLAS 90 offered a possibility to take advantage of compartmentation above freeboard (margin line). The side casing configuration became much more attractive. This is even more evident when lower hold is arranged and considered in damage cases.

Comparing side casings against centre casing in a ro-ro passenger ship of 170 m in overall length with lower hold the difference in required freeboard deck height is between 500-800 mm depending on the size of the hold and the stability criteria applied for the lower hold damages, centre casing configuration requiring higher deck.

Typical arrangement with side casings is presented in figure 1. The side casing arrangement is also beneficial in meeting the Stockholm agreement for water on deck. Depending on the final arrangement on the main deck the number of flood preventing doors can be minimised or even omitted completely.

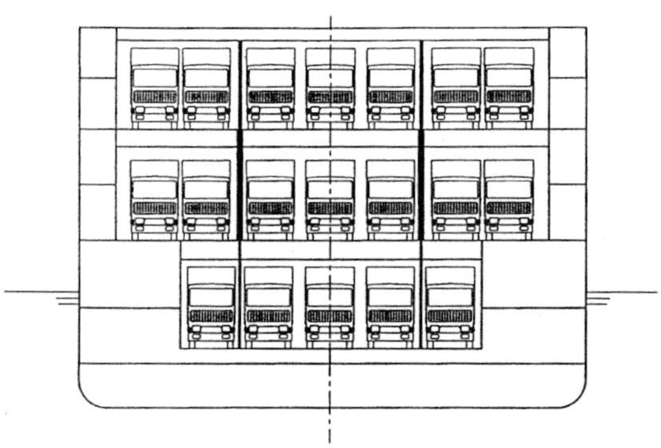

Figure 1: Cross section showing the safety barriers: side casings and longitudinal bulkheads.

RATIONAL CARGO HANDLING

The ro-ro passenger ships of tomorrow must be safer and more efficient to operate on the route and in harbour. This may sound incredible considering the SOLAS 90 and Stockholm agreement together with proposals suggesting that possible cargo movements on vehicle deck must be considered in stability calculations.

Many ro-ro ship accidents occur because the cargo on trailers and the trailers have moved in heavy seas, or because the ships have been incorrectly loaded. A combination of these two factors has caused several capsizings of ro-ro ships.

Human mistakes should not jeopardise the ship's safety. The cargo decks should be designed for rational cargo handling even for short harbour calls and assuming extreme hurry due to eventual delays.

The lane meters on the different ro-ro decks should be available even in short harbour calls. A typical bottleneck is the lower hold. In many ferries the cargo is moved with elevators from the main deck to the lower hold or via a single narrow ramp. Simultaneous loading of all three ro-ro decks is not either always possible leading to a temptation to leave the lower hold empty and raising unnecessarily centre of gravity.

EFFICIENT ARRANGEMENT FOR TRAILER FERRY

This chapter describes our design approach for some recently built and ordered ro-ro trailer ferries.

The design criteria for these vessels have been very straightforward but also very demanding:

- Cargo capacity of 2400-2500 lane meters on two to three decks with minimised main dimensions.
- All lanemeters to be fully available, not theoretical, within the required harbour time, 1 hour only.
- Fully operable ship on year-round service, also in heavy seas and without tug assistance.
- Minimised requirement for lashing
- Minimised maintenance.

The TT-Line ships, 'Robin Hood' and 'Nils Dacke' were the first ones built in accordance with the above basic requirements. They are capable to take trailers on three decks: the lower hold, the main deck and the upper deck. All three decks can be loaded and unloaded simultaneously. Figure 2 shows the principle arrangement.

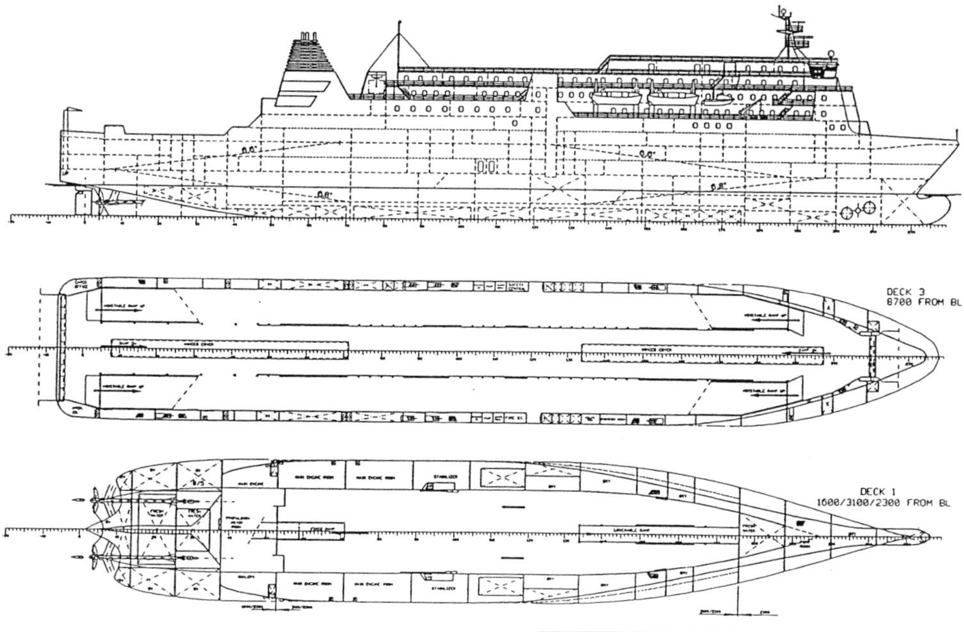

Figure 2. General arrangement of the TT-Line ferries.

Driving on and off the main deck is done through the stern and bow doors. The cargo spaces have side casings and two longitudinal bulkheads. The ship therefore has a double hull both above and below the main deck.

Two ramps lead to the lower hold along the ship's centreline from the main deck. The ramps are placed as close to the bow and stern respectively as possible.

The drive-through principle to the lower hold enables the vehicles to drive straight out, without having to turn or reverse. These are the first ro-ro passenger ships built with the drive-through lower hold leading to a fast and safe loading and unloading procedure. They also have the longest existing lower hold in relation to the ship's length, 56%.

The side casing solution, see figure 1, makes the main deck and upper deck very easy to load and unload. When designing the steel construction, the conclusion was reached that it was no use letting the superstructure rest on heavy steel beams attached to the roof of the main deck and upper hold. Instead it was decided that the superstructure would be held up by two rows of pillars.

In order to attain sufficient air-exchange in the lower hold, air-exchange channels were placed along the pillar rows. A direct development of this was to build two bulkheads instead of pillars along most of the vessel's length with the channels incorporated. When comparing this construction with conventional centre casing design a difference of 300 tons in steel weight was found, side casing configuration being lighter.

The lanes on the two upper decks are divided into two lanes on each outer side and three in the middle. With such subdivision, the fastening of the cargo becomes unnecessary. Only two trailers can be set in motion in the two outer sections of thc ro-ro decks, and in the middle three. Heeling risk because of cargo shift is minimised and has no practical importance. See figure 1.

In order to make the lower hold long enough for drive-through traffic, the machinery was not arranged in a conventional way. A diesel-electric machinery made this possible. The four engines with their generators were placed in four separate engine rooms on both sides of the lower hold. The propeller motors were placed in a separate space astern of the cargo hold.

Aft ramp was led between the electrical propulsion motors starting already at frame 10, 10,80 m forward of transom. Side casings were applied to have the cargo flow down to the lower hold in the middle and to the upper trailer deck on both sides with hoistable ramps, see figure 2.

Instead of an engine room arrangement with two sections, five separate engine rooms are built, which improves safety considerably. With separated engine rooms, all engines are not lost in case of a fire or damage.

The length of the lower hold was still limited in the aft due to the space required for propulsion motors. The next step is to move the propulsion motors outside the hull, into the pod propulsors.

Figure 3 presents engine room and lower hold arrangement for mechanical geared, conventional diesel-electric, and pod-electric machinery configurations. Mechanical arrangement allows 48% of L_{pp} for the lower hold and aft ramp is located between engines. Conventional diesel-electric offers 56% of L_{pp} for the lower hold. But the pod-electric machinery makes it possible to use the maximum within the hull form, 61% of L_{pp}. The gain in the lower hold is 27 m in length, which is 29% increase in the lower hold capacity in comparison with conventional diesel-electric configuration. The transverse position of the aft ramp is not any more dictated by the machinery location but can be optimised to match with the cargo flow into main and upper decks.

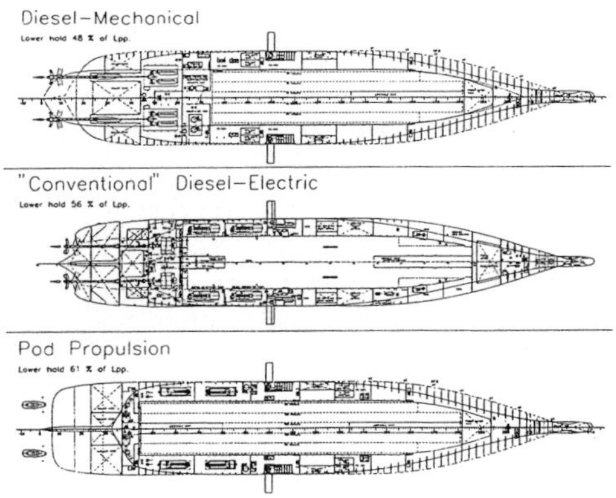

Figure 3. Engine room and lower hold arrangement for three different machinery options on a passenger trailer ferry.

Utilisation of spaces is efficient, there are actually no void spaces left and the volume of machinery spaces is about 20% less in comparison with mechanical propulsion.

DAMAGE STABILITY

Damage stability of the trailer ferry configurations described in the previous chapter, three alternatives in figure 3, has been extensively studied. In principle the B/5 rule is applicable for lower hold ships. However, we decided not to apply the B/5 border but to include the lower hold in all relevant damage cases. This approach is based on overall safety and survivability after even most severe damage cases for any present rule requirements. This is also the principle approach of the new probabilistic proposal of calculating damage stability for ro-ro passenger ships. The focus is on the righting lever (GZ) curve after damage as it presents the actual overall safety and survivability. The damage stability calculations for the maximum lower hold configuration with all two compartment damages (i.e. adjoining compartments aft and fore and/or side compartments) show good GZ-curve capabilities meeting easily all SOLAS 90 requirements.
Figure 4 presents the most severe case lower hold together with the motor room aft and the adjoining side compartment.

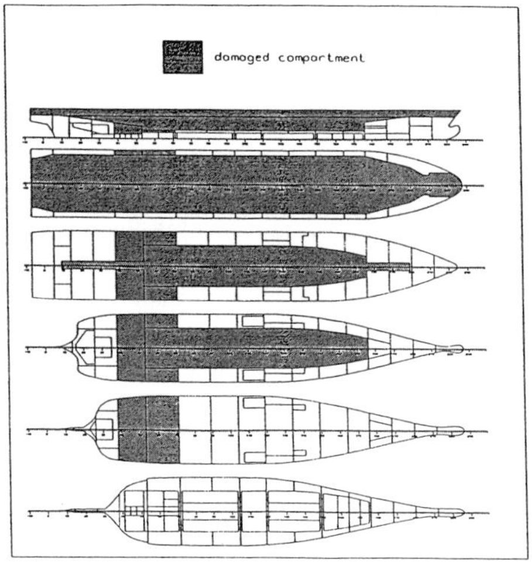

Figure 4. Worst damage case with lower hold damaged.

In addition to the above 'normal' damage cases the following typical and most probable damage situations were studied:

- Three side compartments plus lower hold damage, SOLAS 90 without margin line
- Complete double bottom damage, SOLAS 90 without margin line
- Collision damage extending over 9-11 compartments from bow including lower hold and bulkhead deck, SOLAS 90 without margin line

- Maximum amount of water on deck over three meters corresponding to over 6000 tonnes, simulating an open bow door situation, SOLAS 90.
- Combined lower hold and two side compartments damage plus simultaneously water on deck, survival.

All the above damage cases could be met fulfilling SOLAS 90 final stage criteria, except in some of them the margin line criteria. The lower hold damages with the longest possible hold actually show the best survivability as there is no trim included.
Figure 5 presents one of the above listed special damage cases and stability characteristics at the final stage.

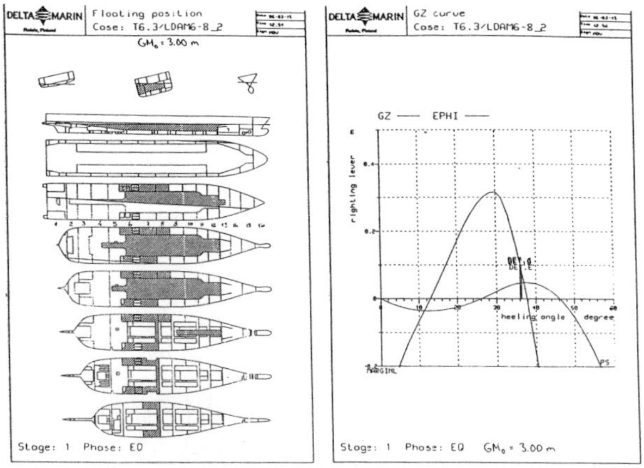

Figure 5. Worst SOLAS damage case together with lower hold flooded.

This means actually that the diesel-electric option with pod propulsion, longest lower hold, shows the best results, see figure 3.
The side casings above bulkhead deck are an essential part of the survivability and according to model tests give a possibility to leave out flood preventing doors on the main deck. The longitudinal bulkheads, see figure 1, within the main deck give also an option to limit the amount of water on the deck if seen necessary depending on operational area.

UPGRADING EXISTING RO-RO PASSENGER FERRIES

The upgrading process of existing ro-ro passenger ferries has been going on already for a few years. First to calculate the A/Amax and to define necessary measures to reach high enough values, and shortly after that to meet the SOLAS 90 (95) requirements and Stockholm agreement.

We have carried out different calculations and studies for more than 50 existing ferries sailing in the Baltic, North Sea and Mediterranean. Most of the vessels require some modifications to meet the set requirements, some of them quite extensive ones.

List of typical modifications can be given:

- closing of eventual openings
- making water/weather tight doors and man-holes
- raising air pipes, passages, and similar
- reduction of asymmetry in flooding with modified tank arrangement and compartmentation
- reduction of flooding volume as above
- making existing side compartments on freeboard deck watertight
- building up new watertight side compartments
- building up longitudinal bulkheads inside B/5 (!)
- using flood preventing doors on freeboard deck
- combinations of above.

External modifications:

- sponsons
- ducktail
- combination of above.

Typical solution is found by starting up with the simplest and cheapest internal modifications and continuing into external ones until satisfactory level has been reached. Some of the internal modifications limit, however, dramatically the operation and/or capacity of the ro-ro deck and are not preferred even though being cheaper and easier to install than the external modifications. Many of the older vessels have also a lack of deadweight due to increased lightweight through the years of operation. The sponson alternative becomes an attractive one, it gives additional displacement, it helps to overcome excessive stern trim problems, it is possible to reduce required ballast in foreship and there is a good gain in stability, both intact and damage.

The deadweight gain is typically between 100-250 tons, depending on the ship and its original hull form. The additional required power according to several model tests is between 7-12%. To reduce this down to 2-8% a ducktail is applied lengthening the waterlines. Figure 6 shows a typical recently built example. The following presents two actual cases.

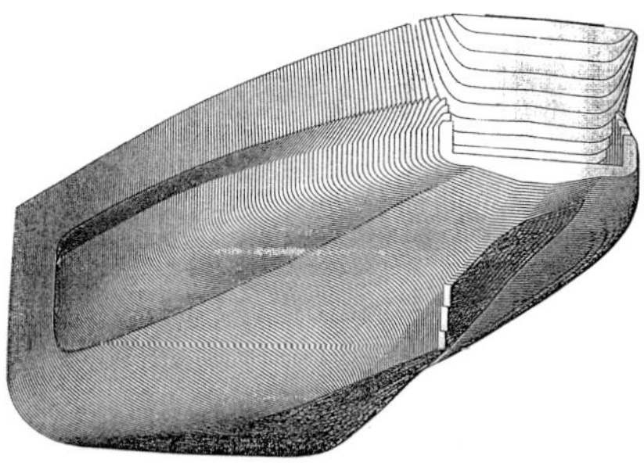

Figure 6. Combined sponson – ducktail arrangement on an existing vessel to increase A/Amax above 0,90.

Case 1:

This is a typical Baltic ferry built in the early eighties. It was originally designed and built to meet the damage stability requirements set by SOLAS 1974. The task was to upgrade the ship to meet both SOLAS 1990 damage stability requirements and the water on deck requirements set by IMO Circ. 1891.

One important feature of this ferry is that the main trailer deck does not have full width but on the sides there are rather wide side casings containing cabins, staircases and storerooms. It was found out in the calculations that all SOLAS 1990 requirements will be met if these side casings will be done watertight. On the port side also a lengthening of the existing side casing with 5 frames was required and thus some lane meters were lost. The water tightness of the side casing required about 20 doors to be changed. And also within the staircases a couple of new steel bulkheads had to be built. (Figure 7 A)

The following step was to cover the water on deck requirements. Calculation method given by IMO Circ. 1891 revealed that two transversal watertight barriers through the main trailer deck were required. This would have caused not only expenses but also significant loss of lane meters and made the loading sequence more difficult.
Also an interesting finding in the calculations was that although two barriers were necessary in order to fulfil requirements, one barrier was not too far away from meeting the requirements. (Figure 7 B)

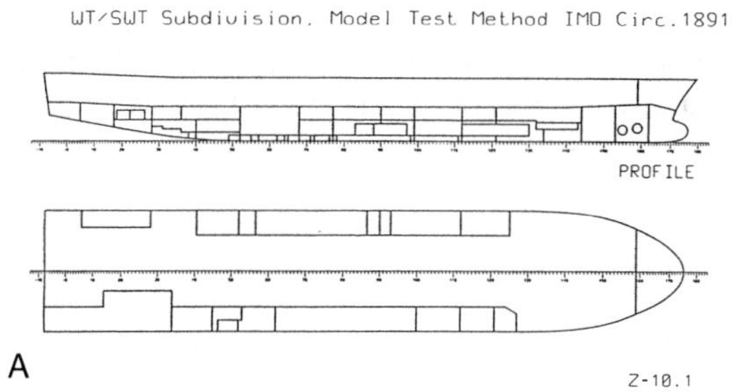

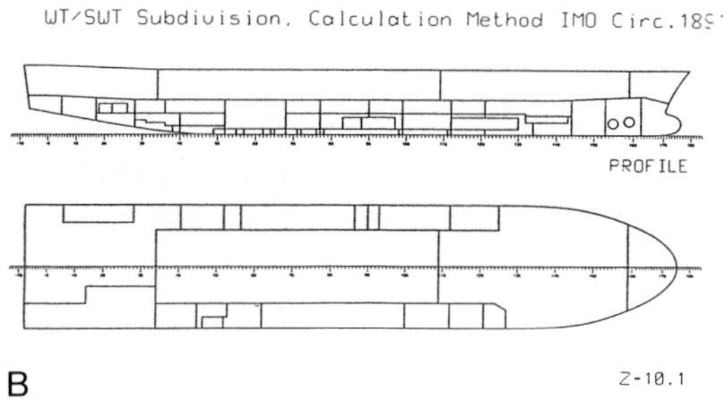

Figure 7.

Based on the above it was decided to perform water on deck model tests according to the model test method described in IMO Circ. 1891. Two worst SOLAS damages were tested with significant wave height of 4 meters, which is the highest required significant wave height meaning unrestricted service area. The outcome of these tests was that water on deck requirements were fulfilled with the same arrangement that was needed to meet the normal SOLAS 90 requirements. So both extra transversal barriers planned on the trailer deck could be skipped. This meant remarkable savings to the operator and fluent loading/unloading sequences also in the future.

In this case it was found out that the model test method was leading to smaller conversion needed on board compared to the straight calculation method. It is not possible to make any general conclusions based on one result but model tests can be recommend as an alternative especially for ferries having a good intact stability and side casings. The ship in question had a GM of 3 meters intact and over 1.5 m when damaged. It was observed during the tests that even though significant amounts of water came on the deck with the waves, the deck also drained very rapidly through the damage because the ship was so stiff.

Case 2:

This is a typical ro-ro ferry built in the early seventies. The A/Amax according to MSC574 was 0.77. The task was to study the possibilities to meet the SOLAS 90 requirements or at least improve the A/Amax ratio so much that the final upgrade could be postponed beyond year 2002.

The main trailer deck has no watertight compartments and the actual GM in design loading condition is 1.69 m. Six different steps including the installation of 320 mm (the width of the web frames) wide watertight internal side casings, duck tail sponson (not widening the original ship), 2,3 or 4 pairs of transversal flood control doors on the trailer deck and external side sponsons (width 1 m) were studied. The conversion weight were estimated to be:

- 48 tonnes for the internal side casings
- 47 tonnes for the duck tail sponson
- 16 tonnes for 1 pair of transversal flood control doors on car deck
- 176 tonnes for external side sponsons

The increased displacement was 51 tonnes for the duck tail sponson and 139 tonnes for the wide side sponsons.

Only three of the calculated combinations were fulfilling the SOLAS 90 requirements in full. The one with 320 mm wide internal side casings, duck tail sponson and 3 pairs internal flood control doors (Figure 8), the other with duck tail sponson and 4 pairs of internal flood control doors, and the last with 1 m wide side casings and 3 pairs of internal flood control doors.

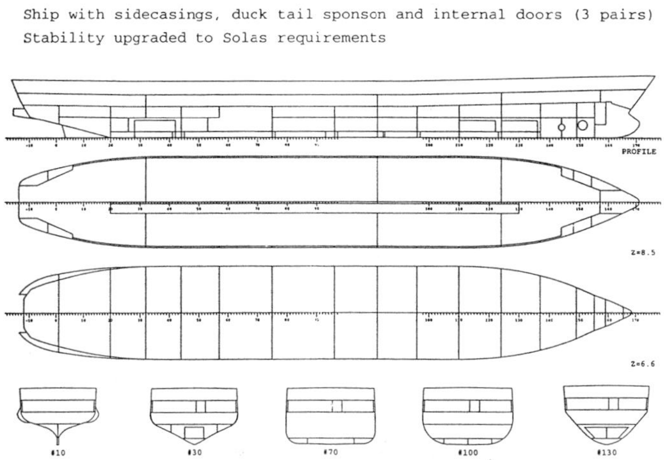

Figure 8.

During the work it became clear that it is utmost important to remain the existing lane capacity and on the other hand it is very important that the service speed could be kept as is today. All the combinations fulfilling the SOLAS damage stability requirements have transversal flood control doors on the trailer deck (2,3 or 4 pairs). These doors reduce trailer capacity. They are normally so called hemi-cyclic doors which can be opened towards both directions, so the door itself does not need free operation space after loading. The last modification which included wide side sponsons would be the best for the cargo capacity and loading operations but on the other hand these wide sponsons (wider than the original ship) always mean some loss in speed.

During the work it was also found out that a step by step conversion is acceptable for the Authorities.

The selected modification was finally a combination where a duck tail sponson was combined with rather short side sponsons (0.6 m wide) making a horse shoe type extension round the aft ship (shown in figure 6). This sponson increased the KM value of the ship by 0.73 m. On the trailer deck existing casings on the aft and forward corners were made watertight. This meant four doors to have watertight sealing and not to be left open during voyage. No new internal side casings and no flood control doors were installed leaving the cargo area totally untouched.

The new A/Amax calculation showed a figure of 0.9372. According to MSC 60/21 the final upgrading must be done before the October 1st, 2005.

CONCLUSIONS

Current design practice of ro-ro passenger ferries is discussed for seaworthiness, damage stability and cargo handling. The importance of the bow flare shape, geometry, for seaworthiness and impact loads is presented with examples of bow flare impact pressures measured in model scale for different bow flare geometries. The developed bow flare estimator gives a simple but reliable tool to validate the bow shape at the early stage of the project. Model tests are recommended especially for extreme hull shapes and/or operational conditions. Special attention should be paid on the bow flare structural strength. Several damages on ferries in heavy weather have the existing dimensioning methods inadequate. Strong recommendation is given to use only DNV (Det Norske Veritas) dimensioning methodology.

Efficient lower cargo hold arrangement for ferries is presented. However, the present damage stability rules and especially their interpretations are not supporting the lower hold arrangement. The studies carried out, however, prove that the lower hold arrangement with side casings on main deck gives a clearly leading to better and higher damage safety than any of the existing arrangements supported by the present deterministic stability rules. The B/5 bulkheads are utilised in the lower hold but damage stability is calculated for lower hold damages as well together with all possible two compartment side and double bottom damages fulfilling the SOLAS 90 requirements. Efficient cargo handling is arranged with drive through ramp arrangement in the lower hold.

Contemporary Ideas on Ship Stability
D. Vassalos, M. Hamamoto, A. Papanikolaou and D. Molyneux (Editors)

A REALISABLE CONCEPT OF A SAFE HAVEN RO-RO DESIGN

Dracos Vassalos

The Ship Stability Research Centre (SSRC), Department of Ship and Marine Technology
University of Strathclyde, Glasgow, UK

ABSTRACT

A considerable amount of effort has been expended, particularly over the recent past, towards enhancing the safety of Ro-Ro vessels. The routes followed include proposals for a stricter regulatory regime, improvements on operational procedures, effective training of personnel onboard and the introduction of more efficient life saving and evacuation appliances and approaches. However attractive (and sometimes necessary) these measures might be, they do not address the root of the problem - namely the ship design concept itself. In this respect, it is particularly alarming to see proposals that undermine the meaningful evolution of the Ro-Ro concept, the most commercially successful ship design. With design-for-safety in mind this paper re-iterates the use of sheer and camber in the design of the Ro-Ro car deck as an efficient means to enhancing survivability drastically and cost-effectively. Two applications of this idea are considered: the first involving a combination of positive sheer with positive camber (PSPC) and the second negative shear with negative camber (NSNC) whilst employing the use of intelligent wash ports (IWP). The impressive enhancement of damage survivability is demonstrated by means of numerical simulation using the suite of software developed at SSRC. The latter is currently being extensively applied by the ferry industry for upgrading, retrofitting and design purposes in the strife of this industry to meet the new demanding survivability standards in the most effective way possible. Following a brief background and a description of the proposed survivability enhancing design ideas, the mathematical/numerical model used to perform the comparative study is briefly explained. The features of the Ro-Ro design and of the damage cases used in the analysis are described next before presenting and discussing the results for the basis ship flat deck (BSFD) design and the two alternatives considered.

KEYWORDS

New Ro-Ro concept designs, deck shear, deck camber, intelligent wash ports, design for survivability.

BACKGROUND

The Ro-Ro concept provides the capability to carry a wide variety of cargoes in the same ship, thus being able to offer a competitive turn-around frequency with minimum port infrastructure or special shore-based equipment. Short sea routes are dominated by Ro-Ro ships with lorries, trailers, train wagons, containers, trade cars and passengers being transferred from the "outer" regions (UK, Ireland, Scandinavia and Finland) to the "main" land (continental Europe). In the Southern Europe corridors, the Ro-Ro freight service is progressively increasing in volume. The case for a long-distance Ro-Ro service to provide a European maritime highway has also been made several times before. This is particularly relevant and important in respect of fast sea transportation where again Ro-Ro ferries play a prominent role. As a result, the world fleet of Ro-Ro ships has steadily increased over the last 15 years to some 5,000. Over the same period there has been an encouraging reduction in the annual vessel casualty rate. However, the large number of serious casualties for this ship type and the overall loss of life have not shown the same improvement as the casualty rate. The maritime industry is acutely aware of recent shipping casualties involving Ro-Ro ferries, which have resulted in severe loss of life. These led to safety becoming the main concern with Ro-Ro vessels. Standards for Ro-Ro ship configuration, construction and operation have undergone close scrutiny and new legislation has been put into place aimed at improving the safety of these vessels, notably SOLAS '90, (IMO Resolution MSDC.12 (56), 1988) as the new global standard for all existing ferries. However, since the great majority of Ro-Ro passenger ships were designed and built prior to the coming-into-force of SOLAS '90, it is hardly surprising that few of them comply with the new requirements.

Furthermore, concerted action to address the water-on-deck problem in the wake of the *Estonia* tragedy led IMO to set up a panel of experts to consider the issues carefully and make suitable recommendations. Following considerable deliberations and debate, a new requirement for damage stability has been agreed among North West European Nations to account for the risk of accumulation of water on the Ro-Ro deck. This new requirement, known as the *Stockholm Agreement* (IMO Resolution 14, 1995), demands that vessels satisfy SOLAS '90 standards (allowing only for a minor relaxation) with, in addition, a constant height of water on deck. The net effect of these developments in legislation is a massive increase in survivability standards to a level many believe to be unattainable without destroying the very concept industry is extremely keen to defend.

Deriving from this, haphazard attempts to improve Ro-Ro safety by introducing ineffective survivability enhancement devices must give way to rational approaches. The ingenuity of designers must be called upon, and be nurtured, to pave the way towards practical designs for cost-effective safety, in order to ensure both the survival and a meaningful evolution of Ro-Ro ships in the future. Attempting to demonstrate that simple, cost-effective ideas, capable of ensuring the survivability of Ro-Ro vessels whilst retaining the Ro-Ro concept intact are there to be discovered, this paper features two configurations of the main deck, which if optimised could render Ro-Ro ships a safe haven. The analysis presented in the following provides ample evidence that such a concept can be realised and demonstrates beyond doubt the survivability effectiveness of one of the most traditional naval architecture design practices.

THE PROPOSED SURVIVABILITY ENHANCING DESIGN IDEAS

Earlier studies have clearly shown that the decisive factor affecting Ro-Ro damage survivability is the water accumulated on the main deck, Vassalos et al (1996). Therefore, any measures to prevent or limit the water accumulation would result in a vessel with enhanced survivability. Should such measures prove effective with the ship damaged at high sea states, it could then be suggested that staying onboard the vessel would be the safest alternative in case of an accident that results in breaching of the hull. This is the idea of a safe haven ship. In the investigation considered here, the level of damage survivability aimed at is $H_s = 4$m over the whole range of feasible loading conditions. This is in accordance with the most severe damage stability requirement currently in force. The idea being advocated here is that of using a curved Ro-Ro deck, rather than a flat deck, (Figure 1), with or without intelligent wash ports as a means of channelling the water on deck to flow out. More specifically, the following two alternatives are examined:

Alternative 1 (PSPC) - Ro-Ro deck with positive sheer and positive camber (Figure 2)

Perceived advantages offered by this idea include:

- In the case of midship damage any water finding its way on the Ro-Ro deck would tend to concentrate in the vicinity of the damage opening because of the fore-and-aft sheer on the deck and flow out.
- In the case of damage forward or aft, the increased freeboard resulting from the deck sheer will ensure that less water reaches the Ro-Ro deck and hence survivability will be improved. Normally, the ensuing trim forward or aft, following respective damages will be conducive to water accumulation towards the vicinity of the damage opening and hence to water egress from the deck.
- Irrespective of the damage location, the presence of positive deck camber potentially provides two additional benefits. Water may flow towards the intact side of the ship resulting in an increased damaged freeboard and hence enhanced survivability. If the ship is inclined towards the damage, the presence of camber in principle impairs water inflow whilst assisting water outflow.

Alternative 2 (NSNC + IWP) – Ro-Ro deck with negative sheer and negative camber together with intelligent wash ports (Figure 3)

Intelligent wash ports are freeing ports with flaps, which passively allow only water outflow, their opening or closing depending on the pressure difference on either side of the flap. The use of these ports has been considered and abandoned on the basis of inconclusive research showing that the overall area of the freeing ports necessary to ensure effective outflow would be too large to offer an attractive solution. The idea put forward here is aimed at minimising the area of opening of the IWP's by utilising again a curved Ro-Ro deck. In a damage scenario resulting in progressive flooding of the Ro-Ro deck, there is a slow build up of water accumulation with the ship heel increasing equally slowly until a point is reached where the heeling effect of the water on deck exceeds the restoring capacity of the vessel. Beyond this point, capsize is inevitable and happens very quickly. Considering the above, if the capacity of IWP's were such as to offset the net inflow of water on deck for the range of loading and environmental conditions the ship is likely to operate in, survivability would be ensured and a safe haven ship could be realised.

Perceived advantages deriving from the idea include:

- Negative deck camber assists in water accumulating near the ship centreline and hence reducing the ship heeling. This is very important, as the damaged freeboard is a critical parameter affecting ship survivability.
- Negative deck sheer assists water flow towards the ship ends where the heeling effect is further reduced due to reduced beam. Additionally, by locating IWP's at the ends water can flow out.
- Negative deck sheer results in increasing damaged freeboards particularly amidships where the ship is the most vulnerable when damaged at this location without having to raise the whole deck which would adversely affect the overall stability of the ship.
- The presence of IWP's would give a Ro-Ro ship a chance in case of accidents similar to the *Herald of Free Enterprise* and the *Estonia* where bow damage with forward speed rendered capsize inevitable and catastrophic.

To assess the survivability and effectiveness of these ideas, use is made of the North West European R&D Project on the "Safety of Passenger/Ro-Ro Vessels" and of the mathematical/numerical models developed at the Department of Ship and Marine Technology of the University of Strathclyde since 1977.

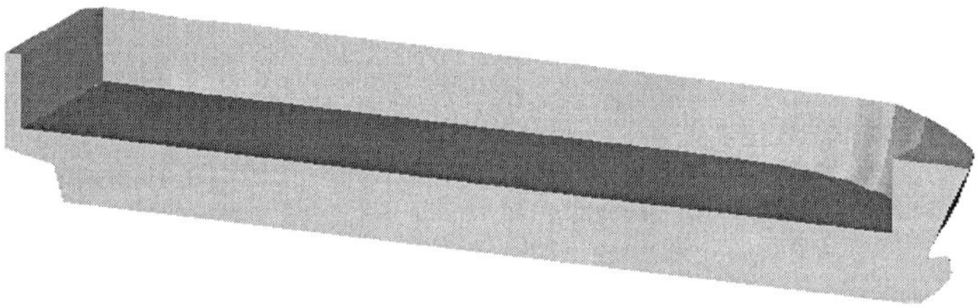

Figure 1: Basis Ship Flat Deck (BSFD)

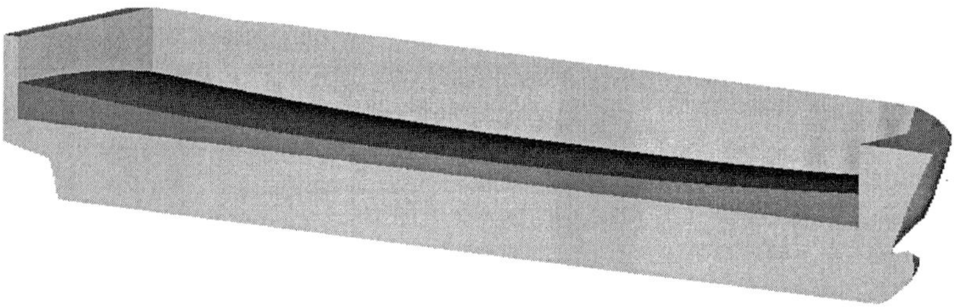

Figure 2: Positive Sheer Positive Camber (PSPC)

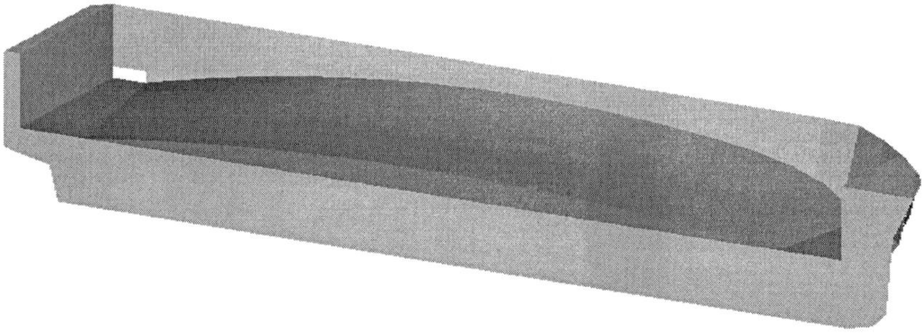

Figure 3: Negative Sheer Negative Camber + Intelligent Wash Ports (NSNC + IWP)

MATHEMATICAL/NUMERICAL MODELS

The generalised models available at SSRC for assessing the damage survivability of Ro-Ro vessels in realistic environments are described by Vassalos (The Water on Deck Problem of Damaged Ro-Ro Ferries). Considerable effort has been expended to ensure the validity of the numerical simulation programs in their ability to predict the capsizal resistance of a damaged vessel in a random sea whilst accounting for progressive flooding, over the whole range of possible applications. These include vessel type and compartmentation (above and below the bulkhead deck), loading condition and operating environment and location and characteristics of damage opening. Such claims have been substantiated by the impressive agreement achieved between theoretical and experimental results spanning a wide range of parameters. Typical results from ten ships, tested through the "Equivalent Safety" route by numerical and physical model experiments are shown in Figure 4 where the agreement between physical model tests and numerical tests is very convincing, Vassalos (1998). This is a clear indication of the ability of the mathematical model used to accurately assess the capsizal resistance of a damaged Ro-Ro vessel subjected to large scale flooding. This derives from the accurate modelling of the dynamic system behaviour in the capsize region. This might at first sound surprising, considering how complex the processes involved are, but can be easily explained by the hydrostatically dominated nature of the capsize phenomenon relating to an extensively flooded vessel. With the exception of very few cases the results from both approaches are identical. In general the agreement between numerical and experimental results has reached a level where any discrepancy of more than 0.25m in critical Hs between the two is considered unacceptable and is normally the cause of a thorough investigation until a satisfactory explanation is found. Typical cases include discrepancies, which are the result of differences in deck permeability of the order of 10% and in heel angle in the order of 0.5^0. The level of confidence in the results of numerical tests is clearly being demonstrated by some ferry owners/operators who proceed with retrofitting plans on the strength of numerical predictions. Efforts are currently under way to collect this and other relevant evidence to prepare a working paper for submission to IMO, aiming for approval to utilise numerical simulation as an alternative route to compliance with damage survivability standards.

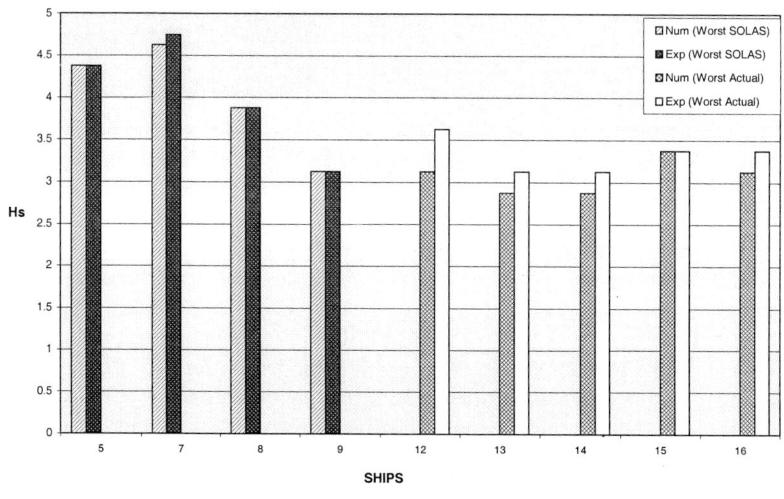

Figure 4: Comparison Between Numerical and Experimental Results – Critical Hs

CASE STUDY

The case study presented here considers as a basis ship, the flat open deck Ro-Ro vessel NORA that is a generic design used in the North West European R&D Project. The two alternatives explained in the foregoing are also described in detail in this section, following a description of NORA. All three alternatives are illustrated in Figure 5.

Basis Ship

The principal design particulars of NORA are shown in Table 1 and the outline design with the original car deck configuration is illustrated in Figure A.1 of Appendix A.

TABLE 1
PRINCIPAL DESIGN PARTICULARS OF NORA

Length, $\mathbf{L_{BP}}$	130.00 m
Beam, **B**	25.50 m
Depth to Car Deck, **D**	8.35 m
Draught, **T**	5.75 m
Displacement, Δ	12,000 tonnes
Block Coefficient, $\mathbf{C_B}$	0.612
Intact **KM**	14.26 m
Intact Freeboard, **F**	2.60 m

Alternative 1 (PSPC)

Details of this arrangement are shown in Figure A.2 of Appendix A. The sheer considered is parabolic in shape with maximum values of 1.0 m at the ends and 0.0 m amidships. The camber is also of parabolic shape with a maximum value of 0.2 m at the centreline of the vessel. This choice was made by taking the ratio of maximum sheer to maximum camber in proportion of the L/B ratio.

Alternative 2 (NSNC + IWP)

Details of this arrangement are shown in Figure A.3 of Appendix A. The negative sheer has now a maximum value of 0.8 m amidships at the side of the car deck reducing to 0.0 m at the location of the IWP. The maximum camber is again 0.2 m but with the negative camber considered here, this would correspond to a drop of 0.2 m along the ship centreline with the deck at side following the shape of the negative sheer. This configuration provides therefore for two flat deck portions along the ship length where the IWP's are located. The freeing ports considered in this case study are located at both sides of the ship at the stern and the bow as illustrated in Figure 5 with dimensions of 10.0 m in length by 0.5 m in height. The 20.0 m reduction in the sheered Ro-Ro deck length is the reason for considering a maximum camber of 0.8 m in this alternative rather than 1.0 m as in alternative 1. No optimisation study has been made to determine the most effective dimensions or location of the IWP's for the damage scenarios considered.

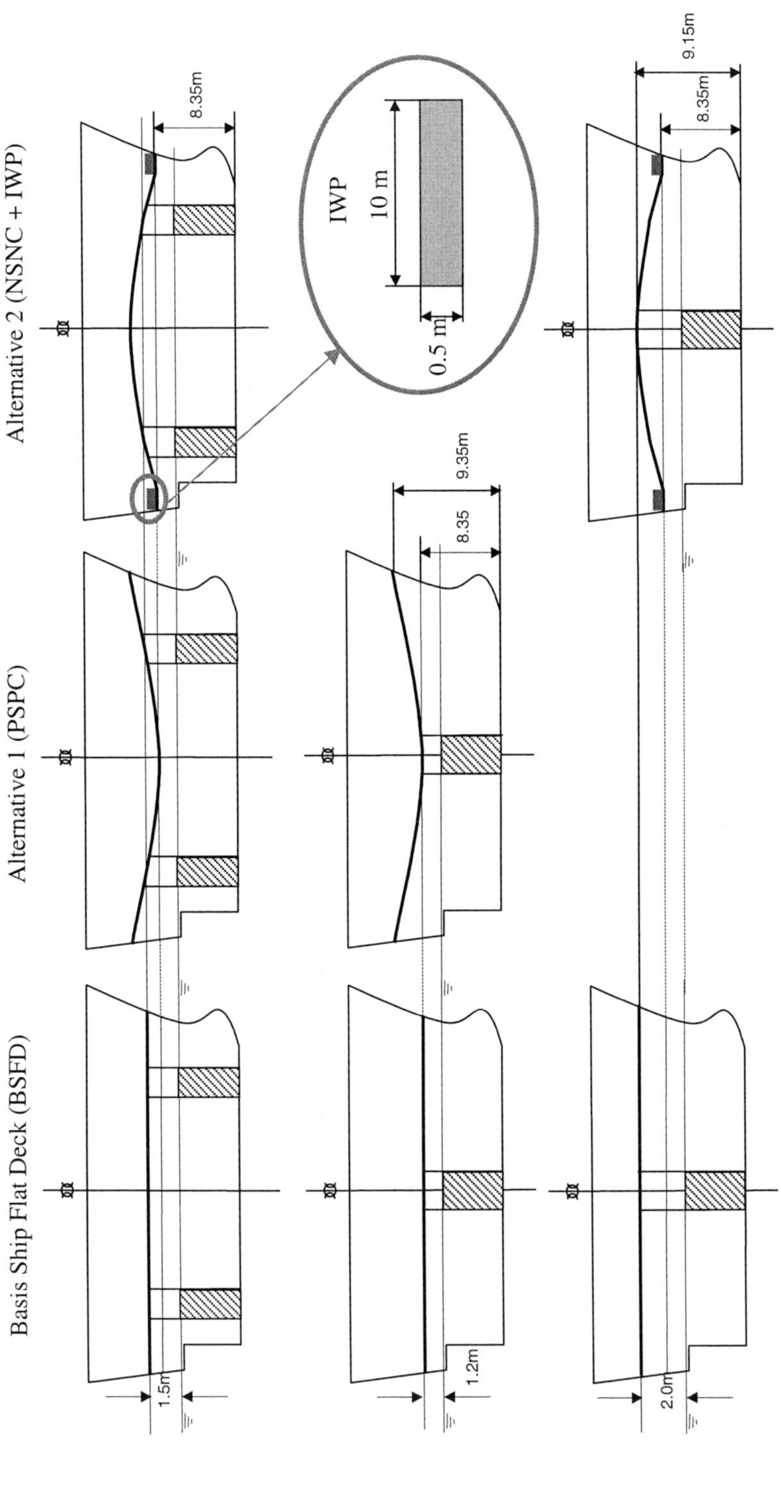

Figure 5: Case Study Alternatives

Particulars of Damage

To evaluate the effectiveness of the proposed ideas in enhancing Ro-Ro damage survivability, it was thought appropriate to consider a relatively low damaged freeboard for the basis ship, for ease of illustration of the survivability enhancing effect of the proposed ideas. To this end, the basis ship damaged freeboard was taken to be 1.2 m. Furthermore to eliminate any bias in the results deriving from the difference in the damage freeboards due to the presence of deck sheer, it was considered appropriate to compare the survivability of the various alternatives at the same damaged freeboard. This was achieved by raising in each case the car deck artificially to the right level. In this respect, it is to be noted that the damaged freeboard for both the fore and aft damages at the location of damage is approximately 1.5 m in both alternatives whilst for the midship damage of alternative 2 the damaged freeboard is 2.0 m. Deriving from the above, the damage cases considered in this case study are given in the Table 2 below referring to all the alternatives and are illustrated in Figure A.4 of Appendix A.

TABLE 2
PARTICULARS OF DAMAGED CASES

Damage Case	Freeboard (m)	Compartment Length (m)	Compartment Location (m)	KM (m)
Aft	1.5	18.00	-47.00 ÷ -29.00	13.71
Midship	1.2	29.20	-22.43 ÷ 10.00	13.45
	2.0	29.20	-13.80 ÷ 0.50	
Forward	1.5	19.30	19.65 ÷ 43.35	13.23

Parametric Investigation

Considering the damage cases described in the foregoing over the range of possible operational and environmental conditions leads to the test matrix shown in Table 3. With the operational KG at 11.5 m, a ± 0.5 m variation was thought to be representative for the KG operational envelop appropriate to this vessel.

TABLE 3
PRINCIPAL DESIGN PARTICULARS OF NORA

	FLAT DECK			ALTERNATIVE 1		ALTERNATIVE 2	
KG (m)	Damaged Freeboard (m)			Damaged Freeboard (m)		Damaged Freeboard (m)	
	1.2 amidships	1.5 fore &aft	2.0 amidships	1.2 amidships	1.5 fore & aft	1.5 fore & aft	2.0 amidships
11.0	X	X	X	X	X	X	X
11.5	X	X	X	X	X	X	X
12.0	X	X	X	X	X	X	X

Wave Environment

The wave environment used in the numerical simulations is representative of the North Sea and is modelled by using a JONSWAP spectrum as shown in the table below.

TABLE 4
PRINCIPAL DESIGN PARTICULARS OF NORA

Significant Wave Height, $\mathbf{H_s}$ (m)	Peak Period, $\mathbf{T_p}$ (s)	Zero-crossing Period, $\mathbf{T_0}$ (s)
1.0	4.00	3.13
1.5	4.90	3.83
2.0	5.66	4.42
2.5	6.33	4.95
3.0	6.93	5.42
4.0	8.00	6.25
5.0	8.95	6.99

$$H_s/L_p = 0.04\ (L_p = 25\,Hs);\quad T_p = \sqrt{\frac{2\pi\,Lp}{g}}\ (T_p = 4H_s^{1/2});\quad T_0 = \frac{T_P}{1.279}$$

Numerical survivability tests have been undertaken for a significant wave height resolution of 0.25 m. Limiting Hs in the derived results represents the maximum sea state that can be survived repeatedly in each damaged case considered. The norm adopted in presenting the results of numerical simulations and model experiments is to provide a capsize region rather than a capsize boundary to correctly reflect the fact that, because of the random nature of all the parameters determining a capsize event, a single boundary curve does not exist.

RESULTS AND DISCUSSION

The results are presented as survivability bands in the form of critical Hs (i.e. significant wave height characterising a limited sea state from survivability point of view) versus KG, allowing for comparisons between the basis ship and the two alternatives proposed at the corresponding KG and freeboard.

Flat Deck (BSFD) Vs Alternative 1 (PSPC)

Midship Damage (Figure 6a)

For open deck Ro-Ro vessels, damage amidships is the most onerous and hence it constitutes the critical damage concerning survivability. This is clearly demonstrated in Figure 6 where the basis ship appears to have very low capsizal resistance, barely managing to survive 3.0 m Hs, even at low KG's. Introducing positive sheer and camber on the deck at levels that could easily be realised, however, results in increasing the damage survivability of the vessel over the required 4.0 m Hs, even at high KG's.

Fore & Aft Damages (Figures 6b & 6c)

For open deck Ro-Ro vessels fore and aft damages are normally less onerous than midship damage, as indicated above. It is interesting, however, to demonstrate that even by increasing the damaged freeboard of the basis ship at a level corresponding to the height of the sheered deck at the damage location, alternative 1 still results in a clear improvement of survivability for both damage cases.

Flat Deck (BSFD) Vs Alternative 2 (NSNC + IWP)

Midship Damage (Figure 7a)

The importance of freeboard in improving damage survivability is clearly demonstrated in Figure 6, where by increasing the damaged freeboard of the basis ship to 2.0 m, the vessel is capable of surviving over 4.0 m sea states almost throughout the range of possible loading conditions. This example explains clearly and justifies in a way the drive inherent in recent criteria and approaches for assessing survivability towards higher freeboards. However the introduction of curved decks as described in alternative 2 together with moderately sized freeing ports improves survivability beyond these levels by an average of 1.0 m Hs over the whole range considered.

Fore & Aft Damages (Figures 7b & 7c)

The potential advantages deriving from alternative 2 are clearly demonstrated in Figure 7 where survival to extreme sea states appears to be realisable, throughout the possible range of interest and well above the survival levels offered by alternative 1.

Figure 6a - MIDSHIP DAMAGE
Freeboard = 1.20 m

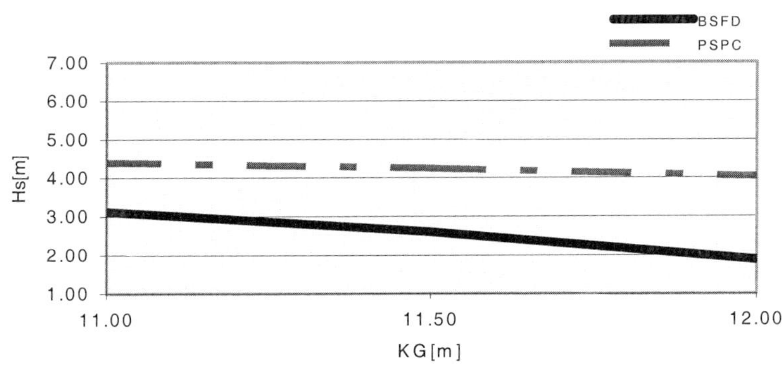

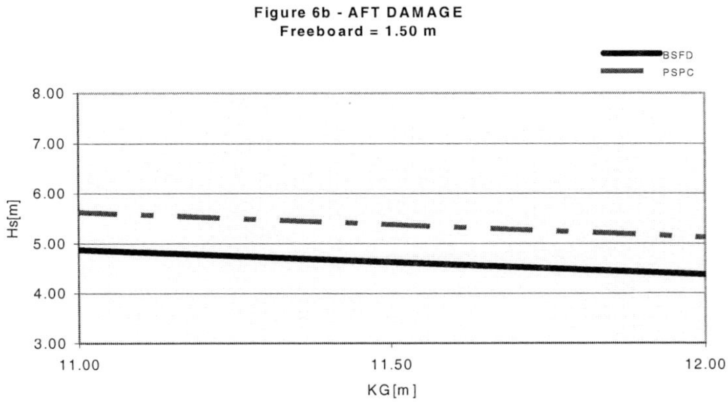

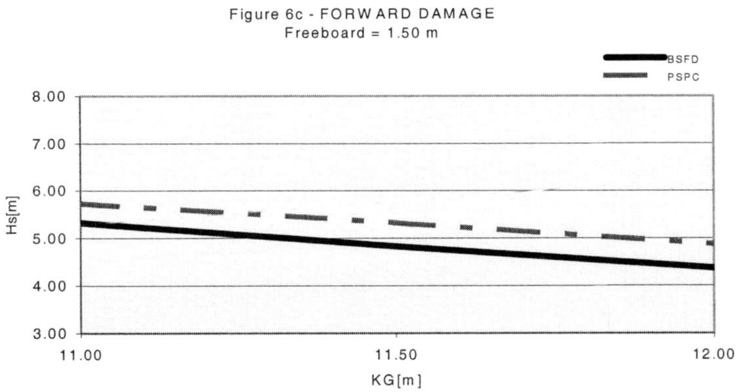

Figure 6: Comparison Between Basis Ship (BSFD) and Alternative 1 (PSPC)

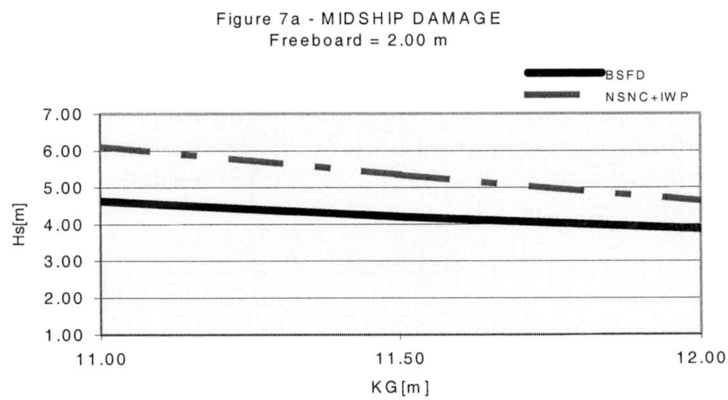

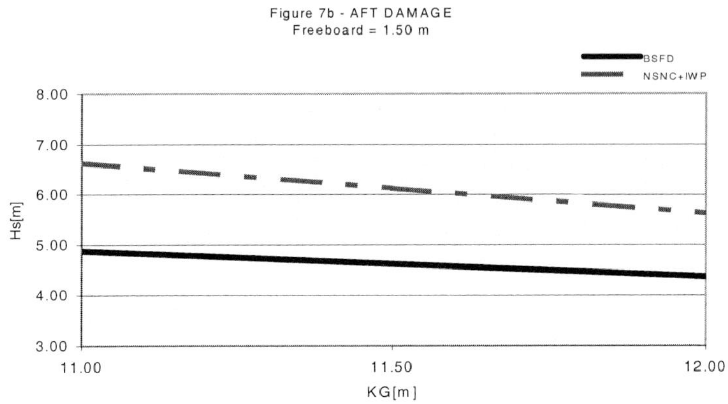

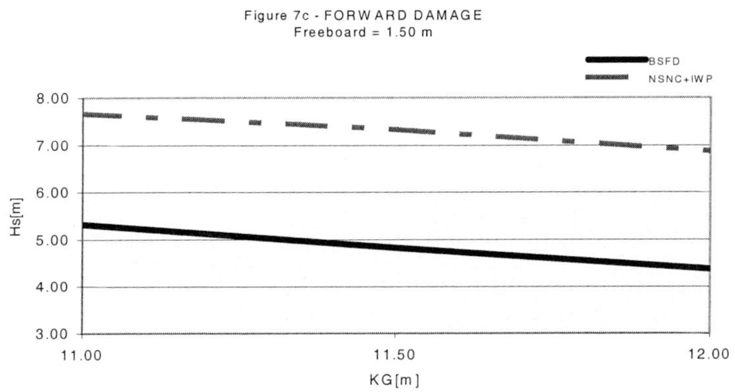

Figure 7: Comparison Between Basis Ship (BSFD) and Alternative 2 (NSNC + IWP)

CONCLUDING REMARKS

Considering that the proposed design alternatives have not been optimally applied and hence the potential improvements on Ro-Ro damage survivability could be even more pronounced renders the results achieved from the introduction of such simple ideas even more impressive. The parametric investigation undertaken in the foregoing leaves little doubt that curved decks optimally designed to resist flooding and assist outflow could help realise Ro-Ro designs which can achieve acceptably high levels of survivability when damaged whilst preserving the flexibility and operational advantages offered by the open undivided Ro-Ro decks.

The results presented in this study help demonstrate to all concerned that cost-effective ship safety cannot be achieved solely by regulations, particularly when the latter derive from lack of understanding, experience or knowledge of the problem at hand.

Give the designers a chance and they will pave the way to safer ships!

References

IMO Resolution MSC.12 (56) (Annex) (1988). Amendments to the International Convention for the Safety of Life at Sea, 1974: Chapter II-1 – Regulation 8. *Adopted on 28 October.*

IMO Resolution 14 (1995). Regional Agreements on Specific Stability Requirements for Ro-Ro Passenger Ships" – (Annex: Stability Requirements Pertaining to the Agreement). *Adopted on 29 November.*

Vassalos, D, Pawlowski, M, and Turan, O. (1996). A Theoretical Investigation on the Capsizal Resistance of Passenger Ro-Ro Vessels and Proposal of Survival Criteria. *Final Report, The Joint North West European Project, University of Strathclyde, Department of Ship and Marine Technology, March.*

Vassalos, D. (1998). Damage Survivability of Passenger/Ro-Ro Vessels by Numerical and Physical Model Testing. *WEMT'98, Rotterdam, May.*

APPENDIX A

DESCRIPTION OF DESIGN ALTERNATIVES AND DAMAGED CASES

Figure A.1: Basis Ship

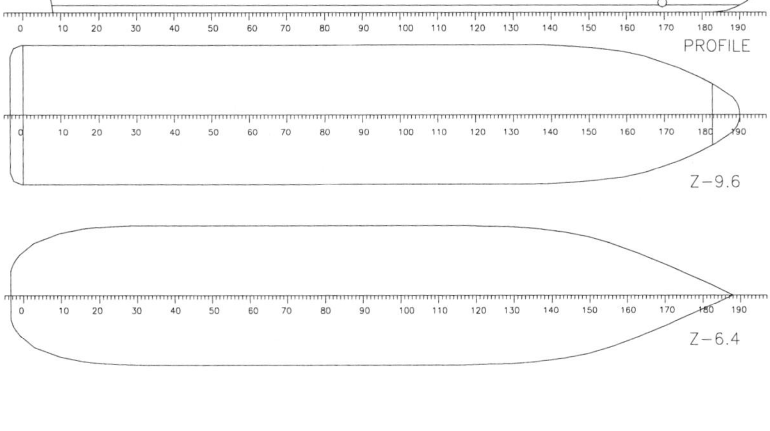

Figure A.2: Alternative 1

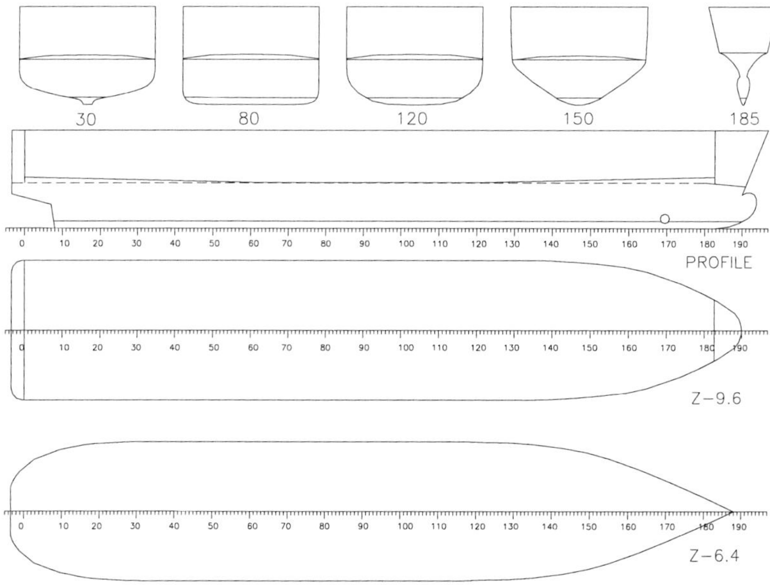

Figure A.3: Alternative 2

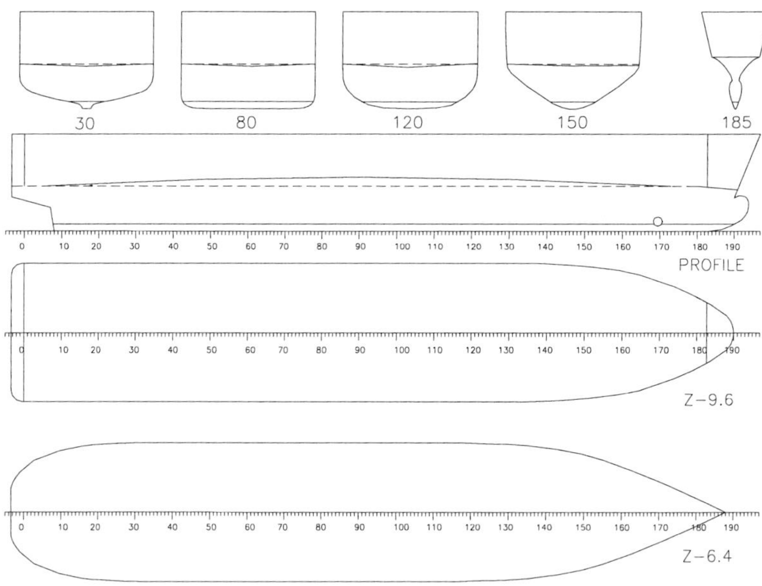

Figure A.4: Damage cases - Aft, Midship and Forward

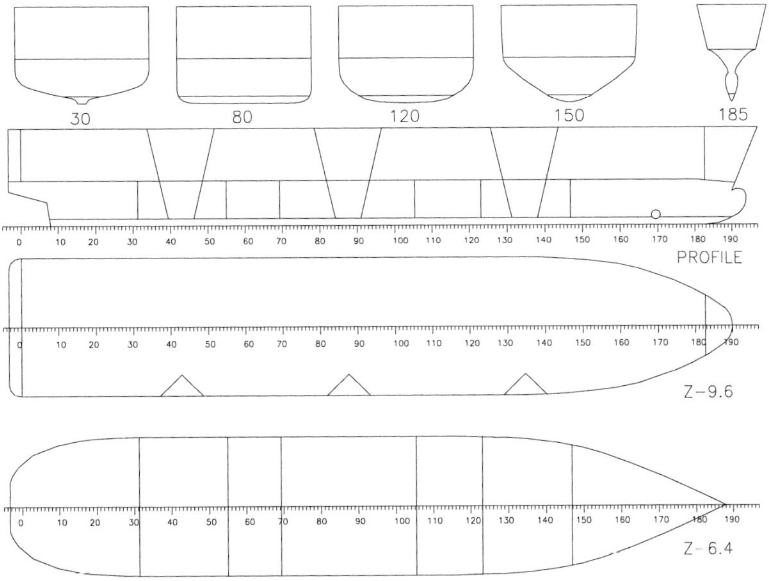

D. Vassalos, M. Hamamoto, A. Papanikolaou and D. Molyneux (Editors)

DESIGN ASPECTS OF SURVIVABILITY OF SURFACE NAVAL AND MERCHANT SHIPS

Apostolos Papanikolaou[1] and Evangelos Boulougouris[2]

[1]Professor, Ship Design Laboratory, National Technical University of Athens, Greece
[2] PhD. cand., Ship Design Laboratory, National Technical University of Athens, Greece

ABSTRACT

The paper addresses design aspects of survivability of merchant and surface naval ships through a common rational methodology, aiming at introducing a new ship design philosophy, namely *design for safety* respectively *design for enhanced survivability*. The envisaged method is based on earlier work of Kurt Wendel (1960) on the evaluation of the damage ship stability by a probabilistic concept and is currently under review by IMO for application to all types of merchant ships in the framework of harmonization of existing stability rules. Recently the outlined method found access, also, into the design process of modern naval ships, especially those characterised by their limited size and increased operational requirements. The methodology aims at supporting early designer decisions, associated with survivability, namely compartmentation and arrangements, which are taken at the preliminary design stage and are very difficult and costly to change, if at all, in latter stages. Therefore, a proper guidance in the preliminary design stage would greatly help the designer to achieve his goals. The paper addresses the fundamental aspects of survivability and introduces into the design process the new probabilistic approach for assessing the damage stability and survivability properties of both naval and merchant ships.

KEYWORDS

Survivability, damage stability, probabilistic stability method, IMCO Res. A. 265, vulnerability, lethality, susceptibility, modelling, simulation, ship design

INTRODUCTION

One might wonder what is common between the design of a naval surface combatant and a merchant ship, particularly a Ro-Ro passenger ship. The answer appears simultaneously trivial and complex: they both *have to* survive in case of damage, for obvious reasons, however under quite different design constraints, external damage threats and environmental-operational conditions. For surface combatants, although the probability of damage is very high and survivability should have been a significant factor in their design, there is, until now, a lack of systematic examination and rational assessment of their survivability at the early design stage. On the other side, following the public outcry after some recent tragic accidents in passenger shipping, particularly *Herald of Free Enterprise (1987)* and *Estonia (1994)*, the need for enhancing the inherent survivability of Ro-Ro passenger ships in case of damage by efficient design measures became obvious.

In the following, a consideration of the possible introduction of a common methodology for the assessment of the survivability of these two totally different ship types is attempted. The proposed method is a generalisation of earlier work of Kurt Wendel (1960) on a probabilistic approach to the damage stability of surface ships. This method was subsequently adopted by SOLAS 1974 through Res. A.265 (VIII), as an alternative to the deterministic SOLAS criteria for passenger ships, and later for the evaluation of the stability of new cargo ships built after 1992. The probabilistic stability approach is currently under review by relevant IMO bodies for application to all types of merchant ships in the framework of harmonisation of existing stability rules.

The Naval Ship Dimension

Modern naval warfare is characterized by highly sophisticated air, surface and underwater weapon systems. In order to accomplish their mission, surface combatants have to carry a large arsenal and a complicate suit of advanced and very expensive electronics. This led to a significant increase of their acquisition and operational cost and eventually reduced the size of the fleets, the various Navies could afford to operate. The obvious need for a high payload to displacement ratio has driven the designers and builders to a reduction of the shell plate thickness for keeping the structural weight as low as possible. This resulted to a shift *from enhanced armour to increased sensor capability*, making naval ship designs today more vulnerable than in the past. It evident that an effective response to the above stated problem is the adoption of a rather new, naval ship design philosophy, namely *Design for Enhanced Survivability.*

Nature makes its creatures adaptive to their environment for survival. In the same way, ships should be designed with an inherent ability to survive in the threat environment they have to operate. For naval ships survivability is the capability to continue to carry out their missions in the combat lethal environment. This is a function of their ability to prevent the enemy from detecting, classifying, targeting, attacking or hitting them. The *inability* to "intercept" any of the above threats is a measure of their *susceptibility*, mathematically expressed by the probability P_H. On the other hand the degree of impairment the ship suffers in case of damage characterizes her *vulnerability*, expressed by $P_{K/H}$ respectively. The product of *susceptibility* and *vulnerability* defines the *killability* P_K of the combatant. It is obvious that *in order to maximise the naval ship's survivability we have to minimise its susceptibility and vulnerability*. Mathematically the above can be expressed by the following global formula:

$$P_K = P_H \times P_{K/H} \tag{1a}$$

Thus the probability of survival S is expressed by:

$$S = 1 - P_K \tag{1b}$$

The susceptibility of a naval combatant is dependent on its *signature* characteristics. Signature reduction measures will decrease the probability of being detected and classified. These include the suppression of the Radar Cross Section (RCS), Infra-red (IR), noise, magnetic and electro-optical signature.

The vulnerability reduction measures must be addressed in the early design phase in order to maximise the results. They include arrangements, redundancy, protection, and equipment hardening as well as damage containment.

If we restrict our analysis only to conventional (high explosive) anti-surface weapons then there are two main damage effects that threat the survival of a combatant: **flooding** and **fire**. Though both are equally essential we will limit our survivability analysis herein only to flooding. The reason is, first of all, that the second aspect (fire) can be effectively performed only at advanced stages of design. Besides for any ship we have to counter a fire onboard, it is assumed that it is assumed staying afloat and upright. Mathematically this means that we consider flooding and fire as *independent* events:

$$P_K[\text{Hit} \cap (\text{Flooding} \cup \text{Fire})] = P_K[(\text{Hit} \cap \text{Flooding}) \cup (\text{Hit} \cap \text{Fire})] = \tag{2a}$$
$$= P_K[\text{Hit} \cap \text{Flooding}] + P_K[\text{Hit} \cap \text{Fire}] - P_K[\text{Hit} \cap \text{Flooding}] \times P_K[\text{Hit} \cap \text{Fire}] \tag{2b}$$

Therefore for the rest of this paper we will be referring to the survivability of naval ships by meaning just *flooding survivability* and by calculating the $P_K[\text{Hit} \cap \text{Flooding}]$.

The Passenger Ship Dimension

Since the early seventies (SOLAS 1974) the International Maritime Organisation (IMO) has adopted probabilistic methodologies for the assessment of survivability of passenger ships. We refer to the regulation A.265 (VIII) of IMO setting an equivalent to the deterministic stability criteria, namely part B of Chapter II of the International Convention for the Safety Of Life At Sea, 1960 (SOLAS 60). The vast number and complexity of the required calculations for a full probabilistic assessment of a ship under consideration was, in those days of limited computer hard- and software, a serious drawback that led to only limited applications to actual ships. This was one serious reason for the further development of the deterministic criteria (SOLAS 90 & SOLAS 95), whereas the results of the probabilistic approach (or a simplified version thereof) were only used as an indicator for the implementation of new regulatory schemes to existing ships (phase-in procedure). However, it is a taken decision of

IMO to formulate[1] and approve a "harmonised" new probabilistic stability framework for all types of ships, possibly by the SOLAS conference in the year 2000[2].

According to IMO Res. A.265, and following the fundamental concept of K. Wendel (1960), there are the following probabilities of events relevant to the ship's damage stability:

1. The probability that a ship compartment or group of compartments **i** *may be flooded (damaged),* $\mathbf{p_i}$.
2. The probability of *survival after flooding* the ship compartment or group of compartments **i** under consideration, $\mathbf{s_i}$.

The total probability of survival is expressed by the attained subdivision index **A** which is given by the sum of the products of $\mathbf{p_i}$, and $\mathbf{s_i}$ for each compartment and compartment group, **i**, along the ship's length:

$$A = \sum_i p_i \cdot s_i \tag{3}$$

The regulations require that this attained subdivision index should be greater than a required subdivision index **R**, which is determined by the number of passengers the ship is carrying and the extent of life-saving equipment onboard. This value is a measure for the ship's probability of surviving a random damage and it is obvious that this value increases with the number of passengers onboard the ship. The factors in the formula determining **R** are so selected to correspond to the mean values of the attained subdivision indexes of a sample of existing ships with *acceptable* stability characteristics. This is, of course, a point for lengthy discussions, because the safety standards of passenger ships have significantly changed over the years, therefore the basis for the evaluation of **R** must be updated to account for these changes.

Increasing the survivability of the Ro-Ro passenger ships requires the specification of the "threats" (or better the "hazards") they have to counter. Similar to naval vessels, passenger ships face mainly two major threats: Flooding and Fire. It is obvious, that for A.265 the survivability is virtually identical to the vulnerability in case of flooding. The regulation does not take into consideration either the *vulnerability* in case of fire or the *susceptibility* of the ship. However, it is formally not difficult and probably advisable to incorporate *susceptibility* into the survivability of Ro-Ro passenger ships, as it has been suggested earlier for naval ships. This concept is more or less adopted in the formulation of the "Safety Assessment", or "Formal Safety Assessment (FSA)", Spouge (1996). This way we can formulate a unified, rational scheme model (methodology) for assessing the survivability of naval surface combatants, passenger Ro-Ro ships and any other type of ship. The only difference would be the type of the anticipated risks each type of ship has to counter. For passenger ships, and

---

[1] see, SLF39, SLF 40, SLF 41, SLF 42

[2] at the time of finalising this paper the date of finalisation and approval of the harmonised stability rules by IMO was still uncertain. In view of the complexity of the subject and pending the results of parallel research work regarding the probabilistic stability concept, it is anticipated that the new harmonised rules will be ready for implementation not earlier than the end of year 2003.

merchant ships in general, we should be considering flooding due to one of the following impact events, Spouge (1996):

1. Ship to ship (collision)
2. Ship to berth/breakwater (contact)
3. Ship to bottom (grounding/stranding).
4. Explosion
5. Terrorist act
6. Material Failure
7. Human Error

The probability of a passenger ship loss in case of flooding or fire is calculated by the formula:

$$P_L\,[\text{Flooding} \cup \text{Fire}] = P_L\,[\text{Flooding}] + P_L\,[\text{Fire}] - P_L\,[\text{Flooding}] \times P_L\,[\text{Fire}] \qquad (4)$$

As has been noted above, we should herein discuss only the probability of loss in case of flooding, namely P_L [Flooding].

OUTLINE OF POSSIBLE SHIP DAMAGE CONSEQUENCES

It is obvious that between the intact condition and the total loss of a ship there are many intermediate stages. Though these stages can be defined in various ways a very common procedure is the one relating them to a *functional hierarchy*, Ball & Calvano (1994). Following this concept, a **naval ship** we may be faced, in descending order, with one of the following damage consequences:

- **Total Kill** when the ship is considered lost entirely because of sinking (foundering) or completely damaged by fire (or other incident).
- **Mobility Kill** if immobilisation or loss of controllability of the ship occurs.
- **Mission Area Kill** if a mission area (e.g. AAW capability) is considered lost for the ship.
- **Primary or Combat System Kill** in case of one or more vital systems of the ship, such as a propulsion engine or a CIWS, are damaged.
- **Hull, Machinery or Electrical (HM&E) Support System Kill** if one or more components supporting a primary/combat system of the ship are damaged (e.g. the cooling water system).

Apparently a *combat system kill* can lead to a *mission area kill* or a *mobility kill* or even a *total kill*. Likewise a *mission area kill* may decrease to a *combat system kill* after the crew makes necessary repairs. Our primary target is to confine damage consequences to the lowest possible level. Accordingly, in case of a **Ro-Ro passenger ship** we may have the following damage scenarios:

- **Loss of stability** (intact or damage ship capsizing)
- **Loss of floatability** (progressive flooding, foundering)
- **Loss of power** (loss of mobility, failure of electrical systems)
- **Loss of controllability**(failure of mechanical, electrical and/or electronic control devices)

Loss of stability and of floatability, though they might both result to the foundering of the ship, they can be addressed separately, because of the different time scales available for evacuation of the ship. With the loss of power and controllability we refer herein to the failure of the ship's main machinery and of vital equipment components, like electrical generators, etc.

SURVIVABILITY PERFORMANCE ANALYSIS FOR NAVAL SHIPS

The effort to develop naval ships of enhanced survivability imposes additional constraints to the naval ship design. In such approaches the *Survivability Performance Analysis (SPA)* shall be an indispensable design tool, to be briefly addressed in the following (see, Boulougouris, 1999).

The SPA analysis is based on the modelling of the event sequence from the enemy's arrival to ship's operational area up to the moment at which a hit might strike the vessel. Thus we have the detection, classification, target acquisition and launch of an enemy attack. The ship's response is to jam, attempt to deceive, or to destroy the enemy's incoming weapons. The probability of ship's detection is a function of the threat's sensor, its range and ship's signature. A rough estimation of the Radar Cross Section (RCS) of a surface ship can be taken from the following formula, Rains (1994) :

$$sigma = 52 \cdot \sqrt{f} \cdot \sqrt[3]{Disp^2} \tag{5}$$

where *sigma* is the ship's RCS in m^2, f is the incident radar frequency in MHz and *Disp* is ship's displacement in tons. The range of ship's detection by the enemy's radar is given by the following equation, acc. to Goddard, Kirkpatrick, Rainey & Ball (1996):

$$R_{\max} = \left[\frac{P_T G^2 \lambda^2 sigma}{(4\pi)^3 P_{\min}} \right]^{1/4} \tag{6}$$

where R_{max} is the maximum detection range, P_T the transmitter's power, $G$the antenna gain, λ is the radar's operating wavelength, *sigma* is the ship's RCS and P_{min} is the minimum detectable received signal from the enemy's sensor. Obviously the lower the RCS of the ship, the closer the enemy has to come for detecting it. An optimisation of the RCS is nowadays possible by application of STEALTH technology. P_{min} depends on the enemy's radar characteristics and also on the environmental conditions. By the later we mean temperature, sea condition as well as jamming. Increase of any of these parameters results to an increase of P_{min} and eventually decrease of the R_{max}.

Assuming that the missile is radar-guided, its course to the target will also depend on ship's RCS. The missile's path depends their Linear Error Probability (LEP). Knowing the missiles' LEP we may assume that the missile's position relatively to ship's profile follows a normal distribution with standard deviation σ, related to the LEP by the following formula,

Przemieniecki (1994):

$$\mathrm{LEP} = 0.6745\sigma. \tag{7}$$

At this phase the ship will try jamming the missile's radar. Because of its higher power, the ship's jamming device will block the missile's radar until it reaches a certain distance from its target. This distance depends on the power ratio between the radar and the jamming device. Once the missile regains a lock on the ship it depends on its aerodynamic characteristics (i.e. maximum turning acceleration and speed) whether it will turn to the vessel's direction. To be successful, the missile's minimum turning radius has to be less than its distance from the ship at that moment, namely, Rains (1994):

$$\text{Missile Radius} = \frac{V_m^2}{N \cdot g} \leq R_{regain} \tag{8}$$

where V_m is the missile's velocity, N the maximum turning acceleration of missile in [g] and g the gravitational acceleration. The range at which the missile will regain a clear picture of the ship's location is given by the formula, Rains (1994):

$$R_{regain} = \sqrt{\frac{P_M}{P_J} \cdot \frac{sigma}{4\pi}} \tag{9}$$

where P_M/P_J power ratio between the missile seeker and the jammer. Thus the effectiveness of jamming can be expressed by the integral of the normal distribution from the ship's either end to a distance R_{max}—R_{regain} towards the centre of the ship.

In order to assess the survivability of a naval ship design we have to estimate the ship's vulnerability and in particular her single hit kill probability. Therefore we have to calculate:

- the probability of "hit of a particular point of the ship".
- the probable "damages extent given a hit at that point".
- the probability of "ship's survival given the hit point and extent".

The impact of a hit can be at any point of the ship's length. The target point of the missile guidance system depends on its type, sensor type and guidance system characteristics. Likewise, the longitudinal point of impact will depend on the shape that the signature of the ship presents to the particular threat sensor. For simplicity and generality, we may assume that the impact location is described by a normal probability distribution with its centre at the ship's centre and a linear error probability (LEP) equal to $0.5 \cdot L_{WL}$.

The damage extent can be taken from a Log-Normal Damage Function. This is given by the following function, Przemieniecki (1994):

$$d(r) = 1 - \int_0^r \frac{1}{\sqrt{2\pi}\beta r} \exp\left[-\frac{\ln^2(r/a)}{2\beta^2} \right] \cdot dr \tag{10}$$

where:

$\alpha = (R_{SK}R_{SS})^{0.5}$

$\beta = \frac{1}{2\sqrt{2}z_{SS}} \ln(\frac{R_{SS}}{R_{SK}})$

R_{SK} "dead-sure kill radius"
R_{SS} "dead-sure surviving radius"

R_{SK} and R_{SS} correspond to 98% and 2% probabilities of damage respectively. Their values can be derived from empirical data for the threat missiles considered. Herein we will assume a R_{SK} diameter equal to the U.S. Navy standards, namely 15%L_{BP}, Surko (1994). The "sure-save" diameter will be taken as 0.24·L, in line with the A.265 IMO-SOLAS regulations for merchant ships. This results in a ratio of R_{SK}/R_{SS} equal to 0.625. The variation of the lognormal damage function is shown in figure 1. Alternatively a diameter depended on warhead's weight can be used.

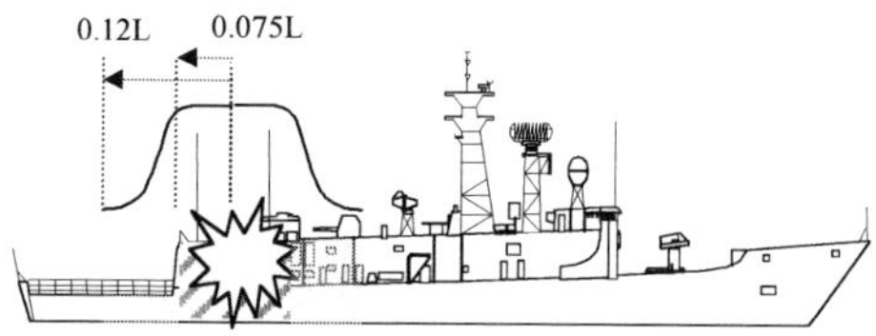

Figure 1: Variation of log-normal damage function

Having defined the first two probabilities namely hit at a particular point and damage extent given the impact point, we are left to define the probability that the ship will survive given the damage location and extent. A rational methodology for this evaluation can be based on the survival criteria of the U.S. Navy, Surko (1994).

The philosophy for transforming these deterministic criteria into a set of rational probabilistic approach criteria will be based on A.265 IMO-SOLAS regulatory concept for merchant ships.

Considering that the basis for the current deterministic U.S. Navy criteria is a significant wave height of 8 ft and aiming at requirements for a specific region of operation, characterised by a sea spectrum, we could calculate the probability that waves will not exceed the U.S. Navy criteria basis-wave height, as this wave height was used as reference for the determination of $\square_{roll}$, namely the roll amplitude due to wave action. As a first step we can propose the following guidance for the formulation of **survival criteria** to be applied in the frame of a probabilistic approach to the survivability of **naval ships** (see, table 1).

It is obvious, that some systematic experimental and theoretical work is needed in order to specify in a more rigorous way the calculation of the ***S*** value for naval ships. In any case, the calculation of the probability distributions of wave exceedence in the area of operation is necessary. For instance, the $P(H_S<8 \text{ ft})= 0.60$ holds for the North Atlantic, but $P(H_S< 8\text{ft})= 0.90$ for the Mediterranean Sea. Thus, a combatant, meeting the U.S. Navy criteria, should have, according to the criteria formulated above, a 60% probability of survival for any 2-compartment damage in the North Atlantic and a 90% probability of survival in the Mediterranean Sea.

TABLE 1

PROPOSED CRITERIA

$S = 1$	$\theta_{roll} = 25$ deg $A_1 \geq 1.4\ A_2$	Wind speed = accord. to DDS-079-1. Min Freeboard $\geq$ 3in + $0.5\times(H_S(0.95)$-8 ft)
$S = P\ (h_S$=8 ft)	The ship meets the current (deterministic) damaged stability criteria of the U.S. Navy.	
$S = 0$	$\theta_{roll} = 10$ deg. $A_1 \leq 1.05\ A_2$	Wind speed $\leq$ 11 knots Margin line immerses.

The above procedure for assessing the damage stability component of the vulnerability could form an important element of an advanced *Survivability Performance Analysis* method for naval ships, as illustrated in the following flowchart.

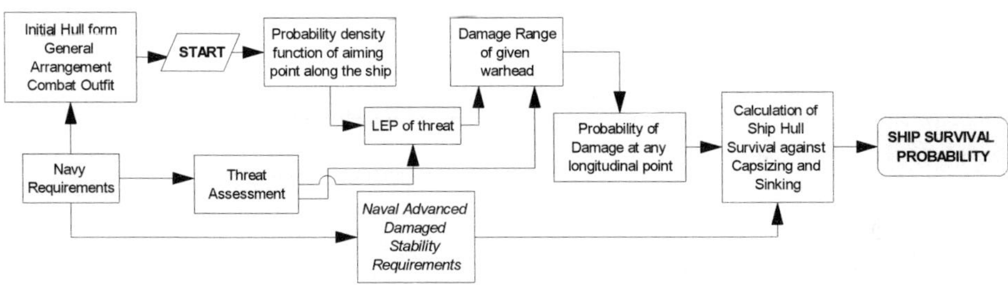

Figure 2: Survivability Performance Analysis for Naval Ships

THE RO-RO SHIP DESIGN PROCEDURE

The existing probabilistic method A.265 is a rational but complicate and non-transparent procedure for assessing the probability of survival of a ship. On the other hand, the ship specific survivability characteristics should be reflected in a clear way in its survivability index for possible design optimisation. By use of modern hardware and software computing tools, an optimisation appears today feasible, as it will be outlined in the following.

As far as the possible risks for passenger ships are concerned, it has been noted at the stage of definition of the various possible damage cases for Ro-Ro vessels that multiple stages of damage, prior to the total loss of ship, can be considered. The probability of loss of controllability, or of power (black out) should be also taken into account. Piping transferring cold water for cooling vital machinery or electrical lines passing through the damaged compartment should be identified and properly included in the calculations. Additionally, the time required to evacuate the ship must be taken into account. This would encourage ship designers to consider arrangements and efficient equipment for the safe evacuation of passengers and crew, beyond the least requirements set by the regulations. Recent work by D. Vassalos et al. (1998) stresses the need for properly assessing the available evacuation time in case of damage and prior to capsizing or foundering. Existing regulations specify the maximum time available for evacuation but no special care is given to consider the ship damage conditions that might affect this time. Large heel or trim may increase the evacuation time significantly especially for children, elderly or persons with mobility problems. Thus damage cases with large heel or significant trim have to receive reduced weights even though the stability requirements are met.

The consideration of the particular significance of the various ship spaces can be incorporated in the regulations by assigning them so-called *"vitality"* coefficients as it has been already done with the variation of space permeability. Thus spaces with vital machinery or equipment will receive more weight in the survivability analysis. This will account for the fact that even though the ship stays afloat, in case of damage, the associated risks are greater compared to the same post-damage condition after damage of non-vital spaces. The Attained Subdivision Coefficient formula should be therefore modified as following:

$$A = \sum_i p_i \cdot s_i \cdot w_i \tag{11}$$

where w_i is a properly defined vitality coefficient for compartment i. The vitality coefficients ($w_i \leq 1.0$) of the various compartments should be in descending order to their significance. As defined above, vitality of course depends on the existence of redundant equipment. Also, if for a damage case involving multiple compartments, both the primary and the emergency equipment are contained in the damage extent, the vitality coefficient has also to be reduced. The product $s_i \cdot w_i$ will therefore express the reduced probability of ship's survival, including functionality, given damage to the specific compartment. The above procedure calls for the introduction of innovative machinery and equipment arrangement concepts (multiple machinery spaces, redundancy of equipment) as has been indicated in recently published work by M. Kanerva (1997). Specific values cannot be proposed in this paper because more work has to be done in this area. It should involve both damage statistics and actual machinery arrangements of existing ships. The above revision of the Attained Subdivision Coefficient formula calling for special consideration of vital spaces can be incorporated in a more general Ro-Ro ship design optimisation scheme, as outlined in the attached chart (Fig. 3). The procedure considers the optimisation with respect to the Local Subdivision Index, as proposed earlier by Sen & Gerigk (1992) and properly modified, as given above to account for the vitality of spaces.

CONCLUSIONS

The paper addressed the survivability of naval and merchant ships (here: Ro-Ro passenger

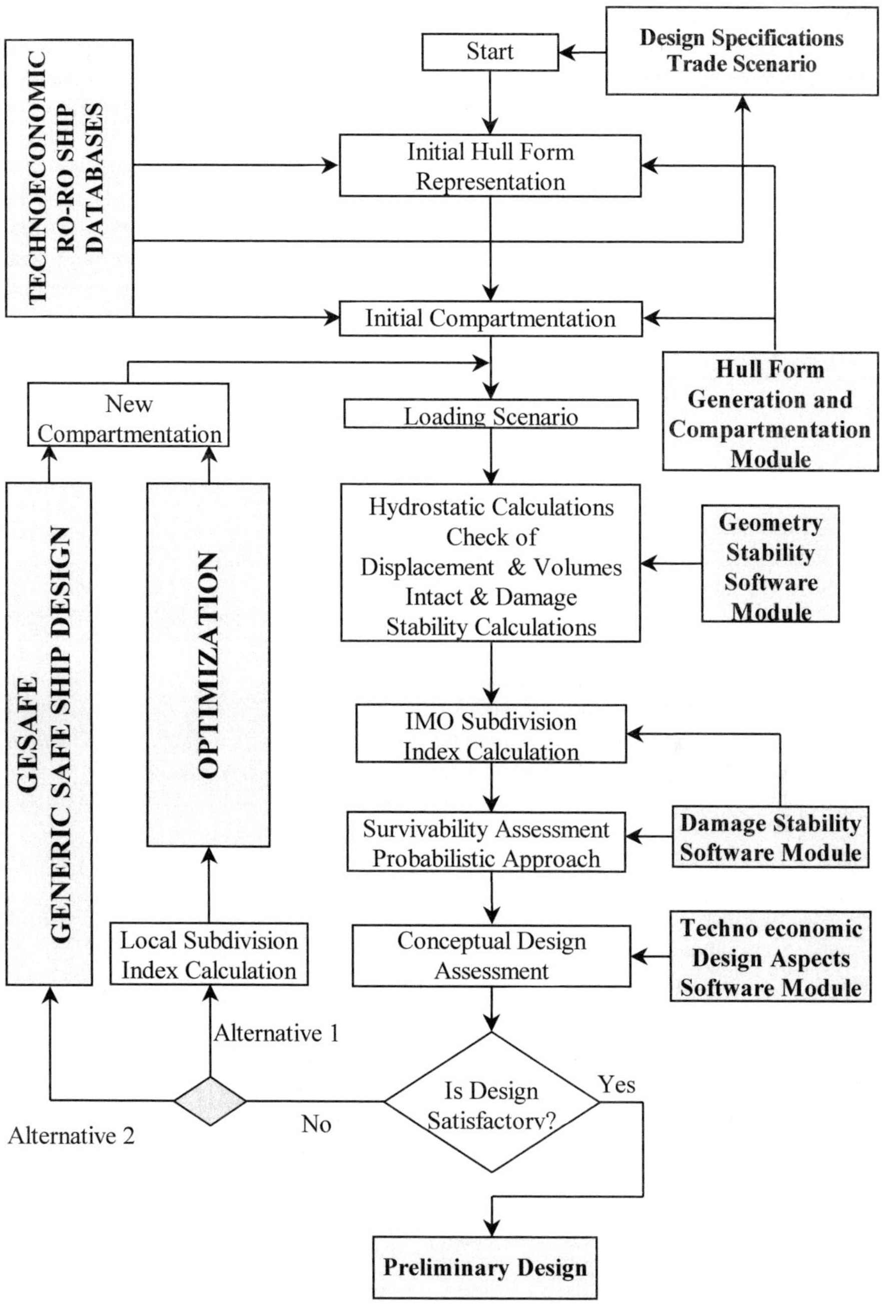

Figure 3: Computer-Aided Safe-Ferry Design Procedure (SAFER-EURORO)

ships) through a common ("harmonized") rational probabilistic procedure. Possible risks for naval and passenger ships have been outlined. Special attention has been paid to the formulation of survival criteria for naval ships, based on the probabilistic stability concept. Following general design concepts of naval ships, a modification of the formula of the Attained Subdivision Index of passenger ships has been proposed accounting for the vitality of specific ship compartments, securing the powering, electric supply and controllability of the ship after damage. The increased importance of modelling and simulation methods in modern ship design requires the development of more rational concepts. Though the above methodology is, at this stage, not complete, it is suggested as a first step to a new design direction. It is of course obvious that much work is still needed to test and to fully verify the proposed procedures. A framework for discussion of the suggested design procedures for Ro-Ro passenger ships is the Thematic Network SAFER-EURORO, Vassalos et. al. (1997), supported by DG XII of the European Commission. Their financial support is herein acknowledged.

References

Ball R. E. & Calvano C. N. (1994). Establishing the Fundamentals of Surface Ship Survivability Design Discipline. *Naval Engineers Journal*, **106:1**, 71-74.

Boulougouris, E. (1999). *Ship Design Optimisation regarding the Survivability after Damage for Ro-Ro Passenger and Naval Ships*. PhD Thesis, Ship Design Laboratory, National Technical University of Athens, *in progress*.

Goddard C. H., Kirkpatrick D. G., Rainey P. G. and Ball J. E. (1996). How much STEALTH?. *Naval Engineers Journal*, **108:3**, 105-116.

Kanerva M. (1997). *Fundamental Rethinking of Passengership Design for Economic Operation by Owners and Construction by Shipyards*. Cruise & Ferry 97, London.

Przemieniecki J. S. (1994). *Mathematical Methods in Defence Analyses*. AIAA, Washington, DC.

Rains D. A. (1994). Methods For Ship Military Effectiveness Analysis. *Naval Engineers Journal*, **106:2,** 126-135.

Sen P. & Gerigk M. (1992). Some Aspects of a Knowledge-Based Expert System for Preliminary Ship Subdivision Design for Safety. *Proc. PRADS'92,* **Vol. 2**, 1187-97.

Spouge J. (1996). Safety assessment of passenger/Ro-Ro vessels. *Int. Seminar n the Safety of Passenger Ro-Ro Vessels*, RINA, London.

Surko S. W. (1994). An Assessment of Current Warship Damaged Stability Criteria. *Naval Engineers Journal*, **106:3**, 120-131.

Vassalos D. et al. (1997), *The Thematic Network SAFER-EURORO, Ro-Ro Design for Safety*. European Commission Thematic Network, DG XII, 1996-2000.

Vassalos D. et al. (1998). *Time Based Survival Criteria for Ro-Ro Vessels*. The Royal Institution of Naval Architects, Spring Meetings.

Wendel K. (1960). Die Wahrscheinlichkeit des Ueberstehens von Verletzungen. *Journal of Ship Technology Research - Schiffstechnik*, **7:36**, 47-61.

Contemporary Ideas on Ship Stability
D. Vassalos, M. Hamamoto, A. Papanikolaou and D. Molyneux (Editors)

A TECHNIQUE FOR ASSESSING THE DYNAMIC STABILITY AND CAPSIZE RESISTANCE OF SHIPS

Martin Renilson[1]

[1]Australian Maritime College, PO Box 986, Launceston, Australia, 7250.

ABSTRACT

Although a number of organisations have developed mathematical models for predicting the extreme behaviour of vessels leading to capsize in severe sea states, a systematic approach to applying these models to determine the required stability against capsize has not yet been agreed.

Three different approaches are possible: running the models for long periods in a range of different irregular waves spectra; running the models in predetermined deterministic wave sequences; and running the models in regular waves. Of course each of the methods has to be applied to a range of ship speeds and headings.

A method is proposed based on regular waves which allows straightforward comparison between different vessels, and with the existing IMO regulations.

The capsize boundaries obtained are distilled into a single plot of limiting KG against Froude Number. It is shown that the maximum KG which will allow the vessel to meet the IMO regulations can also be included on this plot.

KEYWORDS

Ship capsize; Dynamic stability; Stability assessment; Severe seas; Intact ship stability; Following seas; Limiting KG; Mathematical model.

INTRODUCTION

Existing IMO regulations for the intact stability of vessels are based purely on statical concepts utilizing the G_FZ curve and the G_FM value as follows:

Criterion	IMO minimum
Area under the G_FZ curve between 0° and 30°	3.15m-deg
Area under the G_FZ curve between 0° and 40° (or 0° and θf, whichever is less)	5.16m-deg
Area under the G_FZ curve between 30° and 40° (or 30° and θ_f, whichever is less)	1.72m-deg
G_FZ at 30° or greater	0.20m
Angle for maximum G_FZ	25degrees
G_FM	0.15m (0.35m for fishing boats)

Although these criteria do take the non-linear restoring moment into account they are based on statical stability alone and do not consider the dynamics of the behaviour of the vessel in the waves. IMO has adopted criteria to take into account the effect of the wind etc, however all existing criteria are still based on statical rather than dynamical concepts. (IMO 1995)

With the recent advances in knowledge of the behaviour of ships in rough seas and the advent of modern computing the time has come to develop new criteria based fully on dynamical concepts.

MATHEMATICAL MODEL

A complete simulation of a vessel operating in a following sea is a very complex problem in comparison to a vessel operating in calm water. The vessel is subjected to two different hydrodynamic forces simultaneously, namely lifting forces in the horizontal plane acting on the vessel due to plane manoeuvring motion and the forces acting on the vessel arising from the incident wave. In addition, all these forces and moments, as well as the vessel's transverse stability, are functions of the vessel's position in the wave. Using the quasi-static assumption the vessel's heave and pitch can be obtained as functions of its longitudinal position in the wave. Hence the dynamic part of the model reduces to the following four degrees of freedom. The dynamic equations are based on Newton's 2nd law of motion:

$$X = m(\dot{u} - v\dot{\psi}) = X_{wave} + X_{hull} + X_{rudder} + X_{propeller} + X_{wind} \quad (1)$$

$$Y = m(\dot{v} + u\dot{\psi}) = Y_{wave} + Y_{hull} + Y_{rudder} + Y_{propeller} + Y_{wind} \quad (2)$$

$$N = I_{zz}\ddot{\psi} = N_{wave} + N_{hull} + N_{rudder} + N_{propeller} + N_{wind} \quad (3)$$

$$K = I_{xx}\ddot{\phi} = K_{wave} + K_{hull} + K_{rudder} + K_{wind} \quad (4)$$

X_{hull}, Y_{hull}, N_{hull} and K_{hull} are the hull hydrodynamic manoeuvring forces and moments, and can be obtained using empirical, theoretical and experimental methods, Renilson & Tuite (1998).

The wave induced force and moment coefficients, X_{wave}, Y_{wave}, N_{wave} and K_{wave} are based on slender body theory with some simplifications. The wave induced surge force, X_{wave}, is calculated using the Froude-Krylov component of the force. The wave induced sway force, Y_{wave}, yaw moment, N_{wave}, and roll moment, K_{wave}, are calculated using both the Froude-Krylov and diffraction components, Umeda & Renilson, (1994).

The moment induced by the rudder has a significant influence on the behaviour of vessels in following and quartering seas and its modelling requires considerable detail. The rudder forces and moments, X_{rudder}, Y_{rudder}, N_{rudder} and K_{rudder}, are calculated by estimating the actual velocity of water flow acting over the surface of the rudder. The forces within the model are modified to take into account the change in fluid flow over the surface of the rudder due to the presence of the propeller and the wave.

The mathematical model also takes into account the change in GZ relative to the longitudinal position in the wave. This is done hydrostatically and has been validated using semi captive experiments, Tuite & Renilson (1999).

ASSESSMENT OF RESISTANCE TO CAPSIZE

Although a number of mathematical models for predicting the extreme behaviour of vessels leading to capsize in severe sea states have been developed, a systematic approach to applying these models to determine the required stability against capsize has not yet been agreed.

Three different approaches are possible: running the models for long periods in a range of different irregular waves spectra; running the models in predetermined deterministic wave sequences; and running the models in regular waves. Of course each of the methods has to be applied to a range of ship speeds and headings.

Although the use of irregular waves would appear to be the most accurate technique for determining whether a vessel has sufficient resistance against capsize, this approach has three disadvantages which make it impractical as a routine technique. First, capsize is a very rare event, meaning that extremely long run times are necessary to achieve reliable statistics. Second, because of the sensitivity of wave frequencies as well as wave heights to the likelihood of capsize, these long runs must be conducted in a range of different spectra for each speed and heading. Finally, there is the technical difficulty of being confident that the mathematical model, which is usually based on linear superposition, can actually be applied to non-linear motions in irregular waves.

The next approach of using a number of predetermined severe wave trains has a lot of merit as proposed by Tikka and Paulling (1990). Here, however, the difficulty lies in deciding what the predetermined severe wave trains should be. Various proposals have been suggested, see for example Clauss, (2000). Again, the sensitivity of vessel capsize to wave frequencies means this approach has to be adopted with extreme care. It may well be possible to select various standard wave trains which would artificially favour an undesirable vessel feature, and possibly miss a capsize mechanism altogether.

Finally, the approach using regular waves has the advantage of being deterministic, meaning that as the method is not based on a statistical approach each run can be much shorter. On the other hand it is necessary to run the model over a complete range of values of λ/L; h/λ; χ; KG; and Fn.

PROPOSED METHOD

A method is proposed for assessing a vessel's resistance against capsize using regular waves in a manner which allows ready comparison with other vessels and with the existing IMO regulations. Results for regular waves are used as being 'deterministic' they are easier to obtain than those in irregular waves, allowing for ready comparison between different vessels. It has often been demonstrated that in following and quartering seas the results are very similar for regular and irregular waves. See for example Takaishi, (1982).

With regular waves the run lengths can be relatively short as the technique is deterministic and does not depend on a statistical approach. It is also comparatively easy to investigate all combinations of λ/L; h/λ; χ; KG; and Fn.

To simplify the process it is desirable to conduct all the investigations at a constant wave steepness. In order to investigate the relative safety of a vessel it is reasonable to investigate its behaviour in steep waves. A number of existing investigations have utilised different wave steepnesses, making it impossible to compare the results. To overcome this it is proposed that a standard wave steepness of $h/\lambda = 1/15$ be used for all studies of capsizability in future. If this value is adopted as a standard by all investigators in the field then it will be possible to compare different designs on an equal basis. This is necessary to be able to develop standard stability regulations based on rational dynamical concepts.

The first step in the proposed method is to determine the maximum KG value that the vessel can have without capsizing for a fixed λ/L value and Froude Number at the wave steepness of $h/\lambda = 1/15$. This process is repeated for a range of heading angles and the resulting capsize boundary can be represented on a plot of heading angle, χ, against KG as shown for a fictitious vessel in figure 1.

For a given heading angle, if the vessel is loaded with a KG value higher than the boundary shown in figure 1 it will capsize in waves of steepness $h/\lambda = 1/15$ or above, whereas if it has a KG value lower than the boundary it is safe in waves with this steepness.

As it is clearly necessary to load the vessel in such a way that it does not capsize at any heading, the important aspect of this boundary is its minimum value, and this can be considered the limiting KG value for the vessel in this λ/L and at this Froude Number.

As can be seen from figure 1, the limiting value of KG for this case is 0.5m. If the vessel has a greater KG value than this it will capsize at headings other than 20°.

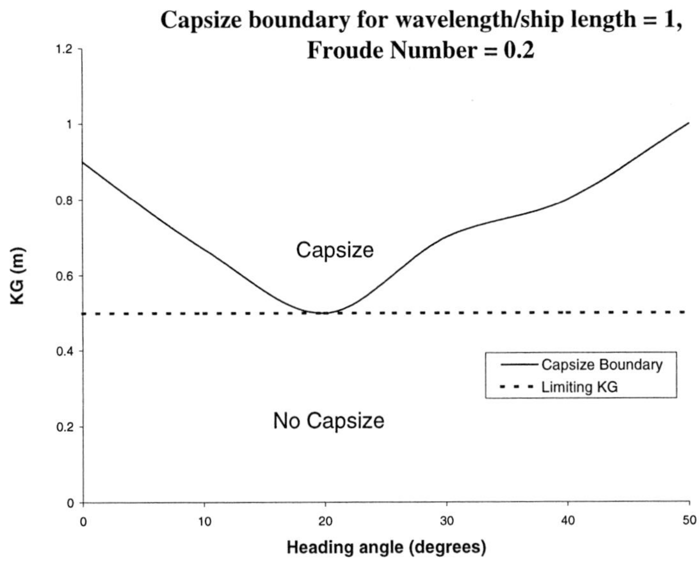

Figure 1: Example of capsize boundary for single λ/L and Fn

Of course, as it is possible to encounter waves with different lengths this exercise must be repeated for a range of wavelengths and in each case the limiting KG value can be obtained as before. These limiting KG values can then be plotted against λ/L as shown in figure 2.

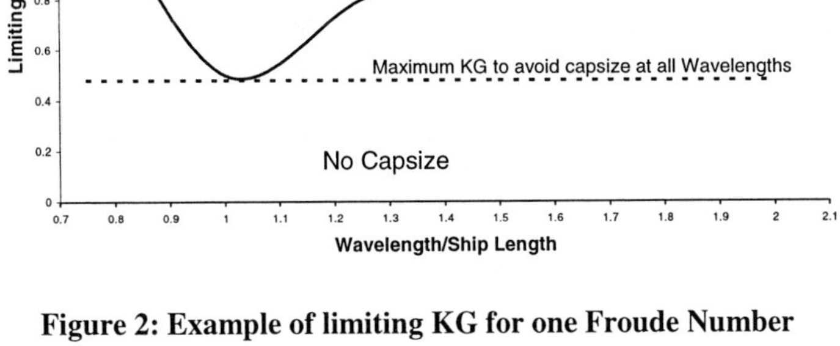

Figure 2: Example of limiting KG for one Froude Number

In figure 2 the capsize boundary indicates the maximum KG which the vessel can be loaded to without capsizing in waves of steepness $h/\lambda = 1/15$ at any heading at that one Froude number. The wavelength/ship length ratio most likely to cause a capsize in waves of steepness $h/\lambda = 1/15$ can easily be determined from this figure. In the case of this fictitious vessel this is 1.03, and at that wavelength/ship length ratio the maximum KG value that the vessel can have without capsizing is 0.48m.

As the ship may encounter waves of any length this means that it is necessary for the KG to be less than 0.48m to guarantee against capsize. This then is the maximum permissible KG for this vessel to travel in following and quartering seas at a Froude number of 0.2.

This can be non-dimensionalised by dividing by the vessel's beam, B, to give KG′ (= KG/B).

This process can then be repeated for the range of speeds, as shown in figure 3. Here the curve is the boundary of the maximum KG′ that the vessel can sustain without capsizing at any heading and any wavelength combination in waves of steepness $h/\lambda = 1/15$. What this means is that if the KG′ value is greater than this for a particular Froude number then there will be at least one combination of heading and wavelength which will cause the vessel to capsize in waves of steepness $h/\lambda = 1/15$. Conversely, if the KG′ value is less than this for a particular Froude number then it will not capsize at any heading in any wavelength in waves of steepness $h/\lambda = 1/15$. This value of KG′ can then be termed the limiting KG′ to prevent capsize at that Froude number.

The maximum KG′ permitted for this fictitious vessel to meet the IMO stability regulations is also shown on figure 3, and it can be seen that at the lower Froude numbers the IMO regulations may be considered too conservative, whereas at the higher Froude numbers they allow a higher KG′ than that permitted by the dynamic approach.

As the dynamic approach is based on the behaviour of the vessel in a wave steepness of $h/\lambda = 1/15$, which was chosen relatively arbitrarily as a 'typical' steep wave, considerable experience will be required with this comparison before it will be possible to discard the statical approach which has been built up over many years. However, if the proposed approach is adopted as uniform by researchers in this field, and the results regularly compared with those from the statical approach then it will be possible to develop the correlation between the approaches. Once this has been done, the dynamical approach can be adopted in favour of the statical approach, and the assessment of vessel stability will be greatly improved.

The final plot, shown in figure 3, is therefore a very good guide to the likelihood of a given vessel capsizing in following and quartering seas, and can reasonably be used to compare the stability of different vessels in a standard manner.

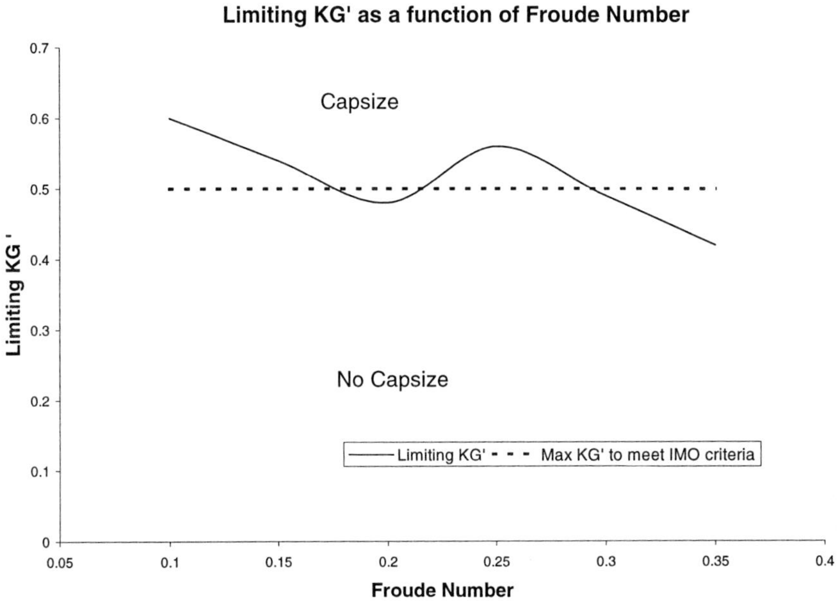

Figure 3: Limiting KG′ as a function of Froude Number

CONCLUDING REMARKS

A simple but comprehensive method for assessing a vessel's stability against capsize has been demonstrated using regular waves.

With this method a vast array of data is distilled into a single plot of limiting KG′ as a function of Froude Number, allowing for straightforward comparison between different vessels, and enabling the designer and operator to easily assess the likelihood of capsize for a given vessel. It is also possible to include the maximum KG which will allow the vessel to meet the IMO criteria on this plot for reference.

It is suggested that this method now be used as standard to compare the likelihood of capsize for different vessel types, and that the results be compared with the maximum KG obtained from the existing IMO regulations in each case.

NOTATION

B	Vessel beam
Fn	Froude number
G_FM	Metacentric height
G_FZ	Righting lever
h	Wave height

I	moment of inertia
K	roll moment
K_{hull}	roll moment induced by the hull
K_{rudder}	roll moment induced by the rudder
K_{wind}	roll moment induced by the wind
K_{wave}	roll moment induced by the waves
KG	Vertical centre of gravity position
KG′	Non-dimensional vertical centre of gravity position (=KG/B)
L	Vessel length
m	mass of ship (physical mass + added mass)
N	yaw moment
N_{hull}	yaw moment induced by hull
$N_{propeller}$	yaw moment induced by propeller
N_{rudder}	yaw moment induced by rudder
N_{wave}	yaw moment induced by waves
N_{wind}	yaw moment induced by wind
u	velocity in the x-direction
$\dot{u}$	acceleration in the x-direction
v	velocity in the y-direction
$\dot{v}$	acceleration in the y-direction
X	surge force
X_{hull}	surge force induced by the hull
$X_{propeller}$	surge force induced by the propeller
X_{rudder}	surge force induced by the rudder
X_{wave}	surge force induced by waves
X_{wind}	surge force induced by the wind
Y	sway force
Y_{hull}	sway force induced by hull
$Y_{propeller}$	sway force induced by propeller
Y_{rudder}	sway force induced by rudder
Y_{wave}	sway force induced by waves
Y_{wind}	sway force induced by wind
θ	Heel angle
θ_f	Angle of downflooding
λ	Wavelength
χ	Vessel heading to waves ($\chi = 0°$ represents following seas)
$\ddot{\phi}$	roll acceleration

REFERENCES

Clauss, G., (2000). Tailor-made Transient Wave Groups for Capsizing Tests, *Seventh International Conference on Stability of Ships and Ocean Vehicles, Launceston, Australia.*

IMO, (1995). *Code on intact stability for all types of ships covered by IMO instruments,* International Maritime Organisation.

Renilson., M. R. and Tuite, A., (1998). Broaching-To: A Proposed Definition and Analysis Method, *Twenty-Fifth American Towing Tank Conference, Iowa City, USA.*

Takaishi, Y. (1982). Consideration on the Dangerous Situations Leading to Capsize of Ships in Waves *Second International Conference on Stability of Ships and Ocean Vehicles, Tokyo, Japan.*

Tikka, K., K., and Paulling, J., R., (1990). Prediction of Critical Wave Conditions For Extreme Vessel Response in Random Seas. *Fourth International Conference on Stability of Ships and Ocean Vehicles, Naples, Italy.*

Tuite, A. and Renilson, M.R. (1999). The Controllability of a Small Vessel Operating in Severe Following Seas. *International Shipbuilding Progress,* **46:446,** pp 129-140.

Umeda N. and Renilson M.R. (1994), *Broaching of a Fishing Vessel in Following and Quartering Seas - Nonlinear Dynamical System Approach.* Fifth International Conference on Stability of Ships and Ocean Vehicles, Melbourne.

ACKNOWLEDGMENTS

The work reported was part funded by the Australian Research Council. The author would also like to thank his colleagues, including in particular Dr Andrew Tuite, for their assistance.

Contemporary Ideas on Ship Stability
D. Vassalos, M. Hamamoto, A. Papanikolaou and D. Molyneux (Editors)

PROBABILITY TO ENCOUNTER HIGH RUN OF WAVES IN THE DANGEROUS ZONE SHOWN ON THE OPERATIONAL GUIDANCE/IMO FOR FOLLOWING/QUARTERING SEA

Y. Takaishi [1], K.Watanabe [2] and K. Masuda [1]

[1] College of Science and Technology, Nihon University
7-24-1 Narashinodai, Funabashi, Chiba 274-8501, JAPAN
[2] Yachiyo Engineering Co., Ltd.
1-10-23 Nakameguro, Meguro, Tokyo 153-8639, JAPAN

ABSTRACT

The time history of encounter waves differs from the wave record at a fixed point, i.e. the groupiness of encounter waves varies according to ship course and speed. The groupiness of ocean waves itself had been estimated statistically by using characteristic values of wave spectrum. This method has been applied to the encounter waves, and the variation of the groupiness has been determined as a function of V/T, i.e. ship speed/wave period, in following and quartering seas. The probability to encounter high run is shown as functions of V/T and number of runs. It is concluded that the groupiness or probability to encounter high run becomes higher in a range of V/T between 1.0 and 2.0 (kt/sec). The danger of encountering such high runs is confirmed by model test results on capsizing.

KEYWORDS

Encounter wave, Following sea, High run, Operational guidance, Probability, Short-crested sea, Wave grouping

INTRODUCTION

One of the dangerous situations for ships navigating in following and quartering seas is to encounter occasionally a group of high waves, i.e. high run. The present author had been explained this phenomena by introducing a concept of the wave energy concentration ratio which is determined on the encounter wave spectrum, and he has proposed the dangerous zone to encounter high run on the V/T-diagram, Takaishi (1994). This idea is now included in the "Guidance to the Master for Avoiding Dangerous Situations in Following and Quartering Seas", which recently has been adopted as the MSC Circular707/IMO (1995).
However, this V/T-diagram gives no quantitative information on the degree of encountering high run. Then, the authors have tried to derive the statistical properties of this encounter wave grouping phenomena in following and quartering seas to quantify the probability of occurrence of high runs in wave group, applying the Longuet-Higgins' method that the groupiness of ocean wave itself depends on the characteristic values of energy spectrum. A band-width parameter, which is calculated for the encounter wave spectrum transformed from the ocean wave spectrum, is used to estimate the probability of wave group length or high run to be appeared in encounter waves of a ship.
The results have been confirmed by the statistical analysis of simulated time series of encounter waves that the probability of occurrence of high runs in following sea becomes higher in the range of $V/T = 1.0 \sim 2.0$ with the highest value at $V/T \cong 1.45$. The influences of directional distribution of wave energy, i.e. the effects of short-crestedness of ocean waves have been also investigated and it has been shown that the increment of encounter wave groupiness is less, as the directional distribution of wave energy becomes wider.
The dangers of encountering such high runs have been demonstrated on the V/T-diagram by using experimental results on capsizing of ship models.

STATISTICAL ANALYSIS OF WAVE GROUP

Longuet-Higgins' Method

The run is the number of successive wave train in a group, all of which exceed a level ρ, as shown in Figure 1. According to Longuet-Higgins (1984), the mean length of wave group, $\bar{l}$, the mean values of wave number and run in a group, $\overline{G}$ and $\overline{H}$, are represented by the following formulae as the functions of the band-width parameter, ν, of the wave spectrum.

$$\bar{l} = \{\mu_2 / (2\pi)\}^{-1/2} (\mu_0 / \rho) e^{\rho^2/(2\mu_0)} \tag{1}$$

$$\overline{G} = (2\pi)^{-1/2} [(1+\nu^2)^{1/2} / \nu] (\mu_0^{1/2} / \rho) e^{\rho^2/(2\mu_0)} \tag{2}$$

$$\overline{H} = (2\pi)^{-1/2} [(1+\nu^2)^{1/2} / \nu] \mu_0^{1/2} / \rho \tag{3}$$

where ν is derived from the n-th moment of the spectrum, as follows,

$$\nu = \sqrt{m_0 m_2 - m_1^2} / m_1 \quad (4)$$

$$\mu_0 = m_0 , \quad \mu_1 = 0 , \quad \mu_2 = m_2 - m_1^2 / m_0 \quad (5)$$

provided that the components of the spectrum in higher frequency range than $1.5f_p$ as well as the lower frequency range than $0.5f_p$ are ignored, where f_p is peak frequency of the spectrum.
The probability density functions of l, G and H are represented by an exponential function, respectively, as follows.

$$p(l) = \bar{l}^{-1} e^{-l/\bar{l}} \quad (6)$$

$$p(G) = \bar{G}^{-1} e^{-G/\bar{G}} \quad (7)$$

$$p(H) = \bar{H}^{-1} e^{-H/\bar{H}} \quad (8)$$

The probability that the run exceeds H is calculated by equation (9).

$$P(H) = \int_H^\infty p(H) dH = e^{-H/\bar{H}} \quad (9)$$

or

$$\ln[P(H)] = -H / H \cong -\sqrt{2\pi / m_0}\, \rho \nu H$$

Namely, the probability, $P(H)$, takes the same value for the same $H/\bar{H}$ value, or ρ. It means that the increased number of run and the increased level of wave height of run have the same effect on the probability.

Application to Encounter Wave

The encounter wave spectrum is obtained by transformation of wave spectrum in wave frequency domain ω into encounter wave frequency domain ω_e, i.e.

$$S(\omega_e, \chi) = S(\omega) / |1 - 2\omega V \cos\chi / g| \quad (10)$$

where V is ship speed, χ is encounter angle, and

$$\omega_e = \omega - \omega^2 V \cos\chi / g \quad (11)$$

The Longuet-Higgins' method is applied to the encounter wave spectrum. The spectrum of Pierson-Moskowitz type is used for calculation of long-crested sea, as shown by the non-dimensional equation (12) and Figure 2.

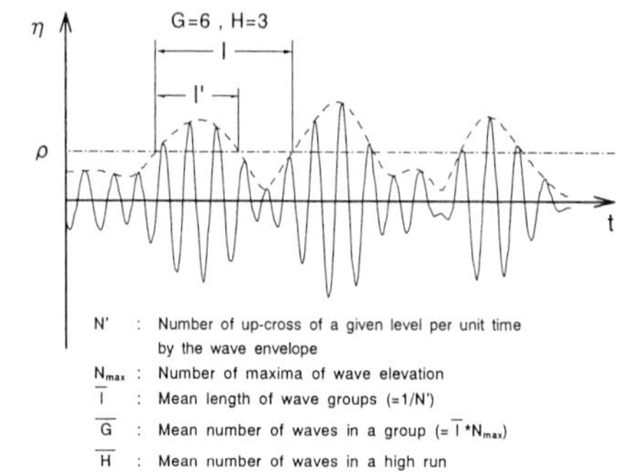

Figure 1 Definition of wave group and high run

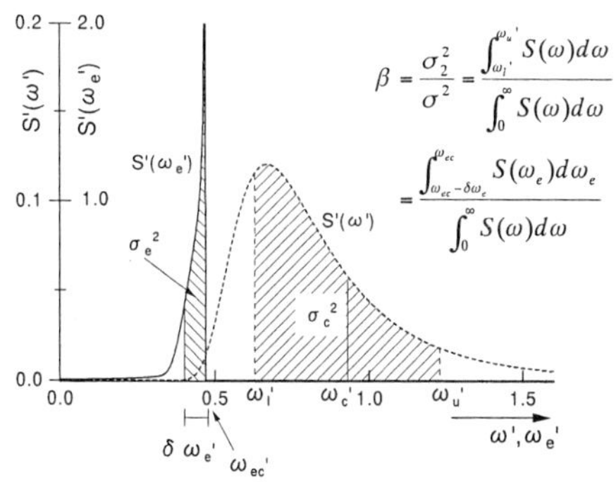

Figure 2 Wave spectrum(P-M type) , encounter wave spectrum and definition of wave energy concentration ratio, β

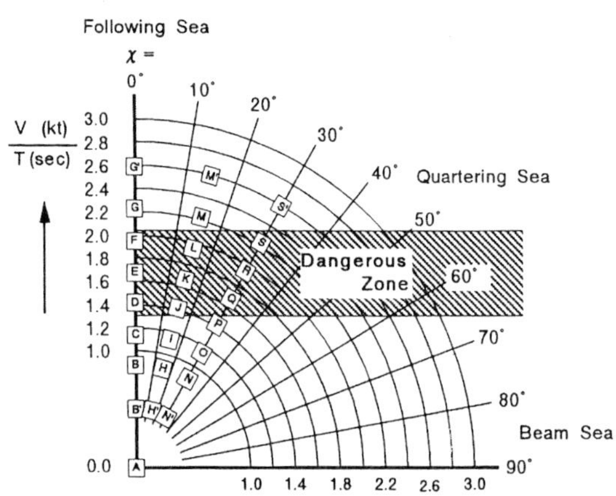

Figure 3 Encounter conditions on V/T-Diagram

$$S'(\omega') = \frac{2\pi S(\omega)}{T_{01} H_{1/3}^2} = 0.11\omega'^{-5} \exp\{-0.44\omega'^{-4}\} \tag{12}$$

where $\omega' = \omega / \omega_0$, $\omega_0 = 2\pi / T_{01}$, T_{01} mean wave period, $H_{1/3}$ the significant wave height. The short-crested wave spectrum is shown by equation (13),

$$S(\omega, \theta_j) = S(\omega) \cdot D(\omega, \theta_j) \tag{13}$$

with the energy distribution function $D(\omega, \theta)$, as equation (14).

$$D(\theta) = \frac{1}{\sqrt{\pi}} \frac{\Gamma(1+n/2)}{\Gamma(1/2+n/2)} \cos^n \theta \tag{14}$$

The encounter wave spectrum in short-crested waves is represented by summing the component wave spectra in various directions, as equation (15),

$$S(\omega_e, \chi) = \sum_{j=1}^{J} S(\omega_{ej}, \chi) = \sum_{j=1}^{J} \frac{S(\omega, \theta_j)\Delta\theta_j}{\left|1 - 2\omega V \cos(\chi - \theta_j)/g\right|} \tag{15}$$

where

$$\omega_{ej} = \omega - \omega^2 V \cos(\chi - \theta_j)/g \tag{16}$$

The band-width parameters of encounter wave spectra have been calculated for navigation conditions as shown in Figure 3, where V / T (kt/sec) is taken as the unique factor to govern the encounter wave grouping. Figures 4, 5 and 6 show the encounter spectra of following sea for various exponents of directional distribution function, $n = \infty, 10$ and 4, respectively. Figures 7 shows the band-width parameter versus V / T.

Then, the characteristic values of encounter wave groupiness such as $\bar{l}$, $\overline{G}$, and $\overline{H}$ are derived by using equations (1), (2) and (3) for the significant wave height, i.e. $\rho / m_0^{1/2} = 2$ or $\rho = H_{1/3} / 2$, as shown in Figures 8 and 9. From these figures it is clearly recognized that the encounter wave groupiness increases remarkably in the range of $V / T = 1.0 \sim 2.0$ and it becomes highest at $V / T = 1.45$.

Statistical Analysis of Encounter Waves Obtained by Simulation

The time series of encounter wave in short-crested sea can be obtained by time-domain simulation as double summation of elemental waves as equation (17),

$$\eta(\mathbf{x}, t) = \sum_{i=1}^{I} \sum_{j=1}^{J} a_{ij} \cos(\omega_i t - \mathbf{k}_{ij} \cdot \mathbf{x} + \varepsilon_{ij}) \tag{17}$$

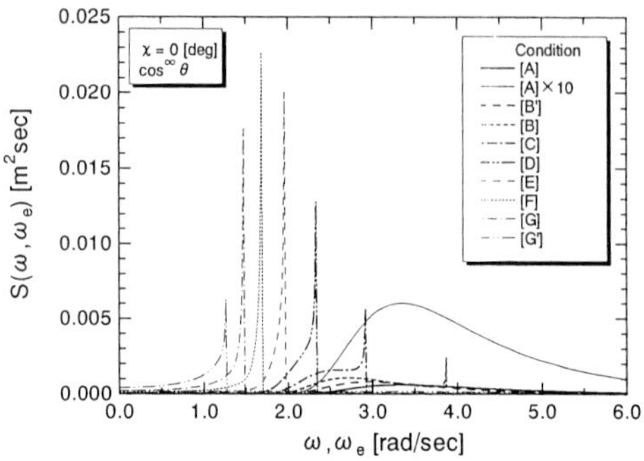

Figure 4 Transformation of wave spectra of long-crested waves $S(\omega)$ into spectra of encounter wave $S(\omega_e)$ in following seas, $\cos^{\infty}\theta$, $\chi = 0$ [deg]

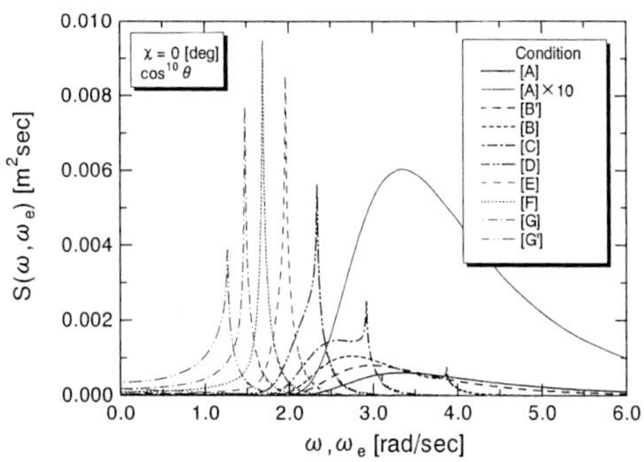

Figure 5 Transformation of wave spectra of short-crested waves $S(\omega,\theta)$ into spectra of encounter wave $S(\omega_e)$ in following seas, $\cos^{10}\theta$, $\chi = 0$ [deg]

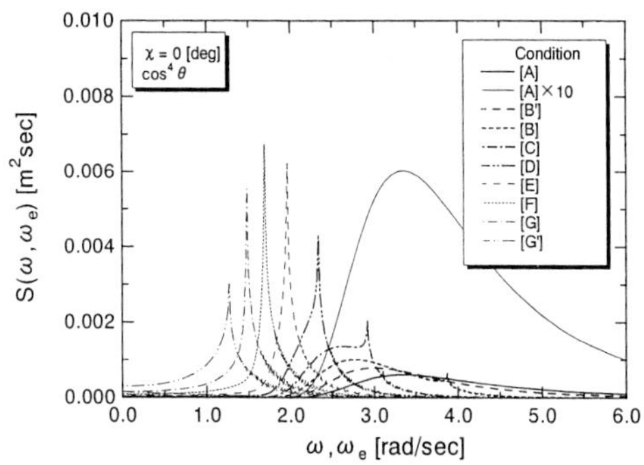

Figure 6 Transformation of wave spectra of short-crested waves $S(\omega,\theta)$ into spectra of encounter wave $S(\omega_e)$ in following seas, $\cos^{4}\theta$, $\chi = 0$ [deg]

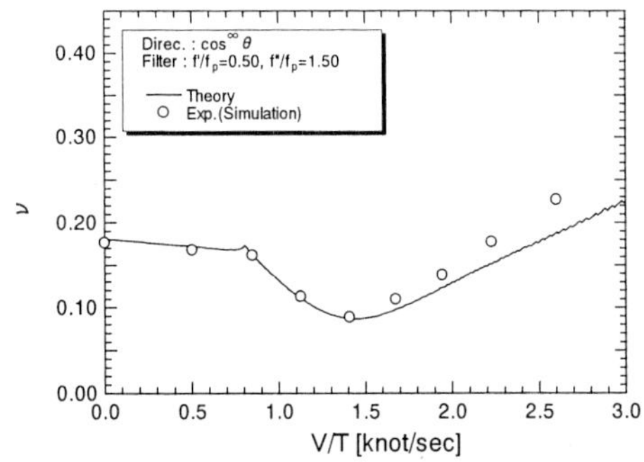

Figure 7 Band width parameter ν vs. V / T in following sea, Long-crested wave, $n = \infty$

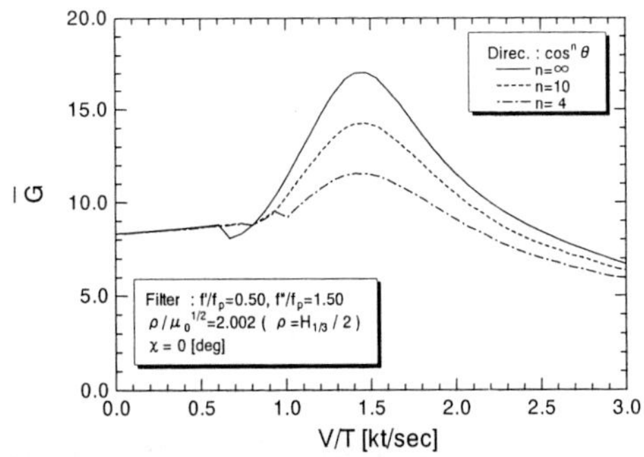

Figure 8 Mean number of waves in a group vs. V / T for the significant wave height

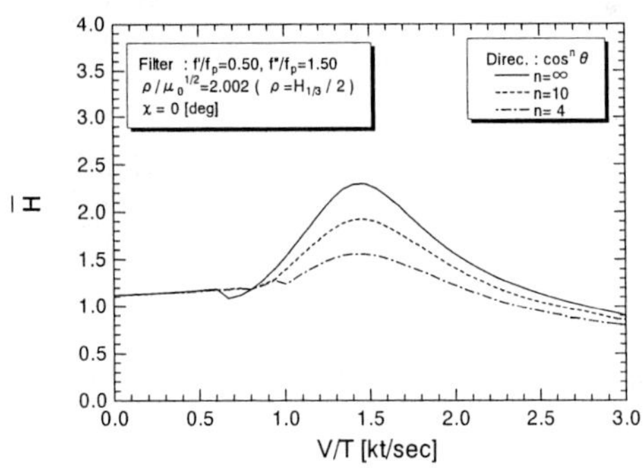

Figure 9 Mean number of waves in a high run exceeding the significant wave height vs. V / T

where

$$
\begin{aligned}
a_{ij} &= \sqrt{2S(\omega_i,\theta_j)\Delta\omega_i\Delta\theta_j} \\
\mathbf{k}_{ij}\cdot\mathbf{x} &= k_i\left(x(t)\cos\theta_j + y(t)\sin\theta_j\right) \\
(x(t), y(t)) &= (x_0 + V\cos\chi\cdot t, y_0 + V\sin\chi\cdot t)
\end{aligned} \tag{18}
$$

The time series of encounter waves in long-crested sea are shown in Figure 10 for various V/T values. The envelopes of wave amplitudes are drawn on the figure.

STASISTICAL PROPERTIES OF ENCOUNTER WAVE GROUP AND EFFECTS OF PARAMETERS

The envelopes of time series of encounter waves have been analysed by statistical manner and the probability that number of run exceeds H has been obtained as shown in Figures 11, 12 as well as Table 1.

The influence of directional distribution of wave spectrum is also shown by chain or broken lines in the figures.
From these results the qualitative features of encounter wave grouping are concluded as follows.

(1) Theoretical calculation and numerical simulation give almost the same tendency that the encounter wave groupiness increases remarkably in the speed range of V/T between 1.0 and 2.0, see Figure 11.
(2) The probability of exceedance, $P(H)$, that the number of waves in a run is greater than H can be represented by a straight line on Figure 12. Therefore, by using Figures 11 and 12, we can estimate $P(H)$ for any conditions of V/T and H.
(3) The increase rate of encounter wave groupiness in the range of $V/T = 1.0\sim2.0$ will become greater, as the value of H increases, see Figure 13.
(4) The increment of encounter wave groupiness in the above-mentioned range of V/T is less, as the directional distribution of wave energy becomes wider.

TABLE 1
PROBABILITY OF RUNS GREATER THAN H OBTAINED BY SIMULATION

Condtion	A		B		C		D		E		F		G	
H	N' = 1557	P(H)	N' = 1089	P(H)	N' = 630	P(H)	N' = 421	P(H)	N' = 387	P(H)	N' = 409	P(H)	N' = 473	P(H)
1	N= 1267	0. 81374	N= 945	0. 86777	N= 580	0. 92064	N= 378	0. 89786	N= 340	0. 87855	N= 344	0. 84108	N= 386	0. 81607
2	385	0. 24727	365	0. 33517	331	0. 52540	233	0. 55344	174	0. 44961	161	0. 39364	139	0. 29387
3	68	0. 04367	57	0. 05234	142	0. 22540	154	0. 36580	101	0. 26098	77	0. 18826	55	0. 11628
4	17	0. 01092	10	0. 00918	48	0. 07619	81	0. 19240	63	0. 16279	36	0. 08802	16	0. 03383
5	2	0. 00129	1	0. 00092	15	0. 02381	42	0. 09976	47	0. 12145	22	0. 05379	6	0. 01269
6	1	0. 00064	0	0. 00000	8	0. 01270	21	0. 04988	32	0. 08269	14	0. 03423	3	0. 00634
7	1	0. 00064	0	0. 00000	3	0. 00476	7	0. 01663	23	0. 05943	11	0. 02690	1	0. 00211
8	0	0. 00000	0	0. 00000	2	0. 00318	5	0. 01188	14	0. 03618	9	0. 02201	0	0. 00000
9	0	0. 00000	0	0. 00000	0	0. 00000	2	0. 00475	8	0. 02067	9	0. 02201	0	0. 00000
10	0	0. 00000	0	0. 00000	0	0. 00000	1	0. 00238	6	0. 01550	5	0. 01223	0	0. 00000
M_0 [m^2sec]	0.0010929		0. 0014062		0. 0013826		0. 00143098		0. 0013334		0. 0011938		0. 00113237	
$H_{1/3}/2$ [m]	0. 064083		0. 073212		0. 073196		0. 0746493		0. 072582		0. 068045		0. 065601	
ν	0. 18299		0. 16334		0. 11256		0. 0859457		0. 10074		0. 11791		0. 15733	

COMPARISON WITH MODEL TEST RESULTS ON CAPSIZING

The capsizing rate of two model ships in following/quartering seas, Kan et al.(1993), are plotted on the V/T-diagram, together with the curve representing the probability to encounter high run, $P(H \geq N)$, where number of run, N, is assumed 4, as shown in Figure 14. From these results, it is clearly shown that the probability to encounter to high runs correlate closely the danger of capsizing events of ships in following/quartering seas. The degree of danger could be assumed to be proportional to the probability, $P(H)$. Note that the capsizing rate for model of larger GM show a better correlation to $P(H)$.

As the results of this study the authors recommend the followings:

(1) The dangerous zone shown in the operational guidance/MSC Circular 707/IMO (1995) may be classified according to the probability, i.e. $P(H)$ values, from the most dangerous zone near V/T=1.5 to the marginal zone lower than 1.0 and higher than 2.0.

(2) The model test in following/quartering seas, i.e. in irregular waves should include such dangerous conditions defined by the mean wave period and ship speed, Takaishi et al. (1996). The V/T-diagram is useful to find such conditions. In performing the model tests in such conditions, it is necessary to let the model to encounter to high runs of waves, i.e. the design of test condition is important. The numerical simulation of encounter waves beforehand the test will be effective. The examples of such model tests were carried out by Umeda et al. (1995) and Hamamoto et al. (1996).

CONCLUSIONS

By statistical analyses of encounter waves both in frequency and time domain, the remarkable increase of groupiness in following/quartering seas has been shown qualitatively as the function of V/T and number of waves in a high run. The degree of danger to encounter high run can be classified in the dangerous zone indicated in the operational guidance, and this result will be useful to evaluate ship stability against capsizing by probabilistic approach.

ACKNOWLEDGEMENT

This work has been performed as a part of the research project RR 24 organized in the Shipbuilding Research Association of Japan. The authors would like to thank advices and discussions by the Committee members, chaired by Prof. M. Fujino, The University of Tokyo.

And this work is originally presented by the present authors, Takaishi et al. (1996), Takaishi et al.(1996) and Takaishi et al.(1997).

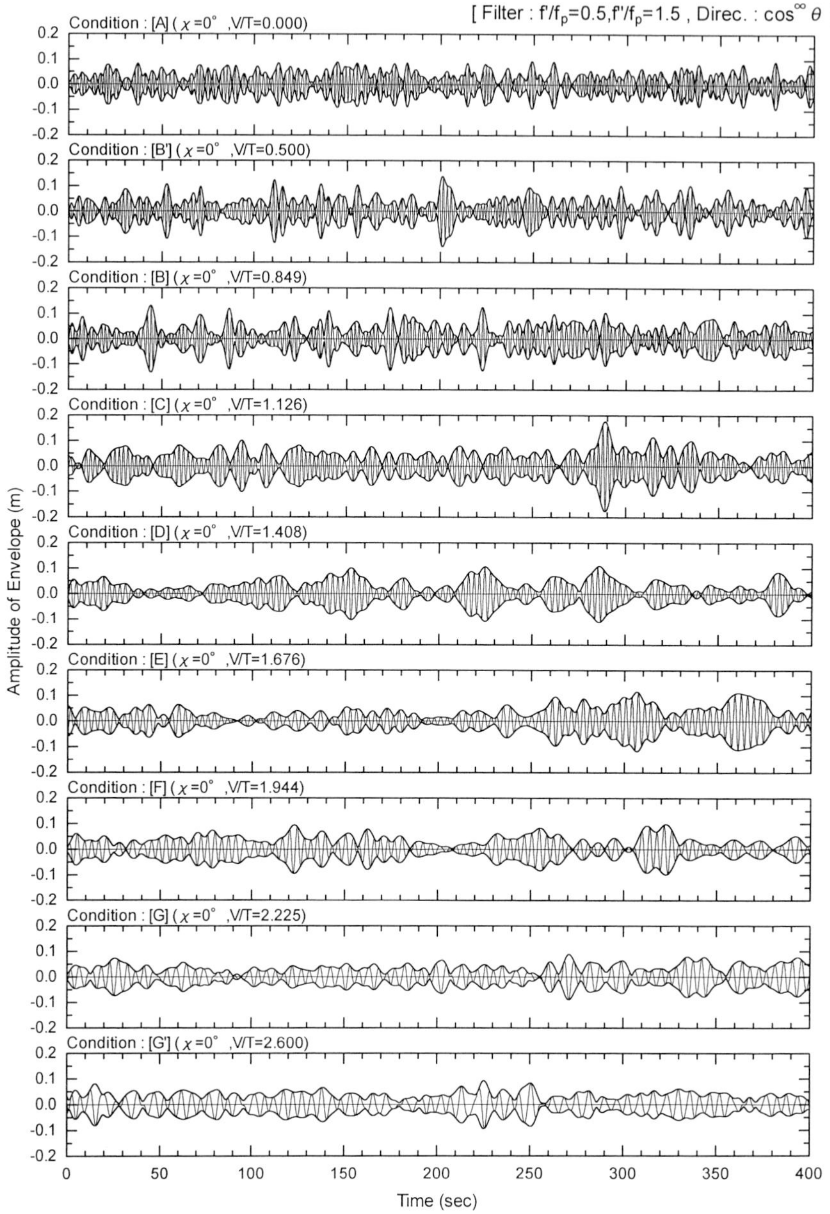

Figure 10 Filtered encounter waves and its envelope in following seas of long-crested waves

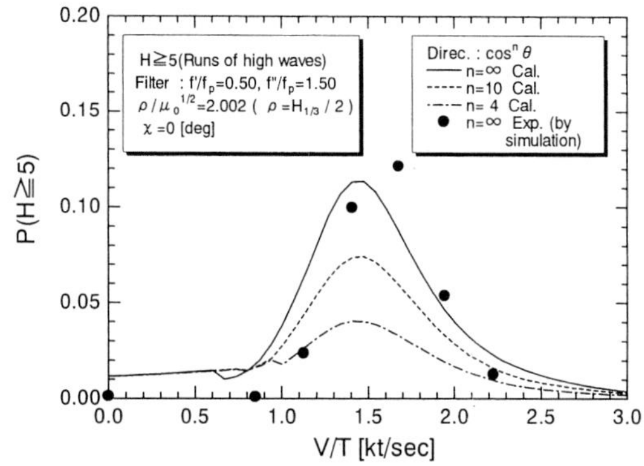

Figure 11 Probability of Runs greater than 5

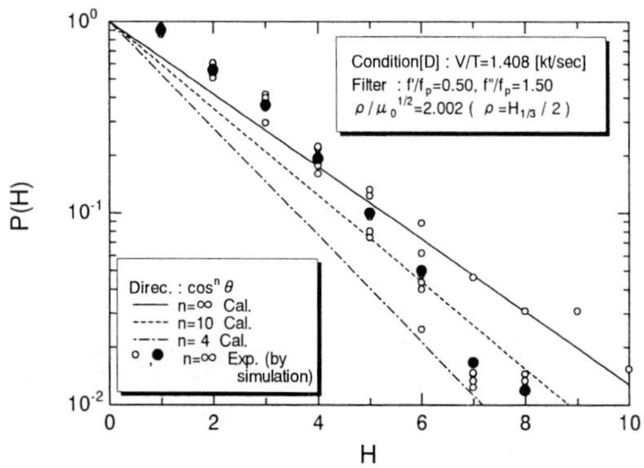

Figure 12 Probability of runs greater than H

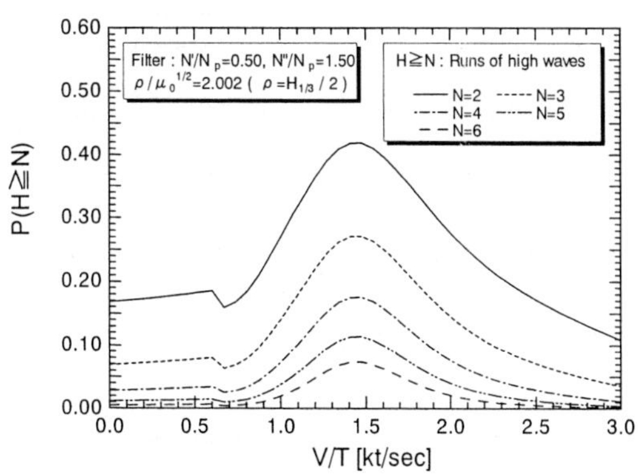

Figure 13 Probability of runs greater than N, $N = 2,3,4,5,6$

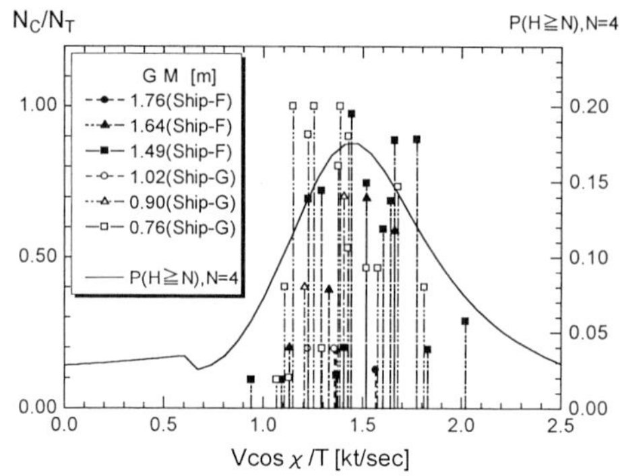

Figure 14 Capsizing rate(N_C / N_T) and probability of runs($P(H \geq N = 4)$) Takaishi et al.(1997)

FIG-14.SMP

REFERENCES

Hamamoto, H. et al. (1996). Model experiments of ship capsize in astern seas, 2nd Report, *J.Society of Naval Architects of Japan* **179**, 79-87.

Kan M. Saruta, T. and Taguchi, H. (1993). Capsizing of a ship in quartering seas (Part 5. Comparative model experiments on mechanism of capsizing. *J. Society of Naval Architects of Japan* **173**, 133-145.

Longuet-Higgins, M.S. (1984). Statistical properties of wave groups in a random sea state, *Phil.Trans.R.Soc. London* **A 312**, 219-250.

MSC/ Circular 707/IMO (1995). Guidance to the master for avoiding dangerous situations in following and quartering seas.

Takaishi, Y.(1994). Dangerous encounter wave conditions for ships navigating in following and quartering seas, *Proc. of 5th STAB, Florida.*

Takaishi, Y., Watanabe, K. and Masuda, K. (1996). Statistical properties of encounter wave grouping phenomena in following and quartering seas, *J.Society of Naval Architects of Japan* **181**, 283-294.

Takaishi, Y., Watanabe, K., Umeda, N. and Masuda, K. (1996). On wave condition for model experiments of ships in following and quartering seas. *J. Kansai Society of Navel Architects,* **Japan 225**, 107-114.

Takaishi, Y. (1996). Probability to encounter high run of waves in the dangerous zone shown in the operational guidance/IMO for following and quartering sea. *Proc. 2nd Workshop on Stability and Operational Safety of Ships, Osaka.* 26-35.

Takaishi, Y., Watanabe, K. and Masuda, K.(1997). Probability to encounter high run of waves in the dangerous zone shown on the operational guidance/IMO for following and quartering sea. *Proc. 6th International Conference on Stability of Ships and Ocean Vehicles, Varna,* **I**. 173-179.

Umeda, N. et al. (1995). Model experiments of ship capsize in astern seas, *J. Society of Naval Architects of Japan* **177**, 207-217.

Contemporary Ideas on Ship Stability
D. Vassalos, M. Hamamoto, A. Papanikolaou and D. Molyneux (Editors)

ONGOING WORK EXAMINING CAPSIZE RISK OF INTACT FRIGATES USING TIME DOMAIN SIMULATION

Kevin McTaggart

Defence Research Establishment Atlantic, P.O. Box 1012,
Dartmouth, Nova Scotia, B2Y 3Z7, Canada

ABSTRACT

This paper describes ongoing efforts to develop methods for predicting capsize risk of intact naval frigates. For a given ship condition, the risk of capsize is considered to be a function of ship speed, heading, significant wave height, and zero-crossing wave period. Implementation of the proposed procedure requires a numerical model for prediction of ship motions in severe seas. The program Fredyn, which uses a time domain strip theory approach, provides an appropriate combination of computational speed and accuracy for predicting capsize of slender naval frigates. When predicting capsize risk in irregular seas, the dependence of capsize occurrence on wave process realization must be considered. This dependence can be modelled by using Gumbel fits of maximum absolute roll angles obtained from several different realizations. The proposed risk analysis procedure can be extended to account for capsize avoidance procedures taken by ship operators.

KEYWORDS

Capsize, frigate, risk, roll, safety, seakeeping, stability.

INTRODUCTION

Risk analysis provides the most rational approach for designing safe ships. Kobylinski (1997) gives an example of risk analysis applied to ship stability, while Mansour et al. (1997) discuss application to structural design. The development of risk analysis approaches is relatively recent, and most ship design continues to be based on rules developed through experience. For stability of warships, the rules developed approximately 40 years ago by Sarchin and Goldberg (1962) continue to be used by several navies.

Traditional design rules suffer from two primary problems. The resulting levels of safety for different ships can be highly variable, leading to insufficient safety in some cases and unreasonable restrictions in other cases. The second primary problem with traditional design rules is that they are not applicable to new designs that differ significantly from the ships upon which the rules were originally based. For example, Sarchin and Goldberg's rules were developed for warships with narrow sterns; however, modern warships have wide transom sterns, making them vulnerable to loss of stability during stern emergence, which can occur while riding on a wave crest. Consequently, modern warships may be more vulnerable to capsize than originally intended by Sarchin and Goldberg.

This paper describes ongoing work to develop rational risk analysis procedures for the stability of intact naval frigates. The present research is being conducted in parallel with Canada's ongoing participation in the Cooperative Research Navies Dynamic Stability Project, which has been described by de Kat (1993) and de Kat et al. (1994).

THEORETICAL APPROACH

Initial work presented by McTaggart (1993, 1995) assumed that if a ship were to capsize in a given year, it would capsize in the most severe annual storm as characterized by significant wave height. The developed capsize wave height approach was based on the following equation:

$$P(C_{annual}) = \sum_{i=1}^{N_{V_s}} \sum_{j=1}^{N_\beta} \sum_{k=1}^{N_{\widetilde{H/\lambda}}} \sum_{l=1}^{N_\chi} \frac{1}{N_\chi} p_{V_s}(V_{s-i})\, p_\beta(\beta_j)\, p_{\widetilde{H/\lambda}}\left(\widetilde{H/\lambda}_k\right) \times Q_{H_{s,annual}}\left(H_C \mid V_s, \beta, \widetilde{H/\lambda}, \chi\right) \quad (1)$$

where $P(C_{annual})$ is annual capsize risk, N_X is the number of discretized values of random variable X, $p_X(X)$ is the probability mass function (PMF) for X, V_s is desired ship speed, β is desired ship heading relative to the waves, $\widetilde{H/\lambda}$ is nominal wave steepness, χ is a seed number for generating random waves, $H_{s,annual}$ is annual maximum significant wave height, $Q_{H_{s,annual}}(H_{s,annual})$ is the exceedence probability of significant wave height for the worst annual storm (typically three hour duration), and $H_C|V_s, \beta, \widetilde{H/\lambda}, \chi$ is the minimum significant wave height which causes capsize given $V_s, \beta, \widetilde{H/\lambda}$, and χ. Equation (1) assumes that if a capsize occurs at significant wave height H_C then capsize will occur for all wave heights greater than H_C. The nominal steepness of the worst annual storm is given by:

$$\widetilde{H/\lambda} = H_s \frac{2\,\pi}{g\,T_p^2} \quad (2)$$

where T_p is peak wave period and g is gravitational acceleration. Storm data from the Canadian East Coast (Canadian Climate Centre, 1991) indicate that a Gumbel distribution provides a good fit for $H_{s,annual}$ and a Weibull distribution provides a good fit for $\widetilde{H/\lambda}$.

Key assumptions of Equation (1) are that capsize risk increases with significant wave height and that nominal wave steepness is of secondary importance. Subsequent work has shown that this assumption is too simplistic, and that the joint distribution of wave height and wave period must be considered; thus, it is inappropriate to consider only extreme annual storms based on significant wave height.

Present work is examining risk of capsize for all possible seaways. For a ship in a seaway of duration D (e.g. one hour), the probability of capsize is:

$$P(C_D) = \sum_{i=1}^{N_{V_s}} \sum_{j=1}^{N_\beta} \sum_{k=1}^{N_{H_s}} \sum_{l=1}^{N_{T_z}} p_{V_s}(V_{s-i})\, p_\beta(\beta_j)\, p_{H_s,T_z}(H_{s-k}, T_{z-l})\, P(C_D|V_s, \beta, H_s, T_z) \quad (3)$$

where T_z is zero-crossing wave period, which is used for compatibility with BMT Global Wave Statistics (1986). Similar expressions have been presented by Kobylinski (1997) and Dahle and Myrhaug (1994, 1996). An important assumption of the above equation is that desired ship speed and heading are independent of wave conditions. This assumption is conservative because ship operators will alter speed and course to reduce capsize risk.

Once the probability of capsize for duration D has been computed using Equation (3), the associated annual probability of capsize can be computed as follows:

$$P(C_{annual}) = 1 - [1 - P(C_D)]^{1\ year/D} \quad (4)$$

NUMERICAL MODELLING OF SHIP CAPSIZE

A risk analysis approach such at that described above requires an accurate and efficient method for predicting ship capsize in a seaway. The present method uses the time domain program Fredyn, as described by de Kat (1993), de Kat et al. (1994), and Hooft and Pieffers (1988). Fredyn considers forces from waves, wind, and ship maneuvering. Strip theory gives fast computation times and acceptable accuracy for naval frigates. It is uncertain whether Fredyn can provide acceptable accuracy for low length/beam vessels such as small fishing boats. For a typical simulation in waves, the program runs approximately 15 times faster than real time on a 300 MHz Pentium II personal computer.

When the Cooperative Research Navies Dynamic Stability Project commenced in 1990, it was hoped that simplified numerical models could be developed for ship capsize. For example, a simplified numerical model would likely be adequate for predicting ship capsize at zero speed in beam seas. Subsequent research revealed that capsize modes for frigates typically require modelling of all six degrees of freedom. Frigates appear to be vulnerable to capsize in following seas through loss of static stability while riding on a wave crest, parametric excitation, and broaching.

Validation and improvement of the Fredyn program is an ongoing process. For frigate motions in extreme conditions, a lack of useful experimental data for validation prompted a series of model tests described by de Kat and Thomas (1998a,1998b) Fredyn appears to give very good agreement with experiments for capsize of a frigate in waves.

SHIP CAPSIZE RISK FOR GIVEN CONDITIONS

The application of Equation (3) for prediction of capsize risk requires a suitable method to determine capsize risk for given conditions $P(C_D|V_s, \beta, H_s, T_z)$. For a Fredyn simulation of a ship in irregular waves, a random phase approach is used to generate a wave realization. The wave realization is dependent upon a seed number χ provided as input for generation of random wave phases. The occurrence of capsize can be highly dependent on the input seed number. Upon initial consideration, it would seem appropriate to determine $P(C_D|V_s, \beta, H_s, T_z)$ using N_s simulations with different seed numbers as follows:

$$P(C_D|V_s, \beta, H_s, T_z) = \frac{N_C}{N_s} \tag{5}$$

where N_C is the number of simulations for which capsize occurs. The main disadvantage of this approach is that a large number of simulations can be required to obtain accurate estimates of $P(C_D|V_s, \beta, H_s, T_z)$. A much more efficient and useful approach is to apply a statistical fit to the maximum roll angles $\phi_{max,D}$ from N_s simulations of duration D. Fitted distribution parameters can be selected to obtain good agreement with sample exceedence probabilities given by:

$$Q_{\phi_{max,D}|V_s,\beta,H_s,T_z}(\phi^i_{max,D}|V_s, \beta, H_s, T_z) = \frac{i}{N_s + 1} \tag{6}$$

where $Q_X(X)$ is exceedence probability for quantity X and i denotes the rank in descending order from N_s simulations. Figures 1 and 2 show fitted Gumbel distributions for a frigate in following seas with significant wave heights of 10 m and 12 m. The Gumbel distribution parameters were estimated using a least squares procedure given by Thoft-Christensen and Baker (1982). The results indicate that a Gumbel distribution provides a very good fit to the conditional probability $Q_{\phi_{max,D}|V_s,\beta,H_s,T_z}(\phi_{max,D}|V_s, \beta, H_s, T_z)$. The clustering of results in Figure 2 at a roll angle of 90 degrees occurs because a Fredyn simulation terminates when roll angle reaches 90 degrees.

Once the conditional probability distribution $Q_{\phi_{max,D}|V_s,\beta,H_s,T_z}(\phi_{max,D}|V_s, \beta, H_s, T_z)$ has been estimated for given conditions, the associated conditional capsize probability is given by:

$$P(C_D|V_s, \beta, H_s, T_z) = Q_{\phi_{max,D}|V_s,\beta,H_s,T_z}(\phi_C|V_s, \beta, H_s, T_z) \tag{7}$$

where ϕ_C is the capsize roll angle for the given ship. For naval ships, the capsize roll angle is often taken as being 70 degrees, which is considered to be the nominal angle for onset of downflooding.

Work in progress suggests that only ten simulations are required to obtain good estimates for the conditional distribution $Q_{\phi_{max,D}|V_s,\beta,H_s,T_z}(\phi_{max,D}|V_s, \beta, H_s, T_z)$. An added benefit of using a fitted distribution for each seaway is that Equation (3) can be revised to give exceedence probabilities for all maximum roll angles in all seaways as follows:

$$\begin{aligned} Q_{\phi_{max,D}}(\phi_{max,D}) = & \sum_{i=1}^{N_{V_s}} \sum_{j=1}^{N_\beta} \sum_{k=1}^{N_{H_s}} \sum_{l=1}^{N_{T_z}} p_{V_s}(V_{s-i}) p_\beta(\beta_j)\, p_{H_s,T_z}(H_{s-k}, T_{z-l}) \\ & \times Q_{\phi_{max,D}|V_s,\beta,H_s,T_z}(\phi_{max,D}|V_s, \beta, H_s, T_z) \end{aligned} \tag{8}$$

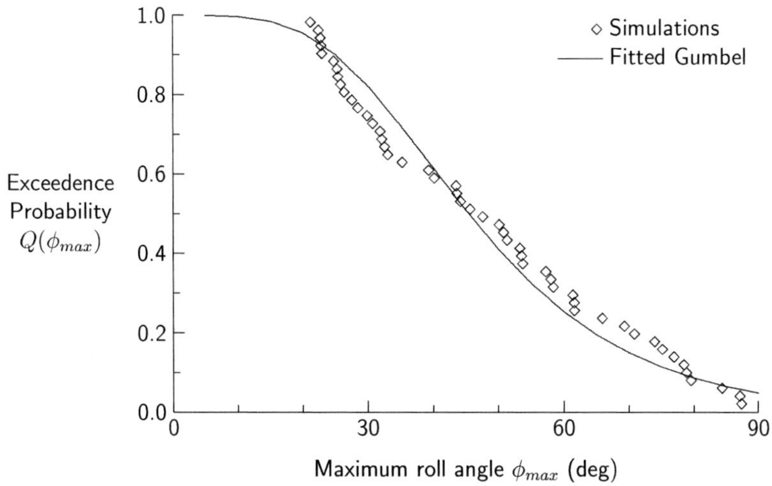

Figure 1: Roll Exceedence Probability for Frigate in One Hour Seaway, $V_s = 10$ knots, $\beta = 30$ degrees, $H_s = 10$ m, $T_z = 10$ s

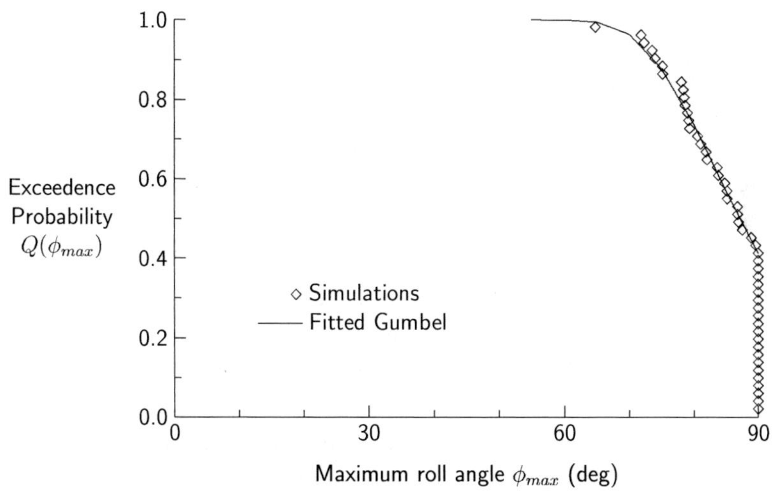

Figure 2: Roll Exceedence Probability for Frigate in One Hour Seaway, $V_s = 10$ knots, $\beta = 30$ degrees, $H_s = 12$ m, $T_z = 10$ s

INFLUENCE OF CAPSIZE AVOIDANCE BY OPERATOR

The risk analysis approach presented thus far does not consider efforts by the operator to reduce capsize risk. Weather routing and altering of course are two possible methods of reducing capsize risk. Faulkner and Williams (1997) suggest that weather routing will not always reduce exposure to severe conditions. For a warship, mission requirements can provide a significant impediment to avoiding severe conditions.

Numerical simulations and experiments have shown that modern naval frigates are most susceptible to capsize when travelling at high speed in high following seas. Figure 3 taken from McTaggart (1995) shows the conditional probability of heading given capsize, and indicates that capsize risk can be greatly reduced by altering course. Fortunately, the quantitative results are consistent with operator practice of altering course to head seas in severe conditions.

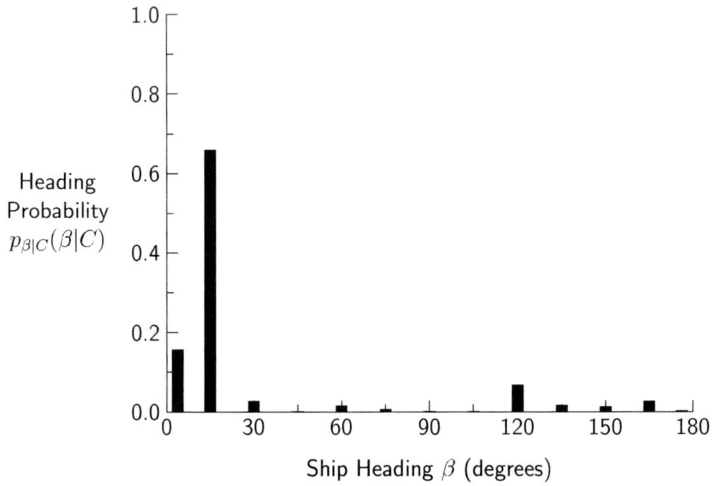

Figure 3: Conditional Probability of Ship Heading Given Capsize

Various researchers are examining ship operational records to determine operator actions in severe conditions. For example, Glen et al. (1998) have recently completed a study for the Ship Structures Committee. Using operational profile data, it should be relatively simple to incorporate capsize avoidance behaviour into risk predictions by modifying Equation (3) to include altering of speed and course in response to significant wave height:

$$P(C_D) = \sum_{i=1}^{N_{V_s}} \sum_{j=1}^{N_\beta} \sum_{k=1}^{N_{H_s}} \sum_{l=1}^{N_{T_z}} p_{V_s|H_s}(V_{s-i}|H_{s-k})\, p_{\beta|H_s}(\beta_j|H_{s-k})\, p_{H_s,T_z}(H_{s-k}, T_{z-l}) \times P(C_D|V_s, \beta, H_s, T_z) \quad (9)$$

FUTURE PLANS

Work continues with the presented risk analysis procedure so that it can be used for routine engineering computations. Although the procedure has been implemented, there are several areas for refinement.

The properties of the Gumbel distribution suggest that simulation durations can be a fraction of the nominal seaway duration for which the capsize probability $P(C_D)$ is required. For example, it is possible that 10 simulations of 15 minute duration could be adequate to determine the statistical properties of maximum roll for a given seaway of 60 minute duration, thus permitting a significant reduction in computation time.

When applying the proposed risk analysis procedure, an acceptable level of capsize risk must be specified. Selection of an acceptable risk level is complicated by the fact that any risk analysis procedure gives risk predictions which are different from actual risk levels; thus, a risk analysis procedure usually has to be calibrated using past operational experience. The present proposed risk analysis method will likely be calibrated using operational data for naval ships which have had long service lives without any near capsizes.

As mentioned in the previous section, operator avoidance action will greatly reduce the risk of ship capsize. Equation (9) provides a promising approach for considering such avoidance action, and will likely be implemented as required operational profile data become available.

The ongoing validation and improvement of the Fredyn ship motion program will lead to improved capsize risk predictions. It is likely that Fredyn will eventually be superceded by another time domain code with a more sophisticated treatment of nonlinear and three-dimensional effects.

CONCLUSIONS

A rational method has been developed for predicting capsize risk of intact naval frigates. The method requires an accurate and efficient numerical model of ship motions in severe seas. The program Fredyn fulfills this role for naval frigates. For a ship in an irregular seaway, the occurrence of capsize can be highly dependent on the wave process realization. This effect can be effectively modelled by fitting a Gumbel distribution to maximum roll angles from several different seaway realizations. Future improvements to the risk model will include the effects of actions by ship operators to avoid ship capsize.

REFERENCES

British Maritime Technology Limited. (1986), *Global Wave Statistics*, Unwin Brothers, London.

Canadian Climate Centre. (1991). *Wind/Wave Hindcast Extremes for the East Coast of Canada, Volume 1.* Prepared under contract by MacLaren Plansearch Limited and Oceanweather Inc.

Dahle E. A. and Myrhaug D. (1994). Risk Analysis Applied to Capsize of Smaller Vessels in Breaking Waves. *Transactions, Royal Institution of Naval Architects* **136**.

Dahle E. A. and Myrhaug D. (1996). Capsize Risk of Fishing Vessels. *Schiffstechnik (Ship Technology Research)* **43**, 164–171.

de Kat J.O. (1993). The Development of Survivability Criteria Using Numerical Simulations. *United States Coast Guard Vessel Stability Symposium '93*, New London, Connecticut.

de Kat J.O., Brouwer R., McTaggart K., and Thomas L. (1994). Intact Ship Survivability in Extreme Waves: Criteria from a Research and Navy Perspective. *STAB '94, Fifth International Conference on Stability of Ships and Ocean Vehicles*, Melbourne, Florida.

de Kat J.O. and Thomas W.L. III. (1998a). Broaching and Capsize Model Tests for Validation of Numerical Ship Motion Predictions. *Fourth International Stability Workshop*, St. John's, Newfoundland.

de Kat J.O. and Thomas W.L. III. (1998b). Extreme Rolling, Broaching, and Capsizing-Model Tests and Simulations of a Steered Ship in Waves. *22nd Symposium on Naval Hydrodynamics*, Washington.

Faulkner D. and Williams R.A. (1997). Design for Abnormal Ocean Waves. *Transactions, Royal Institution of Naval Architects* **139**.

Glen I.F., Paterson R.B., and Luznik L. (1998) *Sea Operational Profiles for Structural Reliability Assessments.* Fleet Technology Limited, Kanata. Prepared for Defence Research Establishment Atlantic and Ship Structures Committee.

Hooft J.P. and Pieffers J.B.M. (1988). Maneuverability of Frigates in Waves. *Marine Technology* **25:4**, 262–271.

Kobylinski L.K. (1997). Rational Approach to Ship Safety Requirements. *Second International Conference on Marine Technology*, Szczecin, Poland.

Mansour A.E., Wirsching P.H., Lucket M.D., Plumpton A.M., and Lin Y.H. (1997). Structural Safety of Ships. *Transactions, Society of Naval Architects and Marine Engineers* **105**.

McTaggart K.A. (1993). Risk Analysis of Intact Ship Capsizing. *United States Coast Guard Vessel Stability Symposium '93*, New London, Connecticut.

McTaggart K.A. (1995), Capsize Risk Prediction Including Wind Effects. *Proceedings of the Third Canadian Marine Hydrodynamics and Structures Conference*, Dartmouth, Nova

Scotia.

Sarchin T.H. and Goldberg L.L. (1962). Stability and Buoyancy Criteria for U.S. Naval Surface Ships. *Transactions, Society of Naval Architects and Marine Engineers* **70**.

Thoft-Christensen P. and Baker M.J. (1982). *Structural Reliability and Its Applications*, Springer-Verlag, Berlin.

AUTHOR INDEX

Aihara, K. 411
Akinturk, A. 473
Amagai, K. 137
Armenio, V. 33

Belenky, V.L. 149
Birmingham, R. 389
Bishop, S.R. 89
Blume, P. 237, 265
Boulougouris, E. 553

Calisal, S.M. 473
Chang, B.C. 237
Contento, G. 33, 125
Cotton, B. 89, 101, 499

de Kat, J.O. 69, 249
Dodworth, K. 487

Francescutto, A. 33, 125
Fujiwara, T. 341
Furukawa, Y. 435

Grochowalski, S. 3

Hamamoto, M. 15
Haraguchi, T. 291
Hirayama, T. 365
Hong, S.W. 353
Hong, S.Y. 353
Hutchison, B.L. 187

Ikeda, Y. 199, 225, 401, 449
Ishida, S. 291, 341

Jasionowski, A. 487
Jost, A.E. 265

Kambisseri, R. 199, 225
Kanerva, M. 523
Kang, C.G. 353
Katayama, T. 401, 449
Kendrick, A. 303
Kijima, K. 435, 461
Kimura, N. 137
Komuro, M. 411

Letizia, L. 315

Masuda, K. 575
McTaggart, K. 587
Molyneux, D. 303
Munif, A. 15
Murashige, S. 291, 341, 411

Nagamatsu, T. 461
Nishimura, K. 365

Odabasi, A.Y. 27
Ogawa, Y. 341
Okumura, H. 449

Papanikolaou, A. 279, 553
Peirce, T. 303

Rakhmanin, N.N. 377, 511
Renilson, M. 565
Rudman, M.J. 473

Schindler, M. 213
Shimizu, N. 199
Spanos, D.A. 279
Spyrou, K.J. 47, 101, 111, 499

Takaishi, Y. 575
Taschereau, A. 303
Tasevski, B. 473
Thomas III, W.L. 69
Thompson, J.M.T. 89, 101, 499
Turan, O. 315

Ueno, K. 137
Umeda, N. 57, 423

Vassalos, D. 163, 315, 487, 537
Vilensky, G.V. 511

Washio, Y. 461
Watanabe, I. 341
Watanabe, K. 575
Wong, A. 473

Yamada, T. 411

Zaraphonitis, G. 279
Zhivitsa, S.G. 377